Additional copies of *Flight Instructor Flight Maneuvers and Practical Test Prep* are available from

Gleim Publications, Inc.
P.O. Box 12848 • University Station
Gainesville, Florida 32604
(352) 375-0772
(800) 87-GLEIM or (800) 874-5346
FAX: (352) 375-6940
Internet: www.gleim.com | E-mail: avmarketing@gleim.com

The price is $17.95 (subject to change without notice). Orders must be prepaid. Call us, order online, or use the order form on page 583. Shipping and handling charges apply to all orders. Add applicable sales tax to shipments within Florida.

Gleim Publications, Inc. guarantees the immediate refund of all resalable texts and unopened software and audios purchased directly from Gleim Publications, Inc. if they are returned within 30 days. Shipping and handling charges are nonrefundable. Returns of books purchased from bookstores and other resellers should be made to the respective bookstore or reseller.

REVIEWERS AND CONTRIBUTORS

Harry Kraemer, ATP, CFII, MEI, Master CFI, and Master Ground Instructor, is one of our aviation research consultants. Mr. Kraemer reviewed the content of this book.

Christopher A. Noth, CMEL, AGI, B.S., Embry-Riddle Aeronautical University, is an aviation technical research assistant. Mr. Noth developed and incorporated revisions into the text.

John F. Rebstock, B.S., Fisher School of Accounting, University of Florida, specializes in ensuring that our questions and Knowledge Transfer Outlines are user-friendly. Mr. Rebstock reviewed portions of the manuscript.

The CFIs who have worked with us throughout the years to develop and improve our pilot training materials.

The many FAA employees who helped, in person or by telephone, primarily in Gainesville, Orlando, Oklahoma City, and Washington, D.C.

ACKNOWLEDGMENTS

The photograph on the front cover is of a Cessna Citation Mustang, and it is courtesy of Cessna Aircraft Company.

A PERSONAL THANKS

This manual would not have been possible without the extraordinary effort and dedication of Julie Cutlip, Eileen Nickl, Mumbu Ngugi, and Teresa Soard, who typed the entire manuscript and all revisions; and drafted and laid out the diagrams and illustrations in this book.

The authors also appreciate the production and editorial assistance of Katie Goodrich, Jim Harvin, Jean Marzullo, Shane Rapp, Victoria Rodriguez, and Laura Ter Keurst.

Finally, we appreciate the encouragement and tolerance of our families throughout this project.

AND
PRACTICAL TEST PREP

FOURTH EDITION

by

Irvin N. Gleim, Ph.D., CFII

and

Garrett W. Gleim, CFII

ABOUT THE AUTHORS

Irvin N. Gleim earned his private pilot certificate in 1965 from the Institute of Aviation at the University of Illinois, where he subsequently received his Ph.D. He is a commercial pilot and flight instructor (instrument) with multiengine and seaplane ratings, and is a member of the Aircraft Owners and Pilots Association, American Bonanza Society, Civil Air Patrol, Experimental Aircraft Association, and Seaplane Pilots Association. He is the author of flight maneuvers and practical test prep books for the private, instrument, commercial, and flight instructor certificates/ratings, and study guides for the private/recreational, instrument, commercial, flight/ground instructor, fundamentals of instructing, airline transport pilot, and flight engineer FAA knowledge tests. Three additional pilot training books are *Pilot Handbook*, *Aviation Weather and Weather Services*, and *FAR/AIM*.

Dr. Gleim has also written articles for professional accounting and business law journals, and is the author of widely used review manuals for the CIA exam (Certified Internal Auditor), the CMA exam (Certified Management Accountant), the CFM exam (Certified in Financial Management), the CPA exam (Certified Public Accountant), and the EA exam (IRS Enrolled Agent). He is Professor Emeritus, Fisher School of Accounting, University of Florida, and is a CFM, CIA, CMA, and CPA.

Garrett W. Gleim earned his private pilot certificate in 1997 in a Piper Super Cub. He is a commercial pilot (single and multi-engine), ground instructor (advanced and instrument) and flight instructor (instrument and multi-engine), and is a member of the Aircraft Owners and Pilots Association and National Association of Flight Instructors. He is the author of study guides for the private/recreational, instrument, commercial, flight/ground instructor, fundamentals of instructing, and airline transport pilot FAA knowledge tests. He received a Bachelor of Science in Economics from The Wharton School, University of Pennsylvania. He is also a CPA (not in public practice).

Gleim Publications, Inc.
P.O. Box 12848 · University Station
Gainesville, Florida 32604

(352) 375-0772
(800) 87-GLEIM or (800) 874-5346
FAX: (352) 375-6940

Internet: www.gleim.com
E-mail: admin@gleim.com

ISSN 1092-4183

ISBN 978-1-58194-567-1

First Printing: March 2007

This is the first printing of the fourth edition of
*Flight Instructor Flight Maneuvers
and Practical Test Prep*
Please e-mail update@gleim.com with **FIFM 4-1**
in the subject or text. You will receive our current
update as a reply. Updates are available until the
next edition is published.

EXAMPLE:

To:	update@gleim.com
From:	your e-mail address
Subject:	**FIFM 4-1**

CAUTION: This book is an academic presentation for training purposes only. Under NO circumstances can it be used as a substitute for your Pilot's Operating Handbook or FAA-approved *Airplane Flight Manual*. **You must fly and operate your airplane in accordance with your Pilot's Operating Handbook or FAA-approved *Airplane Flight Manual*.**

HELP !!

This is the Fourth Edition, designed specifically for pilots who aspire to become flight instructors. Please send any corrections and suggestions for subsequent editions to the author, c/o Gleim Publications, Inc. The last page in this book has been reserved for you to make comments and suggestions. It can be torn out and mailed to Gleim Publications, Inc.

NOTE: UPDATES

Send e-mail to update@gleim.com as described at the top right of this page, and visit our Internet site for the latest updates and information on all of our products. To continue providing our customers with first-rate service, we request that questions about our books and software be sent to us via mail, e-mail, or fax. The appropriate staff member will give each question thorough consideration and a prompt response. Questions concerning orders, prices, shipments, or payments will be handled via telephone by our competent and courteous customer service staff.

TABLE OF CONTENTS

This book contains a systematic discussion and explanation of the FAA's Practical Test Standards (Airplane), which will assist you in (1) preparing for and (2) successfully completing your FAA Practical Test!

PREFACE

This book will prepare you to pass your FLIGHT INSTRUCTOR FAA PRACTICAL TEST! In addition, this book will assist you and your flight instructor in planning and organizing your flight training.

We provide you with the easiest, fastest, and least expensive means of successfully completing your flight instructor FAA practical test. We have reorganized and integrated the relevant FAA books, manuals, and documents (including the latest Practical Test Standards).

The flight instructor practical test is a rigorous test of both concept knowledge and motor skills. This book explains all of the knowledge that your instructor and FAA examiner will expect you to demonstrate and discuss with him/her. Previously, flight instructor candidates had only the FAA PTS "reprints" to study. Now you have the PTS followed by a thorough explanation of each task and a step-by-step description of each flight maneuver. Thus, through careful organization and presentation, we will decrease your preparation time, effort, and frustration, **and** increase your knowledge and understanding.

As an additional feature of this book, we have listed some of the common errors made by pilots in executing each flight maneuver or operation. You will be aware of *what not to do*. We all learn by our mistakes, but our *common error* list provides you with an opportunity to learn from the mistakes of others.

Most books create additional work for the user. In contrast, *Flight Instructor Flight Maneuvers and Practical Test Prep* facilitates your effort; i.e., it is easy to use. The outline format, numerous illustrations and diagrams, type styles, indentations, and line spacing are designed to improve readability. Concepts are often presented as phrases rather than complete sentences.

Relatedly, our outline format frequently has an "a" without a "b" or a "1" without a "2." While this violates some journalistic *rules of style*, it is consistent with your cognitive processes. This book was designed, written, and formatted to facilitate your learning and understanding. Another similar counterproductive "rule" is *not to write in your books*. We urge you to mark up this book to facilitate your learning and understanding.

We are confident this book will facilitate speedy completion of your flight training and practical test. We also wish you the very best in subsequent flying and in obtaining additional ratings and certificates. If you have *not* passed your flight instructor FAA knowledge test and do *not* have *Flight Instructor Pilot FAA Knowledge Test* (another book with a red cover) or *FAA Test Prep* CD-Rom, please order today. Almost everything you need to pass the FAA's knowledge and practical tests for the flight instructor certificate is available from Gleim in our Flight Instructor Pilot Kit and our Online Ground School. If your FBO, flight school, or aviation bookstore is out of stock, call (800) 874-5346 or visit www.gleim.com.

We encourage your suggestions, comments, and corrections for future printings and editions. The last page of this book has been designed to help you note corrections and suggestions throughout your preparation process. Please use it, tear it out, and mail it to us. Thank you.

Enjoy Flying -- Safely!

Irvin N. Gleim
Garrett W. Gleim

March 2007

PART I:
GENERAL INFORMATION

Part I (Study Units 1 through 5) of this book provides general information to assist you in obtaining your flight instructor certificate:

Part II consists of Study Units I through XIV, which provide an extensive explanation of each of the 72 tasks required of those taking the flight instructor FAA practical test in a single-engine airplane (land). Part II is followed by Appendix A, FAA Flight Instructor Practical Test Standards, which is a reprint of all 72 tasks in one location.

Flight Instructor Flight Maneuvers and Practical Test Prep is one of six related books for obtaining your flight instructor certificate. The other five are

1. *Flight/Ground Instructor FAA Knowledge Test*
2. *Fundamentals of Instructing FAA Knowledge Test*
3. *Aviation Weather and Weather Services*
4. *Pilot Handbook*
5. *FAR/AIM*

This book assumes that you have these five companion books available. You will be referred to them as appropriate.

Flight/Ground Instructor FAA Knowledge Test contains all of the FAA's 850-plus airplane-related questions and organizes them into logical topics called subunits. The book consists of 121 subunits, which are grouped into 10 study units. Each study unit begins with a brief, user-friendly outline of what you need to know, and answer explanations are provided next to each question. This book will transfer knowledge to you and give you the confidence to do well on the FAA flight instructor (airplane) knowledge test.

Fundamentals of Instructing FAA Knowledge Test contains all of the FAA's 160 questions organized into 30 subunits, which are grouped into six study units. This book will also transfer knowledge to you and give you confidence to do well on the FAA knowledge test.

Aviation Weather and Weather Services combines all of the information from the FAA's *Aviation Weather* (AC 00-6A), *Aviation Weather Services* (AC 00-45E), and numerous other FAA publications into one easy-to-understand book. It will help you study all aspects of aviation weather and provide you with a single reference book.

Pilot Handbook is a complete text and reference for all pilots. Aerodynamics, airplane systems, airspace, and navigation systems are among the topics explained in *Pilot Handbook*.

FAR/AIM is an essential part of every flight instructor's library. Gleim's *FAR/AIM* is an easy-to-read reference book containing all of the Federal Aviation Regulations (FARs) applicable to general aviation flying, plus the full text of the FAA's *Aeronautical Information Manual (AIM)*.

RECAP OF REQUIREMENTS TO OBTAIN A FLIGHT INSTRUCTOR CERTIFICATE

1. Be at least 18 years old.

2. Be able to read, write, and converse fluently in English.

3. Hold a commercial pilot certificate with an instrument rating or an airline transport pilot (ATP) certificate.

4. Hold at least a current third-class FAA medical certificate.

5. Receive and log ground training (such as studying *Fundamentals of Instructing FAA Knowledge Test*) on the following subjects:

   a. The Learning Process
   b. Barriers to Learning
   c. Human Behavior and Effective Communication
   d. Teaching Methods
   e. Planning Instructional Activity
   f. Critique and Evaluation

6. Receive and log ground training (such as studying this book, *Flight/Ground Instructor FAA Knowledge Test*, *Pilot Handbook*, and *Aviation Weather and Weather Services*). Subjects include

   a. Airplanes and Aerodynamics
   b. Airplane Performance
   c. Airplane Instruments, Engines, and Systems
   d. Airports, Airspace, and ATC
   e. Weight and Balance
   f. Aviation Weather
   g. Federal Aviation Regulations
   h. Navigation
   i. Flight Maneuvers
   j. Aeromedical Factors and Aeronautical Decision Making (ADM)

7. Pass the fundamentals of instructing (FOI) and flight instructor knowledge (written) test with a score of 70% or better.

   NOTE: An instructor endorsement is not required for the FOI or any instructor knowledge test.

8. Receive flight instruction from a CFI who has held a flight instructor certificate during the 24 months immediately preceding giving you instruction and who has given at least 200 hr. of flight instruction as a CFI. Your CFI will endorse your logbook certifying that you are proficient to pass the practical test on the areas of operations explained in Part II, Study Units I through XIV, of this book.

9. Obtain a logbook endorsement from a CFI who has provided you with spin entry, spin, and spin recovery training in an airplane certificated for spins and has found that you are competent and possess instructional proficiency in those training areas.

10. Obtain a logbook endorsement from an authorized instructor for the fundamentals of instructing listed in 5. and 6. above. This endorsement is required prior to taking the practical test.

11. Successfully complete a practical test, which will be given as a final exam by an FAA inspector or designated pilot examiner. The practical test will be conducted as specified in Part II of this book.

STUDY UNIT ONE
THE FLIGHT INSTRUCTOR CERTIFICATE

1.1 WHY GET A FLIGHT INSTRUCTOR CERTIFICATE?

A. The flight instructor certificate is commonly the next step after you receive your commercial pilot certificate.

    1. The majority of commercial pilots find their first jobs as certificated flight instructors (CFI).

    2. In addition to producing income, flight instructing also provides much-needed flight time and experience for those aspiring to work with an airline or a corporate flight department.

B. Many professional pilots, after gaining experience and hours instructing, will move on to other types of flying. Some, however, will find a rewarding and challenging career in flight instruction.

    1. Flight instructing provides the unique experience of being able to introduce the wonders of flying to others.

    2. Thus, flight instruction is a legitimate career choice in itself.

    3. Many maintain that the CFI certificate is the most important certificate granted by the FAA.

1.2 WHAT IS A FLIGHT INSTRUCTOR CERTIFICATE?

A. A flight instructor certificate is similar in appearance to your commercial pilot (or ATP) certificate and will allow you to give both flight and ground instruction.

B. The certificate will be sent to you by the FAA upon completion of your training program, the two computerized pilot knowledge tests, and a practical test. A sample flight instructor certificate is reproduced below.

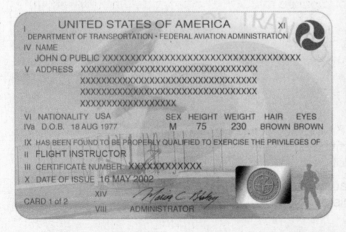

C. Your flight instructor certificate is valid only when it is accompanied by your commercial pilot (or ATP) certificate.

    1. Your flight instructor certificate expires at the end of 2 years. The expiration date will be on your certificate.

1.3 HOW TO GET STARTED

A. **Talk to several people who have recently attained their flight instructor certificate.** Visit your local airport and ask for the names of several people who have just completed their pilot training. When you locate one person, (s)he can usually refer you to another. How did they do it?

    1. Flight training: Airplane? CFI? What period of time? Cost? How structured was the program?

    2. Ask for their advice. How would they do it differently? What do they suggest to you?

    3. What difficulties did they encounter?

B. **Talk to several CFIs.** Tell them you are considering becoming a CFI. Evaluate each as a prospective instructor.

    1. What does each CFI recommend?

    2. What are the projected costs?

    3. What is the rental cost for the training aircraft?

    4. Ask for the names and phone numbers of several persons who recently obtained the flight instructor certificate under his/her direction.

    5. Does the flight instructor's schedule and the schedule of available aircraft fit your schedule?

    6. Where do most initial CFI applicants take the practical test? What is its estimated cost?

        a. You may have to fly to the nearest FAA Flight Standards District Office (FSDO) where an FAA inspector will give you your practical test.

C. Once you have made a preliminary choice of flight instructor and/or FBO, **sit down with your CFI and plan a schedule of flight instruction**.

    1. When and how often you will fly

    2. When you will take the FAA pilot knowledge tests

    3. When you should plan to take your practical test

    4. When and how payments will be made for your instruction

    5. Review, revise, and update the total cost to obtain your commercial certificate (see the next page).

D. **Prepare a tentative written time budget and a written expenditure budget.**

> Flight training:
>
> Dual: _____ hours × $_____ .... $ _____
> Dual complex airplane: _____ hours × $_____ .... $ _____
>
> Ground training: _____ hours × $_____ .... $ _____
>
> FAA Knowledge Tests
> Fundamentals of Instructing (FOI)...................... $ _____
> Flight Instructor--Airplane (FIA)......................... $ _____
>
> Practical test (airplane)... $ _____
>
> Practical test (examiner).. $ _____
>
> Medical exam (for third-class medical)..................... $ _____
>
> This book... $ __19.95__
>
> Gleim's *Fundamentals of Instructing FAA Knowledge Test*.................. $ __14.95__
>
> Gleim's *Flight/Ground Instructor FAA Knowledge Test*..................... $ __19.95__
>
> Gleim's *FAA Test Prep* CD-Rom (CD for Windows 95, 98, or NT 4.0 or higher)*..... $ __59.95__
>
> Gleim's *Aviation Weather and Weather Services*...................... $ __24.95__
>
> Gleim's *Pilot Handbook*... $ __15.95__
>
> Gleim's *FAR/AIM*.. $ __15.95__
>
> Other books:
> One or more section charts.................................. $ _____
> Information manual(s) for your training airplane(s).......... $ _____
>
> TOTAL... $ _____
>
> *Contains both the FOI and flight/ground instructor questions.

E. A final point to consider is the employment possibilities available after you have earned your flight instructor certificate. Some flight schools may give you a job instructing if you complete your training with them. It is important to look ahead to when you will stop paying to fly and start getting paid to fly.

1.4 ADDITIONAL INSTRUCTOR RATINGS

A. This book is geared toward the training, maneuvers, and practical test for initial certification as a Certificated Flight Instructor--Airplane Single-Engine Land. Your flight instructor certificate will indicate airplane single-engine without specifying land or sea. This certificate will allow you to give ground and flight instruction in single-engine airplanes for

1. Recreational pilot certificate

2. Private pilot certificate

3. Commercial pilot certificate

4. Flight instructor certificate (airplane), after being a CFI for 24 months and giving at least 200 hr. of instruction

5. Flight reviews

6. Single-engine airplane checkouts

7. Recurrency training

B. You may apply for additional ratings on your flight instructor certificate. To qualify for these additional ratings, you must

   1. Have a current pilot certificate with the ratings appropriate to the flight instructor rating sought.

   2. Have at least 15 hr. of pilot-in-command time in the category and class of aircraft appropriate to the rating sought.

   3. Pass a knowledge test and practical test as may be prescribed in FAR Part 61 for the rating sought.

C. To give instrument instruction in preparation for the instrument rating, you must become a CFII, or Certificated Flight Instructor -- Instrument (airplane). To qualify, you must

   1. Pass the Instrument Flight Instructor--Airplane knowledge test, which is drawn from the FAA instrument pilot knowledge test bank administered to instrument rating applicants. The FAA test questions appear in Gleim's *Instrument Pilot FAA Knowledge Test.*

   2. Have at least 15 hr. of pilot-in-command time in airplanes.

   3. Pass a practical test that will require you to fly and teach instruments from the right seat.

D. To give instruction in preparation for the multiengine rating or a checkout in a multiengine airplane, you must become a Certificated Multiengine Instructor (MEI). To qualify, you must

   1. Have a multiengine rating on your current commercial pilot certificate. Some pilots only possess private privileges on their multiengine ratings.

   2. Have at least 15 hr. of pilot-in-command time in multiengine airplanes.

   3. Have 5 hr. of pilot-in-command time in the make and model of aircraft before you can instruct in it.

   4. Pass a practical test that will require you to fly the airplane and instruct from the right seat through all normal and emergency maneuvers.

      a. Just like the initial multiengine rating, there is no FAA knowledge test to pass prior to this rating.

E. Beyond the airplane single-engine, there are six additional ratings that can be added to your flight instructor certificate:

   1. Airplane Multiengine
   2. Instrument Airplane
   3. Rotorcraft - Helicopter
   4. Rotorcraft - Gyroplane
   5. Glider
   6. Instrument Helicopter

STUDY UNIT TWO
FLIGHT INSTRUCTION TECHNIQUES

The purpose of this study unit is to assist you in making the transition from student to teacher. The word "teacher" has many different connotations and will therefore be replaced by the words "instructor" and "instructing" (the FAA's terms), which are more appropriate.

While a matter of semantics, we argue that people learn rather than are taught. Instructors are facilitators to explain how to learn, help students organize and interrelate information, demonstrate flight maneuvers, and also provide critiques (feedback) on student performance. Instructors also impart their own feelings, experiences, and styles of flying to their students.

Note that the effort and results are those of the student. Instructors are responsible for directing student effort so that the results are optimized. They can also help motivate students.

Pilots often minimize their own efforts and are used to instructors who are babysitters and/or entertainers. Be honest: To become a proficient pilot is a lot of work, as well as costly. You must develop and commit to a mindset that it is worthwhile (rewarding) to become a safe and proficient pilot.

This study unit and book are largely directed to you as a student preparing for your flight instructor FAA practical test. You should also think of the discussion in terms of you as a flight instructor giving instruction to your student.

2.1 AVIATION INSTRUCTOR'S HANDBOOK (FAA-H-8083-9)

A. In late 1999, the FAA released a new edition of *Aviation Instructor's Handbook (AIH)*. It is a 152-page book designed to provide you with information on learning and instructing and ways to relate that information to students.

    1. The problem with the *AIH*, as with many FAA books, is that there is too much verbiage.

    2. Thus, while there are many good ideas in *AIH*, they generally are not presented and organized so that you can easily commit them to long-term memory in order to master and use them.

B. Gleim's *Fundamentals of Instructing FAA Knowledge Test* thoroughly covers the material presented in AIH but organizes it in a manner to facilitate learning. Thus, the need to purchase the *AIH* is questionable.

    1. *AIH* consists of 11 chapters. The *AIH* table of contents is reproduced on page 8 and *Fundamentals of Instructing FAA Knowledge Test* table of contents is reproduced on page 9.

    2. Much of this information is also presented in Part II, Study Unit I, Fundamentals of Instructing, beginning on page 65 of this book.

C. While you need to learn the fundamentals of instructing to pass the FOI knowledge test, it will not be sufficient background to become an effective CFI. We will try to simplify the process beyond the FOI knowledge test.

    1. First, read and think about the learning process.

        a. How do you learn?
        b. How will your students learn?

    2. Second, study the remainder of this study unit, which discusses ground instruction and flight instruction.

        a. Many teachers, professors, instructors, etc., have never stopped to think about how people (or they) learn! Invest an hour in this worthwhile endeavor by concentrating on this study unit.

FAA
AVIATION INSTRUCTOR'S HANDBOOK
(AIH)

Table of Contents

   1. THE LEARNING PROCESS
   2. HUMAN BEHAVIOR
   3. EFFECTIVE COMMUNICATION
   4. THE TEACHING PROCESS
   5. TEACHING METHODS
   6. CRITIQUE AND EVALUATION
   7. INSTRUCTIONAL AIDS AND TRAINING TECHNOLOGIES
   8. INSTRUCTOR RESPONSIBILITIES AND PROFESSIONALISM
   9. TECHNIQUES OF FLIGHT INSTRUCTION
  10. PLANNING INSTRUCTIONAL ACTIVITY
  11. PROFESSIONAL DEVELOPMENT
      APPENDIX A. SAMPLE TEST ITEMS
      APPENDIX B. INSTRUCTOR ENDORSEMENTS
      REFERENCES
      GLOSSARY
      INDEX

GLEIM'S
FUNDAMENTALS OF INSTRUCTING FAA KNOWLEDGE TEST

Table of Contents

2.2 HOW PILOTS LEARN

A. Learning in the broad sense is the change in behavior, knowledge, or sensorimotor skill as a result of experience, practice, effort, etc. We are interested in the narrower (and aviation-related) definitions of knowledge and the sensorimotor skills of flying an airplane in contrast to other behaviors.

   1. Stated another way, psychologists have defined many categories of learning, such as classical conditioning, trial-and-error learning, sensorimotor learning, verbal learning, concept learning, and rule learning (Wingfield, p. 8)[1]. We are interested in concept and rule learning (ground school) and sensorimotor learning (flight maneuvers).

---

[1] Arthur Wingfield, *Human Learning and Memory: An Introduction*, copyright © 1979 by Harper & Row.

2.   Wingfield (p. 25) sets forth a learning model consisting of three major stages: input, storage, and retrieval. Furthermore, he distinguishes between short-term and long-term memory, as diagramed below.

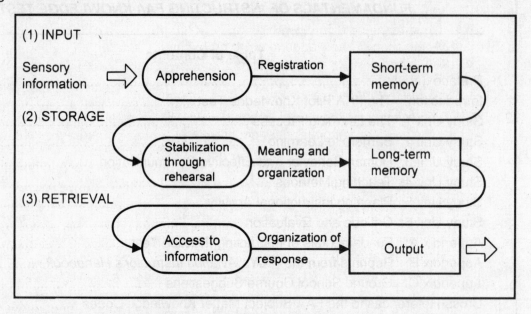

3.   This illustration is an oversimplification because there are several levels or variations of adding meaning and organization to information. Most aviation concepts are multidimensional. Much more can be learned about an airplane by walking around it, looking at it from all sides, getting into it, opening the cowling, etc., than by merely viewing a picture of the airplane, which is two-dimensional.

4.   Aviation concepts can be better understood by examining their multiple aspects. For example, a flight maneuver may be viewed in terms of the

- Objective of the maneuver
- Flight path of the airplane through the maneuver
- Related FARs
- Airplane - make and model
- Aerodynamic theory
- Weather, e.g., wind

- Airspeeds/power settings
- Visual cues - horizon
- Flight instrument cues
- Airplane configuration - flaps and gear
- Altitudes
- Other traffic

5.   The point here is that there are many dimensions to pilot knowledge in general and even to individual simple flight maneuvers.

6.   This multidimensionality describes understanding, i.e., relating many concepts, rules, relationships. The preceding list of multiple aspects of flight maneuvers is incomplete; take a few minutes to pencil in a few additional dimensions in the margin. As you study, try to interrelate data. Ask and answer the question WHY? to all rules, conclusions, and observations.

B. **Levels of Cognitive Ability**

1. In ascending order of complexity, one categorization (Bloom)[2] of the levels of knowledge is

    a. Recall knowledge
    b. Understanding to interpret
    c. Application of knowledge to solve problems
    d. Analytical skills
    e. Synthesis
    f. Ability to evaluate

2. The various levels of knowledge above simple recall coexist and tend to be cumulative. They constitute building blocks of cognitive processes. To interpret, you need some recall knowledge; to solve problems, you must understand to interpret, etc. A discussion of these levels of knowledge and an example of each is presented below and on the next page based on knowledge of Class D airspace and related knowledge.

    a. **Recall knowledge.** The first level is recall knowledge, e.g., definitions of technical terms and sources of information. The FAA pilot knowledge test questions often test this kind of knowledge, which is the most fundamental since it entails basic memorization.

        1) EXAMPLE: Class D airspace is the airspace around an airport at which a control tower is operating. Class D airspace normally extends from the surface to 2,500 ft. AGL within a 4.4-NM radius from the geographic center of the airport. All aircraft operations must be conducted with controller approval/direction.

    b. **Understanding to interpret.** The second level of knowledge is the understanding and interpretation of written and quantitative data. Questions at this level test understanding of concepts, including interrelationships within data. This level of knowledge is also called comprehension.

        1) EXAMPLE: If there is a blue airport symbol on the VFR sectional chart and the letters CT (meaning control tower) followed by a frequency, a pilot must use this communications frequency to talk with the controller in the control tower.

    c. **Application of knowledge to solve problems.** The third level of knowledge is problem solving. Questions at this level examine practical applications of concepts to solve a problem. Unfortunately, some problem solving is based only on recall knowledge.

        1) EXAMPLE: Application of this knowledge occurs when the pilot is actually operating the airplane and, about to enter Class D airspace, determines the control tower frequency and enters it on the radio, lifts the microphone to verbalize a call to the controller, and identifies his/her position to the controller.

    d. **Analytical skills.** The fourth level of knowledge is analytical ability, including identification of cause-and-effect relationships, internal inconsistencies or consistencies, relevant and irrelevant items, and underlying assumptions.

        1) EXAMPLE: The pilot is given a pattern entry and a traffic report regarding an airplane that is 3 mi. ahead. (S)he must enter the assigned pattern with the underlying assumption that (s)he will be able to adjust his/her glide slope properly for landing. (S)he may also disregard the traffic report as unimportant (depending, of course, on both airplanes' relative speeds), since the other airplane will be on the ground well ahead of him/her.

---

[2]Bloom's *Taxonomy of Educational Objectives*, copyright © 1956 by David McKay Company, Inc.

e. **Synthesis.** The fifth level is the ability to put parts together to form a new whole.

   1) EXAMPLE: All of the elements of analysis of the controller's clearance -- watching for other traffic and operating the airplane within its stated limitations in existing weather conditions -- must be synthesized to produce a successful approach to the airport.

f. **Ability to evaluate.** The sixth level is evaluative ability. What is the best (most effective) method (alternative)? Evaluation has in common with analysis and synthesis the consideration of qualitative as well as quantitative variables. Qualitative variables are usually multidimensional and thus cannot be meaningfully quantified or measured on a single dimension.

   1) EXAMPLE: In evaluating the approach to the airport, the pilot may determine that another type of pattern entry may better facilitate safety. (S)he may ask for an amended clearance, or his/her evaluation may indicate that the currently assigned entry is acceptable.

2.3 GROUND TRAINING

A. First and foremost: Ground training is extremely important to facilitate flight training. Each preflight and postflight discussion is as important as the actual flight instruction of each flight training lesson!

   1. Unfortunately, most students and many CFIs incorrectly overemphasize the in-airplane portion of a flight lesson.

      a. The airplane, all of its operating systems, ATC, other traffic, etc., are major distractions from the actual flight maneuver and the aerodynamic theory/factors underlying the maneuver.

      b. This is not to diminish the importance of dealing with operating systems, ATC, other traffic, etc.

   2. Formal ground school classes to support flight training generally do not exist except at aeronautical universities and some Part 141 programs. Most community college, adult education, and FBO ground schools are directed toward FAA pilot knowledge (written) tests.

      a. Ground school classes are important, and if you are fortunate to be able to participate, take advantage of them.

      b. Home study, in addition to pre- and postflight briefings, however, is essential to professional flight training.

B. **The effort and results are those of the student.** Instructors are responsible for directing student effort for optimum results. Prepare for each flight lesson so you know, before the fact, exactly what is going to happen and why. The more you prepare, the better you will do, both in execution of maneuvers and acquisition of knowledge.

C. These six steps are directed to you as a student CFI, but they also apply to your future student pilots and to you as their flight instructor.

   1. At the end of each flight lesson, find out exactly what is planned for the next flight lesson.

   2. At home, begin by reviewing everything that occurred during the last flight lesson -- preflight briefing, flight, and postflight briefing. Make notes on follow-up questions and discussion with your CFI to occur at the beginning of the next preflight briefing.

   3. Study all new flight maneuvers scheduled for the next flight lesson and review those flight maneuvers that warrant additional practice (refer to the appropriate study units in Part II of this book). Make notes on follow-up questions and discussion with your CFI to occur at the beginning of the next preflight briefing.

4.   Before each flight, sit down with your CFI for a preflight briefing. Begin with a review of the last flight lesson. Then focus on the current flight lesson. Go over each maneuver that is to be executed, including maneuvers to be reviewed from previous flight lessons.

5.   During each flight lesson, be diligent about safety (continuously check traffic and say so as you do it). Always use clearing turns. During maneuvers, compare your actual experience with your expectations (based on your prior knowledge from completing your Flight Maneuver Analysis Sheet).

6.   Your postflight briefing should begin with a self-critique by you, followed by evaluation by your CFI. Ask questions until you are satisfied that you have expert knowledge. Finally, develop a clear understanding of the time of your next flight lesson and the maneuvers to be covered then.

2.4 FLIGHT MANEUVER ANALYSIS SHEET (FMAS)

A.   We have developed a method of analyzing and studying flight maneuvers that incorporates 10 variables:

1. Maneuver	5. Altitude(s)	9. Traffic considerations
2. Objective	6. Airspeed(s)	10. Completion standards
3. Flight path	7. Control forces	
4. Power settings(s)	8. Time(s)	

B.   A copy of an FMAS (front and back) appears on pages 14 and 15 for your convenience. When you reproduce the forms for your own use and later for your students, photocopy on the front and back of a single sheet of paper to make the forms more convenient. The front side contains space for analysis of the above variables. The back side contains space for

1.   Make- and model-specific information

   a.   Weight
   b.   Airspeeds
   c.   Fuel
   d.   Center of gravity
   e.   Performance data

2.   Flight instrument review of maneuver

   a.   Attitude indicator .... AI
   b.   Airspeed indicator .... ASI
   c.   Turn coordinator .... TC
   d.   Heading indicator .... HI
   e.   Vertical speed indicator .... VSI
   f.   Altimeter .... ALT

3.   Common errors

C.   You should prepare/study/review an FMAS for each maneuver you intend to perform before each flight lesson. Photocopy the form (front and back) on single sheets of paper. Changes, amplifications, and other notes should be added subsequently. Blank sheets of paper should be attached (stapled) to the FMAS, including self-evaluations, "to do" items, questions for your CFI, etc., for your home study during your flight instruction program. FMASs are also very useful to prepare for the practical test.

1.   A major benefit of the FMAS is preflight lesson preparation. It serves as a means to discuss maneuvers with your CFI (and you with your student) before and after each flight. It emphasizes preflight planning, airplane make and model knowledge, flight instruments, and common errors.

2.   Also, the FMAS helps you, your future students, and pilots in general to focus on the operating characteristics of your/their airplane, including weight and balance. Weight and balance, which includes fuel, should be carefully reviewed prior to each flight.

CFI _____

Student _____

Date _____

GLEIM'S
FLIGHT MANEUVER ANALYSIS SHEET

1. **MANEUVER** _____

2. **OBJECTIVES/PURPOSE** _____
_____

3. **FLIGHT PATH (visual maneuvers)**

4. **POWER SETTINGS**   5. **ALT**   6. **A/S**

MP	RPM	SEGMENT OF MANEUVER		
____	____	a. _____	____	____
____	____	b. _____	____	____
____	____	c. _____	____	____

Pencil in expected indication on each of 6 flight instruments on reverse side.

7. **CONTROL FORCES**

    a. _____
    _____

    b. _____
    _____

    c. _____
    _____

8. **TIME(S), TIMING** _____

9. **TRAFFIC CONSIDERATIONS**   **CLEARING TURNS REQUIRED** _____
    _____

10. **COMPLETION STANDARDS/ATC CONSIDERATIONS** _____
_____
_____

AIRPLANE MAKE/MODEL _____

WEIGHT		AIRSPEEDS	
Gross	_____	V_{SO}	_____
Empty	_____	V_{S1}	_____
Pilot/Pasngrs	_____	V_X	_____
Baggage	_____	V_Y	_____
Fuel (gal × 6)	_____	V_A	_____
		V_{NO}	_____
		V_{NE}	_____

CENTER OF GRAVITY			
Fore Limit	_____	V_{FE}	_____
Aft Limit	_____	V_{LO}	_____
Current CG	_____	V_R	_____

```
┌─────────────────────────────┐
│   ( ASI )  ( AI )  ( ALT )   │
│   ( TC )   ( HI )  ( VSI )   │
└─────────────────────────────┘
```

FUEL

Capacity	L _____gal	R _____gal
Current Estimate	L _____gal	R _____gal
Endurance (Hr.)	_____	
Fuel-Flow -- Cruise (GPH)	_____	

PRIMARY vs. SUPPORTING INSTRUMENTS

(IFR maneuvers) -- instruments: AI, ASI, ALT, TC, HI, VSI, RPM and/or MP
(most relevant to instrument instruction)

	PITCH	BANK	POWER
ENTRY			
primary	_____	_____	_____
supporting	_____	_____	_____
ESTABLISHED			
primary	_____	_____	_____
supporting	_____	_____	_____

PERFORMANCE DATA

	Airspeed	Power* MP	RPM
Takeoff Rotation	_____	_____	_____
Climbout	_____	_____	_____
Cruise Climb	_____	_____	_____
Cruise Level	_____	_____	_____
Cruise Descent	_____	_____	_____
Approach**	_____	_____	_____
Approach to Land (Visual)	_____	_____	_____
Landing Flare	_____	_____	_____

* If you do not have a constant-speed propeller, ignore manifold pressure (MP).
**Approach speed is for holding and performing instrument approaches.

COMMON ERRORS

_____

_____

_____

2.5 FLIGHT TRAINING

A. After you complete your preflight briefing, including a review and critique of your student's FMAS, you should move out to the airplane and observe your student performing the preflight, engine start, taxi, etc. Critique his/her performance, including flight maneuvers to your practice area.

   1. For example, if poor radio technique or inadequate attention to traffic is a problem, explain both the deficiency and corrective action.

   2. Vary takeoffs and landings (normal, soft-field, short-field) and other maneuvers as a cumulative review process throughout the curriculum, e.g., do S-turns across a road after execution of a surprise emergency approach and landing.

B. As you approach the location appropriate to begin the intended flight maneuver, implement the telling and doing technique. It is called the demonstration-performance method and is also used for ground instruction.

   1. **Instructor tells, instructor does.** This is a continuation of preparing your student, which began during the preflight discussion. It is important that your demonstration conform to the explanation as closely as possible. If a deviation occurs, you should point it out and immediately explain why it occurred.

   2. **Student tells, instructor does.** This step assures you and your student that the explanation and demonstration have been adequate and are thoroughly understood. This is a transition to the next step. Frequently, this step is changed to **instructor tells, student does**, which may fit the learning/teaching style of the student/instructor.

   3. **Student tells, student does.** This step is where learning takes place and where performance habits are formed. You must be alert during this step to detect any errors in technique and to prevent the formation of bad habits.

   4. **Student does, instructor evaluates.** During this step, you review what has been covered during the flight, and determine to what extent your student has met the objectives outlined during the preflight discussion.

C. During the first step (instructor tells, instructor does) and second step (student tells, instructor does), you will demonstrate one complete circuit of the maneuver that you are introducing to your student.

   1. Point out the pattern that (s)he is supposed to fly, noting the visual references to be used.

      a. As you correct for any wind, mention the wind direction and how you are correcting for this condition.

      b. Keep your explanation simple and to the point.

   2. This step is a transfer from the preflight discussion to the actual performance of the maneuver.

D. During the third step (student tells, student does), you should instruct and correct your student's errors as each part of the maneuver is being performed.

   1. During this step, you will concentrate on specific items as they are encountered (e.g., wind drift correction errors).

E.   The fourth step (student does, instructor evaluates) is when you take an overall look at your student's performance of the maneuver.

   1.   Your student may have a specific problem (e.g., altitude control) that prevents him/her from seeing the whole maneuver.

      a.   While concentrating to maintain altitude, (s)he fails to plan for the next part of the maneuver.

   2.   You should redirect your student to the basic elements of the maneuver, i.e., planning, orientation, and airplane control.

   3.   Stress that all of the parts are required to complete the maneuver successfully.

F.   Note that each maneuver should be taught based on both outside visual references and reference to flight instruments. This is known as integrated flight instruction.

2.6   INTEGRATED FLIGHT INSTRUCTION

A.   The objective of integrated flight instruction technique is the formation of firm habit patterns for the observance of and reliance on flight instruments by your student.

   1.   The ability to fly in instrument weather is not the objective of this type of primary training, although it does greatly facilitate later instrument flight training.

B.   By requiring your student to perform all flight maneuvers (except those requiring a ground reference) by reference to instruments, as well as by outside references, (s)he will develop from the start the habit of continuously monitoring his/her own and the airplane's performance.

   1.   The early establishment of proper habits of instrument cross-check, instrument interpretation, and airplane control will be of great assistance to your student in gaining competence in the area of operation discussed in Part II, Study Unit XII, "Basic Instrument Maneuvers", starting on page 455, which is included in the Private Pilot Practical Test.

   2.   The habits formed at this time will also give your student a firm foundation for later training for an instrument rating.

C.   During the conduct of integrated flight instruction, you must be especially vigilant for other air traffic while your student is flying by reference to flight instruments. You must guard against having attention diverted to your student's performance for extended periods.

   1.   At the same time, you must ensure that your student develops, from the start of his/her training, the habit of looking for other traffic at all times when (s)he is not flying under simulated instrument conditions.

2.7 RIGHT SEAT VS. LEFT SEAT

A. Until now, you have flown primarily from the left seat. The left seat is used only by convention and probably because we drive on the right side of the road (with the driver on the left side). Thus, the only advantage of the left side is that we are able to transfer the feeling of sitting on the left side of an automobile to the left (rather than right) side of an airplane.

    1. Also, flight instruments are generally positioned for the convenience of the left seat pilot on single-engine aircraft. Thus, you have to concentrate on learning to analyze different visual cues and instrument indications from the right seat.

    2. This viewing error of the instrument indications appearing to be in a different location when viewed from the right seat is called parallax.

B. To continue this convention, your students will be flying from the left side, and you will be on the right side. Therefore, you need to become acquainted with the right side.

    1. **Do not wait** to become acquainted during your first lesson.

    2. Visit the airplane you will be taking lessons in and spend 30+ min. gaining familiarity with the flight, engine, and other controls while sitting in the right seat.

        a. Simulate a takeoff, landing, and several flight maneuvers to get a feel for where the controls are and how they feel from the right seat.

C. After you have done this, you will be ready to fly with your CFI in the left seat.

    1. When you try to fly the airplane with your right hand, if you have not experienced this before, you will probably find that your "feel" for the airplane does not readily transfer over to your right hand.

        a. This is especially evident during your roundout and landing.

    2. Many pilots beginning CFI training find that they must fly for several hours as the sole manipulator of the controls before becoming proficient in the right seat.

STUDY UNIT THREE
FLIGHT INSTRUCTOR RELATED FARs

Technically, FAR is the abbreviation for Federal Acquisition Regulation, and the FAA is pushing 14 CFR as the acronym for Federal Aviation Regulation. CFR stands for Code of Federal Regulations, and the Federal Aviation Regulations are in Title 14. We, however, continue to use the popular and widely used "FAR" acronym.

This study unit contains outlines that paraphrase Subpart H of Part 61, which is specific to flight instructors. Also included are those regulations in Parts 61, 91, and 141 that are especially important to a flight instructor.

Gleim's *FAR/AIM* contains the reprint of all general aviation-related FARs (Parts 1, 43, 61, 67, 71, 73, 91, 97, 103, 105, 119, 135, 137, 141, and 142) and NTSB Part 830. You should have the *FAR/AIM* in your possession and can reference it during your training and on your practical test.

(continued next page)

3.1 PART 1 - DEFINITIONS AND ABBREVIATIONS

This part contains three sections (definitions, abbreviations, and rules of construction). The latter section contains legalese and is not relevant to most pilots. Part 1 - Definitions and Abbreviations of *FAR/AIM* is a seven-page discussion that contains over 160 definitions and over 80 abbreviations.

3.2 PART 61 - CERTIFICATION: PILOTS, FLIGHT INSTRUCTORS, AND GROUND INSTRUCTORS

Part 61 contains nine subparts, labeled A through I.

Subpart A -- General
Subpart B -- Aircraft Ratings and Pilot Authorizations
Subpart C -- Student Pilots
Subpart D -- Recreational Pilots
Subpart E -- Private Pilots
Subpart F -- Commercial Pilots
Subpart G -- Airline Transport Pilots
Subpart H -- Flight Instructors
Subpart I -- Ground Instructors

Subpart H is outlined beginning on page 22. Prior to the outline of Subpart H, we list all of the section titles of Subparts A through F for your review and convenience. Refer as necessary and appropriate to *FAR/AIM* for these FARs. You should be conversant with Part 61 rules, including pilot certification requirements for recreational, private, and commercial pilot certificates with a single-engine airplane rating. You are not required to know any airline transport pilot or ground instructor requirements.

Subpart A -- General

61.1	Applicability and Definitions
61.3	Requirement for Certificates, Ratings, and Authorizations
61.4	Qualification and Approval of Flight Simulators and Flight Training Devices
61.5	Certificates and Ratings Issued under This Part
61.7	Obsolete Certificates and Ratings
61.11	Expired Pilot Certificates and Reissuance
61.13	Issuance of Airman Certificates, Ratings, and Authorizations
61.14	Refusal to Submit to a Drug Test
61.15	Offenses Involving Alcohol or Drugs
61.16	Refusal to Submit to an Alcohol Test or to Furnish Test Results
61.17	Temporary Certificate
61.19	Duration of Pilot and Instructor Certificates
61.21	Duration of a Category II and a Category III Pilot Authorization (for Other than Part 121 or Part 135 Use)
61.23	Medical Certificates: Requirement and Duration
61.25	Change of Name
61.27	Voluntary Surrender or Exchange of Certificate
61.29	Replacement of Lost or Destroyed Airman or Medical Certificate or Knowledge Test Report
61.31	Type Rating Requirements, Additional Training, and Authorization Requirements
61.33	Tests: General Procedure
61.35	Knowledge Test: Prerequisites and Passing Grades
61.37	Knowledge Test: Cheating or Other Unauthorized Conduct
61.39	Prerequisites for Practical Tests
61.41	Flight Instruction Received from Flight Instructors Not Certificated by the FAA
61.43	Practical Tests: General Procedures
61.45	Practical Tests: Required Aircraft and Equipment
61.47	Status of an Examiner Who Is Authorized by the Administrator to Conduct Practical Tests
61.49	Retesting after Failure
61.51	Pilot Logbooks
61.53	Prohibition on Operations during Medical Deficiency
61.55	Second-in-Command Qualifications
61.56	Flight Review
61.57	Recent Flight Experience: Pilot in Command
61.58	Pilot-in-Command Proficiency Check: Operation of Aircraft Requiring More than One Required Pilot Flight Crewmember
61.59	Falsification, Reproduction, or Alteration of Applications, Certificates, Logbooks, Reports, or Records
61.60	Change of Address

Subpart B -- Aircraft Ratings and Special Certificates

Subpart C -- Student Pilots

Subpart D -- Recreational Pilots

Subpart E -- Private Pilots

Subpart F -- Commercial Pilots

Subpart H -- Flight Instructors

61.181 Applicability

1.   This subpart prescribes the requirements for the issuance of flight instructor certificates and ratings, the conditions under which those certificates and ratings are necessary, and the limitations on those certificates and ratings.

61.183 Eligibility Requirements

1. To be eligible for a flight instructor certificate, you must

   a. Be at least 18 years of age.

   b. Read, write, and converse fluently in the English language.

   c. Hold a commercial pilot or airline transport pilot certificate with an aircraft rating appropriate to the flight instructor rating sought.

   d. Hold an instrument rating if applying for an airplane or instrument instructor rating.

   e. Receive a logbook endorsement from an authorized instructor for the fundamentals of instructing listed in FAR 61.185 appropriate to the required knowledge test.

   NOTE: This endorsement is required for your practical test, not the knowledge test.

   f. Pass a knowledge test on each of the areas listed in FAR 61.185.

   g. Pass a practical test on the items listed in FAR 61.187.

   h. Present a logbook endorsement from the CFI who provided spin entry, spin, and spin recovery training in an airplane of the appropriate category that is certificated for spins.

      1) The examiner conducting the practical test may either accept the spin training endorsement or require demonstration of the spin entry, spin, and spin recovery maneuver during the flight portion of the practical test.

61.185 Aeronautical Knowledge

1. You must receive and log ground training from an authorized instructor on the fundamentals of instructing, to include

   a. The learning process
   b. Elements of effective teaching
   c. Student evaluation, quizzing, and testing
   d. Course development
   e. Lesson planning
   f. Classroom instruction techniques

2. You must also have logged ground training from an authorized instructor (AGI or a CFI) in all the subjects in ground training for the recreational, private, and commercial certificates and in all the instrument rating requirements if an instrument instructor certificate is sought.

61.187 Flight Proficiency

1. You must receive and log flight and ground training from an authorized instructor on the areas of operations explained in Part II, Study Units I through XV, of this book.

   a. Your logbook must contain an endorsement from an authorized instructor certifying that you are proficient to pass the practical test.

61.189 Flight Instructor Records

1. Each CFI must sign the logbook of each person to whom (s)he gives flight or ground training.

2. A CFI must maintain a record in a logbook or a separate document of the following:

   a. The name of each person whose logbook or student pilot certificate (s)he has endorsed for solo flight privileges and the date of the endorsement.

   b. The name of each person (s)he has endorsed for a knowledge or practical test, including the kind of test, the date, and the results.

3. These records shall be maintained for a period of at least 3 years.

61.191 Additional Flight Instructor Ratings

1. The holder of a flight instructor certificate who applies for an additional rating on that certificate must

    a. Hold a valid pilot certificate with ratings appropriate to the instructor rating sought

    b. Have at least 15 hr. as pilot in command in the category and class of aircraft appropriate to the rating sought

    c. Pass the knowledge test (if appropriate) and practical test prescribed for the rating sought

61.193 Flight Instructor Privileges

1. A CFI is authorized within the limitations of his/her flight instructor certificate and ratings to give training and endorsements that are required for and relate to

    a. A student pilot certificate

    b. A pilot certificate

    c. A flight instructor certificate

    d. A ground instructor certificate

    e. An aircraft rating

    f. An instrument rating

    g. A flight review, operating privilege, or recency of experience requirement of FAR Part 61

    h. A practical test

    i. A knowledge test

61.195 Flight Instructor Limitations and Qualifications

1. A CFI may not conduct more than 8 hr. of flight training in any 24-consecutive-hour period.

2. A CFI may not conduct flight training in any aircraft for which (s)he does not hold a pilot and flight instructor certificate in the category, class, and type (if a type rating is required) of aircraft.

3. A CFI must hold an instrument rating on his/her flight instructor certificate (i.e., be a CFII) to provide instrument flight training to someone who wants to add an instrument rating to his/her certificate.

4. A CFI may not endorse a

    a. Student pilot certificate unless (s)he has

       1) Given that student the flight training required for solo flight privileges and

       2) Determined that the student is prepared to conduct the flight safely under known circumstances, subject to any limitations listed in the student's logbook that the instructor considers necessary for the safety of the flight.

    b. Student pilot certificate and his/her logbook for a solo cross-country flight, unless the CFI has determined the student's flight preparation, planning, equipment, and proposed procedures are adequate for the proposed flight under the existing conditions and within any limitations listed in the logbook that the instructor considers necessary for the safety of the flight.

    c. Student pilot certificate and his/her logbook for solo flight in Class B airspace or at an airport within Class B airspace unless that CFI has

       1) Given that student ground and flight training in that Class B airspace or at that airport.

       2) Determined that the student is proficient to operate the aircraft safely.

    d.   Logbook of a recreational pilot, unless the CFI has

       1)   Given that pilot the required ground and flight training
       2)   Determined that the pilot is proficient to operate the airplane safely

    e.   Logbook of a pilot for a flight review, unless the CFI has conducted a review of that pilot in accordance with FAR 61.56, Flight Review

    f.   Logbook of a pilot for an instrument proficiency check, unless the CFII has tested that pilot in accordance with the requirements of FAR 61.57(d), Instrument Proficiency Check

5.   A CFI may not give flight training in an aircraft that requires the pilot in command to hold a type rating unless (s)he holds a type rating for that aircraft on his/her pilot certificate.

6.   A CFI may not give training required for the issuance of a certificate or rating in a multiengine airplane unless (s)he has at least 5 flight hours of PIC time in that specific make and model.

7.   A CFI must perform all training in an airplane that has fully functioning dual controls.

    a.   However, only instrument flight training may be given in a single-engine airplane equipped with a single, functioning throwover control wheel when the CFII has determined the flight can be conducted safely, and the person manipulating the controls has at least a private pilot certificate with ASEL ratings.

8.   A CFI who provides flight training for a certificate or rating must provide an airplane that meets the following two requirements:

    a.   The airplane must have at least two pilot stations.

    b.   For a single-place airplane, the presolo flight training must be in an airplane with two pilot stations and of the same category, class, and type, if appropriate.

9.   A CFI who provides training to an initial applicant for a flight instructor certificate must have held a flight instructor certificate for at least 24 months and have given at least 200 hr. of flight training as a CFI.

10.   A CFI shall not make any self-endorsement for a certificate, rating, flight review, authorization, operating privilege, practical test, or knowledge test.

61.197 Renewal of Flight Instructor Certificates

1.   A person who holds a flight instructor certificate that has not expired may renew his/her certificate for an additional 24 calendar months by one of the following methods:

    a.   Pass a practical test for one of the ratings listed on the current flight instructor certificate or for an additional flight instructor rating.

    b.   Present to an FAA inspector

       1)   A record of training students that shows within the preceding 24 calendar months the CFI has endorsed at least five students for a practical test and at least 80% (4 out of 5) passed on the first attempt

       2)   A record that shows the CFI has served as a company check pilot, chief flight instructor, company check airman, or flight instructor in a Part 121 or 135 operation, or in a position involving the regular evaluation of pilots

       3)   A graduation certificate showing the person has successfully completed an approved flight instructor refresher course within the 3 calendar months before the expiration month of his/her certificate

2.   The expiration of a renewed CFI certificate will be 24 calendar months from

    a.   The month the renewal requirements are accomplished or

    b.   The month of expiration of the current CFI certificate, provided

        1)   The renewal requirements are accomplished within the 3 calendar months preceding the current expiration month.

        2)   The successful completion of a flight instructor refresher course was within 3 calendar months preceding the current expiration month.

61.199 Expired Flight Instructor Certificates and Ratings

1.   The holder of an expired CFI certificate may exchange it for a new one with the same ratings by passing the practical test prescribed by the FARs for one of the ratings listed on the expired CFI certificate.

3.3 PART 67 - MEDICAL STANDARDS AND CERTIFICATION

Part 67 contains Subparts A through E listed below, which are in *FAR/AIM*.

Subpart A -- General

67.1	Applicability
67.3	Issue
67.7	Access to the National Driver Register

Subpart B -- First-Class Airman Medical Certificate

67.101	Eligibility
67.103	Eye
67.105	Ear, Nose, Throat, and Equilibrium
67.107	Mental
67.109	Neurologic
67.111	Cardiovascular
67.113	General Medical Condition
67.115	Discretionary Issuance

Subpart C -- Second-Class Airman Medical Certificate

67.201	Eligibility
67.203	Eye
67.205	Ear, Nose, Throat, and Equilibrium
67.207	Mental
67.209	Neurologic
67.211	Cardiovascular
67.213	General Medical Condition
67.215	Discretionary Issuance

Subpart D -- Third-Class Airman Medical Certificate

67.301	Eligibility
67.303	Eye
67.305	Ear, Nose, Throat, and Equilibrium
67.307	Mental
67.309	Neurologic
67.311	Cardiovascular
67.313	General Medical Condition
67.315	Discretionary Issuance

Subpart E -- Certification Procedures

67.401	Special Issuance of Medical Certificates
67.403	Applications, Certificates, Logbooks, Reports, and Records: Falsification, Reproduction, or Alteration; Incorrect Statements
67.405	Medical Examinations: Who May Give
67.407	Delegation of Authority
67.409	Denial of Medical Certificate
67.411	Medical Certificates by Flight Surgeons of Armed Forces
67.413	Medical Records
67.415	Return of Medical Certificate after Suspension or Revocation

3.4 PART 91 - GENERAL OPERATING AND FLIGHT RULES

Part 91 contains Subparts A through J. You should be conversant with Subparts A through E.

Subpart A -- General

Subpart B -- Flight Rules

Subpart C -- Equipment, Instrument, and Certificate Requirements

Subpart D -- Special Flight Operations

Subpart E -- Maintenance, Preventive Maintenance, and Alterations

Subpart F -- Large and Turbine-Powered Multiengine Airplanes and Fractional Ownership Program Aircraft

Subpart G -- Additional Equipment and Operating Requirements for Large and Transport Category Aircraft

Subpart H -- Foreign Aircraft Operations and Operations of U.S.-Registered Civil Aircraft Outside of the United States

Subpart I -- Operating Noise Limits

Subpart J -- Waivers

Subpart A -- General

91.1	Applicability
91.3	Responsibility and Authority of the Pilot in Command
91.5	Pilot in Command of Aircraft Requiring More than One Required Pilot
91.7	Civil Aircraft Airworthiness
91.9	Civil Aircraft Flight Manual, Marking, and Placard Requirements
91.11	Prohibition against Interference with Crewmembers
91.13	Careless or Reckless Operation
91.15	Dropping Objects
91.17	Alcohol or Drugs
91.19	Carriage of Narcotic Drugs, Marihuana, and Depressant or Stimulant Drugs or Substances
91.21	Portable Electronic Devices
91.23	Truth-in-Leasing Clause Requirement in Leases and Conditional Sales Contracts
91.25	Aviation Safety Reporting Program: Prohibition against Use of Reports for Enforcement Purposes

Subpart B -- Flight Rules

91.101	Applicability
91.103	Preflight Action
91.105	Flight Crewmembers at Stations
91.107	Use of Safety Belts, Shoulder Harnesses, and Child Restraint Systems
91.109	Flight Instruction; Simulated Instrument Flight and Certain Flight Tests
91.111	Operating near Other Aircraft
91.113	Right-of-Way Rules: Except Water Operations
91.115	Right-of-Way Rules: Water Operations
91.117	Aircraft Speed
91.119	Minimum Safe Altitudes: General
91.121	Altimeter Settings
91.123	Compliance with ATC Clearances and Instructions
91.125	ATC Light Signals
91.126	Operating on or in the Vicinity of an Airport in Class G Airspace
91.127	Operating on or in the Vicinity of an Airport in Class E Airspace
91.129	Operations in Class D Airspace
91.130	Operations in Class C Airspace
91.131	Operations in Class B Airspace
91.133	Restricted and Prohibited Areas
91.135	Operations in Class A Airspace
91.137	Temporary Flight Restrictions in the Vicinity of Disaster/Hazard Areas
91.138	Temporary Flight Restrictions in National Disaster Areas in the State of Hawaii
91.139	Emergency Air Traffic Rules
91.141	Flight Restrictions in the Proximity of the Presidential and Other Parties
91.143	Flight Limitation in the Proximity of Space Flight Operations
91.144	Temporary Restriction on Flight Operations during Abnormally High Barometric Pressure Conditions
91.145	Management of Aircraft Operations in the Vicinity of Aerial Demonstrations and Major Sporting Events
91.151	Fuel Requirements for Flight in VFR Conditions
91.153	VFR Flight Plan: Information Required
91.155	Basic VFR Weather Minimums
91.157	Special VFR Weather Minimums
91.159	VFR Cruising Altitude or Flight Level
91.167	Fuel Requirements for Flight in IFR Conditions
91.169	IFR Flight Plan: Information Required
91.171	VOR Equipment Check for IFR Operations
91.173	ATC Clearance and Flight Plan Required
91.175	Takeoff and Landing under IFR
91.177	Minimum Altitudes for IFR Operations
91.179	IFR Cruising Altitude or Flight Level
91.180	Operations Within Airspace Designated as Reduced Vertical Separation Minimum Airspace
91.181	Course to Be Flown
91.183	IFR Radio Communications
91.185	IFR Operations: Two-Way Radio Communications Failure
91.187	Operation under IFR in Controlled Airspace: Malfunction Reports
91.189	Category II and III Operations: General Operating Rules
91.191	Category II and Category III Manual
91.193	Certificate of Authorization for Certain Category II Operations

Subpart C -- Equipment, Instrument, and Certificate Requirements

91.203   Civil Aircraft: Certifications Required
91.205   Powered Civil Aircraft with Standard Category U.S. Airworthiness Certificates: Instrument and
    Equipment Requirements
91.207   Emergency Locator Transmitters
91.209   Aircraft Lights
91.211   Supplemental Oxygen
91.213   Inoperative Instruments and Equipment
91.215   ATC Transponder and Altitude Reporting Equipment and Use
91.217   Data Correspondence between Automatically Reported Pressure Altitude Data and the
    Pilot's Altitude Reference
91.219   Altitude Alerting System or Device: Turbojet-Powered Civil Airplanes
91.221   Traffic Alert and Collision Avoidance System Equipment and Use
91.223   Terrain Awareness and Warning System

Subpart D -- Special Flight Operations

91.303   Aerobatic Flight
91.305   Flight Test Areas
91.307   Parachutes and Parachuting
91.309   Towing: Gliders and Unpowered Ultralight Vehicles
91.311   Towing: Other than under Section 91.309
91.313   Restricted Category Civil Aircraft: Operating Limitations
91.315   Limited Category Civil Aircraft: Operating Limitations
91.317   Provisionally Certificated Civil Aircraft: Operating Limitations
91.319   Aircraft Having Experimental Certificates: Operating Limitations
91.321   Carriage of Candidates in Elections
91.323   Increased Maximum Certificated Weights for Certain Airplanes Operated in Alaska
91.325   Primary Category Aircraft: Operating Limitations
91.327   Aircraft having a Special Airworthiness Certificate in the Light Sport Category: Operating Limitations

Subpart E -- Maintenance, Preventive Maintenance, and Alterations

91.401   Applicability
91.403   General
91.405   Maintenance Required
91.407   Operation after Maintenance, Preventive Maintenance, Rebuilding, or Alteration
91.409   Inspections
91.410   Special Maintentance Program Requirements
91.411   Altimeter System and Altitude Reporting Equipment Tests and Inspections
91.413   ATC Transponder Tests and Inspections
91.415   Changes to Aircraft Inspection Programs
91.417   Maintenance Records
91.419   Transfer of Maintenance Records
91.421   Rebuilt Engine Maintenance Records

Subpart F -- Large and Turbine-Powered Multiengine Airplanes and Fractional Ownership Program Aircraft

91.501   Applicability
91.503   Flying Equipment and Operating Information
91.505   Familiarity with Operating Limitations and Emergency Equipment
91.507   Equipment Requirements: Over-the-Top or Night VFR Operations
91.509   Survival Equipment for Overwater Operations
91.511   Radio Equipment for Overwater Operations
91.513   Emergency Equipment
91.515   Flight Altitude Rules
91.517   Passenger Information
91.519   Passenger Briefing
91.521   Shoulder Harness
91.523   Carry-on Baggage
91.525   Carriage of Cargo
91.527   Operating in Icing Conditions
91.529   Flight Engineer Requirements
91.531   Second-in-Command Requirements
91.533   Flight Attendant Requirements
91.535   Stowage of Food, Beverage, and Passenger Service Equipment during Aircraft Movement on the Surface,
    Takeoff, and Landing

Subpart G -- Additional Equipment and Operating Requirements for Large and Transport Category Aircraft

91.601   Applicability
91.603   Aural Speed Warning Device
91.605   Transport Category Civil Airplane Weight Limitations
91.607   Emergency Exits for Airplanes Carrying Passengers for Hire
91.609   Flight Recorders and Cockpit Voice Recorders
91.611   Authorization for Ferry Flight with One Engine Inoperative
91.613   Materials for Compartment Interiors

Subpart H -- Foreign Aircraft Operations and Operations of U.S.-Registered Civil Aircraft Outside of the United States

91.701   Applicability
91.702   Persons on Board
91.703   Operations of Civil Aircraft of U.S. Registry outside of the United States
91.705   Operations within Airspace Designated as Minimum Navigation Performance Specification Airspace
91.706   Operations within Airspace Designated as Reduced Vertical Separation Minimum Airspace
91.707   Flights between Mexico or Canada and the United States
91.709   Operations to Cuba
91.711   Special Rules for Foreign Civil Aircraft
91.713   Operation of Civil Aircraft of Cuban Registry
91.715   Special Flight Authorizations for Foreign Civil Aircraft

Subpart I -- Operating Noise Limits

91.801   Applicability: Relation to Part 36
91.803   Part 125 Operators: Designation of Applicable Regulations
91.805   Final Compliance: Subsonic Airplanes
91.815   Agricultural and Fire Fighting Airplanes: Noise Operating Limitations
91.817   Civil Aircraft Sonic Boom
91.819   Civil Supersonic Airplanes that Do Not Comply with Part 36
91.821   Civil Supersonic Airplanes: Noise Limits
91.851   Definitions
91.853   Final Compliance: Civil Subsonic Airplanes
91.855   Entry and Nonaddition Rule
91.857   Stage 2 Operations Outside of the 48 Contiguous United States, and Authorization for Maintenance
91.858   Special Flight Authorizations for Non-revenue Stage 2 Operations
91.861   Base Level
91.863   Transfers of Stage 2 Airplanes with Base Level
91.865   Phased Compliance for Operators with Base Level
91.867   Phased Compliance for New Entrants
91.869   Carry-forward Compliance
91.871   Waivers from Interim Compliance Requirements
91.873   Waivers from Final Compliance
91.875   Annual Progress Reports
91.877   Annual Reporting of Hawaiian Operations

Subpart J -- Waivers

91.903   Policy and Procedures
91.905   List of Rules Subject to Waivers

3.5 PART 141 - PILOT SCHOOLS

Subpart A -- General

141.1 Applicability

1.   This part prescribes the requirements for issuing pilot school certificates, provisional pilot school certificates, and associated ratings and the general operating rules for the holders of those certificates and ratings.

141.3 Certificate Required

1.   No one can operate as a certificated pilot school without an FAA certificate.

141.5 Requirements for a Pilot School Certificate

1.   An applicant may be issued a pilot school certificate if the applicant

   a.   Completes the necessary application.
   b.   Holds a provisional pilot school certificate for at least 24 months before applying for a pilot school certificate.
   c.   Meets the applicable requirements of Subparts A through C of Part 141.
   d.   Has trained and recommended at least 10 students for a knowledge or practical test for a pilot, flight, or ground instructor certificate, an additional rating, or any combination, and at least 80% of all tests administered were passed on the first attempt within the preceding 24 months.

141.7 Provisional Pilot School Certificate

1.   An applicant who meets the requirements of Subparts A through C of Part 141 but does not meet the recent training activity requirements of item d. in 141.5 above may be issued a provisional pilot school certificate.

141.9 Examining Authority

1.   An applicant is issued an examining authority for that flight school if (s)he meets the requirements of Part 141 Subpart D, Examining Authority (141.61 through 141.67).

141.11 Pilot School Ratings

1.   Ratings are issued along with each pilot school certificate specifying the courses the school can conduct. Course ratings available are

   a.   Recreational pilot
   b.   Private pilot
   c.   Commercial pilot
   d.   Instrument rating
   e.   Airline transport pilot
   f.   Flight instructor
   g.   Flight instructor instrument
   h.   Ground instructor
   i.   Additional aircraft category or class rating
   j.   Aircraft type rating
   k.   Pilot refresher
   l.   Flight instructor refresher
   m.   Ground instructor refresher
   n.   Agricultural aircraft operations
   o.   Rotorcraft external-load operations
   p.   Special operations
   q.   Test pilot
   r.   Pilot ground school

141.13 Application for Issuance, Amendment, or Renewal

1.   Application for issuance, amendment, or renewal of a certificate must be done in a manner described by the FAA. Two copies of each proposed training course outline are required for issuance or amendment of a certificate or rating.

141.17 Duration of Certificate and Examining Authority

1.   Unless suspended, surrendered, or revoked, a pilot school certificate expires

   a.   At the end of the 24th month after the month in which it was issued or renewed

   b.   When a change in ownership of the school or facilities occurs, unless within 30 days after that date an application is made for an appropriate amendment and no change in the facilities, instructor, personnel, or training course is involved

   c.   Upon notice by the FAA that the school has failed for more than 60 days to maintain the facilities, aircraft, and personnel for at least one of its approved courses

2.   The examining authority on a pilot school certificate expires when that certificate expires or is suspended, revoked, or surrendered.

141.18 Carriage of Narcotic Drugs, Marijuana, and Depressant or Stimulant Drugs or Substances

1.   If a certificate holder permits any of its aircraft to engage in any operation in violation of 91.19, Carriage of Narcotic Drugs, that operation is the basis for suspending or revoking that certificate.

141.19 Display of Certificate

1.   The holder of a pilot school certificate shall display the certificate in a place that is easy for the public to see.

141.21 Inspections

1.   The holder of a pilot school certificate shall allow the FAA to inspect the school and its facilities, equipment, and records to determine its compliance with the regulations.

141.23 Advertising Limitations

1.   Advertising by certificated pilot schools may not be false or misleading.

2.   If a school is certificated, it must differentiate between courses that have been approved by the FAA and those that have not.

3.   A school that moves or whose certificate has expired or been revoked must promptly remove all indications that the school is certificated by the FAA.

141.25 Business Office and Operations Base

1.   Each certificated pilot school shall maintain a principal business office with a mailing address as shown on the school's certificate.

   a.   The business office shall have adequate facilities and equipment to maintain school files and operate the business of the school.

   b.   The office may not be shared with or used by another pilot school.

2.   Each certificated school shall notify the FSDO at least 30 days before changing the location of its business office or base of operations.

3.   Before training can be conducted at a new operating base, the base must have been inspected and any amendments to training courses approved.

141.27 Renewal of Certificates and Ratings

1. Renewal of pilot school certificates and ratings are for 24 months and must be made not less than 30 days before expiration of the current school certificate.

2. Provisional pilot school certificates may not be renewed but rather may be used as a basis for applying for a pilot school certificate not less than 30 days before the provisional certificate expires.

   a. After expiration, a provisional pilot school holder may not reapply for a like certificate for at least 180 days.

Subpart B -- Personnel, Aircraft, and Facilities Requirements

141.31 Applicability

1. This subpart prescribes the personnel, aircraft, and physical facility requirements for a pilot school or a provisional pilot school certificate.

141.33 Personnel

1. An applicant for a pilot school certificate must show that (s)he has adequate personnel and instructors, including a chief instructor for each course of training, who are qualified and competent to perform their duties.

2. Each dispatcher, aircraft handler, line crewman, and serviceman must be instructed in the proper procedures of his/her job.

3. Qualified personnel may be used in more than one position.

4. Each instructor to be used must hold a flight or ground instructor certificate, as appropriate, with ratings for the course of instruction and aircraft used.

5. Each applicant shall designate a chief flight instructor for each course of training who meets the requirements of FAR 141.35, Chief Instructor Qualifications.

6. When necessary, an assistant chief flight instructor shall be designated who will assist and serve in the chief's absence.

7. Under certain conditions, a pilot school may designate a person to be a check instructor.

141.35 Chief Instructor Qualifications

1. Lists required qualifications in detail

141.36 Assistant Chief Instructor Qualifications

1. Lists required qualifications in detail

141.37 Check Instructor Qualifications

1. Lists required qualifications in detail

141.38 Airports

1. Applicants for pilot or provisional pilot school certificates must demonstrate that they have continuous use of the airport at which training flights originate.

2. The airport must have at least one runway or takeoff area that allows training aircraft to make normal takeoffs and landings at full gross weight under normal conditions with adequate safety margins.

3. The airport must have a wind direction indicator that is visible from the ends of each runway at ground level.

4.   The airport must have a traffic direction indicator when the airport does not have an operating control tower and UNICOM advisories are not available.

5.   Each airport used for night training flights must have permanent runway lights.

141.39 Aircraft

1.   An applicant for a pilot school certificate must show that each aircraft used by that school in flight instruction and solo flights meets the following requirements:

   a.   It must be registered as a U.S. civil aircraft.

   b.   It must be certificated in the standard or primary airworthiness category (with some specialized exceptions).

   c.   It must be maintained as a "for hire" aircraft in accordance with the requirements of Part 91.

   d.   For instruction, it must be at least a two-seat aircraft with power and flight controls accessible to both pilots.

   e.   For use in IFR, it must be equipped and maintained for IFR operations.

      1)   This does not apply to simulated instrument flight.

141.41 Flight Simulators, Flight Training Devices, and Training Aids

1.   Applicants for the pilot school certificate or provisional pilot certificate must meet detailed requirements as to the use of and physical arrangements regarding flight simulators and flight training devices.

2.   Each training aid, including audiovisuals and aircraft components, must be accurate and appropriate for the course for which it is listed.

141.43 Pilot Briefing Areas

1.   An applicant for a pilot school certificate must show continuous access to a briefing area to be used for training at the airport where training flights originate.

2.   The training area must provide adequate shelter for students waiting to go on their training flights.

3.   The training area must be equipped to conduct pilot briefings.

4.   Schools approved under FAR Part 141 to teach the instrument rating or commercial pilot courses must have a telephone line available to the Flight Service Station, or the school may be located within easy walking distance of the FSS.

141.45 Ground Training Facilities

1.   Each room, training booth, or other space used for instruction purposes must be heated, lighted, and ventilated to conform to local building, sanitation, and health codes and must also facilitate the purpose for which it is intended.

Subpart C -- Training Course Outline and Curriculum

141.51 Applicability

1.   This subpart prescribes the curriculum and course outline requirements for the issuance of a pilot school or provisional school certificate and ratings.

141.53 Approval Procedures for a Training Course: General

1. Applicants for pilot or provisional pilot school certificates must obtain FAA approval of each training course outline for which a certificate and rating are sought.

2. Applications must be made at least 30 days before any training is conducted under that training course.

141.55 Training Course: Contents

1. This section gives the content measurements for each course outline and each course syllabus that is to be taught.

141.57 Special Curricula

1. Pilot school or provisional pilot school certificate applicants may apply to conduct special courses if these courses will produce results equivalent to those produced by normal Part 141 courses and Part 61 curricula.

Subpart D -- Examining Authority

141.61 Applicability

1. This subpart prescribes the requirements for the issuance of an examining authority to the holder of a pilot school certificate.

2. It also gives the privileges and limitations of that examining authority.

141.63 Examining Authority Qualification Requirements

1. The application for examining authority must be made in the manner prescribed and on the form designed by the FAA.

2. To be eligible, the applicant must hold a pilot school certificate and also show that

    a. It has actively conducted a certificated pilot school for at least 24 months before the date of application.
    b. Within those 24 months, at least 10 students graduated from the course, and, of those 10, at least nine passed the test the first time.

141.65 Privileges

1. The holder of examining authority may recommend graduates of the school's approved courses for pilot certificates and ratings without their taking the appropriate FAA practical test, or knowledge test, or both.

141.67 Limitations and Reports

1. The holder of an examining authority may not recommend a person for a certificate or rating without taking the FAA knowledge or practical test unless

    a. The person has been enrolled in the holder's approved course.
    b. (S)he has successfully completed all of the course at that school.
    c. Each knowledge or practical test taken is comparable in depth and scope to the FAA test.

2. A final ground school test may not be given unless the test has been approved by the FAA, nor may it be given if there is reason to believe that the test has been compromised, i.e., available to participants in advance of the testing.

3. The holder shall submit a copy of the training record for each person it recommends for a certificate or rating.

Subpart E -- Operating Rules

141.71 Applicability

1. This subpart prescribes the operating rules applicable to a certificated pilot school or a provisional pilot school

141.73 Privileges

1. The certificate holder may advertise and conduct approved pilot training courses in accordance with the certificate and ratings held.

2. A school holding examining authority may recommend each graduate for the issuance of a pilot certificate and rating appropriate to the approved course of training without taking the FAA knowledge or practical test from an FAA inspector or designated examiner.

141.75 Aircraft Requirements

1. A pretakeoff checklist, a prelanding checklist, and an operator's handbook (if furnished by the manufacturer) must be carried on each aircraft.

141.77 Limitations

1. The certificate holder may not issue a graduation certificate to a student or recommend him/her for a certificate or rating if (s)he has not completed the specified training.

2. To graduate, the student must have completed all the curriculum requirements of that course.

3. A student may be credited, but not for more than one-half of the requirement, with previous flight experience or knowledge based upon a proficiency or knowledge test, or both, conducted by the school.

4. Course credits may be transferred from one certificated school to another.

    a. The receiving school shall determine the amount of credit based upon a knowledge or proficiency test, or both.

    b. Credit may not be given unless the other school

    1) Is certificated,
    2) Certifies the type and amount of training,
    3) Gives the results of each stage of training and any tests, and
    4) Follows the approved course, or

        a) The student was enrolled in the approved course while (s)he took the training.

141.79 Flight Training

1. No person other than a CFI who has the appropriate ratings and the qualifications set forth in the course outline may give instruction in an approved course.

2. No student pilot may be authorized to start a solo flight until the flight has been approved by an authorized CFI present at that airport.

3. Each chief flight instructor must complete, at least once every 12 months, an approved syllabus of training consisting of ground or flight training, or both, or a flight instructor refresher course.

4. Each CFI teaching an approved course must pass a flight check given by the chief instructor, assistant chief instructor, or check instructor each 12 months.

5. A CFI may not be used in an approved course until he has been briefed regarding the objectives and standards of the course.

141.81 Ground Training

1. Ground training instructors must have an appropriate flight or ground instructor certificate unless certain additional steps are taken by the chief instructor to compensate for the lack of an appropriate certificate.

141.83 Quality of Training

1. Each certificate holder must comply with the approved training course and must provide training of such quality that eight of the last 10 students passed either of the following tests on their first try:

    a. A test for a pilot certificate and/or rating
    b. A test to determine his/her competence of a completed stage of training in a special preparation course

2. The failure to maintain this quality standard can be used as the basis for suspension or revocation of the certificate.

3. The holder shall allow the FAA to make any checks or tests necessary to determine compliance with this requirement.

141.85 Chief Instructor Responsibilities

1. The chief instructor is responsible for

    a. Certifying all training records, graduation certificates, test reports, and student recommendations
    b. Ensuring that each instructor, prior to his/her teaching an approved course, passes an initial proficiency check and thereafter passes one at least annually
    c. Ensuring that each stage or final test is given to students of approved courses
    d. Maintaining acceptable training techniques, procedures, and standards for the school

2. The chief flight instructor shall be available at the school or by telephone, radio, or other electronic means during the time that instruction is given.

141.87 Change of Chief Instructor

1. The appropriate FSDO office shall be immediately notified if there is a change in the chief instructor for an approved course.

2. The school may continue to operate without a chief instructor for up to 60 days, but, during that time, each stage or final test must be conducted by an FAA inspector or designated pilot examiner.

141.89 Maintenance of Personnel, Facilities, and Equipment

1. Each training airport, aircraft, and facility must continuously meet the certificate holder's approved training course outline and other appropriate requirements.

2. Each instructor or chief instructor must meet the qualifications of each course and of Part 141.

141.91 Satellite Bases

1. A pilot school or provisional pilot school certificate holder may conduct ground or flight training at a base other than its main operations base if

    a. An assistant chief instructor is designated for each satellite base and is available while instruction is being conducted.
    b. The airport facilities and personnel used meet the requirements of Subpart B of Part 141 and the course outline.

   c. The instructors at the satellite base are under direct supervision of the chief flight instructor or assistant flight instructor for the course being taught.

   d. The appropriate FSDO is notified in writing if instruction is conducted there for more than 7 consecutive days.

141.93 Enrollment

1. When a student enrolls in an approved training course, the pilot school or provisional pilot school will provide the student with

   a. A certificate of enrollment indicating the course and the date of enrollment

   b. A copy of the training syllabus

   c. A copy of the safety procedures

      1) Weather minimums for dual and solo flights
      2) Ramp starting and taxiing procedures
      3) Fire precautions and procedures
      4) Redispatch procedures after unscheduled landings on and off airports
      5) Aircraft discrepancies
      6) Securing aircraft after use
      7) Fuel reserves for local and cross-country flights
      8) Avoidance of other aircraft in flight and on the ground
      9) Minimum altitude limitations and simulated emergency landing instructions
      10) Description and use of assigned practice areas

2. The pilot school must maintain a monthly listing of persons enrolled in each training course offered by the school.

141.95 Graduation Certificate

1. Each pilot school or provisional pilot school shall issue a graduation certificate to each student completing the approved course of training.

2. The certificate shall include

   a. The name of the school and the number of the school certificate

   b. The name of the graduate to whom it was issued

   c. The course of training for which it was issued

   d. The date of graduation

   e. A statement that the student has satisfied each required stage of the approved course of training, including the stage tests

   f. A certification by the chief instructor attesting to the information contained on the certificate

   g. A statement showing the cross-country training that the student received in the course of training

Subpart F -- Records

141.101 Training Records

1. Each pilot school or provisional pilot school shall establish and maintain current and accurate records of participation and accomplishment for all students. These records will include

   a. The date of enrollment

   b. A chronological log of the student's attendance, subjects covered, and flight operations conducted

   c. The date the student graduated, terminated training, or transferred

2. Upon the student's graduation, termination, or transfer, the records shall be certified by the chief instructor.

3. Records will be maintained for at least one year.

4. Copies of the student's records will be made available to the student upon request.

3.6 NTSB PART 830 - NOTIFICATION AND REPORTING OF AIRCRAFT ACCIDENTS OR INCIDENTS AND OVERDUE AIRCRAFT, AND PRESERVATION OF AIRCRAFT WRECKAGE, MAIL, CARGO, AND RECORDS

The National Transportation Safety Board has its own Federal regulations, and Part 830 concerns aircraft accident reporting. It has five subparts, A through E, and only seven sections, which are outlined in Chapter 4 of *Pilot Handbook*.

Subpart A -- General

830.1 Applicability
830.2 Definitions

Subpart B -- Initial Notification of Aircraft Accidents, Incidents, and Overdue Aircraft

830.5 Immediate notification
830.6 Information to be given in notification

Subpart C -- Preservation of Aircraft Wreckage, Mail, Cargo, and Records

830.10 Preservation of aircraft wreckage, mail, cargo, and records

Subpart D -- Reporting of Aircraft Accidents, Incidents, and Overdue Aircraft

830.15 Reports and statement to be filed

Subpart E -- Reporting of Public Aircraft Accidents and Incidents

830.20 Reports to be filed

3.7 LOGBOOK "SIGN-OFF" ENDORSEMENTS

Logbook Entries and Certificate Endorsements is Task L in Study Unit II, Technical Subject Areas. CFI-recommended logbook entries per the FAA's AC 61-65D, *Certification: Pilots and Flight and Ground Instructors*, are reproduced beginning on page 162 of this book.

Also, the endorsements for the FAA knowledge tests appear in the back of each Gleim FAA Knowledge Test book.

Name: _____

I certify that I have reviewed the above individual's preparation for the FAA Private Pilot--Airplane knowledge test [covering the topics specified in FAR 61.105(b)(1) through (13)] using the *Private Pilot FAA Knowledge Test* book, Test Prep CD-Rom, and/or Online Ground School by Irvin N. Gleim and find him/her competent to pass the knowledge test.

| Signed | Date | Name | CFI Number | Expiration Date |

STUDY UNIT FOUR
YOUR FAA PRACTICAL (FLIGHT) TEST

During 1987, the FAA began taking over testing of initial CFI applicants (i.e., not those adding a rating to an existing CFI certificate). Beginning in April 1988, all initial CFI practical tests were administered by FAA inspectors.

FAA inspectors are full-time FAA employees who administer various types of flight tests and review the paperwork generated by FAA-designated pilot examiners. Years ago, FAA inspectors administered most practical tests, which were given at no cost. For budgetary reasons, the FAA developed a system whereby well-qualified instructors could become FAA-designated pilot examiners and administer FAA practical tests for a fee. Most FAA flight tests are currently administered by FAA-designated examiners.

The FAA took over initial CFI testing to become more involved with the flight instructor community. It was hoped that FAA inspector interaction with new CFIs would improve the quality of flight instruction.

As one would expect, the FAA could not handle the thousands of CFI applicants. Backlogs developed quickly. In July 1988, the FAA began a system of farming out some CFI practical tests to certain designated pilot examiners. This policy continues (and in this book we use the term "FAA inspector/examiner").

At least 4 weeks before you are ready for your practical test, call your nearest FSDO and ask to schedule the test. This procedure varies from FSDO to FSDO. You may be asked to submit your request in writing, or you may be able to do everything by telephone.

Begin by asking whether an FAA inspector will be in your area or whether you will have to incur the expense of flying your airplane to the FSDO (if so, you will save the designated pilot examiner's fee). Alternatively, your FSDO may direct you to a designated pilot examiner. If so, follow the same routine as you did for your instrument and commercial practical tests. In either event, you should learn the name of your FAA inspector/examiner and reconfirm your practical test with your FAA inspector/ examiner a day in advance to avoid any scheduling error.

As soon as your practical test is scheduled, contact one or two individuals who recently took their flight instructor practical test with your FAA inspector/examiner. Your CFI may be able to direct you to others who recently took the CFI practical test. In addition, most CFIs who train CFIs ask their students about their practical tests and thus can give you guidance on what to expect. This book also provides you with much guidance.

Ask each person to explain the routine, length, emphasis, maneuvers, and any peculiarities (i.e., surprises). Obtaining guidance is a very important step because, like all people, examiners are unique. One particular facet of the practical exam may be tremendously important to one FAA inspector/examiner, while another FAA inspector/examiner may emphasize an entirely different area. By gaining this information beforehand, you can focus on the areas of apparent concern to the FAA inspector/examiner. Also, knowing what to expect will relieve some of the apprehension and tension about your practical test.

4.1 FAA PRACTICAL TEST STANDARDS TASKS

A. The intent of the FAA is to structure and standardize practical tests by specifying required tasks and acceptable performance levels to FAA inspectors and FAA-designated pilot examiners. These tasks (procedures and maneuvers) listed in the PTS are mandatory on each practical test unless specified otherwise.

B. The 72 tasks for the flight instructor certificate (single-engine airplane land) are listed on the next page in 14 areas of operation as organized by the FAA. To the right of each task is the page number on which the discussion begins in Study Units I through XIV.

   1. The 25 tasks that can be completed away from the airplane are indicated on the next page as "oral" and are termed "knowledge only" tasks by the FAA (Study Units I through IV).

   2. The 47 tasks that are usually completed in the airplane are indicated "flight" and are termed "knowledge and skill" tasks by the FAA.

   3. Your FAA inspector/examiner is not required to test you on all tasks.

      a. Generally, at least one task in each area of operation is required.
      b. However, some tasks are mandatory.

C. In Area of Operation XI, Slow Flight, Stalls, and Spins, you will be required to perform at least one proficiency stall, one demonstration stall, and spins. Spins (marked ***) may be excluded if you have a logbook endorsement, signed by the flight instructor who conducted your spin training, certifying your instructional competence in spin entries, spins, and spin recoveries.

   1. This exception is permitted because many CFI applicants take their entire practical test in a high-performance airplane, in which it may be dangerous and even prohibited to spin.

      a. Thus, CFIs are delegated the responsibility for training and testing CFI applicants in spin proficiency.

   2. However, despite the exception, if a CFI applicant fails the practical test due to deficiencies of knowledge or skills relating to stall awareness, spin entry, spins, or spin recovery techniques, (s)he must during the retest satisfactorily demonstrate both knowledge and skill in these areas in an airplane of the appropriate category that is certificated for spins.

*At least one task will be selected by your FAA inspector/examiner.

**Required task.

***May be satisfied by CFI logbook sign-off.

NOTE: In the PTS format, the FAA has done away with reference to "oral tests" and "flight tests." The current FAA position is that all tasks require oral examining about the applicant's knowledge. Nonetheless, we feel it is useful to separate the "knowledge only" tasks from the "knowledge and skill" tasks.

4.2 FORMAT OF PTS TASKS

A. Each of the FAA's 72 flight instructor tasks listed on page 41 is presented in a shaded box in Study Units I through XIV, similar to Task III.A. reproduced below.

III.A. TASK: CERTIFICATES AND DOCUMENTS

   REFERENCES: 14 CFR Parts 43, 61, and 91; FAA-H-8083-3, AC 61-23/FAA-H-8083-25; FAA-S-8081-12, FAA-S-8081-14; POH/AFM.

Objective. To determine that the applicant exhibits instructional knowledge of the elements related to certificates and documents by describing:

1. The training requirements for the issuance of a recreational, private, and commercial pilot certificate.

2. The privileges and limitations of pilot certificates and ratings at recreational, private, and commercial levels.

3. Class and duration of medical certificates.

4. Recent pilot flight experience requirements.

5. Required entries in pilot logbook or flight record.

---

1. The task number is followed by the title.

2. The reference list identifies the FAA publication(s) that describe(s) the task.

   a. Our discussion of each task is based on the FAA reference list. Note, however, that we will refer you to *Pilot Handbook* for further discussion of specific topics.

   b. A listing of the FAA references used in the PTS is on page 64.

3. Next, the task has "**Objective.** To determine that the applicant . . . ," followed by "Exhibits instructional knowledge . . ." of aviation concepts, "Demonstrates and simultaneously explains . . ." various maneuvers, and "Analyzes and corrects simulated common errors . . ." of various maneuvers.

B. Each task in this book is followed by the following general format:

I.   General information

   1.   The FAA's objective and/or rationale for this task
   2.   A list of Gleim's *Pilot Handbook* chapters and/or modules that provide additional discussion of the task, as appropriate
   3.   Any general discussion relevant to the task

II.   Comprehensive discussion of each concept or item listed in the FAA's task

---

4.3 AIRPLANE AND EQUIPMENT REQUIREMENTS

A. You are required to provide an appropriate and airworthy airplane for the practical test. The airplane must be equipped for, and its operating limitations must not prohibit, the areas of operations required on the test.

B. The takeoff and landing maneuvers and appropriate emergency procedures must be accomplished in a complex airplane. A complex airplane is defined as an airplane that is equipped with

1. Retractable gear
2. Adjustable flaps
3. Controllable pitch propeller

C. You may provide a complex airplane for the entire test or elect to provide another airplane for those tasks that do not require a complex airplane. Your CFI can discuss the pros and cons of using two airplanes for training and the practical test. Regardless of the airplane used, you are required to meet the flight instructor knowledge and skill standards throughout the entire test.

4.4 WHAT TO TAKE TO YOUR PRACTICAL TEST

A. You should ensure that you are completely prepared to begin your practical test before you meet your examiner. If you are unprepared, the test will become time-consuming and awkward for you and the examiner as you search for items that should have been located beforehand.

B. The following checklist from the FAA's Private Pilot Practical Test Standards should be reviewed with your instructor both 1 week and 1 day before your scheduled practical test:

1. Acceptable Airplane with Dual Controls

   a. Aircraft Documents

      1) Before your practical test, we suggest that you gather the following documents in one place so that you do not waste time looking for them once the test has started.

         a) Airworthiness Certificate
         b) Registration Certificate
         c) Operating Limitations

            i) Review the information in your Pilot's Operating Handbook (FAA-Approved *Airplane Flight Manual*)

               • Before the practical test, perform a weight and balance computation based upon the anticipated conditions on the day of your flight;

               • Determine the takeoff and landing distances for that day's flight; and

               • List your airplane's V-speeds on an index card for quick reference.

   b. Aircraft Maintenance Records

      1) Logbook Record of Airworthiness Inspections and AD Compliance

         a) Paperclip or otherwise indicate the location of the following items in your aircraft's airframe, powerplant, and propeller (if applicable) logbooks and other records.

            i) Most-recent Annual Inspection (must be within the preceding 12 calendar months)

            ii) Most-recent 100-hr. Inspection (must be within the preceding 100 hours of flight time if the airplane is operated for compensation or hire)

            iii) ELT Battery Due Date (must be after the date of the practical test)

            iv) Most-recent Transponder Test and Inspection (must be within the preceding 24 calendar months)

            v) Most-recent Altimeter and Static System Test and Inspection (must be within the preceding 24 calendar months)

            vi) Most-recent VOR Receiver Check (must be within the preceding 30 days)

            vii) Records of compliance with each one-time and recurring Airworthiness Directive (AD) (records of compliance with recurring ADs must be within the appropriate interval)

2.   Personal Equipment

    a.   View-Limiting Device
    b.   Current Aeronautical Charts, including appropriate Airport Diagrams
    c.   Computer and Plotter
    d.   Flight Plan Form
    e.   Flight Logs
    f.   Current *AIM, Airport/Facility Directory*, and Appropriate Publications (e.g., FARs)
    g.   A calculator to perform any weight and balance or performance computations

3.   Personal Records

    a.   Before your practical test, we suggest that you gather the following items in a folder or other container.

        1)   Identification -- photo/signature ID
        2)   Pilot Certificate (i.e., student or recreational)
        3)   Current Medical Certificate
        4)   Completed Application for an Airman Certificate and/or Rating (FAA Form 8710-1)
        5)   Airman Computer Test Report (pilot knowledge test grade report)
        6)   Logbook with Instructor's Endorsement for your Private Pilot Practical Test

            a)   Paperclip or otherwise mark the page with your practical test endorsement.
            b)   Total the last page's times in pencil so your application can be compared at a glance.

        7)   Notice of Disapproval (only if you previously failed your practical test)
        8)   Approved School Graduation Certificate (if applicable)
        9)   Examiner's Fee
        10)   Letter of Discontinuance (only if you discontinued a test previously for reasons other than unsatisfactory performance)

4.5 PRACTICAL TEST APPLICATION FORM

A. Prior to your practical test, you and your instructor will complete FAA Form 8710-1 (which appears on pages 47 and 48), and your instructor will sign the top of the back side of the form.

    1. An explanation on how to complete the form is attached to the original, and we have reproduced it on the next page.

        a. The form is not largely self-explanatory.

        b. The FAA requires dates to be presented as three groups of digits separated by hyphens. Years should be shown with four digits. For example, January 15, 2007 should only be expressed as 01-15-2007.

    2. Do not go to your practical test without FAA Form 8710-1 properly filled out; remind your CFI about it as you schedule your practical test.

B. If you are enrolled in a Part 141 flight school, the Air Agency Recommendation block of information on the back side may be completed by the chief instructor of your Part 141 flight school. (S)he, rather than a designated examiner or FAA inspector, will administer the practical test if examining authority has been granted to your flight school.

C. Your examiner or Part 141 flight school chief instructor will forward this and other required forms (listed on the bottom of the back side) to the nearest FSDO for review and approval.

    1. Then they will be sent to Oklahoma City. From there, your permanent flight instructor certificate will be issued and mailed to you.

    2. However, you will be issued a temporary certificate when you successfully complete the practical test (see Subunit 4.9, Your Temporary Flight Instructor Certificate, on page 52).

AIRMAN CERTIFICATE AND/OR RATING APPLICATION
INSTRUCTIONS FOR COMPLETING FAA FORM 8710-1

I. APPLICATION INFORMATION. *Check appropriate blocks(s).*

Block A. Name. Enter legal name. Use no more than one middle name for record purposes. Do not change the name on subsequent applications unless it is done in accordance with 14 CFR Section 61.25. If you do not have a middle name, enter "NMN". If you have a middle initial only, indicate "Initial only." If you are a Jr., or a II, or III, so indicate. If you have an FAA certificate, the name on the application should be the same as the name on the certificate unless you have had it changed in accordance with 14 CFR Section 61.25.

Block B. Social Security Number. Optional: See supplemental Information Privacy Act. Do not leave blank: Use only **US Social Security Number**. Enter either "SSN" or the words "Do not Use" or "None." SSN's are not shown on certificates.

Block C. Date of Birth. Check for accuracy. Enter eight digits; Use numeric characters, i.e., 07-09-1925 instead of July 9, 1925. Check to see that DOB is the same as it is on the medical certificate.

Block D. Place of Birth. If you were born in the USA, enter the city and state where you were born. If the city is unknown, enter the county and state. If you were born outside the USA, enter the name of the city and country where you were born.

Block E. Permanent Mailing Address. Enter residence number and street, P.O. Box or rural route number in the top part of the block above the line. The City, State, and ZIP code go in the bottom part of the block below the line. Check for accuracy. Make sure the numbers are not transposed. FAA policy requires that you use your permanent mailing address. **Justification must be provided on a separate sheet of paper signed and submitted with the application when a PO Box or rural route number is used in place of your permanent physical address. A map or directions must be provided if a physical address is unavailable.**

Block F. Citizenship. Check USA if applicable. If not, enter the country where you are a citizen.

Block G. Do you read, speak, write and understand the English language? Check yes or no.

Block H. Height. Enter your height in inches. Example: 5'8" would be entered as 68 in. No fractions, use whole inches only.

Block I. Weight. Enter your weight in pounds. No fractions, use whole pounds only.

Block J. Hair. Spell out the color of your hair. If bald, enter "Bald." Color should be listed as black, red, brown, blond, or gray. If you wear a wig or toupee, enter the color of your hair under the wig or toupee.

Block K. Eyes. Spell out the color of your eyes. The color should be listed as blue, brown, black, hazel, green, or gray.

Block L. Sex. Check male or female.

Block M. Do You Now Hold or Have You Ever Held An FAA Pilot Certificate? Check yes or no. (NOTE: A student pilot certificate is a "Pilot Certificate.")

Block N. Grade of Pilot Certificate. Enter the grade of pilot certificate (i.e., Student, Recreational, Private, Commercial, or ATP). Do NOT enter flight instructor certificate information.

Block O. Certificate Number. Enter the number as it appears on your pilot certificate.

Block P. Date Issued. Enter the date your pilot certificate was issued.

Block Q. Do You Now Hold A Medical Certificate? Check yes or no. If yes, complete Blocks R, S, and T.

Block R. Class of Certificate. Enter the class as shown on the medical certificate, i.e., 1st, 2nd, or 3rd class.

Block S. Date Issued. Enter the date your medical certificate was issued.

Block T. Name of Examiner. Enter the name as shown on medical certificate.

Block U. Narcotics, Drugs. Check appropriate block. Only check "Yes" if you have actually been convicted. If you have been charged with a violation which has not been adjudicated, check ."No".

Block V. Date of Final Conviction. If block "U" was checked "Yes" give the date of final conviction.

II. CERTIFICATE OR RATING APPLIED FOR ON BASIS OF:
Block A. Completion of Required Test.
1. AIRCRAFT TO BE USED. (If flight test required) – Enter the make and model of each aircraft used. If simulator or FTD, indicate.
2. TOTAL TIME IN THIS AIRCRAFT (Hrs.) – (a) Enter the total Flight Time in each make and model. (b) Pilot-In-Command Flight Time - In each make and model.

Block B. Military Competence Obtained In. Enter your branch of service, date rated as a military pilot, your rank, or grade and service number. In block 4a or 4b, enter the make and model of each military aircraft used to qualify (as appropriate).

Block C. Graduate of Approved Course.
1. NAME AND LOCATION OF TRAINING AGENCY/CENTER. As shown on the graduation certificate. Be sure the location is entered.
2. AGENCY SCHOOL/CENTER CERTIFICATION NUMBER. As shown on the graduation certificate. Indicate if 142 training center.
3. CURRICULUM FROM WHICH GRADUATED. As shown on the graduation certificate.
4. DATE. Date of graduation from indicated course. Approved course graduate must also complete Block "A" COMPLETION OF REQUIRED TEST.

Block D. Holder of Foreign License Issued By.
1. COUNTRY. Country which issued the license.
2. GRADE OF LICENSE. Grade of license issued, i.e., private, commercial, etc.
3. NUMBER. Number which appears on the license.
4. RATINGS. All ratings that appear on the license.

Block E. Completion of Air Carrier's Approved Training Program.
1. Name of Air Carrier.
2. Date program was completed.
3. Identify the Training Curriculum.

III. RECORD OF PILOT TIME. The minimum pilot experience required by the appropriate regulation must be entered. It is recommended, however, that ALL pilot time be entered. If decimal points are used, be sure they are legible. Night flying must be entered when required. You should fill in the blocks that apply and ignore the blocks that do not. Second In Command "SIC" time used may be entered in the appropriate blocks. Flight Simulator, Flight Training Device and PCATD time may be entered in the boxes provided. Total, Instruction received, and Instrument Time should be entered in the top, middle, or bottom of the boxes provided as appropriate.

IV. HAVE YOU FAILED A TEST FOR THIS CERTIFICATE OR RATING? Check appropriate block.

V. APPLICANT'S CERTIFICATION.
A. SIGNATURE. The way you normally sign your name.
B. DATE. The date you sign the application.

TYPE OR PRINT ALL ENTRIES IN INK   Form Approved OMB No: 2120-0021

DEPARTMENT OF TRANSPORTATION
FEDERAL AVIATION ADMINISTRATION

Airman Certificate and/or Rating Application

I Application Information   ☐ Student   ☐ Recreational   ☐ Private   ☐ Commercial   ☐ Airline Transport   ☐ Instrument

☐ Additional Rating   ☐ Airplane Single-Engine   ☐ Airplane Multiengine   ☐ Rotorcraft   ☐ Balloon   ☐ Airship   ☐ Glider   ☐ Powered-Lift

☐ Flight Instructor ____ Initial ____ Renewal ____ Reinstatement   ☐ Additional Instructor Rating   ☐ Ground Instructor

☐ Medical Flight Test   ☐ Reexamination   ☐ Reissuance of _____ certificate   Other _____

A. Name (Last, First, Middle)		B. SSN (US Only)	C. Date of Birth Month Day Year	D. Place of Birth

E. Address	F. Citizenship ☐ USA ☐ Other ____	Specify	G. Do you read, speak, write, & understand the English language? ☐ Yes ☐ No

City, State, Zip Code	H. Height	I. Weight	J. Hair	K. Eyes	L. Sex ☐ Male ☐ Female

M. Do you now hold, or have you ever held an FAA Pilot Certificate? ☐ Yes ☐ No	N. Grade Pilot Certificate	O. Certificate Number	P. Date Issued

Q. Do you hold a Medical Certificate? ☐ Yes ☐ No	R. Class of Certificate	S. Date Issued	T. Name of Examiner

U. Have you ever been convicted for violation of any Federal or State statutes relating to narcotic drugs, marijuana, or depressant or stimulant drugs or substances? ☐ Yes ☐ No	V. Date of Final Conviction

II. Certificate or Rating Applied For on Basis of:

☐ A. Completion of Required Test	1. Aircraft to be used (if flight test required)	2a. Total time in this aircraft / SIM / FTD hours	2b. Pilot in command hours
☐ B. Military Competence Obtained In	1. Service	2. Date Rated	3. Rank or Grade and Service Number
	4a. Flown 10 hours PIC in last 12 months in the following Military Aircraft.	4b. US Military PIC & Instrument check in last 12 months (List Aircraft)	
☐ C. Graduate of Approved Course	1. Name and Location of Training Agency or Training Center	1a. Certification Number	
	2. Curriculum From Which Graduated	3. Date	
☐ D. Holder of Foreign License Issued By	1. Country	2. Grade of License	3. Number
	4. Ratings		
☐ E. Completion of Air Carrier's Approved Training Program	1. Name of Air Carrier	2. Date	3. Which Curriculum ☐ Initial ☐ Upgrade ☐ Transition

III RECORD OF PILOT TIME (Do not write in the shaded areas.)

	Total	Instruction Received	Solo	Pilot in Command (PIC)	Cross Country Instruction Received	Cross Country Solo	Cross Country PIC	Instrument	Night Instruction Received	Night Take-off/ Landings	Night PIC	Night Take-Off/ Landing PIC	Number of Flights	Number of Aero-Tows	Number of Ground Launches	Number of Powered Launches
Airplanes				PIC / SIC			PIC / SIC				PIC / SIC	PIC / SIC				
Rotor-craft				PIC / SIC			PIC / SIC				PIC / SIC	PIC / SIC				
Powered Lift				PIC / SIC			PIC / SIC				PIC / SIC	PIC / SIC				
Gliders																
Lighter Than Air																
Simulator Training Device																
PCATD																

IV. Have you failed a test for this certificate or rating? ☐ Yes ☐ No

V. Applicants's Certification -- I certify that all statements and answers provided by me on this application form are complete and true to the best of my knowledge and I agree that they are to be considered as part of the basis for issuance of any FAA certificate to me. I have also read and understand the Privacy Act statement that accompanies this form.

Signature of Applicant	Date

FAA Form 8710-1 (4-00) Supersedes Previous Edition   NSN: 0052-00-682-5007

Instructor's Recommendation

I have personally instructed the applicant and consider this person ready to take the test.

Date	Instructor's Signature (Print Name & Sign)	Certificate No:	Certificate Expires

Air Agency's Recommendation

The applicant has successfully completed our _____ course, and is recommended for certification or rating without further _____ test.

Date	Agency Name and Number	Officials Signature
		Title

Designated Examiner or Airman Certification Representative Report

☐ Student Pilot Certificate Issued (Copy attached)

☐ I have personally reviewed this applicant's pilot logbook and/or training record, and certify that the individual meets the pertinent requirements of 14 CFR Part 61 for the certificate or rating sought.

☐ I have personally reviewed this applicant's graduation certificate, and found it to be appropriate and in order, and have returned the certificate.

☐ I have personally tested and/or verified this applicant in accordance with pertinent procedures and standards with the result indicated below.

☐ Approved -- Temporary Certificate Issued (Original Attached)

☐ Disapproved -- Disapproval Notice Issued (Original Attached)

Location of Test (Facility, City, State)	Duration of Test		
	Ground	Simulator/FTD	Flight

Certificate or Rating for Which Tested	Type(s) of Aircraft Used	Registration No.(s)

Date	Examiner's Signature (Print Name & Sign)	Certificate No.	Designation No.	Designation Expires

Evaluator's Record (Use For ATP Certificate and/or Type Ratings)

	Inspector	Examiner	Signature and Certificate Number	Date
Oral	☐	☐	_____	_____
Approved Simulator/Training Device Check	☐	☐	_____	_____
Aircraft Flight Check	☐	☐	_____	_____
Advanced Qualification Program	☐	☐	_____	_____

Aviation Safety Inspector or Technician Report

I have personally tested this applicant in accordance with or have otherwise verified that this applicant complies with pertinent procedures, standards, policies, and or necessary requirements with the result indicated below.

☐ Approved -- Temporary Certificate Issued (Original Attached)   ☐ Disapproved -- Disapproval Notice Issued (Original Attached)

Location of Test (Facility, City, State)	Duration of Test		
	Ground	Simulator/FTD	Flight

Certificate or Rating for Which Tested	Type(s) of Aircraft Used	Registration No.(s)

☐ Student Pilot Certificate Issued   ☐ Certificate or Rating Based on   ☐ Flight Instructor   ☐ Ground Instructor

☐ Examiner's Recommendation   ☐ Military Competence   ☐ Renewal

 ☐ Accepted   ☐ Rejected   ☐ Foreign License   ☐ Reinstatement

☐ Reissue or Exchange of Pilot Certificate   ☐ Approved Course Graduate   Instructor Renewal Based on

☐ Special Medical test conducted -- report forwarded   ☐ Other Approved FAA Qualification Criteria   ☐ Activity   ☐ Training Course

 to Aeromedical Certification Branch, AAM-330   ☐ Test   ☐ Duties and Responsibilities

Training Course (FIRC) Name	Graduation Certificate No.	Date

Date	Inspector's Signature (Print Name & Sign)	Certificate No.	FAA District Office

Attachments:

☐ Student Pilot Certificate (Copy)

☐ Knowledge Test Report

☐ Temporary Airman Certificate

☐ Notice of Disapproval

☐ Superseded Airman Certificate

☐ Airman's Identification (ID)

Form of ID _____

Number _____

Expiration Date _____

Telephone Number _____

ID:

Name: _____

Date of Birth: _____

Certificate Number: _____

E-Mail Address _____

FAA Form 8710-1 (4-00) Supersedes Previous Edition   NSN: 0052-00-682-5007

4.6 AUTHORIZATION TO TAKE THE PRACTICAL TEST

A. Before applicants for the flight instructor certificate take the practical test, FAR 61.187 requires them to have logged instruction from an authorized flight instructor in the areas of operations, which are discussed in Part II, Study Units I through XIV, of this book.

B. An authorized flight instructor is a person who has held a flight instructor certificate during the 24 months preceding the date the instruction is given and who has given at least 200 hr. of flight instruction as a certificated flight instructor.

C. Your logbook must contain the following endorsements from your flight instructor.

     1. Logbook endorsement for ground and flight proficiency/practical test: FAR 61.183(g) and 61.187(a) and (b)

         I certify that (First name, MI, Last name) has received the required training of Sec. 61.187(b). I have determined he/she is prepared for the CFI-(aircraft category and class) practical test.

         Date _____ Signature _____ CFI No. _____ Expiration Date _____

     2. Logbook endorsement for fundamentals of instructing: FAR 61.183(d) and 61.185(a)(1)

         I certify that (First name, MI, Last name) has received the required training of Sec. 61.185(a).

         Date _____ Signature _____ CFI No. _____ Expiration Date _____

     3. Logbook endorsement for spin training: FAR 61.183(i)(1)

         I certify that (First name, MI, Last name) has received the required training of FAR 61.183(i). I have determined that he/she is competent and proficient on instructional skills for training stall awareness, spin entry, spins, and spin recovery procedures.

         Date _____ Signature _____ CFI No. _____ Expiration Date _____

4.7 ORAL PORTION OF THE PRACTICAL TEST

A. Your practical test will begin in the FAA inspector's office at your local FSDO or at your examiner's office.

     1. You should have with you

         a. This book
         b. The POH for the airplane(s) (including weight and balance data)
         c. Gleim's *FAR/AIM*
         d. All of the items listed on pages 43 and 44

     2. Your FAA inspector/examiner will probably begin by reviewing your paperwork (FAA Form 8710-1, Airman Computer Test Report, logbook sign-off, etc.).

         a. If you are taking your practical test with an FAA-designated pilot examiner, (s)he will be collecting payment for his/her services.

3.   Typically, your FAA inspector/examiner will begin with questions about fundamentals of instructing followed by discussions on technical subjects. Then you will be asked to prepare a lesson on a maneuver to be performed in flight.

   a.   At times you are required to perform the role of the instructor while your FAA inspector/examiner performs the role of your student.

      1)   This is done in your lesson presentation.

      2)   Ensure that you know when you are expected to be in this role. If you are not sure, ask your FAA inspector/examiner.

4.   During the practical test, you will be evaluated on your instructional knowledge.

   a.   The term **instructional knowledge** means the instructor applicant is capable of using the appropriate reference to provide the "application or correlative level of knowledge" of a subject matter topic, procedure, or maneuver.

   b.   It also means that your discussions, explanations, and descriptions should follow the recommended teaching procedures and techniques explained in Study Unit I, Fundamentals of Instructing, beginning on page 65, which is a summary of the FAA's *Aviation Instructor's Handbook* (FAA-H-8083-9).

5.   As your FAA inspector/examiner asks you questions, follow the guidelines listed below:

   a.   Attempt to position yourself in a discussion mode with him/her rather than being interrogated by the FAA inspector/examiner.

   b.   Be respectful but do not be intimidated. Both you and your FAA inspector/examiner are professionals.

   c.   Draw on your knowledge from this book and other books, your CFI, and your prior experience.

   d.   If you do not know an answer, try to explain how you would research the answer.

6.   Be confident that you will do well. You are a good pilot. You have thoroughly prepared for this discussion by studying the subsequent pages and have worked diligently with your CFI.

B.   After you discuss various aspects of the 25 "knowledge only" tasks, you will move out to your airplane to begin your flight test, which consists of selected tasks from the 47 "knowledge and skill" tasks.

1.   If possible and appropriate in the circumstances, thoroughly preflight your airplane just before you go to your FAA inspector/examiner's office.

2.   As you and your examiner approach your airplane, explain that you have already preflighted the airplane (explain any possible problems and how you resolved them).

   a.   Your FAA inspector/examiner may want you to perform a preflight inspection.

3.   Volunteer to answer any questions.

4.   Make sure you walk around the airplane to observe any possible damage by ramp vehicles or other aircraft while you were in your examiner's office.

5.   As you enter the airplane, make sure that your cockpit is organized and you feel in control of your charts, clock, navigation logs, etc.

4.8 FLIGHT PORTION OF THE PRACTICAL TEST

A. Throughout the flight portion of the practical test, your FAA inspector/examiner will evaluate your ability to demonstrate and simultaneously explain procedures and maneuvers and to give flight instruction to students at various stages of flight training and levels of experience.

    1. You should be explaining what, why, and how you are performing various tasks.

    2. Stress the importance of collision avoidance, checklist usage, stall awareness, distractions, airplane control, sound judgment, etc.

B. Your FAA inspector/examiner will normally spend more time on the performance of the maneuver that you presented during the oral portion of the practical test.

C. The objective of a task that involves pilot skill (i.e., flight maneuver) consists of four parts. Those four parts jointly determine that you exhibit:

    1. Instructional knowledge of the elements of a task through descriptions, explanations, and simulated instruction

    2. Instructional knowledge of common errors related to a task, including their recognition, analysis, and correction

    3. The ability to demonstrate and simultaneously explain the key elements of a task. The task demonstration must be to the commercial pilot standards, and the teaching techniques and procedures should conform to those set forth in the FAA's *Aviation Instructor's Handbook* (FAA-H-8083-9) and the *Airplane Flying Handbook* (FAA-H-8083-3).

    4. The ability to analyze and correct common errors related to a task

D. Your FAA inspector/examiner may ask you to teach a maneuver that (s)he then performs to allow you to demonstrate the evaluation process. Any botched maneuver on your part should be turned into an opportunity to demonstrate knowledge of common errors and evaluation techniques.

E. Throughout the flight portion of the practical test, your FAA inspector/examiner will evaluate your ability to use good **aeronautical decision-making procedures** in order to identify risks.

    1. Your FAA inspector/examiner will develop scenarios that incorporate as many TASKs as possible to evaluate your **risk management** in making safe aeronautical decisions.

        a. For example, your FAA inspector/examiner may develop a scenario that incorporates weather decisions and performance planning.

        b. The scenarios should be realistic and within the capabilities of the aircraft used for the practical test.

    2. Your ability to utilize all the assets available in making a risk analysis to determine the safest course of action is essential for satisfactory performance.

F. **Single-Pilot Resource Management** refers to the effective use of ALL available resources: human resources, hardware, and information.

    1. It is similar to Crew Resource Management (CRM) procedures that are being emphasized in multi-crewmember operations except that only one crewmember (the pilot) is involved.

    2. Human resources "... includes all other groups routinely working with the pilot who are involved in decisions that are required to operate a flight safely.

        a. These groups include, but are not limited to: dispatchers, weather briefers, maintenance personnel, and air traffic controllers."

    3. Pilot Resource Management is not a single TASK; it is a set of skill competencies that must be evident in all TASKs in the practical test as applied to single-pilot operation.

4.9 YOUR TEMPORARY FLIGHT INSTRUCTOR CERTIFICATE

A. When you successfully complete your practical test, your FAA inspector/examiner will prepare a temporary flight instructor certificate similar to the one illustrated below.

    1. The temporary certificate is valid for 120 days.

B. Your permanent certificate will be sent to you directly from the FAA Aeronautical Center in Oklahoma City in about 60 to 90 days.

    1. If you do not receive your permanent certificate within 120 days, your FAA inspector/ examiner can arrange an extension of your temporary certificate (i.e., issue you another temporary certificate).

**I. UNITED STATES OF AMERICA
DEPARTMENT OF TRANSPORTATION_FEDERAL AVIATION ADMINISTRATION**

II. **TEMPORARY AIRMAN CERTIFICATE**

III. CERTIFICATE NO.

THIS CERTIFIES THAT   IV.

V.

DATE OF BIRTH	HEIGHT	WEIGHT	HAIR	EYES	SEX	NATIONALITY	VI.
	IN.						

IX. has been found to be properly qualified and is hereby authorized in accordance with the conditions of issuance on the reverse of this certificate to exercise the privileges of

RATINGS AND LIMITATIONS

XII.

XIII.

THIS IS ☐ AN ORIGINAL ISSUANCE ☐ A REISSUANCE
OF THIS GRADE OF CERTIFICATE   DATE OF SUPERSEDED AIRMAN CERTIFICATE

VII. AIRMAN'S SIGNATURE

BY DIRECTION OF THE ADMINISTRATOR

EXAMINER'S DESIGNATION NO. OR INSPECTOR'S REG. NO.

X. DATE OF ISSUANCE   X. SIGNATURE OF EXAMINER OR INSPECTOR

DATE DESIGNATION EXPIRES

FAA Form 8060-4 (4-69) Supersedes Previous Edition

4.10 FAILURE ON THE PRACTICAL TEST

A. The majority of applicants pass their flight instructor practical test the first time, and virtually all who experienced difficulty on their first attempt pass the second time.

B. If, in the judgment of the FAA inspector/examiner, you do not meet the standards of performance of any task performed, the applicable area of operation is considered unsatisfactory, and thus the practical test is failed.

    1. You or your FAA inspector/examiner may discontinue the test at anytime when an area of operation is failed.

        a. The test will be continued only with your consent.

    2. You will be given credit for those areas of operation satisfactorily performed.

        a. During the retest, you may be asked to perform any task, including those previously considered satisfactory.

    3. Specific reasons for disqualification are

        a. Failure to perform a procedure or maneuver to the commercial pilot skill level (or, in the case of the Flight Instructor - Instrument applicant, to the instrument pilot skill level) while giving effective flight instruction

        b. Failure to provide an effective instructional explanation (i.e., clear, concise, technically accurate, and complete, with no prompting from the FAA inspector/examiner) while demonstrating a procedure or maneuver

        c. Any action or lack of action by the applicant that requires corrective intervention by the FAA inspector/examiner to maintain safe flight

        d. Failure to use proper and effective visual scanning techniques to clear the area before and while performing maneuvers

    4. When on the ground, your FAA inspector/examiner will complete the Notice of Disapproval of Application, FAA Form 8060-5, which appears on the next page, and will indicate the areas necessary for reexamination.

C. You should do the following:

    1. Indicate your intent to work with your instructor on your deficiencies.

    2. Inquire about rescheduling the next practical test.

        a. Many examiners have a reduced fee for a retake (FAA inspectors do not charge for their services).

3.   Inquire about having your flight instructor discuss your proficiencies and deficiencies with the FAA inspector/examiner.

UNITED STATES OF AMERICA
DEPARTMENT OF TRANSPORTATION-FEDERAL AVIATION ADMINISTRATION

NOTICE OF DISAPPROVAL OF APPLICATION

NOTE
PRESENT THIS FORM
UPON APPLICATION
FOR REEXAMINATION

NAME AND ADDRESS OF APPLICANT

CERTIFICATE OR RATING SOUGHT

On the date shown, you failed the examination indicated below

☐ FLIGHT   ☐ ORAL   ☐ PRACTICAL

AIRCRAFT USED (Make and Model)

FLT. TIME RECORDED IN LOGBOOK

PILOT-IN-COMM. OR SOLO	INSTRUMENT	DUAL

UPON REAPPLICATION YOU WILL BE REEXAMINED ON THE FOLLOWING:

I have personally tested this applicant and deem his performance unsatisfactory for the issuance of the certificate or rating sought.

DATE OF EXAMINATION	SIGNATURE OF EXAMINER OR INSPECTOR	DESIGNATION OR OFFICE NO.

FAA Form 8060-5 (5-80)

4.11 DISCONTINUANCE

A.   It may become necessary for you to discontinue your practical test for reasons other than unsatisfactory performance, such as:

1.   Equipment failure
2.   Weather
3.   Illness

B.   When a practical test is discontinued for reasons other than unsatisfactory performance, FAA Form 8700-1, Airman Certificate and/or Rating Application, and, if applicable, the Airman Knowledge Test Report, will be returned to you.

C.   At that time the examiner will prepare, sign, and issue a Letter of Discontinuance to you.

1.   The Letter of Discontinuance should identify the AREAS OF OPERATION and their associated TASKs of the practical test that were successfully completed.

2.   Be advised that the Letter of Discontinuance must be presented to the examiner when the practical test is resumed, and it must be made part of the certification file.

4.12 FLIGHT INSTRUCTOR SPECIAL EMPHASIS PROGRAM

A. The Flight Instructor Special Emphasis Program was created by the Orlando Flight Standards District Office (FSDO) in 1998 in order to increase FAA involvement in flight instructor operations and certification occurring within the FSDO's geographical area.

1. The objective of the program is to reduce the overall occurrence of accidents and incidents within the district, particularly those involving CFIs.

    a. A summary of the program and its findings is available at http://www.faa.gov/about/office_org/field_offices/fsdo/orl/local_more/cfi_program.

    b. An airworthiness checklist, designed to assist you in documenting your aircraft's airworthiness, is also available at this link.

B. As one of the major elements of this program, all initial CFI practical tests in the Orlando district are conducted by the FAA.

1. Based on the considerable experience gained through this program, the Orlando FSDO has compiled a list of common weak areas observed on initial CFI practical tests.

2. This list (also available at the link above) has been reproduced for your use on page 517. Next to each item, we have provided you with the location of information about the topic, either in this book or in Gleim's *Pilot Handbook* (indicated by "PH").

END OF STUDY UNIT

STUDY UNIT FIVE
SPECIAL EMPHASIS AREAS

5.1 GENERAL INFORMATION

A. Your FAA inspector/examiner will place special emphasis on areas of aircraft operation considered critical to flight safety.

B. You should familiarize yourself with ways to ensure safety in all phases of flight.

C. These areas should be emphasized throughout the oral and flight portions of your test and in your flight instruction.

D. These areas are

1. Positive aircraft control
2. Positive exchange of the flight controls procedure
3. Stall/spin awareness
4. Collision avoidance
5. Wake turbulence avoidance
6. Land and hold short operations (LAHSO)
7. Runway incursion avoidance
8. CFIT
9. ADM and risk management
10. Wire strike avoidance
11. Checklist usage

5.2 SPECIAL EMPHASIS AREAS EXPLAINED

A. **Positive aircraft control**

1. In common terms, positive aircraft control means that you are always "in control" of the airplane.

    a. During all maneuvers, the aircraft should appropriately respond to all of your control inputs in a manner that ensures safe flight.

    b. You should always be trying to "lead the airplane," anticipating the outcome of your actions and planning ahead.

B. **Positive exchange of the flight controls procedure**

1. Before starting the aircraft, be sure to state that you will be using the positive exchange of flight controls procedures.

    a. This procedure is designed to ensure that all parties know who will be actively controlling the airplane at any given point in time.

    b. This positive exchange ensures that someone is always controlling the airplane and that only one person at a time will be making control inputs unless otherwise required.

2. There are three verbal steps in exchanging the flight controls. Consider an example in which you pass the flight controls to the FAA inspector/examiner.

    a. You state, "You have the flight controls."
    b. The inspector/examiner responds, "I have the flight controls."
    c. You reaffirm, "You have the flight controls."

C. Stall/spin awareness

1. You should be aware and demonstrate your awareness of possible conditions that could lead to the aircraft stalling or spinning.

2. You should emphasize those conditions in which students most often accidentally encounter stalls and/or spins, such as:

    a. Just after takeoff (from steep climb angles)
    b. Just before touchdown (as in flat, slow approaches)
    c. While practicing stalls

D. Collision avoidance

1. **Scanning** the sky for other aircraft is a key factor in collision avoidance. You and your student (or left-seat passenger) should scan continuously to cover all areas of the sky visible from the cockpit.

    a. You must develop an effective scanning technique that maximizes your visual capabilities.

    b. Effective scanning is accomplished with a series of short, regularly-spaced eye movements that bring successive areas of the sky into the central visual field.

        1) Each eye movement should not exceed 10°.
        2) Each area should be observed for at least 1 second to enable detection.

    c. Visual tasks inside the cabin should represent no more than 1/4 to 1/3 of the scan time outside or no more than 4 to 5 seconds on the instrument panel for every 16 seconds outside.

2. Specific Techniques

    a. Determining relative altitude -- Use the horizon as a reference point. If you see another aircraft above the horizon, it is probably on a higher flight path. If it appears to be below the horizon, it is probably flying at a lower altitude.

    b. Taking appropriate action -- You must be familiar with the rules of right-of-way so that, if an aircraft is on an obvious collision course, you can take the appropriate evasive action.

    c. Considering multiple threats -- The decision to climb, descend, or turn is a matter of personal judgment, but you should anticipate that the other pilot also may be making a quick maneuver. Watch the other aircraft during the maneuver, but begin your scanning again immediately. There may be even more aircraft in the area!

    d. Observing collision course targets -- Any aircraft that appears to have no relative motion and stays in one scan quadrant is likely to be on a collision course. Also, if a target shows no lateral or vertical motion, but it increases in size, take evasive action.

    e. Recognizing high-hazard areas

        1) Airways, VORs, and airport traffic areas are places where aircraft tend to cluster.
        2) Remember that most collisions occur on days when the weather is good.

    f. Practicing cockpit management -- Study maps, checklists, and manuals BEFORE flight, along with other proper preflight planning (e.g., noting necessary radio frequencies). Also, organizing cockpit materials can reduce the time you need to look at them during flight, permitting more scan time.

    g. Improving windshield conditions -- Dirty or bug-smeared windshields can greatly reduce your ability to see other aircraft. Keep a clean windshield.

    h. Considering visibility conditions -- Smoke, haze, dust, rain, and flying toward the sun can also greatly reduce the ability to detect other aircraft.

    i.   Being aware of visual obstructions in the cockpit.

        1)   You may need to move your head to see around blind spots caused by fixed aircraft structures, such as door posts, wings, etc. It may even be occasionally necessary to maneuver your airplane (e.g., lift a wing) to facilitate seeing.

        2)   Check that curtains and other cockpit objects (e.g., maps that glare on the windshield) are removed and stowed during flight.

    j.   Using lights.

        1)   Day or night, exterior lights can greatly increase the visibility of any aircraft.

        2)   Keep interior lights low at night so that you can see out in the dark.

    k.   Requesting ATC support -- ATC facilities often provide radar traffic advisories (e.g., flight following) on a workload-permitting basis. Use this support whenever possible or when required.

        1)   Nevertheless, being in a radar environment (i.e., where traffic is separated by radar) still requires vigilance to avoid collisions. Radar does not relieve you of the responsibility to see and avoid other aircraft.

E. **Wake turbulence avoidance**

    1.   You should memorize and practice the following wake turbulence avoidance procedures:

        a.   **Landing behind a larger aircraft that is landing on the same runway** -- Stay at or above the larger aircraft's final approach flight path. Note the aircraft's touchdown point and land beyond it.

        b.   **Landing behind a larger aircraft that is landing on a parallel runway closer than 2,500 ft. to your runway** -- Consider possible vortex drift to your runway. Stay at or above the larger aircraft's final approach path and note its touchdown point.

        c.   **Landing behind a larger aircraft that is landing on a crossing runway** -- Cross above the larger aircraft's flight path.

        d.   **Landing behind a larger aircraft departing on the same runway** -- Note the larger aircraft's rotation point. Land well prior to the rotation point.

        e.   **Landing behind a larger aircraft departing on a crossing runway** -- Note the larger aircraft's rotation point.

            1)   If the larger aircraft rotates past the intersection, continue your approach and land prior to the intersection.

            2)   If the larger aircraft rotates prior to the intersection, avoid flight below the larger aircraft's flight path.

               a)   Unless your landing is assured well before reaching the intersection, abandon the approach.

        f.   **Departing behind a larger aircraft taking off** -- Note the larger aircraft's rotation point. You should rotate prior to the larger aircraft's rotation point. Continue to climb above and stay upwind of the larger aircraft's climb path until turning clear of its wake.

            1)   Avoid subsequent headings that will cross below and behind a larger aircraft.

            2)   Be alert for any critical takeoff situation that could lead to a vortex encounter.

        g.   **Intersection takeoffs on the same runway** -- Be alert to adjacent larger aircraft operations, particularly upwind of your runway. If intersection takeoff clearance is received, avoid a subsequent heading that will cross below a larger aircraft's path.

h. **Departing or landing after a larger aircraft has executed a low approach, a missed approach, or a touch-and-go landing** -- Vortices settle and move laterally near the ground, so the vortex hazard may exist along the runway and in your flight path.

   1) Ensure that an interval of at least 2 minutes has elapsed before your takeoff or landing.

i. **En route VFR** -- Avoid flight below and behind a larger aircraft's path. If you observe a larger aircraft above and on the same track as your airplane (meeting or overtaking), adjust your position laterally, preferably upwind.

F. Land and Hold Short Operations (LAHSO)

1. Land and hold short operations (LAHSO) take place at some airports with an operating control tower in order to increase airport capacity and improve the flow of traffic.

   a. LAHSO requires that you land and hold short of an intersecting runway, an intersecting taxiway, or some other designated point on a runway.

2. Before accepting a clearance to land and hold short, you must determine that you can safely land and stop within the available landing distance (ALD).

3. Student pilots should not participate in the LAHSO program.

   a. Accordingly, if you are issued and choose to accept a LAHSO clearance during the flight portion of your practical test, you should be the flying pilot.

4. You, as pilot in command, have the final authority to accept or decline any LAHSO clearance.

   a. You are expected to decline a LAHSO clearance if you determine it will compromise safety.

5. You should receive a LAHSO clearance only when there is a minimum ceiling of 1,000 ft. and visibility of 3 SM.

   a. The intent of having basic VFR weather conditions is to allow pilots to maintain visual contact with other aircraft and ground vehicle operations.

G. Runway incursion avoidance

1. The potential for runway incidents and accidents can be reduced through adequate planning, coordination, and communication. The following are some practices to help prevent a runway incursion:

   a. Read back all runway/taxiway crossing and/or hold instructions.
   b. Review airport layouts prior to taxi, before landing if able, and while taxiing as needed.
   c. Review NOTAMs for up-to-date information.
   d. Be familiar with airport signs and markings.
   e. Request progressive taxi from ATC when unsure of taxi route.
   f. Check for traffic before crossing or entering any runway or taxiway.
   g. Make sure aircraft position and taxi lights are on whenever aircraft is moving.
   h. When landing, clear the active runway in a timely fashion.
   i. Use proper phraseology and good radio discipline at all times.
   j. Write down complex taxi instructions.

H. Controlled Flight into Terrain (CFIT)

1. Controlled flight into terrain (CFIT) can be defined as an event in which a normally functioning aircraft is inadvertently flown into terrain, water, or obstacles, often without prior knowledge by the crew.

   a. Inform your students that, due to the physical forces involved, the vast majority of CFIT accidents are fatal to one or more of the aircraft's occupants.

   b. Point out that the element of surprise is also a factor in the lethality of CFIT accidents because it prevents the crew from taking actions (such as reducing airspeed or modifying the flight path) to minimize the impact forces.

   c. Accordingly, you must aggressively emphasize **avoidance** of the scenarios that lead to CFIT accidents.

2. Explain that, while CFIT accidents are typically associated with IFR operations in mountainous areas, they can happen to aircraft operating under IFR or VFR, over all kinds of terrain, and at any time of day or night.

   a. Point out that, while many factors can contribute to a CFIT accident, one causal factor common to most such accidents is **the crew's loss of situational awareness**.

   b. You should therefore teach your students to devote maximum possible attention to maintaining situational awareness during all phases of flight.

      1) Emphasize that the risk of losing situational awareness is greatest in conditions of darkness or reduced visibility and during times of high workload.

      2) Without a clearly discernible horizon, even large deviations from a desired heading, attitude, altitude, or course may not be immediately obvious.

   c. Under high-workload conditions, a pilot's instrument scan can break down, allowing such changes to go unnoticed and setting the stage for a CFIT event.

   d. Point out, however, that times of low workload can also be hazardous because they can lead to boredom and complacency at a time when the pilot should be monitoring the progress of the flight and preparing for the upcoming phases.

I. Aeronautical Decision Making (ADM) and risk management

1. **ADM** is a systematic approach to the mental process used by aircraft pilots to consistently determine the best course of action in response to a given set of circumstances. It can be practiced using the DECIDE model.

   a. **D**etect. The decision maker detects the fact that change has occurred.

   b. **E**stimate. The decision maker estimates the need to counter or react to the change.

   c. **C**hoose. The decision maker chooses a desirable outcome (in terms of success) for the flight.

   d. **I**dentify. The decision maker identifies actions that could successfully control the change.

   e. **D**o. The decision maker takes the necessary action.

   f. **E**valuate. The decision maker evaluates the effect(s) of his/her action countering the change.

   g. The six elements of the DECIDE model should be treated as a continuous loop. If a pilot practices the DECIDE model in all decision making, its use can become very natural and result in better decisions being made under all types of situations.

2.   **Risk management** is the part of the decision-making process that relies on situational awareness, problem recognition, and good judgment to reduce risks associated with flight.

   a.   There are four **risk elements** involved in decisions made during a flight: the pilot in command, the airplane, the environment, and the operation. In decision making, each risk element is evaluated to attain an accurate perception of circumstances.

   1)   **Pilot.** Consider factors such as competency, condition of health, mental and emotional state, level of fatigue, and many other variables.

   2)   **Airplane.** Assess performance, equipment, and airworthiness.

   3)   **Environment.** Consider a range of factors not related to pilot or airplane: weather, air traffic control, NAVAIDs, terrain, takeoff and landing areas, and surrounding obstacles.

   4)   **Operation.** Assessing factors relating to pilot, airplane, and environment is largely influenced by the purpose of the operation. Decisions should be made in the context of why the flight is being made, how critical it is to maintain the schedule, and whether or not the trip is worth the risks.

J.   **Wire strike avoidance**

1.   While low-level flying is, by its nature, dangerous, the possibility of wire strike makes it even more risky.

   a.   The advent of cellular towers and other technologies has led to the regular construction of many new towers.

   b.   Some towers are held in place by guide wires attached to the surface.

2.   Proper planning can help pilots avoid wire strike. This planning should include the use of:

   a.   Applicable sectional and terminal area charts
   b.   *Airport/Facilities Directories* (*A/FDs*)
   c.   Minimum en route and/or approach altitudes
   d.   Notices to Airmen (NOTAMs)

   1)   NOTAMs are especially important because they contain timely information about new construction that may not be included in printed publications.

3.   Pilots must also be cognizant of the risks of wire strikes when choosing areas for emergency landings/ditching.

   a.   Power lines are often found along roads and on the edges of fields.

K.   **Checklist usage**

1.   The proper use of checklists is important to:

   a.   Keep the pilot on task.
   b.   Ensure the pilot does not miss any steps important to safety.
   c.   Provide a smooth, consistent application of both normal and emergency procedures.

2.   You must set an example for your students through your checklist usage.

3.   Ensure that the checklist is a **tool**, not a crutch.

   a.   You should be able to function without the checklist.
   b.   The checklist is designed to "back you up," not remind you how to do things.

END OF STUDY UNIT

This is the end of Part I. Part II consists of Study Units I through XIV. Each study unit covers one Area of Operation in the Flight Instructor Practical Test Standards.

PART II:
FAA PRACTICAL TEST STANDARDS AND FLIGHT MANEUVERS: DISCUSSED AND EXPLAINED

Part II of this book (Study Units I through XIV) provides an in-depth discussion of the Flight Instructor Practical Test Standards (PTS). Each of the 14 areas of operation with its related task(s) is presented in a separate study unit.

		No. of Tasks	No. of Pages
I.	Fundamentals of Instructing.	7	30
II.	Technical Subject Areas.	12	80
III.	Preflight Preparation.	5	22
IV.	Preflight Lesson on a Maneuver to be Performed in Flight.	1	8
V.	Preflight Procedures.	5	34
VI.	Airport Operations.	3	16
VII.	Takeoffs, Landings, and Go-Arounds.	9	76
VIII.	Fundamentals of Flight.	4	20
IX.	Performance Maneuvers.	4	26
X.	Ground Reference Maneuvers.	4	30
XI.	Slow Flight, Stalls, and Spins.	8	48
XII.	Basic Instrument Maneuvers.	5	32
XIII.	Emergency Operations.	4	24
XIV.	Postflight Procedures.	1	7
		72	453

Each task, reproduced verbatim from the PTS, appears in a shaded box within each study unit. General discussion is presented under "A. General Information." This is followed by "B. Task Objectives," which is a detailed discussion of each element of the FAA's task, including the common errors for each "knowledge and skill" task (e.g., flight maneuver).

The practical test is passed if, in the judgment of your FAA inspector/examiner, you demonstrate satisfactory performance with regard to

1. Knowledge of the fundamentals of instructing

2. Knowledge of the technical subject areas

3. Knowledge of the CFI's responsibilities concerning the pilot certification process

4. Knowledge of the CFI's responsibilities concerning logbook entries and pilot certificate endorsements

5. Ability to demonstrate the selected procedures and maneuvers to at least the commercial pilot skill level while giving effective instruction

6. Competence in teaching the selected procedures and maneuvers

7. Competence in describing, recognizing, analyzing, and correcting common errors simulated by your FAA inspector/examiner

8. Knowledge of the development and effective use of a course of training, a syllabus, and a lesson plan

Be confident. You have prepared diligently and are better prepared and more skilled than the average CFI applicant. Treat your FAA inspector/examiner as a good student, and constantly instruct him/her on the entire process. Emphasize safety throughout.

1. Constantly scan for other aircraft. As appropriate, say "check traffic" and point out traffic to your FAA inspector/examiner when you see it.

2. Constantly talk to your FAA inspector/examiner. Explain what you are doing, why, and how you will accomplish these flight maneuvers. If you feel you have nothing to say, talk about scanning for traffic and how it could affect your operations.

3. Remember to speak loudly, clearly, and at a rate that can be easily understood.

4. Your flying will not be perfect. If there are discrepancies, point them out immediately and explain why they occurred. Remember that a CFI does not hide his/her mistakes but openly points them out with an explanation of the errors that were made.

Each task has an FAA reference list that identifies the publication(s) that describe(s) the task. Our discussion is based on the current issue of these references. The following FAA references are used in the Flight Instructor PTS:

FAR Part 61 -- Certification: Pilots, Flight Instructors, and Ground Instructors
FAR Part 91 -- General Operating and Flight Rules
NTSB Part 830 -- Notification and Reporting of Aircraft Accidents and Incidents
AC 00-2 -- *Advisory Circular Checklist*
AC 00-6 -- *Aviation Weather*
AC 00-45 -- *Aviation Weather Services*
AC 61-23/FAA-H-8083-25 -- *Pilot's Handbook of Aeronautical Knowledge*
FAA-H-8083-15 -- *Instrument Flying Handbook*
AC 61-65 -- *Certification: Pilots and Flight Instructors*
AC 61-67 -- *Stall and Spin Awareness Training*
AC 61-84 -- *Role of Preflight Preparation*
AC 90-48 -- *Pilot's Role in Collision Avoidance*
AC 91-13 -- *Cold Weather Operation of Aircraft*
FAA-H-8083-1 -- *Aircraft Weight and Balance Handbook* (formerly AC 91-23 -- *Pilot's Weight and Balance Handbook*)
FAA-H-8083-3 -- *Airplane Flying Handbook* (formerly AC 61-21 -- *Flight Training Handbook*)
FAA-H-8083-9 -- *Aviation Instructor's Handbook* (formerly AC 60-14 -- *Aviation Instructor's Handbook*)
FAA-S-8081-14 -- *Instrument Rating Practical Test Standards*
FAA-S-8081-12 -- *Commercial Pilot Practical Test Standards*
FAA-S-8081-14 -- *Private Pilot Practical Test Standards*
FAA/ASY-20 95/001 -- *Airport Markings, Signs, and Selected Surface Lighting*
AIM -- *Aeronautical Information Manual*
A/FD -- *Airport/Facility Directory*
NOTAMs -- Notices to Airmen
POH/AFM -- Pilot's Operating Handbook (FAA-approved *Airplane Flight Manual*)

STUDY UNIT I
FUNDAMENTALS OF INSTRUCTING

This study unit explains the seven tasks (A-G) of Fundamentals of Instructing. These tasks are "knowledge only." Your FAA inspector/examiner is required to test you on at least Task F, Flight Instructor Characteristics and Responsibilities. (S)he may test you on additional tasks in this area of operation; therefore, you must be prepared for all seven.

NOTE: The FAA released its *Aviation Instructor's Handbook* (FAA-H-8083-9), a 152-page book, in late 1999. We believe, however, that this study unit and your Gleim *FOI FAA Knowledge Test* book are all you need for this topic (Fundamentals of Instructing).

THE LEARNING PROCESS

I.A. TASK: THE LEARNING PROCESS
REFERENCE: FAA-H-8083-9.

Objective. To determine that the applicant exhibits instructional knowledge of the elements of the learning process by describing:

1. Learning theory.
2. Characteristics of learning.
3. Principles of learning.
4. Levels of learning.
5. Learning physical skills.
6. Memory.
7. Transfer of learning.

A. General Information

1. The objective of this task is to measure your understanding of the elements of the learning process and how they relate to instructing.

B. Task Elements

1. **Learning Theory**

a. **Learning** can be defined as a change in behavior or perception as a result of experience.

1) The behavior change can be physical and overt (a better glide path, for instance) or psychological and attitudinal (better motivation, more acute perceptions and insights).

2. **Characteristics of learning**

    a.   Learning is purposeful.

        1)   Each student has specific purposes and goals. Students learn from any activity that tends to further these purposes.

        2)   Thus, in the process of learning, the student's purpose is of great significance.

    b.   Learning comes through experience.

        1)   Knowledge cannot be poured into the student's head. The student can learn only from individual experiences.

        2)   The learning of a physical skill requires actual experience in performing that skill.

            a)   Student pilots learn to fly an airplane only if their experience includes flying.

    c.   Learning is multifaceted.

        1)   The learning process may include any (or all) of the following elements: verbal, conceptual, perceptual, motor skills, emotional, and problem solving.

        2)   Learning is multifaceted in another sense. While learning the subject at hand, the student may be learning other useful things as well. This learning, although called "incidental," can have a significant impact on the student's total development.

    d.   Learning is an active process.

        1)   Students do not soak up knowledge like a sponge absorbs water.

        2)   For students to learn, they must react and respond to the material -- either outwardly or inwardly, physically or emotionally.

        3)   If learning is a process of changing behavior, clearly that process must be an active one.

    e.   Learning style is a concept that can play an important role in improving instruction and student success. This concept recognizes that how a person learns is dependent on that person's background and personality, as well as on the instructional methods used.

        1)   Instructors who can recognize student learning style differences and associated problems will be much more effective than those who do not understand this concept.

3. **Principles of learning**

    a.   **Principle of readiness**

        1)   Individuals learn best when they are ready to learn.

            a)   Readiness implies a degree of single-mindedness and eagerness.
            b)   Getting students ready to learn is the instructor's job.

        2)   If students have a strong purpose, a clear objective, and a well-fixed reason for learning something, they will make more progress than if they lack motivation.

            a)   When students are ready to learn, they will meet the instructor at least halfway, which will make the instructor's job easier.

        3)   There are other factors that the instructor will not be able to control (e.g., a student's personal problem) which may cause the student to have little interest in learning.

b. **Principle of exercise**

1) This principle states that those things most often repeated are best remembered.

   a) This principle is the basis of practice and drill.

2) Students will not learn to perform crosswind landings on one instructional flight.

   a) They learn by applying what they have been told and shown.
   b) Every time practice occurs, learning continues.

3) The instructor must provide opportunities for students to practice or repeat and must see that this process is directed toward a goal.

c. **Principle of effect**

1) Learning is strengthened when accompanied by a pleasant or satisfying feeling, and that learning is weakened when associated with an unpleasant feeling.

2) An experience that produces feelings of defeat, frustration, anger, confusion, or futility is unpleasant for the student.

   a) EXAMPLE: If an instructor attempts to teach landings during the first flight lesson, the student is likely to feel inferior and to be dissatisfied.

3) Impressing students with the difficulty of a flight maneuver can make the teaching task difficult.

   a) Usually it is better to tell students that a maneuver, although difficult, is within their capability to understand or perform.

4) Whatever the learning situation, it should contain elements that affect the students positively and give them feelings of satisfaction.

d. **Principle of primacy**

1) Primacy, the state of being first, often creates a strong, almost unshakable, impression.

   a) For an instructor, this means that what is taught must be right the first time.

2) If a student does not learn the proper use of the rudder control, the instructor will have a difficult task of unteaching the bad habits and reteaching the correct ones.

3) The first experience should be positive and functional and lay the foundation for all that is to follow.

e. **Principle of intensity**

1) A student will learn more from the real thing than from a substitute.

   a) A student is likely to gain greater understanding of stalls by performing them than from merely reading about them.

2) A vivid, dramatic, or exciting learning experience teaches more than a routine or boring experience.

f. **Principle of recency**

1) Things most recently learned are best remembered.

   a) Conversely, the further a student is removed in time from a new fact or understanding, the more difficult it is to remember it.

2) Instructors recognize the principle of recency when planning a summary for a ground lesson or a postflight critique.

   a) The instructor repeats, restates, or reemphasizes important matters at the end of a lesson to make sure a student remembers them.

4. **Levels of learning**

    a. Rote

        1) Rote learning, the lowest level of learning, is the ability to repeat back something that has been taught without understanding or being able to apply what has been learned.

        2) EXAMPLE: A flight instructor tells a student pilot to enter a turn by banking the airplane with aileron control and applying sufficient rudder pressure in the same direction to prevent slipping or skidding.

            a) A student who can repeat this instruction has learned by rote.

    b. Understanding

        1) At this level, the student cannot only repeat what has been taught but also understands the principles and theory behind the knowledge.

        2) EXAMPLE: With proper instruction on the effect and use of the flight controls and experience in their use in straight flight, the student can develop old and new perceptions into an insight on how to make a turn; i.e., the student develops an understanding of how to turn the airplane.

    c. Application

        1) At this level, the student not only understands the theory but can also apply what has been learned and perform it correctly.

        2) EXAMPLE: When a student understands the procedures for entering a turn, has had turns demonstrated, and has practiced turn entries until consistency has been achieved in an acceptable performance level, the student has developed the skill to apply what has been taught.

        3) Application is a major level of learning and one at which an instructor is too often willing to stop.

            a) Discontinuing instruction on one element of piloting performance and directing further instruction exclusively to other elements violates the building block concept of instruction by failing to apply what has been learned to future lessons.

    d. Correlation

        1) At this level, the student is able to associate various learned elements with other segments or blocks of learning or accomplishment.

            a) This level of learning should be the objective of all instruction.

        2) EXAMPLE: The student who has achieved this level of learning in turn entries has developed the ability to correlate the elements of turn entries with the performance of such combined piloting operations as those required for the performance of ground reference maneuvers.

e. Besides the four basic levels of learning, the following three domains of learning consider what is to be learned (i.e., knowledge only, a change in attitude, a physical skill, or a combination of knowledge and physical skill). For each of the three domains, we list the hierarchy of educational objective levels (from the least complex to most complex) and a series of action verbs for each of these levels.

   1) **Cognitive domain** (knowledge).

      a) Knowledge - describe, identify, name, point to, recognize, or recall
      b) Comprehension - convert, explain, locate, report, restate, or select
      c) Application - compute, demonstrate, employ, operate, or solve
      d) Analysis - compare, discriminate, distinguish, or separate
      e) Synthesis - compile, compose, design, reconstruct, or formulate
      f) Evaluation - assess, evaluate, interpret, judge, rate, score, or write

   2) **Affective domain** (attitudes, beliefs, and values).

      a) Receiving - ask, choose, give, locate, select, rely, or use
      b) Responding - conform, greet, help, perform, recite, or write
      c) Valuing - appreciate, follow, join, justify, show concern, or share
      d) Organization - accept responsibility, adhere, defend, or formulate
      e) Characterization - assess, delegate, practice, influence, revise, or maintain

   3) **Psychomotor domain** (physical skills).

      a) Perception - choose, detect, identify, isolate, or compare
      b) Set - Begin, move, react, respond, state, or select
      c) Guided response - assemble, build, calibrate, fix, grind, or mend
      d) Mechanism - same as above except with greater proficiency
      e) Complex overt response - same as above except more highly coordinated
      f) Adaptation - adapt, alter, change, rearrange, reorganize, or revise
      g) Origination - combine, compose, construct, design, or originate

f. These additional levels of learning can be tied to the practical test standards to show the level of knowledge or skill required for a particular task.

5. **Learning physical skills**

  a. Physical skills involve more than muscles.

   1) Concepts of how to perform a skill are developed and attitudes are changed.

  b. Desire to learn

   1) A student who has a desire to learn will learn more quickly and show more rapid progress in improving the skill.

   2) The instructor should build on the student's natural enthusiasm.

  c. Patterns to follow

   1) The best way to prepare a student to perform a task is to provide a clear, step-by-step example.

   2) In flight, the instructor provides the demonstration, emphasizing the steps and techniques, to provide the student with a clear impression of what to do.

  d. Performance of the skill

   1) The student must practice the new skill so as to develop coordination between muscles and visual and tactile senses.

   2) As a student gains proficiency in a skill, verbal instructions mean more.

     a) A long, detailed explanation is confusing before the student begins performing; specific comments are more meaningful and useful after the skill has been partially mastered.

e.   Knowledge of results

    1)   The instructor provides a helpful and often critical function in making certain that students are aware of their progress.

        a)   A student should know when his/her performance is right and when it is wrong.

    2)   Students should be told of their progress as soon after the performance as possible, for they should not be allowed to practice mistakes.

    3)   One way to make students aware of their progress is to repeat a demonstration, showing them the standard against which they can compare their performance.

f.   The pattern of progress

    1)   Learning typically follows a pattern that, if shown on a graph, would be called the learning curve.

        a)   The first part of the curve indicates rapid early improvement.

        b)   Then the curve levels off. This is called the learning plateau.

        c)   The student may stay on the learning plateau for significant periods of effort.

g.   Duration and organization of a lesson

    1)   A primary consideration in planning for student performance is the length of time devoted to practice.

        a)   A beginning student reaches a point at which additional practice is not only unproductive but may even be harmful.

            i)   When that point is reached, errors increase and motivation declines.

    2)   Depending on the skill being learned, the practice period (and perhaps the instructional period) may need to be divided into segments rather than to remain one continuous segment.

h.   Evaluation vs. critique

    1)   In the initial stages, practical suggestions are more valuable to the student than a grade.

    2)   Early evaluation provides a check on the effectiveness of the instruction and a basis on which to predict eventual student learning proficiency.

    3)   The observations on which the evaluations are based can also identify the student's strengths and weaknesses, a prerequisite for making constructive criticism.

i.   Application of skill

    1)   The student must learn the skill so well that it becomes easy, even habitual, to perform it.

    2)   The student must recognize the types of situations in which it is appropriate to use the skill.

6.   **Memory**

a.   Memory is an integral part of the learning process. Memory is a multi-stage process that includes the following three parts:

    1)   The **sensory register** receives input from the five senses and quickly processes it according to the individual's preconceived concept of what is important.

        a)   This is a selective process where the sensory register is set to recognize certain stimuli and immediately transmit them to the working memory for action.

2) The **working or short-term memory** receives information from the sensory register where it may temporarily remain or rapidly fade, depending on the individual's priorities.

    a) If the information is determined by the individual to be important enough to remember, it must be coded in some way for transmittal to long-term memory.

        i)   The coding process may include the use of acronyms, the chronology of events, images, semantics, or an individually developed structure based on past experiences.

        ii) It requires putting the information in context.

    b) Developing a logical strategy for coding information is a significant step in the learning process.

3) **Long-term memory** is where information is stored for future use.

    a) If the initial coding in the working or short-term memory is not properly accomplished, recall will be distorted and it may be impossible.

    b) Long-term memory is a reconstruction, not a pure recall of events or information and is subject to limitations such as time, biases, and in many cases, personal inaccuracies.

b. The following are three theories of forgetting:

    1) The theory of **disuse** states that a person forgets those things that are not used. Students are saddened by the small amount of actual data retained several years after graduation.

    2) The theory of **interference** holds that people forget because new experiences overshadow the original learning experience. In other words, new or similar subsequent events can displace facts learned previously.

    3) The theory of **repression** states that some forgetting is due to the submersion of ideas into the subconscious mind. Material that is unpleasant or produces anxiety is forgotten by the individual, although not intentionally. It is a subconscious and protective response.

c. The instructor should teach thoroughly and with meaning. Material thoroughly learned is highly resistant to forgetting.

d. Retention (remembering) is encouraged/stimulated by the following teaching methods:

    1) Positive feedback, such as praise, recognition, or other types of reward

        a) Responses that give a pleasurable return tend to be repeated.

    2) Meaningful repetition (application)

        a) Three to four repetitions provide the maximum effect, after which the possibility of retention and the rate of learning rapidly decrease.

    3) Recall is promoted by association.

        a) Each bit of information or action that is associated with something to be learned tends to facilitate later recall by the student.

    4) Favorable attitudes aid retention.

        a) People learn and remember only what they are motivated to know.
        b) The most effective motivation is based on positive or rewarding objectives.

    5) Learning with all the senses is most effective.

        a) When several senses respond together, a fuller understanding and greater chance of recall is achieved.

7.   **Transfer of learning**

   a.   The student may be either aided or hindered by things learned previously. This process is called **transfer of learning**.

   b.   Positive transfer occurs when the learning of one skill aids in learning another skill.

   1)   EXAMPLE: Flying rectangular patterns aids in flying traffic patterns.

   c.   Negative transfer occurs when a previously learned skill interferes with learning a new skill.

   1)   EXAMPLE: A student may try to steer a taxiing plane with the control yoke in the same manner as (s)he drives a car.

   2)   Negative transfer thus agrees with the interference theory of forgetting.

   d.   The significance of transfer of learning for the instructor is that the students can be helped to achieve it by following the guidelines below.

   1)   Plan for transfer as a primary objective. The chance for success is increased if the instructor deliberately plans to achieve it.

   2)   Make certain that the students understand that what is to be learned can be applied to other situations. Prepare them to seek other applications.

   3)   Maintain high-order learning standards. The more thoroughly the students understand the material, the more likely they are to see its relationship to new situations.

   a)   Overlearning may even be appropriate.
   b)   Avoid unnecessary rote learning, since it does not foster transfer.

   4)   Provide meaningful learning experiences that build students' confidence in their ability to transfer learning.

   a)   This suggests activities that challenge them to exercise their imagination and ingenuity in applying their knowledge and skills.

   5)   Use instructional material that helps form valid concepts and generalizations. Use materials that make relationships clear.

END OF TASK

HUMAN BEHAVIOR AND EFFECTIVE COMMUNICATION

I.B. TASK: HUMAN BEHAVIOR AND EFFECTIVE COMMUNICATION

REFERENCE: FAA-H-8083-9.

Objective. To determine that the applicant exhibits instructional knowledge of the elements of the teaching process by describing:

1. Human behavior-

a. Control of human behavior.
b. Human needs.
c. Defense mechanisms.
d. The flight instructor as a practical psychologist.

2. Effective communication-

a. Basic elements of communication.
b. Barriers of effective communication.
c. Developing communication skills.

A. General Information

1. The objective of this task is to determine your instructional knowledge of the elements of the teaching process.

B. Task Elements

1. **Exhibit your instructional knowledge by explaining the following elements of human behavior.**

a. **Control of human behavior**

1) The instructor directs and controls the behavior of the students and guides them toward a goal.

a) This process involves directing students' actions and modifying their behavior.

2) The relationship between the instructor and the students has a profound impact on how much the students learn. To students, the instructor is a symbol of authority.

3) Students expect the instructor to exercise certain controls, and they recognize and submit to authority as a valid means of control.

a) The instructor's challenge is to know what controls are best for what circumstances.

4) The instructor should create an environment that enables students to help themselves.

b. **Human needs**

1) Human needs can be organized into a series of levels. The pyramid of human needs has been suggested by Abraham Maslow. For instance, physical needs must be satisfied before so-called "higher" needs can be used as motivators. He suggests that needs must be satisfied in the following ascending order:

a) **Physical needs** pertain to food, rest, exercise, and protection from the elements. Until these needs are satisfied to a reasonable degree, a student cannot concentrate on learning.

b) **Safety needs** include protection against danger, threat, and deprivation.

c) **Social needs** are the needs to belong and to associate with other people.

d) **Egoistic needs** will usually have a direct influence on the student-instructor relationship. Egoistic needs are of two kinds:

i) Relating to one's self-esteem: needs for self-confidence, independence, achievement, and knowledge

ii) Relating to one's reputation: needs for status, appreciation, and the deserved respect of one's fellow beings

e)   **Self-fulfillment needs** are at the top of the hierarchy of human needs. These are the needs for realizing one's own potentialities, for continued development, and for being creative.

   i)   This need of a student should offer the greatest challenge to an instructor.

   ii)   Helping students to realize self-fulfillment is perhaps the most worthwhile accomplishment for an instructor.

2)   You should strive to help your student satisfy these needs in a manner that will ensure a healthy environment for learning.

c.   **Defense mechanisms**

1)   Certain behavior patterns are called defense mechanisms because they are subconscious defenses against the reality of unpleasant situations. People use these defenses to soften feelings of failure, alleviate feelings of guilt, and protect feelings of personal worth and adequacy. Defense mechanisms can be hindrances to learning.

   a)   **Rationalization.** When a person cannot accept the real reasons for his/her own behavior, this device permits the substitution of excuses for reasons. Rationalization is a subconscious technique for justifying actions that otherwise would be unacceptable.

   b)   **Flight.** Students escape from a frustrating situation by taking physical or mental flight.

      i)   To flee physically, students may develop symptoms or ailments that give them excuses for removing themselves from the frustration.

      ii)   More frequent than physical flight is mental flight or daydreaming.

   c)   **Aggression.** A person can avoid a frustrating situation by means of aggressive behavior. Shouting at and accusing others are typical defense mechanisms. Social pressure usually forces student aggressiveness into more subtle forms. Typically, they may

      i)   Ask irrelevant questions
      ii)   Refuse to participate in class activities
      iii) Disrupt activities

   d)   **Resignation.** Students become so frustrated that they lose interest and give up.

      i)   They may no longer believe it profitable or even possible to work further.

      ii)   They accept defeat.

      iii) Resignation usually occurs when the student has completed early lessons without grasping the fundamentals and then becomes bewildered and lost in the advanced phase.

   e)   **Compensation.** With compensation, students often attempt to disguise the presence of a weak or undesirable quality by emphasizing a more positive one.

      i)   They also may try to reduce tension by accepting and developing a less preferred but more attainable objective instead of a more preferred but less attainable objective.

   f)   **Projection.** With projection, students relegate the blame for their own shortcomings, mistakes, and transgressions to others or attribute their motives, desires, characteristics, and impulses to others.

      i)   When students say, "Everyone will cheat on a test if given the chance," they are projecting.

g) **Denial of reality.** Occasionally, students may ignore or refuse to acknowledge disagreeable realities. They may turn away from unpleasant sights, refuse to discuss unpopular topics, or reject criticism.

h) **Reaction formation.** Sometimes individuals protect themselves from dangerous desires by not only repressing them, but by actually developing conscious attitudes and behavioral patterns that are just the opposite.

2) Since defense mechanisms involve some self-deception and distortion of reality, defense mechanisms lead to poor pilot decision making.

d. **The flight instructor as a practical psychologist**

1) At times, the flight instructor must act as a psychologist. Students feel numerous different emotions, and it is the instructor's duty to react in a positive way to those emotions. The following is a list of tips the instructor can follow to aid in this process:

a) Keep students motivated. Students gain more from wanting to learn than from being forced to learn.

b) Keep students informed. Students feel insecure when they do not know what is expected of them or what is going to happen.

c) Approach students as individuals. When instructors limit their thinking to the whole group without considering the individuals who make up that group, their effort is directed at an average personality that really fits no one.

d) Give credit when due. When students do something extremely well, they wish their abilities and efforts to be noticed. Otherwise, they become frustrated.

e) Criticize constructively. Although it is important to give praise and credit when deserved, it is equally important to identify mistakes and failures.

f) Be consistent. If the same thing is acceptable one day and not acceptable the next, the student becomes confused.

g) Admit errors. No one, including the student, expects an instructor to be perfect. The instructor can win the respect of students by honestly acknowledging mistakes.

2. **Exhibit your instructional knowledge by explaining the following elements of effective communication.**

a. **Basic elements of communication**

1) Communication is a two-way transmittal of ideas, information, or feelings. Effective communication is measured by the similarity between what is transmitted and what is received. The three basic elements of communication are the

a) **Source.** The source of communication is the sender of information. As an instructor, it is your responsibility to send information in a way that will be understood by the student.

b) **Symbols.** Information and ideas are transmitted through symbols and codes. This can be in the form of words, gestures, and expressions. A successful instructor combines all possible forms of symbols to effectively convey the information.

c) **Receiver.** The receiver is the final link to communication. (S)he processes the symbols that the source sends. The instructor must be aware that each student's background and individual experiences have a major influence on how information is received.

b. **Barriers of effective communication**

1) Whenever an instructor conveys information or ideas to the student, he or she must be aware of the barriers to effective communication. They are:

   a) **Lack of common experience.** This is the greatest barrier to effective communication. Aviation, by its vary nature, has many technical words and subjects. The student may not understand what is being taught because they don't have the background the instructor does.

   b) **Confusion between the symbol and the symbolized object.** This happens when a word or symbol is confused with another word or symbol, thereby leading to a misunderstanding.

   c) **Abstractions.** An abstraction is any word or symbol that does not elicit a specific image from the receiver.

   d) **Interference.** Often, effective communication lies on the shoulders of the instructor. However, there are times when an outside source provides a barrier to effective communication. These outside sources can be environmental, physiological, or psychological.

c. **Developing communication skills**

1) Like any other skill, communication skills must be developed. This can be done by role playing, instructing, and taking more instruction.

END OF TASK

THE TEACHING PROCESS

I.C. TASK: THE TEACHING PROCESS

REFERENCE: FAA-H-8083-9.

Objective. To determine that the applicant exhibits instructional knowledge of the elements of the teaching process by describing:

1. Preparation of a lesson for a ground or flight instructional period.

2. Presentation methods.

3. Application, by the student, of the material or procedure presented.

4. Review and evaluation of student performance.

A. General Information

1. The objective of this task is to determine your instructional knowledge of the elements of the teaching process.

B. Task Elements

1. **Preparation of a lesson for a ground or flight instructional period**

   a. For each lesson or instructional period, the instructor must determine the material to be covered, the objectives of the lesson, and the goals to be attained.

   b. The preparation for a lesson should also include home study or other special preparation by the student.

   c. As part of the preparation, the instructor should make certain that all necessary supplies, materials, and equipment are readily available and that the equipment is operating properly.

   d. The instructor's preparation should include actual reference to the syllabus for the course involved and a study of objectives.

   e. The instructor must develop a detailed written lesson plan if the instructional period is to be effective.

      1) A lesson plan is the instructor's statement of lesson objectives, the procedures and facilities to be used in presenting it, the specific goals to be attained, and the means to be used for evaluating the results achieved.

   f. One good way to write lesson plans is to begin by formulating performance-based objectives.

      1) Performance-based objectives are used to set measurable, reasonable standards that describe the desired performance of the student.

      2) The objectives must be written. If they are not written, they become subject to the fallibility of recall, interpretation, or loss of specificity with time.

   g. Performance-based objectives consist of the following three parts:

      1) Description of the skill or behavior explains the desired outcome of the instruction.

      2) Conditions are necessary to specifically explain the rules under which the skill or behavior is demonstrated. Information, such as equipment, tools, reference material, and limiting parameters, should be included.

         a) EXAMPLE: Using sectional charts, a flight computer, and a Cessna 172, navigate from point A to point B while maintaining standard hemispheric altitudes.

3)   Criteria is a list of standards that measure the accomplishment of the objective. The criteria should be stated so that there is no question whether the objective has been met.

   a)   Using the example in item g.2) on the previous page, the revised objective may now read, "Using a sectional chart and a flight computer, plan a flight and fly from point A to point B in a Cessna 172. Arrival at point B should be within five minutes of planned arrival time, and cruise altitude should be maintained within 200 ft. during the en-route phase of the flight."

2.   **Presentation methods**

   a.   The instructor's presentation of the knowledge and skills make up the lesson. The choice of the method of presentation is determined by the nature of the subject matter and the objective in teaching it.

   b.   The lecture method is used primarily to introduce students to new subject material. It is also valuable for

      1)   Summarizing ideas
      2)   Showing relationships between theory and practice
      3)   Reemphasizing main points

   c.   The guided discussion method is used by the instructor to draw out what the students know by bringing about discussion through skillful use of questions.

   d.   The demonstration/performance method is used extensively for flight training. It is done in the following steps:

      1)   Instructor tells; instructor does.
      2)   Student tells; instructor does.
      3)   Student tells; student does.
      4)   Instructor supervises and evaluates.

3.   **Application, by the student, of the material or procedure presented**

   a.   In a classroom situation, the student may be asked to explain the new material or to perform a maneuver or operation.

      1)   EXAMPLE: At the end of a classroom period on the flight computer, the student may be asked to work a flight planning problem involving the computation of groundspeed, drift correction, and estimated time en route.

   b.   The application step involves the student's performance of a procedure that has been explained and demonstrated by the instructor.

   c.   During flight instruction, portions of the instructor's explanation and demonstration activity are usually alternated with portions of the student's performance activity.

      1)   It is rare that the instructor completes an explanation and a demonstration and then allows the student to accomplish performance activities without interruptions for corrections and further demonstrations.

      2)   It is very important that the student perform the maneuver the right way the first few times since this is when habits are established.

         a)   Bad habits are hard to correct.

4. **Review and evaluation of student performance**

    a. Review and evaluation are integral parts of each classroom, ground, or flight lesson. Before the end of the instructional session, the instructor should recapitulate what has been covered during the session and require the students to demonstrate the extent to which the lesson objectives have been met.

        1) The instructor's evaluation may be informal and noted only for use in planning the next lesson for the students, or it may be recorded to certify the students' progress in the course.

        2) Students should be aware of their progress. Their advances and deficiencies should be noted at the conclusion of the lesson.

            a) The failure of the instructor to assure that students are aware of their progress, or lack of it, may impose a barrier between instructor and students, which may make further instruction more difficult.

    b. In flight training, the instructor must remember that it is difficult for the student to obtain a clear picture of his/her progress, since (s)he has little opportunity for a direct comparison with others.

        1) Only the instructor can provide a realistic evaluation of performance and progress.

    c. Each lesson should include a review and an evaluation of things previously learned.

        1) If the evaluation reveals a deficiency or fault in the knowledge or performances on which the present lesson is based, it must be corrected before a new lesson can begin.

        2) If deficiencies or faults not associated with the present lesson are revealed, they should be carefully noted and pointed out.

            a) Such corrective measures as are practicable within the limitations of the situation should be taken immediately.

            b) More thorough remedial actions must be included in future lesson plans.

    d. The evaluation of student performance and accomplishment during a lesson should be based on the objectives and goals established in the instructor's lesson plan.

END OF TASK

TEACHING METHODS

I.D. TASK: TEACHING METHODS

REFERENCE: FAA-H-8083-9.

Objective.   To determine that the applicant exhibits instructional knowledge of the elements of teaching methods by describing:

1.   Material organization.
2.   The lecture method.
3.   The cooperative or group learning method.

4.   The guided discussion method.
5.   The demonstration-performance method.
6.   Computer-based training method.

A.   General Information

    1.   The objective of this task is to determine your instructional knowledge of the elements of teaching methods.

B.   Task Elements

    1.   **Material organization**

        a.   Regardless of the teaching method used, an instructor must properly organize the material. One effective way to organize a lesson is introduction, development, and conclusion.

        b.   The introduction should serve several purposes:

            1)   To establish common ground between the instructor and the students

            2)   To capture and hold the attention of the group

            3)   To indicate what is to be covered during the presentation and relate this coverage to the entire course

            4)   To point out specific benefits the student can expect from the learning

            5)   To establish a receptive attitude toward the subject and lead into the lesson development

        c.   The instructor must logically organize the material to show the relationships of the main points. The instructor usually shows these primary relationships by developing the main points from

            1)   The past to the present
            2)   The simple to the complex
            3)   The known to the unknown
            4)   The most frequently used to the least frequently used

        d.   An effective conclusion retraces the important elements of the lesson and relates them to the objective.

            1)   This review and wrap-up of ideas reinforces the student's learning and improves the retention of what has been learned.

            2)   No new ideas should be introduced in the conclusion because, at this point, they are likely to confuse the students.

2. **The lecture method**

   a.   The lecture is used primarily to introduce students to new subject material. It is also valuable for

       1)   Summarizing ideas
       2)   Showing relationships between theory and practice
       3)   Reemphasizing main points

   b.   There are four types of lectures:

       1)   The illustrated talk, in which the speaker relies heavily on visual aids to convey his ideas to the listeners

       2)   The briefing, in which the speaker presents a concise array of facts to the listeners, who do not expect elaboration or supporting material

       3)   The formal speech, in which the speaker's purpose is to inform, to persuade, or to entertain

       4)   The teaching lecture, for which the instructor must plan and deliver an oral presentation in a manner that helps the students reach the desired learning outcomes

   c.   The lecturer does not receive direct reaction (either words or actions) from the students.

       1)   Thus, the instructor must be alert for subtle responses from the class and adjust the lesson accordingly.

          a)   Each class is made up of individuals who may react differently from another group to the same material.

       2)   The instructor must recognize that the lecture method is least useful for evaluating student performance.

   d.   As with course and lesson plan development, there are four steps to developing a lecture:

       1)   Establish the objective and desired outcomes.
       2)   Research the subject.
       3)   Organize the material.
       4)   Plan productive classroom activities.

   e.   The lecture can be formal or informal.

       1)   A formal lecture provides no accurate means for checking student learning.

       2)   Characteristically, the more informal a lecture, the more students will participate, ask questions, etc.

3. **The cooperative or group learning method**

   a.   Cooperative or group learning method is an instructional strategy that organizes students into small groups so they can work together to maximize their own and each other's learning.

   b.   The most significant characteristic of group learning is that it continually requires active participation of the student in the learning process.

4.   **The guided discussion method**

    a.   Fundamentally, the guided discussion method of teaching is the reverse of the lecture method. The instructor aims to draw out what the students know rather than spend the class period telling them what (s)he knows.

    b.   In the guided discussion, learning is produced through the skillful use of questions to bring about discussion that will develop an understanding of the subject.

    c.   Each question, in order to be effective, should

        1)   Have a specific purpose
        2)   Be clear in meaning
        3)   Contain a single idea
        4)   Stimulate thought
        5)   Require definite answers
        6)   Relate to previously taught information

    d.   Avoid questions that require short, categorical answers, such as "yes," "no," "green," "white," "one," or "four." Open-ended questions usually begin with "how" or "why."

    e.   To encourage enthusiasm and stimulate discussion, the instructor should create a relaxed, informal atmosphere.

    f.   When it appears that the students have adequately discussed the ideas supporting this particular part of the lesson, the instructor should summarize what they have accomplished.

        1)   The interim summary can be made immediately after the discussion of each learning outcome.

        2)   A summary consolidates what students have learned, emphasizes how much they know already, and points out any aspects they missed.

        3)   Thus, it prepares the way for further fruitful discussion.

    g.   The more intense the discussion and the greater the participation, the more effective the learning will be.

    h.   The students can reach the desired learning objectives only if they have some knowledge to exchange with each other.

        1)   Students without some background in a subject should not be asked to discuss that subject.

5.   **The demonstration-performance method**

    a.   The demonstration/performance method of instruction has five essential phases:

        1)   Instructor explanations
        2)   Instructor demonstrations
        3)   Student performance, first by explaining as the instructor does, then by explaining and doing it by him/herself
        4)   Instructor supervision
        5)   Instructor evaluations

    b.   The student performance and instructor supervision are performed concurrently.

    c.   When the instructor is demonstrating, it is important that the demonstration conform to the explanation as closely as possible.

        1)   If a deviation occurs, the instructor should point it out and account for it.

        2)   To downplay or ignore the discrepancy creates confusion and undermines student confidence in the teacher.

        3)   Acknowledging and correcting the performance demonstrates how to deal with an unexpected situation calmly and with good judgment.

6. **Computer-based training method**

    a. Computer-based training (CBT), sometimes called computer-based instruction (CBI), uses a personal computer (PC) as a training device.

        1) One of the major advantages of CBT is that students can progress at a rate that is comfortable for them. Additionally, a student is able to access the CBT at their own convenience.

    b. An excellent example of CBT is Gleim's *FAA Test Prep* CD-Rom for preparation of FAA knowledge tests.

    c. Computers are used for training at many different levels, such as

        1) Flight simulators
        2) Flight training devices (FTDs)
        3) Personal computer-based aviation training devices (PCATDs)

    d. While computers provide many training advantages, they also have limitations.

        1) Improper or excessive use of CBT should be avoided.

        2) CBT should not be used by the instructor as stand-alone training any more than a textbook or video.

    e. In teaching flight students, CBT programs can be used by the instructor as simply another form of reference for students to study.

        1) The instructor must continue to monitor and evaluate the progress of a student as usual.

END OF TASK

CRITIQUE AND EVALUATION

I.E. TASK: CRITIQUE AND EVALUATION

REFERENCE: FAA-H-8083-9.

Objective. To determine that the applicant exhibits instructional knowledge of the elements of evaluation by describing:

1. Critique-

  a. Purpose and characteristics of an effective critique.

  b. Methods and ground rules for a critique.

2. Evaluation-

  a. Characteristics of effective oral questions and what types to avoid.

  b. Responses to student questions.

  c. Characteristics and development of effective written questions.

  d. Characteristics and uses of performance test, specifically, the FAA practical test standards.

A. General Information

    1. The objective of this task is to determine your instructional knowledge of the elements of critique and evaluation.

B. Task Elements

    1. **Exhibit your instructional knowledge by explaining the following elements of critiques.**

        a. **Purpose and characteristics of an effective critique**

            1) A critique is provided to a student as a means to better their performance. Critiques must be given for every lesson so the student knows where they stand and how they can improve on their performance. A good critique is:

                a) **Objective.** The instructor's personal feelings about a particular student must never be expressed during a critique.

                b) **Flexible.** Every student behaves in a different manner. An effective critique must be flexible enough to fit the particular student and situation.

                c) **Acceptable.** A student must accept the instructor to accept the instructor's critique. The critique must be fair and sincere for the student to accept it.

                d) **Comprehensive.** A comprehensive critique must contain both positive and negative aspects of the lesson. It does not necessarily have to be long, but it must be complete.

                e) **Constructive.** General negative criticism must never be part of an effective critique. However, praise without merit offers no help to the student.

                f) **Organized.** There must be some sort of organization if a critique is to be fully understood by the student.

                g) **Thoughtful.** An instructor must understand the student's situation and react accordingly to it. Ridicule and anger have no place in a critique.

                h) **Specific.** Specific instructions, information, and suggestions make an effective critique.

b. **Methods and ground rules for a critique**

1) Though the instructor is always responsible for the critique, there are ways in which the critique can bring in other students and elements to make it more interesting, and possibly more effective. For example,

a) The instructor/student critique
b) The student-led critique
c) The small group critique
d) An individual critique led by another student
e) The self critique
f) The written critique

2) The following is a list of ground rules that every critique should follow:

a) Never extend the critique into time allocated for something else.

b) Avoid trying to cover too much.

c) Provide a summary at the end of every critique to stress the main points.

d) Avoid absolute statements.

e) Avoid controversies and try to remain neutral with group factions.

f) Don't allow yourself to be put in a position of having to defend your critique.

g) If a critique is written, make sure it agrees with the oral portion.

2. **Exhibit your instructional knowledge by explaining the following elements of evaluation.**

a. **Characteristics of effective oral questions and what types to avoid**

1) Characteristics of effective oral questions:

a) Each question must have only one correct answer.

b) Each question must apply to the subject being taught.

c) Each question should be brief, concise, clear, and definite.

d) Each question should center on only one idea that is limited to who, what, where, when, how, or why -- not a combination.

e) Each question should present a challenge. Easy questions do not stimulate learning.

2) Types of oral questions to avoid:

a) Trick questions. The students will soon feel they are in a battle of wits with the instructor. The significance of the subject being taught will be lost.

b) Irrelevant questions. Introduction of unrelated facts only obscures the learning process and retards student progress.

c) Questions requiring yes and no answers. Questions should be framed so that the answers are specific and factual. One-word answers may be the product of a good guess.

d) Questions that are too general. Questions such as "Do you understand?" or "Have you any questions?" fail to elicit answers from students that are adequate for accurate evaluations.

i) A student's assurance of understanding or his/her lack of questions provides no evidence of comprehension or even knowledge of the subject under discussion.

b.   **Responses to student questions**

1)   Quizzing goes both ways. When answering student questions, the instructor needs to clearly understand the question before attempting an answer.

2)   The instructor should display interest in the student's questions and give as direct and accurate an answer as possible.

3)   If a student's question is too advanced for the particular lesson and confusion may result from a complete and complicated answer, the instructor may

a)   Carefully explain that the question was good and pertinent,

b)   Explain that to answer would unnecessarily complicate the learning task at hand, and

c)   Advise the student to reintroduce the question later at the appropriate point in training or meet outside class for a more complete discussion.

4)   Occasionally, a student will ask a question the instructor cannot answer. The best course is to freely admit not knowing the answer.

a)   The instructor should then promise to find out the answer or offer to help the student look it up in appropriate references.

c.   **Characteristics and development of effective written questions**

1)   Written test questions fall into two general categories: supply and selection.

2)   Supply-type questions require the student to furnish a response in the form of a word, sentence, or paragraph.

a)   Supply-type test items

i)   Require students to organize their thoughts and ideas,
ii)   Demand the ability to express ideas, and
iii)   Are subjective.

- Thus, they cannot be graded uniformly, which is their main disadvantage.
- The same test graded by different instructors probably would be assigned different scores.

b)   Supply-type test items take longer for students to answer and for instructors to grade, which is another disadvantage.

3)   Selection-type questions include items for which two or more alternative responses are provided.

a)   True-false and multiple-choice type questions are prime examples of selection-type questions.

b)   Selection-type test items

i)   Are highly objective, and
ii)   Allow direct comparison of students' accomplishments. For example, it is possible to compare student performance

- Within the same class
- Between classes
- Under different instructors

4)   Test question attributes

a)   Reliability. A reliable written test is one that yields consistent results.

b)   Validity. A written test is valid when it actually measures what it is supposed to measure and nothing else.

      c) Usability. A written test that is usable is

           i) Easy to give

           ii) Written in a type size large enough for the students to read easily

           iii) Worded clearly and concisely -- both the directions for taking the test and the test items themselves

           iv) Constructed with graphs, charts, and illustrations that are appropriate to the test items and clearly drawn

           v) Easily graded

      d) Comprehensiveness. A written test must completely sample whatever is being measured.

      e) Discrimination. A discriminating written test will detect small differences in knowledge.

d. **Characteristics and uses of performance test, specifically, the FAA practical test standards**

    1) Performance testing is desirable for evaluating training that involves an operation, a procedure, or a process.

      a) The FAA Practical Test Standards (PTSs) are an example of performance testing.

    2) Federal Aviation Regulations (FARs) specify areas in which knowledge and skill must be demonstrated by the applicant before the issuance of a pilot certificate or rating.

      a) FARs provide the flexibility to permit the FAA to publish PTSs containing specific tasks (procedures and maneuvers) in which pilot competency must be demonstrated.

         i) The FAA will add, delete, or revise tasks whenever it is determined that changes are needed in the interest of safety.

      b) Adherence to provisions of the FARs and the PTSs is mandatory for the evaluation of pilot applicants.

    3) An appropriately rated flight instructor is responsible for training the student to the acceptable standards as outlined in the objective of each task within the appropriate PTS.

      a) The flight instructor must certify that the applicant is able to perform safely as a pilot at the appropriate skill level and is competent to pass the required practical test for the certificate or rating sought.

    4) The examiner who conducts the practical test is responsible for determining that the applicant meets standards outlined in the objective of each task within the appropriate PTS.

      a) For each task that involves "knowledge only" elements, the examiner will orally quiz the applicant on those elements.

      b) For each task that involves both "knowledge and skill" elements, the examiner will orally quiz the applicant regarding knowledge elements and ask the applicant to perform the skill elements.

      c) The examiner will determine that the applicant's knowledge and skill meet the objective in all required tasks.

         i) Oral questioning may be used at any time during the practical test.

END OF TASK

FLIGHT INSTRUCTOR CHARACTERISTICS AND RESPONSIBILITIES

I.F. TASK: FLIGHT INSTRUCTOR CHARACTERISTICS AND RESPONSIBILITIES

   REFERENCE: FAA-H-8083-9.

Objective.   To determine that the applicant exhibits instructional knowledge of the elements of flight instructor characteristics and responsibilities by describing:

1. Aviation instructor responsibilities in-

   a. Providing adequate instruction.

   b. Establishing standards of performance.

   c. Emphasizing the positive.

2. Flight instructor responsibilities in-

   a. Providing student evaluation and supervision.

   b. Preparing practical test recommendations and endorsements.

   c. Determining requirements for conducting additional training and endorsement requirements.

3. Professionalism as an instructor by-

   a. Explaining important personal characteristics.

   b. Describing methods to minimize student frustration.

A.   General Information

   1.   The objective of this task is to determine your instructional knowledge of the elements of flight instructor characteristics and responsibilities.

B.   Task Elements

   1.   **Exhibit your instructional knowledge by explaining the following elements of aviation instructor responsibilities.**

      a.   **Providing adequate instruction**

         1)   It is the flight instructor's duty to provide adequate instruction to the student. The instructor can be effective by having a genuine interest in the positive outcome of the student's training.

      b.   **Establishing standards of performance**

         1)   A high standard of training and performance is essential to student success. Instructors who allow standards to be compromised in the interest of becoming more friendly with the student are doing a disservice to the student and to themselves. Students are generally resentful towards low, easily achieved standards.

      c.   **Emphasizing the positive**

         1)   Students must have a positive image of themselves and of aviation, especially early on in their training, to be successful.

            a)   An instructor must focus on the positive aspects of aviation and flight training for a student to feel comfortable.

2. **Exhibit your instructional knowledge by explaining the following elements of aviation instructor responsibilities.**

   a. **Providing student evaluation and supervision**

      1) It is important for the instructor to consistently evaluate the student's performance.

         a) Evaluation can be done in many forms and at various times during a lesson. For example, an instructor may evaluate a student's performance of a maneuver immediately after (s)he performs it or at the end of the lesson when the student will not be distracted by flying the airplane.

         b) Student evaluation should be made on a regular basis so the student understands his/her progress in the training process.

            i) It is not enough for only the instructor to be aware of the student's progress. The student must be kept informed of his/her progress so as not to become discouraged.

      2) The flight instructor is ultimately responsible for the safety of the student.

         a) Supervision applies to every aspect of training, both dual and solo.

            i) Before a student may fly solo, the flight instructor must be completely confident that the student will practice safe solo operations.

   b. **Preparing practical test recommendations and endorsements**

      1) The flight instructor has a heavy responsibility with regards to practical test recommendations and endorsements.

         a) If an instructor recommends a student for a practical test, they are attesting to the fact that they trained the student on all the applicable sections in the PTS, and they feel the student is competent to pass the practical test.

         b) If a student is found deficient in an area during the practical test, the instructor is accountable, not the student.

      2) A flight instructor who endorses a student is ultimately responsible for that student's safety.

         a) Endorsing a student before (s)he is ready for that endorsement is a breach of faith with the student.

         b) As a flight instructor, you are required to maintain a record of all endorsements for a period of three years. This record has to include the name of the person receiving the endorsement, the date, and the type of the endorsement. In addition, the type, date, and result must be kept for all written and practical test endorsements.

   c. **Determining requirements for conducting additional training and endorsement requirements**

      1) Information regarding endorsements, flight reviews, instrument proficiency checks, and transitions to other aircraft can be found in AC 61-98, *Currency and Additional Qualification Requirements for Certificated Pilots.*

         a) The flight review is an important part of maintaining currency. It is the flight instructor's responsibility to endorse someone for a successful flight review only when that person has performed satisfactorily.

            i) The flight review should be consistent with the current regulations and standards.

      2) A common duty of many flight instructors is aircraft checkouts. These are often required for insurance purposes, and the flight instructor must make sure that the checkout is complete and safe.

3.   **Exhibit your instructional knowledge by explaining the following elements of professionalism as an instructor.**

  a.   **Explaining important personal characteristics**

    1)   A flight instructor must be a professional at all times.

      a)   **Sincerity.** Honesty and sincerity are important traits of a flight instructor. If you do not know the answer to a question, simply admit it and volunteer to find the answer.

      b)   **Acceptance of the student.** It is the instructor's job to teach the student how to fly, not to change or degrade the student. An instructor must be fully accepting of who a student is.

      c)   **Personal appearance and habits.** An aviation instructor is a professional and should act and dress accordingly. Anything less will seem insulting to the student.

      d)   **Demeanor.** Instructors must be consistent, kind, and positive with their students.

      e)   **Safety practices and accident prevention.** Students often emulate the way their instructor flies. Therefore, an instructor must fly safely at all times. In addition, the instructor must stress accident prevention at all times during the training.

      f)   **Proper language.** The instructor must use appropriate language at all times with students.

  b.   **Describing methods to minimize student frustration**

    1)   Instructors must minimize student frustration if the student is expected to learn the material well. This can be done by:

      a)   Motivating students,
      b)   Keeping students informed,
      c)   Approaching students as individuals,
      d)   Giving credit when credit is due,
      e)   Criticizing constructively,
      f)   Being consistent, and
      g)   Admitting errors.

END OF TASK

PLANNING INSTRUCTIONAL ACTIVITY

I.G. TASK: PLANNING INSTRUCTIONAL ACTIVITY REFERENCE: FAA-H-8083-9. **Objective.** To determine that the applicant exhibits instructional knowledge of the elements of planning instructional activity by describing: 1. Developing objectives and standards for course of training. 2. Theory of building blocks of learning. 3. Requirements for developing a training syllabus. 4. Purpose and characteristics of a lesson plan.

A.   General Information

1.   The objective of this task is to determine your instructional knowledge of the elements related to the planning of instructional activity.

   a.   Your FAA inspector/examiner is required to test you on this task.

B.   Task Elements

1.   **Developing objectives and standards for a course of training**

   a.   Determine the objective(s) of the course. **What** is to be learned?

      1)   Why?
      2)   How?
      3)   When?

   b.   What are the completion standards?

   c.   Any instructional activity must be competently planned and organized if it is to achieve the objectives of the course.

      1)   First, you must determine the standards and objectives of the course.

      2)   Second, you must break the overall objectives into a syllabus. A syllabus is the list of segments or blocks of learning to be mastered in the course.

         a)   Check that each block is a truly necessary part of the total objective.

            i)   Extraneous tasks or material can be distracting.

         b)   Arrange the blocks or segments logically:

            i)   Smallest or most basic first.
            ii)   Known to unknown.
            iii)   The blocks of learning identified during a training activity grow progressively smaller in scope as complexity increases.

2.   **Theory of building blocks of learning**

   a.   The building blocks of learning must be introduced after the overall training objectives have been established.

      1)   When determining which blocks to include, it is important to leave out unnecessary blocks, as they detract from the overall objective.

   b.   Blocks should be treated as integral parts of the structure, not isolated subjects.

   c.   Using the building block method of flight instruction provides students with a boost in self-confidence.

      1)   When a student is aware that they have mastered a block, it gives them confidence to move on to the next one.

3.   **Requirements for developing a training syllabus**

   a.   The form of the syllabus may vary, but it is always an abstract or digest of the course of training. It consists of the blocks of learning to be completed in the most efficient order.

   b.   A training syllabus must be flexible and should primarily be used as a guide.

      1)   The order of training can and should be altered, when necessary, to suit the progress of the student and the demands of special circumstances.

   c.   In departing from the order prescribed by the syllabus, the instructor must consider the relationships of the blocks of learning affected.

      1)   It is preferable to skip to a completely different part of the syllabus when the conduct of a scheduled lesson is impossible rather than to proceed to the next block, which may be predicated completely on skills to be developed during the lesson being postponed.

4.   **Purpose and characteristics of a lesson plan**

   a.   A lesson plan should be prepared for each ground school and flight period regardless of instructor experience or student level.

      1)   A so-called "mental outline" is not a lesson plan.
      2)   To be effective, the lesson plan must be in writing.

   b.   Purpose of the lesson plan

      1)   Lesson plans are designed to assure that each student receives the best possible instruction under the existing conditions.

      2)   Lesson plans help instructors keep a constant check on their own activity, as well as that of their students.

      3)   The development of lesson plans by instructors signifies that they have taught the lessons to themselves prior to attempting to teach the lessons to students.

      4)   An adequate lesson plan, when properly used, should

         a)   Assure a wise selection of material and the elimination of unimportant details

         b)   Make certain that due consideration is given to each part of the lesson

         c)   Aid the instructor in presenting the material in a suitable sequence for efficient learning

         d)   Provide an outline of the teaching procedure to be used

         e)   Serve as a means of relating the lesson to the course objectives of training

         f)   Give the inexperienced instructor confidence

         g)   Promote uniformity of instruction regardless of the instructor or the date on which the lesson is given

c.　Characteristics of a well-planned lesson

1)　Unity. Each lesson should be a unified segment of instruction. A lesson is concerned with certain limited objectives, which are stated in terms of desired student learning outcomes.

2)　Content. Each lesson should contain new material. The new facts, principles, procedures, or skills should be related to the lesson previously presented.

　　a)　A short review of earlier lessons is usually necessary, notably in flight training.

3)　Scope. Each lesson should be reasonable in scope. A person can master only a few principles or skills at a time, the number depending on complexity.

　　a)　Presenting too much material in a lesson results in confusion; presenting too little material results in inefficiency.

4)　Practicality. Each lesson should be planned in terms of the conditions under which the training is to be conducted.

　　a)　Lesson plans conducted in an airplane or ground trainer will differ from those conducted in a classroom.

5)　Relation to course of training. Each lesson should be planned and taught so that its relation to the course objectives are clear to each student.

6)　Instructional steps. Every lesson, when adequately developed, falls logically into the four steps of the teaching process, i.e., preparation, presentation, application, and review and evaluation.

7)　Flexibility. Although the lesson plan provides an outline and sequence for the training to be conducted, a degree of flexibility should be incorporated.

　　a)　EXAMPLE: The outline of content may include blank spaces for add-on material, if required.

d.　Any lesson plan should contain the following items:

1)　Lesson objective. The objective of the lesson should be clearly stated in terms of desired student learning outcomes.

　　a)　The objective is the reason for the lesson, i.e., what the instructor expects the student to know or do at the completion of the lesson.

2)　Elements involved. This item is a statement of the elements of knowledge and skill necessary for the fulfillment of the lesson objective.

　　a)　The necessary elements may include both those elements previously learned and those to be introduced during this lesson.

3)　Schedule. The instructor should estimate the amount of time to be spent on a particular lesson and also the approximate time to be devoted to the presentation of the elements of that lesson.

4)　Equipment. This list includes all instructional materials and training aids required to teach the lesson.

5)　Instructor's actions. This item is a statement of the instructor's proposed procedures for presenting the elements of knowledge and performance involved in the lesson.

6)　Student's actions. This item is a statement of desired responses to instruction.

7)　Completion standards. These standards supply the basis for determining how well the student has met the lesson objective in terms of knowledge and skill.

END OF TASK -- END OF STUDY UNIT

STUDY UNIT II
TECHNICAL SUBJECT AREAS

This study unit explains the 12 tasks (A-L) of Technical Subject Areas. These tasks are knowledge only. Your FAA inspector/examiner is required to test you on at least Task L, Logbook Entries and Certificate Endorsements, and one other task; therefore, you must be prepared for all 12.

In late 1999, the FAA published *Airplane Flying Handbook* (FAA-H-8083-3) to replace AC 61-21, *Flight Training Handbook*. The changes included bringing the text up to date, removing some information (e.g., chapter on principles of flight and performance characteristics), and adding some additional information on transition to other types of airplanes (i.e., tailwheel airplanes).

AEROMEDICAL FACTORS

II.A. TASK: AEROMEDICAL FACTORS

  REFERENCES: FAA-H-8083; FAA-S-8081, FAA-S-8081-14; AIM.

Objective. To determine that the applicant exhibits instructional knowledge of the elements related to aeromedical factors by describing:

1. How to obtain an appropriate medical certificate.

2. How to obtain a medical certificate in the event of a possible medical deficiency.

3. The causes, symptoms, effects, and corrective action of the following medical factors:

   a. Hypoxia.
   b. Hyperventilation.
   c. Middle ear and sinus problems.
   d. Spatial disorientation.
   e. Motion sickness.
   f. Carbon monoxide poisoning.
   g. Fatigue and stress.
   h. Dehydration.

4. The effects of alcohol and drugs, and their relationship to flight safety.

5. The effect of nitrogen excesses during scuba dives and how this affects pilots and passengers during flight.

A. General Information

   1. The objective of this task is to determine your instructional knowledge of the elements related to aeromedical factors.

   2. Pilot personal checklist

      a. Aircraft accident statistics show that pilots should conduct preflight checklists on themselves as well as their aircraft. Pilot impairment contributes to many more accidents than do failures of aircraft systems.

      b. I'M SAFE -- I am NOT impaired by

         I llness
         M edication

         S tress
         A lcohol
         F atigue
         E motion

B. Task Elements

   1. **Describe how to obtain an appropriate medical certificate**

      a. FAA medical certificates are issued after a routine medical examination which may be administered only by FAA-designated doctors called Aviation Medical Examiners (AMEs).

      b. The FAA publishes a directory that lists all authorized aviation medical examiners by name and address. Copies of this directory are kept at all FSDOs, FSSs, and other FAA offices.

      c. The student pilot must request a combined medical/student pilot certificate, which functions as a student pilot certificate once it is signed by the AME.

   2. **Describe how to obtain a medical certificate in the event of a possible medical deficiency**

      a. As a CFI, you should encourage a person considering flight training to obtain an appropriate medical certificate before training is started.

1) Even if a person has a medical deficiency, a medical certificate can be issued in many cases.

   a) Operating limitations may be imposed, depending on the nature of the deficiency.

b. Medical deficiency questions should be addressed to an AME, who frequently consults with the FAA Flight Surgeon's Office in the FAA Aeronautical Center in Oklahoma City. Their telephone number is (405) 954-4821.

   1) You should be aware that medical deficiencies will not necessarily prohibit activity as a pilot of an airplane.

   2) Advise such a person to obtain assistance from an AME and the local FSDO.

      a) This assistance is available only when requested by the person seeking the medical certificate.

3. **Describe the causes, symptoms, effects, and corrective action of the following medical factors.**

   a. **Hypoxia** is a state of oxygen deficiency in the body sufficient to impair functions of the brain and other organs.

      1) Although a deterioration in night vision occurs at a cabin pressure altitude as low as 5,000 ft. MSL, other significant effects of altitude hypoxia usually do not occur below 12,000 ft. in the normal, healthy pilot.

      2) From 12,000 to 15,000 ft. MSL (without supplemental oxygen), judgment, memory, alertness, coordination, and ability to make calculations are impaired.

      3) Headache, drowsiness, dizziness, and either a sense of well-being (euphoria) or belligerence occur.

      4) The effects of hypoxia appear more quickly with increased altitude.

         a) Pilot performance can seriously deteriorate within 15 min. at 15,000 ft. MSL.

         b) At altitudes above 15,000 ft. MSL, the periphery of the visual field turns gray. Only central vision remains (tunnel vision).

         c) A blue color (cyanosis) develops in the fingernails and lips.

         d) A person will lose the ability to take corrective and protective action in 20 to 30 min. at 18,000 ft. MSL.

            i) This loss of ability happens in 5 to 12 min. at 20,000 ft. MSL, followed soon by unconsciousness.

      5) Significant effects of hypoxia can occur at even lower altitudes given one or more of the following factors:

         a) Carbon monoxide inhaled in smoking or from exhaust fumes

         b) Small amounts of alcohol and low doses of certain drugs (e.g., antihistamines, tranquilizers, sedatives, and analgesics)

         c) Extreme heat or cold, fever, and/or anxiety

      6) Hypoxia is prevented by understanding the factors that reduce your tolerance to altitude and by using supplemental oxygen above 10,000 ft. MSL during the day and above 5,000 ft. MSL at night.

         a) Corrective action if hypoxia is suspected or recognized includes

            i) Use of supplemental oxygen
            ii) An emergency descent to a lower altitude

b. **Hyperventilation**, which is an abnormal increase in the volume of air breathed in and out of the lungs, can occur subconsciously when you encounter a stressful situation in flight.

1) This abnormal breathing flushes from your lungs and blood much of the carbon dioxide your system needs to maintain the proper degree of blood acidity.

a) The resulting chemical imbalance in the body produces dizziness, tingling of the fingers and toes, hot and cold sensations, drowsiness, nausea, and suffocation. Often you may react to these symptoms with even greater hyperventilation.

2) It is important to realize that early symptoms of hyperventilation and hypoxia are similar. Also, hyperventilation and hypoxia can occur at the same time.

3) The symptoms of hyperventilation subside within a few minutes after the rate and depth of breathing are consciously brought back under control.

a) This can be hastened by controlled breathing in and out of a paper bag held over the nose and mouth. Also, talking, singing, or counting aloud often helps.

c. **Middle ear and sinus problems.** As the cabin pressure decreases during ascent, the expanding air in the middle ear pushes the eustachian tube open and escapes down to the nasal passages, thus equalizing ear pressure with the cabin pressure.

1) Either an upper respiratory infection (e.g., a cold or sore throat) or nasal allergies can produce enough congestion around the eustachian tube to make equalization difficult if not impossible.

2) The difference in pressure between the middle ear and the airplane's cabin can build to a level that will hold the eustachian tube closed. This problem, commonly referred to as **ear block**, produces severe ear pain and loss of hearing that can last from several hours to several days.

a) Rupture of the ear drum can occur in flight or after landing.
b) Fluid can accumulate in the middle ear and become infected.

3) During ascent and descent, air pressure in the sinuses equalizes with aircraft cabin pressure through small openings that connect the sinuses to the nasal passages.

a) Either an upper respiratory infection (e.g., a cold or sinusitis) or nasal allergies can produce enough congestion around one or more of these small openings to slow equalization.

4) As the difference in pressure between the sinus and the cabin mounts, the opening may become plugged, resulting in **sinus block**. A sinus block, experienced most frequently during descent, can occur in the frontal sinuses, located above each eyebrow, or in the maxillary sinuses, located in each upper cheek.

a) It usually produces excruciating pain over the sinus area.
b) A maxillary sinus block can also make the upper teeth ache.
c) Bloody mucus may discharge from the nasal passages.

5) Middle ear and sinus problems are prevented by not flying with an upper respiratory infection or nasal allergic condition.

a) Adequate protection is not provided by decongestant spray or drops to reduce congestion around the eustachian tubes or the sinus openings.

b) Oral decongestants have side effects that can significantly impair pilot performance.

    d.   **Spatial disorientation** is a state of temporary spatial confusion resulting from misleading information sent to the brain by various sensory organs. To a pilot this means simply the inability to tell "which way is up."

        1)   Sight, the semicircular canals of the inner ear, and pressure-sensitive nerve endings (located mainly in the muscles and tendons) are used to maintain spatial orientation.

            a)   However, during periods of limited visibility, conflicting information among these senses makes a pilot susceptible to spatial disorientation.

        2)   The brain relies primarily on sight when there is conflicting information.

            a)   When outside references are limited due to limited visibility and/or darkness, a pilot must rely on the flight instruments for information.

        3)   Spatial disorientation can be corrected by relying on and believing the airplane's instruments or by focusing on reliable, fixed points on the ground.

    e.   **Motion sickness** is caused by continued stimulation of the tiny portion of the inner ear which controls the pilot's sense of balance.

        1)   The symptoms are progressive.

            a)   First, the desire for food is lost.

            b)   Then, saliva collects in the mouth, and the person begins to perspire freely.

            c)   Eventually, the person becomes nauseated and disoriented.

            d)   The head aches and there may be a tendency to vomit.

        2)   Pilots who are susceptible to motion sickness should not take the preventive drugs available over the counter or by prescription.

            a)   These drugs may make the pilot feel drowsy or depress his/her brain functions in other ways.

            b)   Research has shown that most motion sickness drugs cause a temporary deterioration of navigational skills or other tasks demanding keen judgment.

        3)   If suffering from motion sickness, the pilot should

            a)   Open the air vents.
            b)   Loosen clothing.
             c)   Use supplemental oxygen, if available.
            d)   Keep the eyes on a point outside the airplane.
            e)   Avoid unnecessary head movements.
            f)   Cancel the flight and land as soon as possible.

    f.   **Carbon monoxide** is a colorless, odorless, and tasteless gas contained in exhaust fumes and tobacco smoke.

        1)   When inhaled even in minute quantities over a period of time, it can significantly reduce the ability of the blood to carry oxygen.

        2)   Consequently, effects of hypoxia occur.

        3)   Most heaters in light aircraft work by air flowing over the exhaust manifold.

            a)   Using these heaters when exhaust fumes are escaping through manifold cracks and seals is responsible every year for both nonfatal and fatal aircraft accidents from carbon monoxide poisoning.

4)   A pilot who detects the odor of exhaust or experiences symptoms of headache, drowsiness, or dizziness while using the heater should suspect carbon monoxide poisoning and immediately shut off the heater and open the air vents.

    a)   If symptoms are severe or continue after landing, medical treatment should be sought.

g.   **Fatigue and stress** continue to be two of the most treacherous hazards to flight safety. Fatigue is best described as acute (short-term) or chronic (long-term).

1)   **Acute fatigue** is the everyday tiredness felt after long periods of physical or mental strain, e.g., 8 hr. of flight instruction.

    a)   Coordination and alertness can be reduced.

    b)   Acute fatigue is prevented by adequate rest and sleep, as well as regular exercise and proper nutrition.

2)   **Chronic fatigue** occurs when there is not enough time for full recovery between episodes of acute fatigue.

    a)   Performance continues to fall off, and judgment becomes impaired.
    b)   Recovery from chronic fatigue requires a prolonged period of rest.

3)   Stress is defined as the body's response to demands made upon it by everyday living. In flying, there is physical (noise, etc.), physiological (fatigue, lack of sleep, etc.), and psychological (personal or work related problems) stress.

4)   Stress can be helpful in small doses, but extremely harmful in large doses.

    a)   The body produces adrenaline and increases your heart rate to deal with a situation. Too much of this can seriously degrade your decision making ability.

5)   To avoid stress in the cockpit, make sure you fly only when you are at a low stress level. Physical fitness, and proper rest and nutrition are key to reducing stress.

h.   **Dehydration** is the lack of adequate body fluids for the body to carry on normal functions at an optimal level.

1)   Dehydration occurs by either inadequate intake of fluids or loss of fluids through perspiration, vomiting, diarrhea, and excessive urination.

    a)   Vomiting, diarrhea, and excessive urination are separate health problems that usually preclude piloting activities.

    b)   As the atmosphere becomes thinner, it also contains less moisture and more body fluids are lost.

2)   Losses of only a few percent of body fluids can adversely affect both mental and physical process.

3)   On all extended flights, carry water or other suitable liquids to consume as appropriate, i.e., to satisfy thirst.

    a)   Do NOT over-consume so as to require otherwise unnecessary landings to urinate.

4)   Advise and monitor passengers regarding the need to consume appropriate amounts of liquids.

4. **The effects of alcohol and drugs, and their relationship to flight safety**

   a. There is only one safe rule to follow with respect to combining flying and drinking -- **DON'T**.

      1) As little as 1 oz. of liquor, 1 bottle of beer, or 4 oz. of wine can impair flying skills.

         a) Even after the body has completely destroyed a moderate amount of alcohol, a pilot can still be impaired for many hours by hangover.

         b) Alcohol also renders a pilot much more susceptible to disorientation and hypoxia.

      2) The FARs prohibit pilots from performing cockpit duties within 8 hr. after drinking any alcoholic beverage or while under the influence of alcohol.

         a) An excellent rule is to allow at least 12 to 24 hr. "from bottle to throttle," depending on how much alcohol was consumed.

   b. Pilot performance can be seriously impaired by over-the-counter medications.

      1) Many medications have primary or side effects that may impair judgment, memory, alertness, coordination, vision, and the ability to make calculations.

      2) Any medication that depresses the nervous system (i.e., sedative, tranquilizer, antihistamine) can make a pilot more susceptible to hypoxia.

      3) The safest rule is not to fly while taking any medication, unless approved by the FAA.

5. **The effect of nitrogen excesses during scuba dives and how this affects pilots and passengers during flight**

   a. A pilot or passenger who intends to fly after scuba diving should allow the body sufficient time to rid itself of excess nitrogen absorbed during diving.

      1) If this is not done, decompression sickness due to evolved gas (bubbles in the bloodstream) can occur at low altitudes and create a serious in-flight emergency.

   b. The recommended waiting time before flight altitudes of up to 8,000 ft. is at least 12 hr. after a dive which has not required controlled ascent (non-decompression diving).

      1) A person should allow at least 24 hr. after diving which has required controlled ascent (decompression diving).

      2) The waiting time before flight altitudes about 8,000 ft. should be at least 24 hr. after any scuba diving.

   c. These recommended altitudes are actual flight altitudes above mean sea level (MSL) and not pressurized cabin altitudes. The recommendations taken into consideration the risk of decompression of the airplane during flight.

END OF TASK

VISUAL SCANNING AND COLLISION AVOIDANCE

II.B. TASK: VISUAL SCANNING AND COLLISION AVOIDANCE

REFERENCES: AC 61-23/FAA-H-8083-25; AC 90-48/FAA-H-8083-3; AIM.

Objective. To determine that the applicant exhibits instructional knowledge of the elements of visual scanning and collision avoidance by describing:

1. Relationship between a pilot's physical condition and vision.

2. Environmental conditions that degrade vision.

3. Vestibular and visual illusions.

4. "See and avoid" concept.

5. Proper visual scanning procedure.

6. Relationship between poor visual scanning habits and increased collision risk.

7. Proper clearing procedures.

8. Importance of knowing aircraft blind spots.

9. Relationship between aircraft speed differential and collision risk.

10. Situations that involve the greatest collision risk.

A.   General Information

1.   The objective of this task is to determine your instructional knowledge of the elements of visual scanning and collision avoidance.

B.   Task Elements

1.   **Relationship between a pilot's physical condition and vision**

a.   Anything that may affect a pilot's physical or mental condition (e.g., illness, medication, stress, alcohol, fatigue, emotion, hypoxia, etc.) will reduce that pilot's visual acuity.

2.   **Environmental conditions that degrade vision**

a.   Under conditions of dim illumination, small print and colors on aeronautical charts and airplane instruments become unreadable unless adequate cockpit lighting is available.

b.   Dark adaptation is impaired by exposure to cabin pressure altitudes above 5,000 ft., carbon monoxide inhaled in smoking and from exhaust fumes, deficiency of Vitamin A in the diet, and by prolonged exposure to bright sunlight.

c.   Since any degree of dark adaptation is lost within a few seconds of viewing a bright light, the pilot should close one eye when using a white light to preserve some degree of night vision.

d.   Excessive illumination (especially from light reflected off the canopy, surfaces inside the airplane, clouds, water, snow, and desert terrain) can produce glare, with uncomfortable squinting, watering of the eyes, and even temporary blindness.

e.   Dirty or bug-smeared windshields can greatly reduce the ability of pilots to see other aircraft. Keep a clean windshield.

f.   Smoke, haze, dust, rain, and flying towards the sun can also greatly reduce the ability to detect aircraft.

3. **Vestibular and visual illusions**
   a. Certain visual scenes encountered in flight can create illusions of motion and position.
      1) **The leans.** An abrupt correction of a banked attitude, which has been entered too slowly to stimulate the motion sensing system in the inner ear, can create the illusion of banking in the opposite direction. The disoriented pilot will roll the aircraft back into its original dangerous attitude, or if level flight is maintained, will feel compelled to lean in the perceived vertical plane until this illusion subsides.
      2) **Coriolis illusion.** An abrupt head movement in a prolonged constant-rate turn that has ceased stimulating the motion sensing system can create the illusion of rotation or movement in an entirely different axis. The disoriented pilot will maneuver the aircraft into a dangerous attitude in an attempt to stop rotation. This most overwhelming of all illusions in flight may be prevented by not making sudden, extreme head movements, particularly while making prolonged constant-rate turns under IFR conditions.
      3) **Graveyard spin.** A proper recovery from a spin that has ceased stimulating the motion sensing system can create the illusion of spinning in the opposite direction. The disoriented pilot will return the aircraft to its original spin.
      4) **Graveyard spiral.** An observed loss of altitude during a coordinated constant-rate turn that has ceased stimulating the motion sensing system can create the illusion of being in a descent with the wings level. The disoriented pilot will pull back on the controls, tightening the spiral and increasing the loss of altitude.
      5) **Somatogravic illusion.** A rapid acceleration during takeoff can create the illusion of being in a nose-up attitude. The disoriented pilot will push the aircraft into a nose-low, or dive, attitude. A rapid deceleration by a quick reduction of the throttles can have the opposite effect, with the disoriented pilot pulling the aircraft into a nose-up, or stall, attitude.
      6) **Inversion illusion.** An abrupt change from climb to straight-and-level flight can create the illusion of tumbling backwards. The disoriented pilot will push the aircraft abruptly into a nose-low attitude, possibly intensifying this illusion.
      7) **Elevator illusion.** An abrupt upward vertical acceleration, usually caused by an updraft, can create the illusion of being in a climb. The disoriented pilot will push the aircraft into a nose-low attitude. An abrupt downward vertical acceleration, usually caused by a downdraft, has the opposite effect, with the disoriented pilot pulling the aircraft into a nose-up attitude.
      8) **False horizon.** Sloping cloud formations, an obscured horizon, a dark scene spread with ground lights and stars, and certain geometric patterns of ground light can create illusions of not being aligned correctly with the actual horizon. The disoriented pilot will place the airplane in a dangerous attitude.
      9) **Autokinesis.** In the dark, a static light will appear to move about when stared at for several seconds. The disoriented pilot will lose control of the airplane in attempting to align it with the light.
      10) **Flicker vertigo.** A flickering light or shadow at a constant frequency of 4 to 20 times per second may cause dizziness, nausea, and in extreme cases convulsions and unconsciousness. When flying toward the sun, the propeller can cause a flickering effect, especially at reduced power settings. A slight change in propeller RPM usually provides the pilot with relief from the flickering effect.

b. Various surface features and atmospheric conditions encountered during an approach to landing can create optical illusions of incorrect height above and distance from the runway threshold.

1) **Runway width illusion.** A narrower-than-usual runway can create the illusion that the airplane is at a higher altitude than it actually is. The pilot who does not recognize this illusion will fly a lower approach, with the risk of striking objects along the approach path or landing short. A wider-than-usual runway can have the opposite effect, with the risk of leveling out too high and landing hard or overshooting the runway.

2) **Runway and terrain slopes illusion.** An upsloping runway, upsloping terrain, or both, can create the illusion that the airplane is at a higher altitude than it actually is. The pilot who does not recognize this illusion will fly a lower approach. A downsloping runway, downsloping approach terrain, or both, can have the opposite effect.

3) **Featureless terrain illusion.** An absence of ground features, as when landing over water, darkened areas, and terrain made featureless by snow, can create the illusion that the airplane is at a higher altitude than it actually is. The pilot who does not recognize this illusion will fly a lower approach.

4) **Atmospheric illusions.** Rain on the windscreen can create the illusion of greater height, and atmospheric haze the illusion of being at a greater distance from the runway. The pilot who does not recognize these illusions will fly a lower approach. Penetration of fog can create the illusion of pitching up. The pilot who does not recognize this illusion will steepen the approach, often quite abruptly.

5) **Ground lighting illusions.** Lights along a straight path, such as a road, and even lights on moving trains can be mistaken for runway and approach lights. Bright runway and approach lighting systems, especially where few lights illuminate the surrounding terrain, may create the illusion of less distance to the runway. The pilot who does not recognize this illusion will fly a higher approach. Conversely, the pilot overflying terrain which has few lights to provide height cues may make a lower than normal approach.

4. **"See and avoid" concept**

a. FAR 91.113 sets forth the concept of "see and avoid." This rule requires that vigilance shall be maintained at all times, by each person operating an airplane, regardless of whether the operation is conducted under IFR or VFR.

1) Remember that most midair accidents and reported near midair collision incidents occur during good VFR weather conditions and during the hours of daylight.

2) Pilots should also keep in mind their responsibility for continuously maintaining a vigilant lookout regardless of the type of airplane being flown.

5. **Proper visual scanning procedure**

a. While the eyes can observe an approximate 200° arc of the horizon at one glance, only a very small center area called the fovea, in the rear of the eye, has the ability to send clear, sharply focused messages to the brain.

b. All other visual information that is not processed directly through the fovea will be of less detail. An aircraft at a distance of 7 mi. which appears in sharp focus within the foveal center of vision would have to be as close as 7/10 of a mile in order to be recognized if it were outside of foveal vision.

c. Because the eyes can focus only on this narrow viewing area, effective scanning is accomplished with a series of short, regularly spaced eye movements that bring successive areas of the sky into the central visual field.

d.   Each movement should not exceed 10°, and each area should be observed for at least one second to enable detection.

e.   Peripheral vision can be most useful in spotting collision threats from other aircraft. Each time a scan is stopped and the eyes are refocused, the peripheral vision takes on more importance because it is through this element that movement is detected.

   1)   Apparent movement is almost always the first perception of a collision threat and probably the most important because it is the discovery of a threat that triggers the events leading to proper evasive action.

   2)   It is essential to remember, however, that if another aircraft appears to have no relative motion, it is likely to be on a collision course with you.

      a)   If the other aircraft shows no lateral or vertical motion but is increasing in size, take immediate evasive action.

f.   Visual search at night depends almost entirely on peripheral vision.

   1)   In order to perceive a very dim lighted object in a certain direction, the pilot should not look directly at the object but should scan the area adjacent to it.

   2)   Short stops, of a few seconds, in each scan will help to detect the light and its movement.

g.   Lack of brightness and color contrast in daytime and conflicting ground lights at night increase the difficulty of detecting other aircraft.

h.   Pilots are reminded of the requirement to move one's head in order to search around the physical obstructions, such as door and window posts.

   1)   The doorpost can cover a considerable amount of sky, and a small head movement may uncover an area which could be concealing an airplane.

6.   **Relationship between poor visual scanning habits and increased collision risk**

a.   Poor visual scanning unnecessarily increases risks of midair collisions. You and your students should be conversant with and practice proper techniques.

   1)   Ensure early in your student's training that (s)he does not fixate on the flight instruments but uses them to fine-tune the visual references.

b.   Effective scanning also helps avoid "empty-field myopia." This condition usually occurs when flying above the clouds or in a haze layer that provides nothing specific to focus on outside the aircraft. This causes the eyes to relax and seek a comfortable focal distance which may range from 10 to 30 ft. For the pilot, this means looking without seeing, which is dangerous.

7.   **Proper clearing procedures**

a.   Prior to taxiing onto a runway or landing area in preparation for takeoff, pilots should scan the approach areas for possible landing traffic, executing appropriate clearing maneuvers to provide a clear view of the approach areas.

b.   During climbs and descents in flight conditions that permit visual detection of other traffic, pilots should execute gentle banks left and right at a frequency that permits continuous visual scanning of the airspace about them.

c.   Sustained periods of straight-and-level flight in conditions that permit visual detection of other traffic should be broken at intervals with appropriate clearing procedures to provide effective visual scanning.

d.   Entries into traffic patterns while descending create specific collision hazards and should be avoided.

e.   All pilots should emphasize the need for sustained vigilance in the vicinity of VORs and airway intersections due to the convergence of traffic.

f.   Operators of pilot training programs are urged to adopt the following practices:

1)   Pilots undergoing flight instruction at all levels should be requested to verbalize clearing procedures (call out, "clear," "left," "right," "above," or "below") to instill and sustain the habit of vigilance during maneuvering.

2)   In a high-wing airplane, momentarily raise the wing in the direction of the intended turn and look.

3)   In a low-wing airplane, momentarily lower the wing in the direction of the intended turn and look.

4)   Appropriate clearing turns (i.e., a 180° turn or two 90° turns in opposite directions) should precede the execution of all maneuvers including chandelles, lazy eights, stalls, slow flight, climbs, straight-and-level, spins, and other combination maneuvers.

8.   **Importance of knowing aircraft blind spots**

a.   Pilots need to move their heads to see around blind spots caused by fixed aircraft structures, such as door posts, wings, etc. It will be necessary at times to maneuver the aircraft (e.g., lift a wing) to facilitate seeing.

b.   Pilots must ensure curtains and other cockpit objects (e.g., maps on glare shield) are removed and stowed during flight.

9.   **Relationship between aircraft speed differential and collision risk**

a.   The performance capabilities of many airplanes in both speed and rates of climb/descent result in high closure rates, limiting the time available for detection, decision, and evasive action.

b.   EXAMPLE: Two aircraft are approaching head-on and visual detection is made at 3 NM.

1)   If the aircraft are converging at a speed of 200 kt., each pilot has 54 sec. to react to avoid a collision.

2)   If the aircraft are converging at a speed of 300 kt., each pilot has 36 sec. to react to avoid a collision.

c.   Studies have shown that the minimum time it takes for a pilot to spot the traffic, identify it, realize it is a collision threat, react, and have the airplane respond is at least 12.5 sec.

10.   **Situations that involve the greatest collision risk**

a.   While a collision can happen anywhere, there are recognized high hazard areas.

1)   Airways, especially near VOR's, and airports are places where aircraft tend to cluster.

2)   Remember that most collisions occur during days when the weather is good. Being in a "radar environment" still requires vigilance to avoid collisions.

b.   Most collisions occurring en route generally are at or below 8,000 ft. and within 25 mi. of an airport.

END OF TASK

PRINCIPLES OF FLIGHT

II.C. TASK: PRINCIPLES OF FLIGHT

REFERENCES: FAA-H-8083-3; AC 61-23/FAA-S-8081, FAA-S-8083-25.

Objective. To determine that the applicant exhibits instructional knowledge of the elements of principles of flight by describing:

1. Airfoil design characteristics.
2. Airplane stability and controllability.
3. Turning tendency (Torque effect).
4. Load factors in airplane design.
5. Wingtip vortices and precautions to be taken.

A. General Information

1. The objective of this task is to determine your instructional knowledge of the elements of the principles of flight.

2. See Chapter 1, Airplanes and Aerodynamics, in *Pilot Handbook* for discussion relevant to each of the 5 elements in this task.

1.1	Definitions	5 pp.
1.2	The Airplane	5 pp.
1.3	Axes of Rotation	1 p.
1.4	Flight Controls and Control Surfaces	5 pp.
1.5	Forces Acting on the Airplane in Flight	6 pp.
1.6	Dynamics of the Airplane in Flight	4 pp.
1.7	Ground Effect	2 pp.
1.8	How Airplanes Turn	2 pp.
1.9	Torque (Left-Turning Tendency)	5 pp.
1.10	Airplane Stability	8 pp.
1.11	Loads and Load Factors	6 pp.
1.12	Stalls and Spins	7 pp.

B. Task Elements

1. **Airfoil design characteristics**

   a. Airplanes are designed to meet objectives, e.g., fly fast or slowly, carry freight or passengers, etc. The airplane's objective(s) determine(s) size, shape, airfoil design, the number of pilots required, etc.

      1) Similarly, airfoils vary in design-based objectives of lift, drag, desired airspeeds, etc.

   b. For definitions of airfoil and aerodynamic terms, see Module 1.1, Definitions (5 pp.), in *Pilot Handbook*.

2. **Airplane stability and controllability**

   a. Stability is the inherent ability of the airplane, after its equilibrium (i.e., steady flight) is disturbed, to return to its original position. In other words, a stable airplane will tend to return to the original condition of flight if disturbed by a force such as turbulent air. This tendency is primarily an airplane design characteristic.

      1) A stable airplane is easy to fly.

      2) Nevertheless, you cannot depend entirely on stability to return the airplane to the original condition. Even in the most stable airplanes, some conditions will require the use of airplane controls to return the airplane to the desired attitude. Less effort is needed to control the airplane, however, because of the inherent stability.

3) The two types of stability are static and dynamic. Within each type are categories called positive, neutral, and negative stability.

    a) Since stability is desired around all three axes of an airplane, it can be classified as longitudinal, lateral, or vertical.

b. **Static stability** is the initial tendency that the airplane displays after its equilibrium is disturbed.

1) Positive static stability is the initial tendency of the airplane to return to the original attitude, or equilibrium, after being disturbed.

2) Neutral static stability is the initial tendency of the airplane to remain in a new condition (attitude) after its equilibrium has been disturbed.

3) Negative static stability is the initial tendency of the airplane to continue away from the original equilibrium (attitude) after being disturbed.

4) Positive static stability is the most desirable characteristic because the airplane will initially attempt to move to its original trimmed attitude.

c. **Dynamic stability** is the overall tendency that the airplane displays after its equilibrium is disturbed. It is determined by its oscillation tendency after the initial displacement.

1) Positive dynamic stability is the overall tendency of the airplane to return to its original attitude directly or through a series of decreasing oscillations.

2) Neutral dynamic stability is the overall tendency of the airplane to attempt to return to its original attitude, but the oscillations do not increase or decrease in magnitude as time passes.

3) Negative dynamic stability is the overall tendency of the airplane to attempt to return to its original attitude, but the oscillations increase in magnitude as time progresses.

4) Thus, the most desirable combination of the types of stability is a combination of positive static stability with positive dynamic stability.

    a) This combination will tend to return the airplane to its original attitude (or equilibrium).

d. See Module 1.10, Airplane Stability (8 pp.), in *Pilot Handbook*.

e. **Maneuverability** is the quality of an airplane that permits it to be maneuvered easily and to withstand the stresses imposed by maneuvers. It is governed by the airplane's weight, inertia, size and location of flight controls, structural strength, and powerplant. It too is an airplane design characteristic.

f. **Controllability** is the capability of an airplane to respond to the pilot's control, especially with regard to flight path and attitude. It is the quality of the airplane's response to the pilot's control application when maneuvering the airplane, regardless of its stability characteristics.

3. **Turning tendency (Torque effect)**

a. By definition, torque is a force or combination of forces that produces or tends to produce a yawing or rotating motion of an airplane.

1) Torque is made up of four elements that cause or produce a yawing or rotating motion around at least one of the airplane's three axes. These four elements are

    a) Torque reaction from engine and propeller
    b) Corkscrewing effect of the slipstream
    c) Gyroscopic action of the propeller
    d) Asymmetrical loading of the propeller (P factor)

b. Torque (commonly referred to as left-turning tendency) is corrected by applying sufficient rudder pressure (normally right rudder pressure) to maintain coordinated flight.

   1) Torque also produces a rolling motion which is corrected by aileron pressure.

c. See Module 1.9, Torque (Left-Turning Tendency) (5 pp.), in *Pilot Handbook*.

4. **Load factors in airplane design**

   a. Any force applied to deflect an airplane from a straight line produces a stress on its structure. The amount of this force is called load factor.

      1) **Load factor** is the ratio of the total load supported by the airplane's wings to the actual weight of the airplane and its contents:

$$Load\ factor = \frac{Total\ load\ supported\ by\ the\ wings}{Total\ weight\ of\ the\ airplane}$$

      2) A positive load occurs when back pressure is applied to the elevator, causing centrifugal force to act in the same direction as weight.

         a) A negative load occurs when forward pressure is applied to the elevator control, causing centrifugal force to act in a direction opposite to that of weight.

   b. To be certified by the FAA, the structural strength (maximum allowable load factor) of airplanes must conform with prescribed standards set forth by Federal Aviation Regulations. Airplanes are classified as to strength and operational use by means of the category system. Most general aviation airplanes are classified in one or more of the following categories.

      1) The normal category has a maximum limit load factor of 3.8 positive Gs and 1.52 negative Gs.

         a) The limit load factor is the highest (both positive and negative) load factor that can be expected in normal operations under various situations. This load factor can be sustained without causing permanent deformation or structural damage to the airplane.

      2) The utility category has a maximum limit load factor of 4.4 positive Gs and 1.76 negative Gs.

      3) The acrobatic category has a maximum limit load factor of 6.0 positive Gs and 3.0 negative Gs.

      4) This system indicates what operations can be performed in a given airplane without exceeding the load limit. Pilots are cautioned to operate the airplane within the load limit for which the airplane is designed so as to enhance safety and still benefit from the intended use of the airplane.

      5) It should be noted that an airplane's structure is designed to support a certain total load.

         a) It is vital that maximum gross weight limits as well as load factor limits be strictly observed.

   c. The velocity/load factor (or V-g) chart shows the flight operating strength of an airplane that is valid for a certain weight and altitude.

      1) The V-g chart presents the allowable combination of airspeeds and load factors (i.e., the flight envelope) for safe operation.

   d. See Module 1.11, Loads and Load Factors (6 pp.), in *Pilot Handbook*.

5.   **Wingtip vortices and precautions to be taken**

a.   Lift is generated by the pressure differential between the upper and lower wing surfaces. The lower pressure occurs over the upper wing surface. The higher pressure occurs under the wing.

   1)   This pressure differential triggers a roll-up of the airflow behind the wing.

      a)   It results in swirling air masses trailing downstream of the wingtips.

   2)   After the roll-up is completed, the wake consists of two counter-rotating cylindrical vortices.

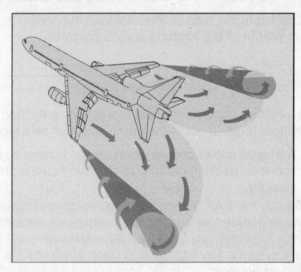

b.   The strength of an airplane's wingtip vortices is governed by the weight, speed, and wing shape of the generating aircraft.

   1)   The angle of attack of the wing directly affects the strength of its vortex.

      a)   As weight increases, angle of attack increases.

      b)   A wing in the clean configuration (flaps retracted) has a greater angle of attack than when flaps are extended.

      c)   As airspeed decreases, angle of attack increases.

   2)   Thus, the greatest vortex strength occurs when the generating aircraft is HEAVY, CLEAN, and SLOW, e.g., during landing and especially during takeoff.

   3)   Wake turbulence presents a hazard to any aircraft that is significantly lighter than the generating aircraft.

      a)   An airplane encountering such wake turbulence could incur major structural damage while in flight.

      b)   The usual hazard is associated with induced rolling which can exceed the rolling capability of the encountering aircraft.

         i)   That is, an airplane may be uncontrollable in the wake turbulence of a large transport airplane.

c.   See Module 3.7, Wake Turbulence (7 pp.), in Chapter 3, Airports, Air Traffic Control, and Airspace, in *Pilot Handbook*.

END OF TASK

AIRPLANE FLIGHT CONTROLS

II.D. TASK: AIRPLANE FLIGHT CONTROLS

REFERENCES: FAA-H-8083-3; AC 61-23/FAA-H-8083-25.

Objective. To determine that the applicant exhibits instructional knowledge of the elements related to the airplane flight controls by describing the purpose, location, direction of movement, effect, and proper procedure for use of the:

1. Primary flight controls.   **3.** Wing flaps.

2. Trim control(s).

A.   General Information

1.   The objective of this task is to determine your instructional knowledge of the elements related to the primary flight controls, trim control(s), and wing flaps.

B.   Task Elements

1.   **Primary flight controls**

a.   The airplane's attitude (i.e., rotation about the three axes) is controlled by deflection of the primary flight controls. These are hinged, movable surfaces attached to the trailing edges of the wing and vertical and horizontal stabilizers. When deflected, these surfaces change the camber (curvature) and angle of attack of the wing or stabilizer and thus change its lift and drag characteristics.

1)   The pilot operates the flight controls through connecting linkage to the rudder pedals and the control yoke.

a)   The control yoke may be either a wheel or a stick.

b.   The **elevator** is attached to the rear of the horizontal stabilizer. The elevator is used to control pitch (i.e., rotation about the airplane's lateral axis) and is controlled by pushing or pulling the control yoke.

1)   The angle of attack of the entire horizontal stabilizer is adjusted by raising or lowering the elevator.

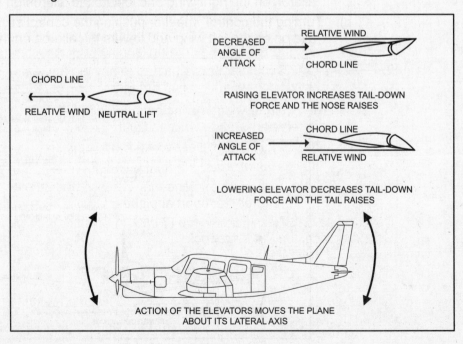

2) Applying back pressure on the control yoke (i.e., the yoke being pulled toward the pilot) raises the elevator. The raised elevator increases the horizontal stabilizer's negative angle of attack and consequently increases the downward tail force. The tail is forced down, increasing the airplane's pitch attitude and thus the angle of attack of the wings.

3) Applying forward pressure to the control yoke (i.e., pushing it forward) lowers the elevator. The lowered elevator decreases the horizontal stabilizer's negative angle of attack and consequently decreases the downward force on the tail. The tail rises, decreasing the airplane's pitch attitude and thus the angle of attack of the wings.

4) Some airplanes, such as the Piper Warrior, have a movable horizontal surface called a **stabilator**, which combines the horizontal stabilizer and the elevator. When the control yoke is moved, the stabilator is moved to raise or lower its leading edge, thus changing its angle of attack and amount of lift.

   a) An **antiservo tab** is attached to the trailing edge of the stabilator and moves in the same direction as the trailing edge of the stabilator.

   b) The movement of the antiservo tab causes the tab to be deflected into the slipstream, providing a resistance to the movement of the stabilator that can be felt on the control yoke.

   i) Without this resistance, the control pressure from the stabilator would be so light that the pilot would move the control yoke too far (overcontrol) for the desired result.

   c) The antiservo tab can also be used as a trim tab.

c. The **ailerons** (French for "little wings") are located on the rear of each wing near the wingtips. The ailerons are used to control roll (i.e., rotation about the longitudinal axis).

1) The ailerons are interconnected in the control system to operate simultaneously in opposite directions from each other. As the aileron on one wing is deflected downward, the aileron on the opposite wing is deflected upward.

   a) Turning the control wheel or pushing the control stick to the right raises the aileron on the right wing and lowers the aileron on the left wing.

   b) Turning the control wheel or pushing the control stick to the left raises the aileron on the left wing and lowers the aileron on the right wing.

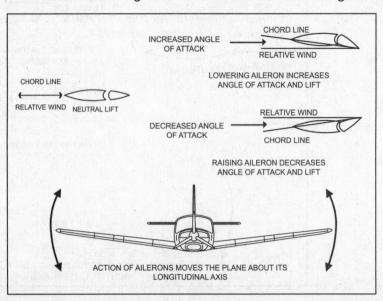

2) When an aileron is lowered, the angle of attack on that wing is increased, which increases the lift. When an aileron is raised, the angle of attack is decreased, which decreases the lift. By this means, the airplane is able to roll (bank) to any desired bank angle.

3) The airplane turns primarily due to banking of the wings, which produces horizontal lift. With wings level, all lift is perpendicular to the Earth. With wings banked, the lift has a horizontal component as well as a vertical component.

    a) The horizontal component (i.e., when the wings are lifting sideways as well as up) counteracts the centrifugal force pulling the airplane straight ahead.

d. The **rudder** is attached to the rear of the vertical stabilizer. Controlled by the rudder pedals, the rudder is used to control yaw (i.e., rotation about the airplane's vertical axis).

1) When the rudder is deflected to one side (left or right), it protrudes into the airflow, causing a horizontal force to be exerted in the opposite direction.

2) This pushes the tail of the airplane in that direction and yaws the nose in the desired direction.

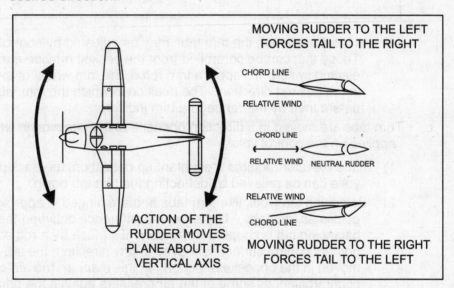

3) The primary purpose of the rudder in flight is to counteract the effect of adverse yaw and to help provide directional control of the airplane.

2.   **Trim control(s)**

a.   Trim devices are secondary flight controls which balance the airplane in flight by reducing the control yoke forces.

b.   A trim tab is a small, adjustable hinged surface, located on the trailing edge of the elevator, aileron, or rudder control surface.

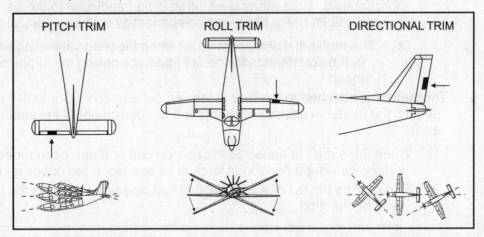

1)   In most airplanes, the trim tabs may be adjusted by controls in the cockpit. Those that can be controlled from the cockpit provide a trim control wheel or electric switch. To apply a trim force, the trim wheel or switch must be moved in the desired direction. The position in which the trim tab is set can usually be determined by reference to a trim indicator.

c.   Trim tabs are moved in a direction opposite to the direction in which pressure is being applied on the control yoke.

1)   If the elevator requires a constant up deflection, the back pressure on the control yoke can be relieved by deflecting the trim tab down.

2)   Balancing tabs look like trim tabs and are hinged in approximately the same places as trim tabs. The essential difference between the two is that the balancing tab is coupled to the control surface by a rod so that, when the primary control surface is moved in any direction, the tab automatically is moved in the opposite direction. In this manner, the airflow striking the tab counterbalances some of the air pressure against the primary control surface and enables the pilot to more readily move and hold it in position.

3)   Servo tabs, sometimes referred to as flight tabs, are used primarily on large airplanes. They aid the pilot in moving the control surface and in holding it in the desired position. Only the servo tab moves in response to movement of the pilot's flight control, and the force of the airflow on the servo tab then moves the primary control surface.

d.   The force of the airflow striking the tab causes the primary control surface to be deflected to a position that will correct the unbalanced condition of the airplane.

e.   The trim tab should not be used to position the primary control. Rather, control pressure should be used on the control yoke or rudder pedals to position the primary control; then the trim tab should be adjusted to relieve the control pressure.

3.   **Wing flaps**

   a.   Wing flaps are used on most airplanes to increase both lift and drag.

   1)   Flaps have three important functions:

   a)   First, they permit a slower landing speed, which decreases the required landing distance.

   b)   Second, they permit a comparatively steeper angle of descent without an increase in speed, which makes it possible to clear obstacles safely when making a landing approach to a short runway.

   c)   Third, they may also be used to shorten the takeoff distance and provide a steeper climb path.

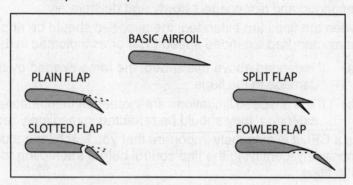

   b.   There are four types of wing flaps.

   1)   Plain flap -- a portion of the trailing edge of the wing on a hinged pivot

   a)   Allows the flap to be moved downward, changing the chord line, angle of attack, and camber of the wing

   2)   Split flap -- a hinged portion of only the bottom surface of the wing

   a)   Increases the angle of attack by changing the chord line
   b)   Creates the least change in pitching moment

   3)   Slotted flap -- a portion of the trailing edge similar to a plain flap but with a gap between the trailing edge of the wing and the leading edge of the flap

   a)   Permits air to pass through and delays the airflow separation along the top of the wing, thus reducing stall speed

   4)   Fowler flap -- a portion of the trailing edge that not only tilts downward but also slides rearward on tracks

   a)   Increases the angle of attack, wing camber, and wing area
   b)   Provides additional lift without significantly increasing drag
   c)   Provides the greatest amount of lift with the least amount of drag, and creates the greatest change in pitching moment

   c.   Wing flaps have different systems of control.

   1)   Most wing flaps are hinged near the trailing edges of the wings, inboard of the ailerons.

   2)   They are controllable by the pilot either manually, electrically, or hydraulically.

   3)   When they are in the up (retracted) position, they fit flush with the wings and serve as part of the wing's trailing edge.

   4)   When in the down (extended) position, the flaps pivot downward from the hinge points to various angles ranging up to 30 ° to 40 ° from the wing.

   d.   Wing flaps have a noticeable effect on aircraft control.

   1)   Wing flaps increase the wing camber (curvature) and angle of attack (when extended), thereby providing greater lift and more drag so that the airplane can descend or climb at a steeper angle or a slower airspeed.

   2)   With a constant power setting while maintaining level flight, the airspeed will be lower with flaps extended because of the drag they create.

   a)   If power is adjusted to maintain a constant airspeed while in level flight, the airplane's pitch attitude will usually be lower with flaps extended.

   e.   It is important to use the correct procedure when using wing flaps.

   1)   Flaps should be used in accordance with the airplane's POH. They should be engaged and disengaged slowly and deliberately.

   2)   When the flaps are extended, the airspeed should be at or below the airplane's maximum flap extended speed (V_{FE}) or as indicated in the POH.

   a)   If extended above this speed, the force exerted by the airflow may result in damage to the flaps.

   b)   If the airspeed limitations are exceeded unintentionally with the flaps extended, they should be retracted immediately regardless of airspeed.

   3)   As a CFI, it is extremely important that you teach your students to form a habit of positively identifying the flap control before attempting to use it to raise or lower the flaps.

   a)   Identification of the flap control prevents inadvertent landing gear operation in a complex airplane with the two controls (i.e., flap and landing gear) in relatively close proximity.

END OF TASK

AIRPLANE WEIGHT AND BALANCE

> **II.E. TASK: AIRPLANE WEIGHT AND BALANCE**
>
> REFERENCES: FAA-H-8083-1; FAA-H-8083-3; AC 61-23/FAA-H-8083-25.
>
> **Objective.** To determine that the applicant exhibits instructional knowledge of the elements of airplane weight and balance by describing:
>
> 1.   Weight and balance terms.
>
> 2.   Effect of weight and balance on performance.
>
> 3.   Methods of weight and balance control.
>
> 4.   Determinations of total weight and center of gravity and the changes that occur when adding, removing, or shifting weight.

A.   General Information

    1.   The objective of this task is to determine your instructional knowledge of the elements of airplane weight and balance.

    2.   In late 1999, the FAA published *Aircraft Weight and Balance Handbook* (FAA-H-8083-1), which replaced *Pilot's Weight and Balance Handbook* (AC 91-23A, dated 1977). The scope of the changes included updating the weight and balance terms (to those found in your airplane's POH) and extensive editing to make the book easier to read.

    3.   For additional reading, see Modules 5.11-5.18 of Chapter 5, Airplane Performance and Weight and Balance, in *Pilot Handbook* for a 12-page discussion on weight balance.

B.   Task Elements

    1.   **Weight and balance terms**

        a.   **Reference Datum** -- an imaginary vertical plane (or line) from which all horizontal distances are measured for balance purposes. The reference datum may be located anywhere the airplane manufacturer chooses.

            1)   One popular location for the reference datum is a specified distance forward of the airplane, measured in inches from some point such as the leading edge of the wing or the engine firewall.

        b.   **Station** -- a location along the airplane fuselage usually given in terms of distance from the reference datum.

        c.   **Arm** -- the horizontal distance, usually measured in inches, from the reference datum to the center of gravity (CG) of an item.

            1)   Arms ahead of the reference datum are negative (–), and those behind the reference datum are positive (+).

            2)   When the reference datum is ahead of the airplane, all of the arms are positive and computational errors are minimized.

        d.   **Moment** -- a force that causes or tries to cause an object to rotate. Moment is the product of the weight (in pounds) of an item multiplied by its arm (in inches). Moments are generally expressed in pound-inches (lb.-in.).

        e.   **Moment Index** -- the moment divided by a reduction number such as 100 or 1,000 to make the moment value smaller and reduce the chance of mathematical errors in computing the center of gravity.

        f.   **Center of Gravity (CG)** -- the point at which an airplane would balance if it were possible to suspend it at that point. The distance of the CG from the reference datum is determined by dividing the total moment by the total weight of the airplane.

        g.   **CG Limits** -- the extreme (forward and aft) center of gravity locations within which the airplane must be operated at a given weight.

        h.   **Usable Fuel** -- the fuel available for flight planning.

        i.   **Unusable Fuel** -- the fuel remaining in the airplane's fuel system after a runout test has been completed in accordance with the FARs.

j. **Standard Empty Weight** -- the weight of the airframe, engines, and all items that have fixed locations and are permanently installed in the airplane. Standard empty weight includes unusable fuel, full operating fluids, and full oil.

k. **Basic Empty Weight** -- the standard empty weight plus any optional equipment that has been installed.

l. **Payload** -- the weight of occupants, cargo, and baggage.

m. **Useful Load** -- the difference between takeoff weight (or ramp weight if applicable) and basic empty weight.

n. **Maximum Ramp Weight** -- the maximum weight approved for ground maneuvers. It includes the weight of start, taxi, and runup fuel.

o. **Maximum Takeoff Weight** -- the maximum weight approved for the start of the takeoff run.

p. **Maximum Landing Weight** -- the maximum weight approved for the landing touchdown.

q. **Maximum Zero Fuel Weight** -- the maximum weight exclusive of usable fuel.

r. **Standard Weights** -- established for numerous items involved in weight and balance computations.

   1) Gasoline is 6 lb./U.S. gal.
   2) Oil is 7.5 lb./U.S. gal.
   3) Water is 8.35 lb./U.S. gal.

2. **Effect of weight and balance on performance**

   a. Effects of weight on flight performance

      1) Increased weight reduces the flight performance of an airplane in almost every respect. The most important performance deficiencies of the overloaded airplane are

         a) Higher takeoff speed required
         b) Longer takeoff run required
         c) Reduced rate and angle of climb
         d) Shorter range
         e) Reduced cruising speed
         f) Reduced maneuverability
         g) Higher stalling speed
         h) Higher landing speed required
         i) Longer landing roll required

   b. Effects of weight on airplane structure

      1) An airplane is certified to be able to withstand certain loads placed on its structure.

         a) As long as gross weight and load factor limits are observed, the total load on the airplane will remain within limits.

         b) If the maximum gross weight is exceeded, load factors well within the load factor limits can cause structural damage.

      2) Structural failures from overloading may be dramatic and catastrophic, but more often they affect structural components gradually in a way which is difficult to detect.

         a) The results of habitual overloading tend to be cumulative and may result in structural failure later during completely normal operations.

         b) Overloading can also accelerate metal fatigue.

   c. Effects of balance on flight performance

      1) The CG location affects the total load placed on the wings in flight.

2) With a forward CG, a greater downward force on the tail is required to maintain level cruising flight.

    a) The total lift required from the wing is increased.

    b) Thus, the wing flies at a higher angle of attack, which results in more drag and a higher indicated stall speed.

3) With an aft CG, less downward force on the tail is required, resulting in less lift required by the wing.

    a) Thus, the wing flies at a lower angle of attack with less drag and a higher cruise speed.

  d. Effect of balance on stability

1) In general, an airplane becomes less stable and controllable as the CG moves aft.

    a) The elevator has a shorter arm (i.e., distance) from the CG and requires greater deflection to produce the same result.

    b) Recovery from a stall is more difficult because the airplane's tendency to pitch down is reduced.

    c) If the CG is moved beyond the aft limit, stall and spin recovery may become impossible.

2) As the CG moves forward, the airplane becomes more nose-heavy.

    a) If the CG is moved beyond the forward limit, the elevator may no longer be able to hold the nose up, particularly at low airspeeds, e.g., takeoff, landing, and power-off glides.

3. **Methods of weight and balance control**

  a. The pilot is responsible for the management of weight and balance control.

1) The pilot has control over the loading and fuel management within the established limits for the particular airplane.

  b. There are various methods used to determine weight and balance conditions. These include

1) Center of gravity calculations
2) Center of gravity graphs
3) Center of gravity tables

4. **Determination of total weight and center of gravity and the changes that occur when adding, removing, or shifting weight**

  a. You must be able to instruct your student (FAA inspector/examiner) to properly compute the total weight and CG location for takeoff and landing.

1) During your practical test, you will be using the weight and balance data for your airplane.

WEIGHT CHANGE AND WEIGHT SHIFT COMPUTATIONS

A. The FAA provides two formulas for weight change and one formula for weight shift. They are not reproduced here because the following weight change and weight shift formula is much simpler and intuitively appealing. It is adapted from a class handout developed by Dr. Melville R. Byington at Embry-Riddle Aeronautical University (used with permission).

B. **Basic Theory.** At issue is the question: **If the CG started out there and certain changes occurred, where is it now?** It can be answered directly using a SINGLE, UNIVERSAL, UNCOMPLICATED FORMULA.

1.   At any time, the CG is simply the sum of all moments divided by the sum of all weights.

$$CG = \frac{\sum M}{\sum W}$$

2.   Since CG was known at some previous (#1) loading condition (with moment = M_1 and weight = W_1), it is logical that the previous CG become the point of departure. Due to weight addition, removal, or shift, the moment has changed by some amount, DM. The total weight has also changed if, and only if, weight has been added or removed. Therefore, the current CG is merely the current total moment divided by the current total weight. In equation format,

$$CG = Current\ moment/Current\ weight\ becomes\ CG = \frac{M^1 \pm \Delta M}{W^1 \pm \Delta W}$$

3.   This UNIVERSAL FORMULA will accommodate any weight change and/or CG shift problem. Before proceeding, certain conventions deserve review:

   a.   Any weight added causes a + moment change (weight removed is –).
   b.   Weight **shifted** rearward causes a + moment change (forward is –).
   c.   A weight **shift** changes only the moment ($\Delta W = 0$).

C.   EXAMPLES:

1.   An airplane takes off at 6,230 lb. with a CG location at station 79.0. What is the location of the CG after 50 gal. (300 lb.) of fuel has been consumed from station 87.0?

$$CG = \frac{M^1 \pm \Delta M}{W^1 \pm \Delta W} = \frac{6,230(79) - 300(87)}{6,230 - 300} = 78.6\ in.$$

2.   An airplane takes off at 3,000 lb. with CG at station 60. Since takeoff, 25 gal. (150 lb.) of fuel has been consumed. Fuel cell CG is station 65. After takeoff, a 200-lb. passenger moved from station 50 to station 90. Find the resulting CG.

$$CG = \frac{M^1 \pm \Delta M}{W^1 \pm \Delta W} = \frac{3,000(60) - 150(65) + 200(90 - 50)}{3,000 - 150} = 62.54\ in.$$

3.   Gross weight of an airplane is 10,000 lb. 500 lb. of cargo is shifted 50 in. How far does the CG shift? (Note original CG and direction of shift are unspecified. Since datum is undefined, why not define it, temporarily, as the initial CG location, even though it is unknown? This causes M_1 to become zero. Incidentally, the direction of CG shift corresponds precisely to the direction of the weight shift.)

$$CG = \frac{M^1 \pm \Delta M}{W^1 \pm \Delta W} = \frac{500 \times 50}{10,000} = 2.5\ in.$$

END OF TASK

NAVIGATION AND FLIGHT PLANNING

II.F. TASK: NAVIGATION AND FLIGHT PLANNING

  REFERENCES: FAA-H-8083-3; AC 61-23/FAA-S-8083-25.

Objective. To determine that the applicant exhibits instructional knowledge of the elements of navigation and flight planning by describing:

1. Terms used in navigation.

2. Features of aeronautical charts.

3. Importance of using the proper and current aeronautical charts.

4. Method of plotting a course, selection of fuel stops and alternates, and appropriate actions in the event of unforeseen situations.

5. Fundamentals of pilotage and dead reckoning.

6. Fundamentals of radio navigation.

7. Diversion to an alternate.

8. Lost procedures.

9. Computation of fuel consumption.

10. Importance of preparing and properly using a flight log.

11. Importance of a weather check and the use of good judgment in making a "go/no-go" decision.

12. Purpose of, and procedure used in, filing a flight plan.

A. General Information

  1. The objective of this task is to determine your instructional knowledge of the elements of navigation and flight planning.

B. Task Elements

  1. **Terms used in navigation**

    a. **Calibrated airspeed (CAS)** -- indicated airspeed corrected for instrument position and error

    b. **Compass heading** -- magnetic heading corrected for deviation

    c. **Density altitude** -- pressure altitude corrected for nonstandard temperature

    d. **Deviation** -- magnetic anomaly that affects the compass

    e. **Groundspeed** -- actual speed of the airplane in relation to the ground

    f. **Indicated airspeed (IAS)** -- airspeed read off the airspeed indicator

    g. **Magnetic course** -- true course corrected for magnetic variation

    h. **Magnetic heading** -- magnetic course corrected for wind (direction and speed)

    i. **Pressure altitude** -- altitude shown when the altimeter is set to 29.92

    j. **Standard pressure (sea level)** -- 29.92"

    k. **Standard temperature (sea level)** -- 15 °C

    l. **True airspeed** -- actual speed relative to the surrounding air (calibrated airspeed corrected for density altitude)

    m. **True course** -- course over ground relative to true north

    n. **True heading** -- true course corrected for wind (direction and speed)

    o. **Variation** -- angular difference between true north and magnetic north

2. **Features of aeronautical charts**

   a. The topographical information featured on sectional and VFR terminal charts portrays surface elevation levels and a great number of visual checkpoints used for VFR flight.

      1) Checkpoints include populated areas, drainage, roads, railroads, and other distinctive landmarks.

   b. The aeronautical information on sectional and VFR terminal charts includes visual and radio aids to navigation, airports, controlled airspace, restricted areas, obstructions, and related data.

   c. You should be very familiar with the features on the aeronautical charts and how to use the legend.

   d. See Module 9.3, Sectional Chart Symbology, in *Pilot Handbook* for a six-page discussion with illustrations.

3. **Importance of using the proper and current aeronautical charts**

   a. It is vitally important that you and your students check the publication date on each aeronautical chart to be used. Obsolete charts should be discarded and replaced by new editions. This is important because revisions in aeronautical information occur constantly.

      1) Revisions may include changes in radio frequencies, new obstructions, temporary or permanent closing of certain runways and airports, and other temporary or permanent hazards to flight.

   b. The National Ocean Survey (NOS) publishes and sells aeronautical charts of the United States and foreign areas. The type of charts most commonly used by pilots flying VFR include the following:

      1) Sectional charts are normally used for VFR navigation, and we will refer to this type of chart in this task. The scale on sectional charts is 1:500,000 (1 in. = 6.86 NM).

      2) VFR terminal area charts depict the airspace designated as Class B airspace. The information found on these charts is similar to that found on sectional charts. They exist for large metropolitan areas such as Atlanta and New York. The scale on terminal charts is 1:250,000 (1 in. = 3.43 NM).

      3) Both the sectional and VFR terminal area charts are revised semiannually.

4. **Method of plotting a course, selection of fuel stops and alternatives, and appropriate actions in the event of unforeseen situations**

   a. By using the straightedge of the plotter, a course line should be drawn from the departure airport to the destination on the sectional chart.

      1) The line should be dark enough to read easily but light enough not to obscure any chart information.

   b. Fuel stops are based on personal comfort and, at minimum, on regulatory requirements.

      1) FAR 91.151 requires that there be enough fuel to fly to the first point of intended landing and, assuming normal cruise power setting, to fly for at least

         a) 30 min. during the day

         b) 45 min. at night

c. Once the course is drawn on the sectional chart, survey the route of flight.

1) Look for available alternate airports en route.

2) Look at the type of terrain, e.g., mountains, swamps, large bodies of water, that would have an impact if an off-airport landing became necessary.

3) Your student should form a habit of mentally preparing for any type of emergency situation and the appropriate action to be taken during the flight.

4) Ensure that the route of flight does not penetrate any restricted (if in use) or prohibited airspace.

5. **Fundamentals of pilotage and dead reckoning**

a. Pilotage is the action of flying cross-country using only a sectional chart to fly from one visible landmark to another.

1) Pilotage becomes difficult in areas lacking prominent landmarks or under conditions of low visibility.

2) During a flight, the pilot will use pilotage in conjunction with dead reckoning to verify his/her calculations and keep track of the airplane's position.

b. **Dead reckoning** is the navigation of the airplane solely by means of computations based on true airspeed, course, heading, wind direction and speed, groundspeed, and elapsed time.

1) Simply, dead reckoning is a system of determining where the airplane should be on the basis of where it has been.

a) Literally, it is deduced reckoning, which is where the term originated, i.e., ded. or "dead" reckoning.

c. Ideally, a VFR pilot uses both pilotage and dead reckoning.

1) Radio navigation should be added to the above two, so that a pilot uses all three.

2) Thus, on cross-country flights, a pilot should start with dead reckoning and confirm with pilotage and radio navigation.

6. **Fundamentals of radio navigation**

a. Radio navigation is a method of navigation by which the pilot follows a predetermined flight path over the Earth's surface by using the properties of radio waves.

1) Radio navigation is done by ground stations or satellites in space transmitting signals to navigation radio receivers installed in the airplane.

2) The pilot determines and controls ground track on the basis of the instrument indications.

b. See Chapter 10, Navigation Systems, in *Pilot Handbook* for a 26-page discussion (with illustrations) on basic radio principles, VOR, DME, ADF, RNAV, LORAN, and GPS.

7. **Diversion to an alternate airport**

a. On cross-country flights, a pilot may need to divert from his/her original plan and land at another (i.e., alternate) airport. Reasons for a diversion may include

1) Low fuel
2) Bad weather
3) Pilot or passenger fatigue, illness, etc.
4) Airplane system or equipment malfunction
5) Any other reason that the pilot decides to divert to an alternate airport

b. Procedures for diverting

1) Confirm your present position on your sectional chart.
2) Select an appropriate alternate airport and estimate a heading to that airport.
3) Write down the time and turn to your new heading.
4) Use a straightedge to draw a new course line on your chart.
5) Refine your heading by using pilotage and making maximum use of available radio navigation aids.
6) Compute new estimated groundspeed, arrival time, and fuel consumption to your alternate airport.

c. See Module 11.5, Diversion to an Alternate Airport (2 pp.), in *Pilot Handbook*.

8. **Lost procedures**

a. Nobody wants to get lost, especially in an airplane, but all pilots occasionally find themselves disoriented. Instruct your student to recognize disorientation quickly and implement corrective action to become reoriented.

b. Steps to avoid becoming lost

1) Always know where you are.
2) Plan ahead and know what the next landmark will be and look for it.

a) Similarly, anticipate the indication of the radio navigation systems.

3) If the radio navigation systems or the visual observations of landmarks do not confirm your expectations, become concerned and take action.

c. When lost, you should

1) Select the best course of action.
2) Maintain the original or an appropriate heading and climb, if necessary.
3) Identify the nearest concentration of prominent landmarks.
4) Use all available radio navigation systems/facilities, or contact ATC for assistance.
5) Plan a precautionary landing if deteriorating weather and/or fuel exhaustion is imminent.

d. See Module 11.6, Lost Procedures (3 pp.), in *Pilot Handbook*.

9. **Computation of fuel consumption**

a. You may compute either fuel burned, fuel consumption rate, or time remaining.

1) The computations are

a) *Fuel burned = Fuel consumption rate × Time*
b) *Time (available) = Fuel to burn ÷ Fuel consumption rate*
c) *Fuel consumption rate = Fuel burned ÷ Time*

b. See Module 9.13, Fuel Computations (3 pp.), in *Pilot Handbook*.

10. **Importance of preparing and properly using a flight (navigation) log**

a. The navigation log will assist a pilot in planning and conducting a cross-country flight.

1) The navigation log allows a pilot to list pertinent information in a logical sequence.
2) During flight, the navigation log is a written record that allows a pilot to track the progress of the flight.

b. See Chapter 11, Cross-Country Flight Planning, in *Pilot Handbook*, for discussion on the use of a navigation log.

11. **Importance of a weather check and the use of good judgment in making a "go/no-go" decision**

   a. See Task III.B., Weather Information, beginning on page 178 for a discussion of the importance of a weather check, means of obtaining weather, and factors to be considered in making a go/no-go decision.

   b. Flight Service Stations (FSSs) are the primary source for obtaining preflight briefings and in-flight weather information.

   1) There are four basic types of briefings to meet your needs:

      a) Standard briefing

         i) A standard briefing should be requested any time you are planning a flight and have not received a previous briefing.

      b) Abbreviated briefing

         i) Request an abbreviated briefing when you need information to supplement mass disseminated data (e.g., TIBS) or to update a previous briefing, or when you need only one or two specific items.

      c) Outlook briefing

         i) Request an outlook briefing whenever your proposed time of departure is 6 hr. or more from the time of the briefing.

      d) In-flight briefing

         i) In situations in which you need to obtain a preflight briefing or an update by radio, you should contact the nearest FSS to obtain this information.

         ii) After communications have been established, advise the FSS of the type of briefing you require.

         iii) You may be advised to shift to the Flight Watch frequency (122.0) when conditions indicate that it would be advantageous.

   c. See Module 8.1, Flight Service Station (FSS), in *Pilot Handbook*, for a three-page discussion of the information and order of presentation of various pilot briefings.

12. **Purpose of, and procedure used in, filing a flight plan**

   a. VFR flight plans are not mandatory but are highly recommended as a safety precaution. In the event that a pilot does not reach his/her destination and close the flight plan as planned, the FAA will institute a search for him/her. The search will begin 30 min. after (s)he was scheduled to reach his/her destination.

   b. FSS will accept flight plans filed in person, over the telephone, or by radio.

   1) Flight plans can also be filed using DUATS.

   c. See Module 11.2, VFR Flight Plan (2 pp.), in *Pilot Handbook*.

END OF TASK

NIGHT OPERATIONS

II.G. TASK: NIGHT OPERATIONS

REFERENCES: FAA-H-8083-3; AC 61-23/FAA-H-8083-25; FAA-S-8081-12, FAA-S-8081-14; AIM.

Objective. To determine that the applicant exhibits instructional knowledge of the elements of night operations by describing:

1. Factors related to night vision.
2. Disorientation and night optical illusions.
3. Proper adjustment of interior lights.
4. Importance of having a flashlight with a red lens.
5. Night preflight inspection.
6. Engine starting procedures, including use of position and anticollision lights prior to start.
7. Taxiing and orientation on an airport.
8. Takeoff and climb-out.

9. In-flight orientation.
10. Importance of verifying the airplane's attitude by reference to flight instruments.
11. Night emergencies procedures.
12. Traffic patterns.
13. Approaches and landings with and without landing lights.
14. Go-arounds.

A. General Information

1. The objective of this task is to determine your instructional knowledge of the elements of night operations.

B. Task Elements

1. **Factors related to night vision**

a. Understanding how the eye operates at night is an important factor in night flying.

1) Two types of light-sensitive nerve endings called cones and rods, located at the back of the eye, or retina, transmit messages to the brain via the optic nerve.

a) The cones are at the center of vision (i.e., directly behind the pupil). When they stop working in the dark, a night blind spot develops at the center of vision.

i) Since the rods function in the dark and are located around the cones, off-center viewing (i.e., seeing an object by looking to one side of it, not directly at it) is possible.

b. Night adaptation

1) When entering a dark area, the pupils of the eyes enlarge to receive as much of the available light as possible.

2) After approximately 5 to 10 min. during which time the cones become adjusted to the dim light, your eyes will become 100 times more sensitive than they were before you entered the dark area.

a) In fact, the cones stop working altogether in semidarkness.

b) Since the rods can still function in light of 1/5,000 the intensity at which the cones cease to function, they are used for night vision.

3) After about 30 min., the rods will be fully adjusted to darkness and become about 100,000 times more sensitive to light than they were in the lighted area.

4) The rods need more time to adjust to darkness than the cones do to bright light. Your eyes become adapted to sunlight in 10 sec., whereas they need 30 min. to fully adjust to a dark night.

    c. Good vision depends on your physical condition. Fatigue, colds, vitamin deficiency, alcohol, stimulants, smoking, or medication can seriously impair your vision.

        1) EXAMPLE: Smoking lowers the sensitivity of the eyes and reduces night vision by approximately 20%.

    d. See Module 6.11, Vision (3 pp.), in *Pilot Handbook*.

2. **Disorientation and night optical illusions**

    a. Various visual scenes encountered during a night flight can create illusions of motion and position. The best way to cope with these illusions is by using and trusting your flight instruments. Some of the illusions encountered at night are

        1) False horizon
        2) Autokinesis
        3) Featureless terrain
        4) Runway slopes
        5) Ground lighting

    b. These illusions are explained in Task II.B., Visual Scanning and Collision Avoidance, beginning on page 102.

3. **Proper adjustment of interior lights**

    a. The cockpit lights should be adjusted to a minimum brightness that will allow the pilot to read the instruments and switches without hindering outside vision.

        1) Low-intensity lighting will eliminate light reflections on the windshield and windows which can obstruct outside vision.

    b. Generally, the pilot will be able to dim the interior lights gradually throughout the course of a flight as his/her night vision improves.

4. **Importance of having a flashlight with a red lens**

    a. At least one reliable flashlight should be considered standard equipment on all night flights.

        1) A "D" cell size flashlight with a bulb-switching mechanism that can be used to select white or red light is preferable.

            a) Use the white light while performing the preflight visual inspection of the airplane. Use the red light for performing cockpit operations.

            b) Since the red light is nonglaring, it will not impair night vision.

        2) Some pilots prefer two flashlights, one with a white light for preflight and the other a penlight type with a red light.

            a) The penlight can be suspended around the neck to keep it readily available during flight.

            b) CAUTION: If a red light is used for reading an aeronautical chart, the red features of the chart will not be visible to you.

5. **Night preflight inspection**

    a. Required equipment for VFR flight at night (FAR 91.205)

        1) Airspeed indicator

        2) Altimeter

        3) Magnetic direction indicator (compass)

        4) Tachometer for each engine

        5) Oil pressure gauge for each engine using a pressure system

        6) Temperature gauge for each liquid-cooled engine

        7) Oil temperature gauge for each air-cooled engine

        8) Manifold pressure gauge for each altitude engine

        9) Fuel gauge indicating the quantity of fuel in each tank

        10) Landing gear position indicator if the aircraft has a retractable landing gear

        11) Approved flotation gear for each occupant and one pyrotechnic signaling device if the aircraft is operated for hire over water beyond power-off gliding distance from shore

        12) Safety belt with approved metal-to-metal latching device for each occupant

        13) For small civil airplanes manufactured after July 18, 1978, an approved shoulder harness for each front seat

        14) An emergency locator transmitter (ELT), if required by FAR 91.207

        15) Approved position (navigation) lights

        16) Approved aviation red or white anticollision light system on all U.S.-registered civil aircraft

        17) If the aircraft is operated for hire, one electric landing light

        18) An adequate source of electricity for all electrical and radio equipment

        19) A set of spare fuses or three spare fuses for each kind required which are accessible to the pilot in flight

            a) Not applicable if your airplane is equipped with resettable circuit breakers

    b. Though not required by FAR 91.205 for VFR night flight, it is recommended, because of the limited outside visual references during night flight, that other flight instruments and equipment supplement those required. These include

        1) Attitude indicator

        2) Sensitive altimeter adjustable for barometric pressure

        3) Individual instrument lights

        4) Adequate cockpit illumination

        5) Landing light

        6) Wingtip strobe lights

    c. A thorough preflight inspection of your airplane (both interior and exterior) is necessary, as in any flight.

        1) Since you may do the preflight inspection at night, you must use your flashlight to illuminate the areas you are inspecting. You should take your time and look at each item carefully.

    d. All airplane lights should be turned on and checked (visually) for operation.

        1) Position lights can be checked for loose connections by tapping the light fixture while the light is on.

            a) If the lights blink while being tapped, further investigation (by a qualified mechanic) to find the cause should be initiated.

e.   All personal materials and equipment should be checked to ensure that you have everything and that all equipment is functioning properly.

  1)   It is very disconcerting to find, at the time of need, that a flashlight, for example, does not work.

f.   Finally, the parking ramp should be checked prior to entering your airplane. During the day, it is easy to see stepladders, chuckholes, stray wheel chocks, and other obstructions, but at night it is more difficult. A check of the area can prevent taxiing mishaps.

6.   **Engine starting procedures, including use of position and anticollision lights prior to start**

a.   Extra caution should be taken at night to ensure that the propeller area is clear.

  1)   This can be accomplished by turning on your airplane's rotating beacon (anticollision light) or by flashing other airplane lights to alert any person nearby to remain clear of the propeller.

  2)   Also, orally announce "clear prop" and wait a few seconds before engaging the starter.

  3)   Think safety.

b.   To avoid excessive drain of electrical current from the battery, keep all unnecessary equipment off until after the engine has been started.

  1)   Once the engine has been started, turn on the airplane's position lights.

7.   **Taxiing and orientation on an airport**

a.   Due to your restricted vision at night, taxi speeds should be reduced. Never taxi faster than a speed that would allow you to stop within the distance illuminated by your landing light.

  1)   Continuous use of the landing light with the low power settings normally used for taxiing may place an excessive burden on your airplane's electrical system.

  2)   Overheating of the landing light bulb may become a problem because of inadequate airflow to carry away the excessive heat generated.

    a)   It is recommended generally that the landing lights be used only intermittently while taxiing, but sufficiently to ensure that the taxiway is clear.

  3)   Be sure to avoid using the wingtip strobes and landing light in the vicinity of other aircraft.

    a)   This vicinity includes the runup area while someone else is landing.
    b)   Lights can be distracting and potentially blinding to a pilot.
    c)   You would expect others to show courtesy when using lights.

b.   Use the checklist in your POH to perform the before-takeoff check.

  1)   During the day, unintended forward movement of your airplane can be easily detected during the runup.

    a)   At night, the airplane may creep forward without being noticed, unless you are alert to this possibility.

    b)   Thus, it is important to lock the brakes during the runup and be attentive to any unintentional forward movement.

c.   Orientation is maintained by airport familiarity (through study of an airport diagram) and by using taxiway markings, lights, and signs.

8. **Takeoff and climb-out**

    a.   Before taxiing onto the active runway for takeoff, you should exercise extreme caution to prevent conflict with other aircraft.

        1)   At controlled airports, where ATC issues the clearance for takeoff, it is recommended that you check the final approach course for approaching aircraft.

        2)   At uncontrolled airports, it is recommended that you make a slow 360 ° turn in the same direction as the flow of air traffic while closely searching for other traffic.

    b.   After ensuring that the final approach and runway are clear of other traffic, you should line up your airplane with the centerline of the runway.

        1)   If the runway has no painted centerline, you should use the runway lighting and align your airplane between and parallel to the two rows of runway edge lights.

        2)   Your landing light and strobe lights (if applicable) should be on as you taxi into this position.

        3)   After the airplane is aligned, the heading indicator should be set to correspond to the known runway direction.

    c.   To begin the takeoff, you should release the brakes and smoothly advance the throttle to takeoff power. As your airplane accelerates, it should be kept moving parallel to the runway edge lights. This is best done by looking at the more distant runway lights rather than those close in and to the side.

        1)   At night your perception of runway length and width, airplane speed, and flight attitude will vary. You must monitor your flight instruments more closely; e.g., rotation should occur at the proper V_R based on your airspeed indicator, not your bodily senses.

        2)   As the airspeed reaches V_R, the pitch attitude should be adjusted to that which will establish a normal climb by referring to both visual and instrument references (e.g., lights and the attitude indicator).

        3)   Do not attempt to pull the airplane forcibly off the ground. It is best to let it fly off in the liftoff attitude while you are cross-checking the attitude indicator against any outside visual references that may be available.

    d.   After becoming airborne, you may have difficulty in noting whether the airplane is getting closer to or farther from the surface because of the darkness of the night.

        1)   By cross-checking your flight instruments, ensure that your airplane continues in a positive climb and does not settle back onto the runway.

           a)   A positive climb rate is indicated by the vertical speed indicator and by a gradual but continual increase in the altimeter indication.

           b)   A climb pitch attitude is indicated on the attitude indicator.

        2)   Check the airspeed to ensure that it is well above a stall and is stabilizing at the appropriate climb speed (e.g., V_Y).

        3)   Use the attitude indicator as well as visual references to ensure that the wings are level, and cross-check with the heading indicator to ensure that you are maintaining the correct heading.

           a)   Normally, no turns should be made until you reach a safe maneuvering altitude.

        4)   Your landing light should be turned off after a climb is well established. This is normally completed during the climb checklist.

           a)   The light may become deceptive if it is reflected by any haze, smoke, or fog that might exist in the takeoff climb.

9.   **In-flight orientation**

  a. Never depart at night without a thorough review of your intended flight plan. Courses, distances, and times of each leg should be computed. At night, your attention is needed for aviating, not for navigation planning.

  b. In spite of fewer usable landmarks or checkpoints, night cross-country flights present no particular problem if preplanning is adequate and you continuously monitor position, time estimates, and fuel consumption.

   1) The light patterns of towns are easily identified, especially when surrounded by dark areas.

    a) Large metropolitan areas may be of little meaning until you gain more night flying experience.

   2) Airport rotating beacons, which are installed at various military and civilian airports, are useful checkpoints.

   3) Busy highways marked by car headlights also make good checkpoints.

   4) On moonlit nights, especially in dark areas, you will be able to identify some unlit landmarks.

  c. Crossing large bodies of water on night flights can be potentially hazardous, not only from the standpoint of landing (ditching) in the water, should it become necessary, but also because the horizon may blend in with the water, in which case control of your airplane may become difficult.

   1) During hazy conditions over open water, the horizon will become obscure, and you may experience a loss of spatial orientation.

   2) Even on clear nights, the stars may be reflected on the water surface, appearing as a continuous array of lights and thus making the horizon difficult to identify.

   3) Always include instrument references in your scan.

  d. Lighted runways, buildings, or other objects may cause illusions to you when they are seen from different altitudes.

   1) At 2,000 ft. AGL, a group of lights on an object may be seen individually, while at 5,000 ft. AGL or higher, the same lights can appear to be one solid light mass.

   2) These illusions may become quite acute with altitude changes and, if not overcome, can present problems with respect to approaches to lighted runways.

  e. At night, it is normally difficult to see clouds and restrictions to visibility, particularly on dark nights (i.e., no moonlight) or under an overcast.

   1) You must exercise caution to avoid flying into weather conditions below VFR (i.e., clouds, fog).

   2) Normally, the first indication of flying into restricted visibility conditions is the gradual disappearance of lights on the ground.

    a) If the lights begin to take on an appearance of being surrounded by a "cotton ball" or glow, you should use extreme caution in attempting to fly farther in that same direction.

   3) Remember that, if you must make a descent through any fog, smoke, or haze in order to land, visibility is considerably less when you look horizontally through the restriction than it is when you look straight down through it from above.

   4) You should never attempt a VFR night flight during poor or marginal weather conditions.

f.   Airplane position (navigation) lights are arranged similarly to those of boats and ships.

1)   A red light is positioned on the left wingtip.
2)   A green light is on the right wingtip.
3)   A white light is on the tail.

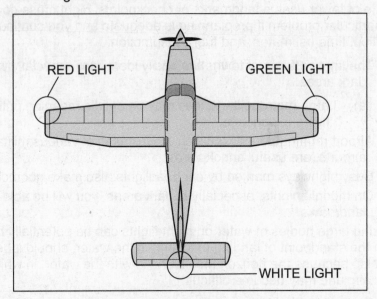

4)   This arrangement provides a means by which you can determine the general direction of movement of other airplanes.

a)   If both a red and a green light are seen, the other airplane is traveling in a general direction toward you.
b)   If only a red light is seen, the airplane is traveling from right to left.
c)   If only a green light is seen, the airplane is traveling from left to right.
d)   Note that the red and green lights cannot be seen from the rear of the airplane.

10. **Importance of verifying the airplane's attitude by reference to flight instruments**

a.   At night, your visual references are limited (and sometimes nonexistent), and you will need to use the flight instruments to a greater degree in controlling the airplane.

1)   This does not mean that you will use only the flight instruments, but that the flight instruments are used more to cross-check the visual references.

11. **Night emergencies procedures**

a.   The greatest electrical load placed upon your airplane's system occurs at night. As with most systems, the greatest chance of failure occurs when the system is being heavily used.

1)   If you suspect a problem, first reduce the electrical load as much as feasible.
2)   The greatest hazard is a total electrical failure during which you lose your position lights, your instrument lights, and radio navigation.
3)   If total failure is suspected, only battery power will be available. You should land at the nearest airport immediately.

    b.   Perhaps your greatest concern about flying a single-engine airplane at night is complete engine failure, even though adverse weather and poor pilot judgment account for most serious accidents.

        1)   If the engine fails at night, the first step is to maintain positive control of your airplane. DO NOT PANIC.

            a)   A normal glide should be established and maintained; turn your airplane toward an airport or away from congested areas.

            b)   A check should be made to determine the cause of the engine failure, including the position of the following:

                i)   Magneto switches
                ii)   Fuel selectors
                iii) Primer

            c)   If possible, correct the malfunction immediately and restart the engine.

        2)   Maintain orientation with the wind to avoid a downwind landing.

        3)   The landing light(s), if equipped, should be checked at altitude and turned on in sufficient time to illuminate the terrain or obstacles along the flight path.

            a)   If the landing light(s) are unusable and outside references are not available, the airplane should be held in level-landing attitude until the ground is contacted.

        4)   Most important of all, positive control of your airplane must be maintained at all times.

            a)   DO NOT allow a stall to occur.

12. **Traffic patterns**

    a.   When you arrive at the airport to enter the traffic pattern and land, it is important that you identify the runway lights and other airport lighting as early as possible.

        1)   If you are unfamiliar with the airport layout, sighting of the runway may be difficult until you are very close-in due to other lighting in the area.

        2)   You should fly towards the airport rotating beacon until you identify the runway lights.

        3)   Your landing light should be on to help other pilots and/or ATC to see you.

    b.   To fly a traffic pattern of the proper size and direction when there is little to see but a group of lights, you must positively identify the runway threshold and runway edge lights.

        1)   Confirm that you are entering the pattern for the proper runway by comparing the runway lights to your heading indicator.

        2)   Once this is done, the location of the approach threshold lights should be known at all times throughout the traffic pattern.

        3)   Fly the traffic pattern as you would during the day.

13. **Approaches and landings with and without landing lights**

    a. Distance may be deceptive at night due to limited lighting conditions, lack of intervening references on the ground, and your inability to compare the size and location of different ground objects. The estimation of altitude and speed may also be impaired.

        1) Consequently, you must use your flight instruments more, especially the altimeter and the airspeed indicator.

        2) Every effort should be made to execute the approach and landing in the same manner as during the day.

        3) Constantly cross-check the altimeter, airspeed indicator, and vertical speed indicator against your airplane's position along the base leg and final approach.

    b. After turning onto the final approach and aligning your airplane between the two rows of runway edge lights, you should note and correct for any wind drift.

        1) Throughout the final approach, power should be used with coordinated pitch changes to provide positive control of your airplane, thus allowing you to accurately adjust airspeed and descent angle.

        2) A lighted visual approach slope indicator (e.g., VASI or PAPI) should be used if available to help maintain the proper approach angle.

    c. The roundout and touchdown should be made in the same manner as day landings. However, your judgment of height, speed, and sink rate may be impaired by the lack of observable objects in the landing area.

        1) You may be aided in determining the proper roundout point if you continue a constant approach descent until your airplane's landing light reflects on the runway and the tire marks on the runway or runway expansion joints can be seen clearly.

            a) At that point, smoothly start the roundout for touchdown and reduce the throttle gradually to idle as your airplane is touching down.

        2) During landings without the use of a landing light or where tire marks on the runway are not identifiable, the roundout may be started when the runway lights at the far end of the runway first appear to be rising higher than your airplane.

            a) This demands a smooth and very timely roundout and requires, in effect, that you "feel" for the runway surface, using power and pitch changes as necessary for the airplane to settle softly onto the runway.

14. **Go-arounds**

    a. Go-arounds always test the pilot's judgment, and even more so at night. Always be prepared to perform the maneuver.

    b. For a detailed discussion on how to perform a go-around, see Task VII.H., Go-Around, beginning on page 303.

END OF TASK

HIGH ALTITUDE OPERATIONS

II.H. TASK: HIGH ALTITUDE OPERATIONS

REFERENCES: 14 CFR Part 91; AC 61-107, FAA-H-8083-3; FAA-S-8081-12; POH/AFM, AIM.

Objective. To determine that the applicant exhibits instructional knowledge of the elements of high altitude operations by describing:

1. Regulatory requirements for use of oxygen.
2. Physiological hazards associated with high altitude operations.
3. Characteristics of a pressurized airplane with various types of supplemental oxygen systems.
4. Importance of "aviators' breathing oxygen."
5. Care and storage of high-pressure oxygen bottles.
6. Problems associated with rapid decompression and corresponding solutions.
7. Fundamental concept of cabin pressurization.
8. Operation of a cabin pressurization system.

A. General Information

1. The objective of this task is to determine your instructional knowledge of the elements of high-altitude operations.

2. FAR 61.31, General Limitations, states that no person may act as pilot in command of a pressurized airplane that has a service ceiling or maximum operating altitude, whichever is lower, above 25,000 ft. MSL unless that person

   a. Has completed the ground and flight training specified in FAR 61.31 (f)(1)(i) and (ii)

   b. Has a logbook endorsement from an authorized instructor (i.e., one that is qualified in a high-altitude airplane) certifying completion of the training

      1) This endorsement is not required if that person can document accomplishment of one of the following prior to April 15, 1991 in an airplane described above:

         a) Served as pilot in command

         b) Completed a pilot proficiency check for a pilot certificate or rating

         c) Completed an official pilot-in-command check by the U.S. Armed Forces

         d) Completed a pilot-in-command proficiency check under FAR Parts 121, 125, or 135 conducted by the FAA or an approved pilot-check airman

      NOTE: For the purposes of this FAR, flight operations conducted above 25,000 ft. MSL are considered to be high altitude.

B. Task Elements

1. **Regulatory requirements for use of oxygen**

   a. You may not operate a civil U.S. registered aircraft

      1) At cabin pressure altitudes above 12,500 ft. MSL up to and including 14,000 ft. MSL unless the required minimum flight crew uses supplemental oxygen for that part of the flight at those altitudes that is longer than 30 min.

      2) At cabin pressure altitudes above 14,000 ft. MSL unless the required minimum flight crew uses supplemental oxygen during the entire time at those altitudes

      3) At cabin pressure altitudes above 15,000 ft. MSL unless each occupant is provided with supplemental oxygen

   b.   You may not operate a civil U.S. registered aircraft with a pressurized cabin

   1)   At flight altitudes above FL 250 unless at least a 10-min. supply of supplemental oxygen is available for each occupant for use in the event that a descent is necessary due to a loss of cabin pressurization

   a)   This is in addition to any oxygen needed to satisfy item a. on the previous page.

   2)   Above FL 350 in a pressurized airplane unless one pilot at the controls of the airplane wears and uses an oxygen mask that is secured and sealed

   a)   The mask must supply oxygen at all times or must automatically supply oxygen when the cabin pressure exceeds 14,000 ft. MSL.

   b)   An exception to this is for flights at or below FL 410. One pilot does not need to wear and use an oxygen mask if both pilots are at the controls and each has a quick-donning type mask that can be placed on the face with one hand from the ready position and secured, sealed, and operating within 5 sec.

   i)   If one pilot is away from the controls, the pilot at the controls must wear and use an oxygen mask that is secured and sealed.

   c.   Supplemental oxygen use is more stringent under FAR Parts 121 and 135.

2.   **Physiological hazards associated with high altitude operations**

   a.   The human body functions normally in the atmospheric area extending from sea level to 12,000 ft. MSL. In this range, brain oxygen saturation is at a level that allows for normal functioning. (Optimal functioning is 96% saturation.)

   1)   At 12,000 ft., brain oxygen saturation is approximately 87%, which begins to approach a level that could affect performance.

   b.   Hypoxia is a lack of sufficient oxygen in the body cells or tissues caused by an inadequate supply of oxygen, inadequate transportation of oxygen, or inability of the body tissues to use oxygen. A description of the four major hypoxia groups follows:

   1)   **Hypoxic (altitude) hypoxia.** Altitude hypoxia poses the greatest potential physiological hazard to a pilot while flying in the high-altitude environment. This type of hypoxia is caused by an insufficient partial pressure of oxygen in the inhaled air resulting from reduced oxygen pressure in the atmosphere at altitude.

   2)   **Histotoxic hypoxia.** This is the inability of the body cells to use oxygen because of impaired cellular respiration. This type of hypoxia, caused by alcohol or drug use, cannot be corrected by using supplemental oxygen because the uptake of oxygen is impaired at the tissue level.

   a)   The only method of avoiding this type of hypoxia is to abstain, before flight, from alcohol or drugs that are not approved by a flight surgeon or an aviation medical examiner.

   3)   **Hypemic (anemic) hypoxia.** This type of hypoxia is defined as a reduction in the oxygen-carrying capacity of the blood. Hypemic hypoxia is caused by a reduction in circulating red blood cells (hemoglobin) or contamination of blood with gases other than oxygen as a result of anemia, carbon monoxide poisoning, or excessive smoking.

4) **Stagnant hypoxia.** This is an oxygen deficiency in the body resulting from poor circulation of the blood because of a failure of the circulatory system to pump blood (and oxygen) to the tissues.

    a) In flight, this type of hypoxia can sometimes be caused by positive pressure breathing for long periods of time or by excessive G-forces.

NOTE: For additional information about hypoxia and its symptoms, refer to Task II.A., Aeromedical Factors, beginning on page 96.

c. Prolonged oxygen use can also be harmful to human health. One hundred percent aviation oxygen can produce toxic symptoms if used for extended periods of time.

1) The symptoms can consist of bronchial cough, fever, vomiting, nervousness, irregular heartbeat, and lowered energy.

2) The sudden supply of pure oxygen following a decompression can often aggravate the symptoms of hypoxia.

3) Therefore, oxygen should be taken gradually, particularly when the body is already suffering from lack of oxygen, to build up the supply in small doses.

d. When nitrogen is inhaled, it dilutes the air we breathe. While most nitrogen is exhaled from the lungs along with carbon dioxide, some nitrogen is absorbed by the body.

1) The nitrogen absorbed into the body tissues does not normally present any problem because it is carried in a liquid state.

2) If the ambient surrounding atmospheric pressure lowers drastically, this nitrogen could change from a liquid and return to its gaseous state in the form of bubbles.

3) These evolving and expanding gases in the body are known as decompression sickness and are divided into two groups.

    a) Trapped gas. Expanding or contracting gas in certain body cavities during altitude changes can result in abdominal pain, toothache, or pain in ears and sinuses if the person is unable to equalize the pressure changes.

        i) Above 25,000 ft., distention can produce particularly severe gastrointestinal pain.

    b) Evolved gas. When the pressure on the body drops sufficiently, nitrogen comes out of solution and forms bubbles, which can have adverse effects on some body tissues. Fatty tissue contains more nitrogen than other tissue, thus making overweight people more susceptible to evolved gas decompression sicknesses.

        i) SCUBA diving will compound this problem because of the compressed air used in the breathing tanks.

e. Vision has a tendency to deteriorate with altitude. A reversal of light distribution at high altitudes (bright clouds below the airplane and darker, blue sky above) can cause a glare inside the cockpit.

1) Glare effects and deteriorated vision are enhanced at night when the body becomes more susceptible to hypoxia.

2) The empty visual field caused by cloudless, blue skies during the day can cause inaccuracies when judging the speed, size, and distance of other aircraft.

3) Sunglasses are recommended to minimize the intensity of the sun's ultraviolet rays at high altitudes.

3. **Characteristics of a pressurized airplane and various types of supplemental oxygen systems**

    a.   Characteristics of a pressurized airplane

        1)   Cabin pressurization is the compression of air in the airplane's cabin to maintain a cabin altitude lower than the actual flight altitude.

            a)   Thus, the need for the full-time use of supplemental oxygen above certain altitudes is eliminated.

        2)   Pressurization in most light airplanes is sent to the cabin from the turbocharger's compressor or from an engine-driven pneumatic pump.

            a)   Since the compression of the air in the turbocharger causes the air to become hot, it is routed through a type of heat exchange unit before it enters the cabin.

            b)   The flow of compressed air into the cabin is regulated by an outflow valve which keeps the pressure constant by releasing excess pressure into the atmosphere.

                i)   This outflow valve also allows the exchange of air from inside to outside of the cabin to eliminate odors and to remove stale air.

        3)   Pressurized airplanes have special structural specifications, such as windows of double thickness and doors with special seals and locking devices, to withstand differential pressures.

            a)   **Differential pressure** is the difference between cabin pressure and atmospheric pressure and is normally expressed in pounds per square inch (psi).

                i)   The higher the airplane climbs, the greater the differential pressure.

            b)   Maximum differential pressure varies among makes and models of airplanes.

                i)   It is not uncommon for there to be differential pressures of 7 psi or for an airplane to be at a flight altitude of 10,000 ft. MSL with a cabin altitude of sea level.

        4)   A pressurized airplane has a cabin pressure control system which provides cabin pressure regulation, pressure relief, vacuum relief, and the means for selecting the desired cabin altitude in the isobaric differential pressure range.

            a)   A cabin pressure regulator, an outflow valve, and a safety valve are used to accomplish these functions.

            b)   In addition, dumping of the cabin pressure is a function of the pressure control system.

        5)   The cabin pressure regulator controls cabin pressure to a selected value in the isobaric range and limits cabin pressure to a preset differential value in the differential range.

            a)   When the airplane reaches the altitude at which the maximum differential pressure is reached, a further increase in airplane altitude will result in an increase in the cabin pressure altitude.

            b)   Differential control is used to prevent the maximum differential pressure, for which the fuselage was designed, from being exceeded.

                i)   This differential pressure is determined by structural strength of the cabin and often by the relationship of the cabin size to the probable areas of rupture, such as window areas and doors.

6) The cabin air pressure safety valve is a combination pressure relief, vacuum relief, and dump valve.

   a) The pressure relief valve prevents cabin pressure from exceeding the maximum differential pressure.

   b) The vacuum relief valve prevents ambient pressure from exceeding cabin pressure by allowing external air to enter the cabin when ambient pressure exceeds cabin pressure.

   c) The dump valve is actuated by the cockpit control switch.

      i) When this switch is positioned to "dump" or "ram," a solenoid valve opens, causing the valve to dump the cabin air to the atmosphere.

7) Several instruments are used in conjunction with the pressurization controller.

   a) The cabin differential pressure gauge indicates the difference between inside and outside pressure.

      i) This gauge should be monitored to assure that the cabin does not exceed the maximum allowable differential pressure.

   b) A cabin altimeter is also provided as a check on the performance of the system.

      i) In some cases, the differential pressure gauge and the cabin altimeter are combined into one instrument.

   c) A third instrument indicates the cabin rate of climb or descent.

   d) A cabin rate-of-climb instrument and a combination cabin altimeter/ differential pressure gauge are shown below.

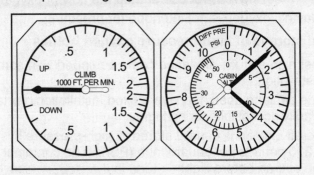

b. Types of supplemental oxygen systems

1) The **continuous flow** oxygen system is the most common system found in general aviation airplanes.

   a) The mask is designed so the oxygen can be diluted with ambient air by allowing the user to exhale around the face piece and comes with a rebreather bag, which allows the individual to reuse part of the exhaled oxygen.

   b) Although certificated up to 41,000 ft., careful attention to system capabilities is required when using this type of system above 25,000 ft.

2)   The **diluter-demand** system increases the usage of the continuous flow system by conserving oxygen at lower altitudes and increasing the oxygen flow at higher altitudes.

    a)   This system is needed above 25,000 ft. and can be used safely up to 40,000 ft.

    b)   This system allows the pilot to select either normal (cabin air and oxygen) or 100% oxygen.

       i)   100% oxygen is normally automatically provided at approximately 30,000 ft.

3)   The **pressure-demand** system is normally installed in high performance turboprop and jet airplanes.

    a)   This system is needed for operations above 40,000 ft., and it delivers pressurized oxygen to your lungs.

    b)   This differs from the two other systems because oxygen is delivered under positive pressure.

       i)   You must exhale against the pressure of the incoming oxygen.

4.   **Importance of aviators' breathing oxygen**

    a.   Aviators' breathing oxygen is specified at 99.5% pure oxygen and not more than .005 mg. of water per liter.

    b.   Medical oxygen contains too much water, which can collect in various parts of the system and freeze.

      1)   Freezing may reduce or stop the flow of oxygen.

    c.   Industrial oxygen is not intended for breathing.

5.   **Care and storage of high-pressure oxygen bottles**

    a.   Most high-altitude airplanes come equipped with some type of fixed oxygen installation.

    b.   If the airplane does not have a fixed installation, portable oxygen equipment must be readily accessible during flight.

    c.   Aircraft oxygen is usually stored in high-pressure system containers of 1,800 to 2,200 pounds per square inch (PSI). The container should be fastened securely in the aircraft before flight.

      1)   When the ambient temperature surrounding an oxygen cylinder decreases, pressure within that cylinder will decrease because pressure varies directly with temperature if the volume of a gas remains constant.

      2)   If a drop in indicated pressure on a supplemental oxygen cylinder is noted, it does not necessarily indicate depletion of the supply. The oxygen may simply have been compacted due to storage of the containers in an unheated area of the aircraft.

    d.   High-pressure oxygen containers should be marked with the PSI tolerance (e.g., 1,800 PSI) before filling the container to that pressure.

    e.   Pilots should be aware of the danger of fire when using oxygen. Materials that are nearly fireproof in ordinary air may be susceptible to burning in oxygen.

      1)   Oils and greases may catch fire if exposed to oxygen and therefore must not be used for sealing the valves and fittings of oxygen equipment.

      2)   Smoking during any kind of oxygen equipment use must also be strictly forbidden.

    f.   Use the recommended procedures for care and storage that are found in the POH.

6. **Problems associated with rapid decompression and corresponding solutions**

    a. **Decompression** is defined as the inability of the airplane's pressurization system to maintain its designed pressure differential. This can be caused by a malfunction in the pressurization system, a malfunction of door or window seals, or structural damage to the airplane.

        1) During rapid decompression, there may be noise and, for a split second, one may feel dazed. The cabin air will fill with fog, dust, or flying debris. Fog occurs due to the rapid drop in temperature and the change of relative humidity. Normally, the ears clear automatically. Belching or passing of intestinal gas may occur. Air will rush from the mouth and nose due to the escape of air from the lungs.

        2) The primary danger of decompression is hypoxia. Unless proper utilization of oxygen equipment is accomplished quickly, unconsciousness may occur in a very short time. The period of useful consciousness is considerably shortened when a person is subjected to a rapid decompression. Oxygen in the lungs is exhaled rapidly. This, in effect, reduces the partial pressure of oxygen in the lungs and thus reduces the pilot's effective performance time by one-third to one-fourth of his/her normal time.

            a) Exposure to windblast and extremely cold temperatures are other hazards faced with a decompression.

        3) An explosive decompression is a change in cabin pressure faster than the lungs can decompress.

            a) Any decompression which occurs in less than 0.5 sec. is considered explosive and potentially dangerous.

            b) To avoid potentially dangerous flying debris in the event of an explosive decompression, all loose items, such as baggage and oxygen cylinders, should be properly secured.

    b. Recovery from all types of decompression is similar. Oxygen masks should be donned, and a rapid (emergency) descent initiated as soon as possible to avoid the onset of hypoxia.

        1) Although top priority in such a situation is reaching a safe altitude, pilots should be aware that cold-shock in piston engines can result from a high-altitude rapid descent, causing cracked cylinders or other engine damage.

        2) The time allowed to make a recovery to a safe altitude before loss of useful consciousness is, of course, much less with an explosive than with a gradual decompression.

   c.   Types of evolved gas decompression sickness

      1)   The **bends**, also known as caisson disease, is characterized by pain in and around the joints. The term "bends" is used because the resultant pain is eased by bending the joints.

         a)   The pain gradually becomes more severe, can eventually become temporarily incapacitating, and can result in collapse.

      2)   The **chokes** refers to a decompression sickness that manifests itself through chest pains and burning sensations, a desire to cough, possible cyanosis, a sensation of suffocation, progressively shallower breathing and, if a descent is not made immediately, collapse and unconsciousness.

      3)   **Paresthesia** is a third type of decompression sickness, characterized by tingling, itching, a red rash, and cold and warm sensations, probably resulting from bubbles in the central nervous system (CNS).

         a)   CNS disturbances can result in visual deficiencies, such as illusionary lines or spots, or a blurred field of vision.

         b)   Some other effects of CNS disturbances are temporary partial paralysis, sensory disorders, slurred speech, and seizures.

   d.   Shock can often result from decompression sicknesses as a form of body protest to disrupted circulation.

      1)   Shock can cause nausea, fainting, dizziness, sweating, and/or loss of consciousness.

      2)   The best treatment for decompression sickness is descent to a lower altitude and landing.

7.   **Fundamental concept of cabin pressurization**

   a.   Cabin pressurization is the compression of air in the airplane's cabin to maintain a cabin altitude lower than the actual flight altitude.

8.   **Operation of a cabin pressurization system**

   a.   If your airplane is equipped with a pressurization system, you must know the normal and emergency operating procedures.

      1)   Follow the procedures explained in your airplane's POH.

END OF TASK

FEDERAL AVIATION REGULATIONS AND PUBLICATIONS

II.I. TASK: **FEDERAL AVIATION REGULATIONS AND PUBLICATIONS**

REFERENCES: 14 CFR Parts 1, 61, 91; NTSB Part 830; AC 00-2, AC 61-23/FAA-H-8083-25; POH/AFM, AIM.

Objective. To determine that the applicant exhibits instructional knowledge of the elements related to Federal Aviation Regulations and publications

1. Availability and method of revision of 14 CFR Parts 1, 61, 91, and NTSB Part 830 by describing--

    a. Purpose.
    b. General content.

2. Availability of flight information publications, advisory circulars, practical test standards, pilot operating handbooks, and FAA-approved airplane flight manuals by describing--

    a. Availability.
    b. Purpose.
    c. General content.

A. General Information

    1. The objective of this task is to determine your instructional knowledge of the elements related to Federal Aviation Regulations (FAR) and publications.

    2. FARs are organized by part, and the most common parts relevant to general aviation pilots and flight instructors are Parts 61 and 91 and NTSB Part 830.

B. Task Elements

    1. **Availability and method of revision of 14 CFR Parts 1, 61, 91, and NTSB Part 830**

        a. **Availability and method of revision**

            1) Availability

                a) Most FSSs, FSDOs, and other FAA offices have complete sets of the FARs available for public use.

                b) The FAA issues AC 00-44, which is a listing of the current publication status, prices (single issue and subscriptions), and order forms for obtaining the FARs.

                c) Gleim's *FAR/AIM* contains all of the FARs applicable to general aviation, with easy-to-use indexes. We provide free updates by e-mail, Internet, FAX, and U.S. mail.

            2) FARs are issued pursuant to and are enforceable under the laws of the United States. In essence, the FAA is authorized to issue and revise rules and regulations in accordance with its executive agency (the FAA is a component of the U.S. Department of Transportation). Notices of Proposed Rulemaking (NPRMs) are published in the *Federal Register* and, after comment periods, they are issued as FARs, i.e., published in the *Federal Register* as final rules. The National Transportation Safety Board (NTSB) has similar rulemaking authority.

b.   **Purpose and general content**

1)   Part 61, Certification: Pilots, Flight Instructors, and Ground Instructors, contains

a)   The requirements for issuing pilot, flight instructor, and ground instructor certificates and ratings

b)   The conditions under which those certificates and ratings are necessary

c)   The privileges and limitations of those certificates and ratings

d)   The following subparts

i)   Aircraft ratings (e.g., instrument) and pilot authorizations
ii)   Student pilots
iii)   Recreational pilots
iv)   Private pilots
v)   Commercial pilots
vi)   Airline transport pilots
vii) Flight instructors
viii) Ground instructors

2)   Part 91, General Operating and Flight Rules, describes rules governing the operation of aircraft.

a)   The subparts are

i)   General
ii)   Flight rules

- General
- Visual
- Instrument

iii)   Maintenance, preventive maintenance, and alterations
iv)   Large and turbine-powered multiengine airplanes
v)   Operating noise limits

3)   The NTSB is a function within the U.S. Department of Transportation (as is the FAA) that has rule-making authority similar to the FAA.

a)   NTSB Part 830 addresses aircraft accidents and is titled, Rules Pertaining to the Notification and Reporting of Aircraft Accidents or Incidents and Overdue Aircraft, and Preservation of Aircraft Wreckage, Mail, Cargo, and Records.

b)   It contains numerous forms and paragraphs, including

i)   Applicability
ii)   Definitions
iii)   Immediate Notification
iv)   Information to Be Given in Notification
v)   Preservation of Aircraft Wreckage, Mail, Cargo, and Records
vi)   Reports and Statements to Be Filed

2.   **Availability, purpose, and general content of aviation related publications**

a.   *Airport/Facility Directory (A/FD)*

1)   The *Airport/Facility Directory (A/FD)* is a Civil Flight Information Publication published and distributed every 8 weeks by the National Aeronautical Charting Office (NACO), a division of the FAA.

a)   It is a directory of all airports, seaplane bases, and heliports open to the public; communications data; navigational facilities; and certain special notices and procedures.

b)   Subscriptions to the publication are for sale by the

> FAA, National Aeronautical Charting Office
> Distribution Division, AJW-3550
> 10201 Good Luck Road
> Glenn Dale, MD 20769-9700
> Telephone: 1-800-638-8972 (within the U.S.)
> 301-436-8301/6990
> 301-436-6829 (FAX)

c)   One of these directories is published for each of seven geographical districts:

    i)   Northwest
    ii)   Southwest
    iii)   North Central
    iv)   South Central
    v)   East Central
    vi)   Northeast
    vii) Southeast

2)   The *A/FD* is a vital publication for cross-country flight planning.

a)   All pertinent information regarding airports, FSS contact information, etc., is contained in the *A/FD*.

3)   Table of contents of each issue

1.	Abbreviations
2.	Legend, Airport/Facility Directory
3.	Airport/Facility Directory
4.	Heliports
5.	Seaplane Bases
6.	Notices
7.	FAA and National Weather Service Telephone Numbers
8.	Air Route Traffic Control Centers
9.	Flight Service Station Communication Frequencies
10.	FSDO Addresses/Telephone Numbers
11.	Preferred IFR Routes
12.	VOR Receiver Check
13.	Parachute Jumping Areas
14.	Aeronautical Chart Bulletin
15.	Tower Enroute Control (TEC)
16.	National Weather Service (NWS) Upper Air Observing Stations
17.	Enroute Flight Advisory Service (EFAS)

b.   ***Aeronautical Information Manual (AIM)***

1)   The *AIM* provides pilots with a vast amount of basic flight information and Air Traffic Control (ATC) procedures in the United States.

a)   It is an 8" × 10 1/2" paperback, with page changes published every 196 days by the FAA.

b)   Subscriptions are available from the Government Printing Office, (202) 512-1800. They accept MasterCard and VISA.

2)   Chapters and section titles

CHAPTER 1. AIR NAVIGATION
Section 1. Navigation Aids
Section 2. Area Navigation (RNAV) and Required Navigation Performance (RNP)

CHAPTER 2. AERONAUTICAL LIGHTING AND OTHER AIRPORT VISUAL AIDS
Section 1. Airport Lighting Aids
Section 2. Air Navigation and Obstruction Lighting
Section 3. Airport Marking Aids and Signs

CHAPTER 3. AIRSPACE
Section 1. General
Section 2. Controlled Airspace
Section 3. Class G Airspace
Section 4. Special Use Airspace
Section 5. Other Airspace Areas

CHAPTER 4. AIR TRAFFIC CONTROL
Section 1. Services Available to Pilots
Section 2. Radio Communications Phraseology and Techniques
Section 3. Airport Operations
Section 4. ATC Clearances/Separations
Section 5. Surveillance Systems
Section 6. Operational Policy/Procedures for Reduced Vertical Separation
            Minimum (RVSM) in the Domestic U.S., Alaska, Offshore Airspace
            and the San Juan Fir

CHAPTER 5. AIR TRAFFIC PROCEDURES
Section 1. Preflight
Section 2. Departure Procedures
Section 3. En Route Procedures
Section 4. Arrival Procedures
Section 5. Pilot/Controller Roles and Responsibilities
Section 6. National Security and Interception Procedures

CHAPTER 6. EMERGENCY PROCEDURES
Section 1. General
Section 2. Emergency Services Available to Pilots
Section 3. Distress and Urgency Procedures
Section 4. Two-Way Radio Communications Failure
Section 5. Aircraft Rescue and Fire Fighting Communications

CHAPTER 7. SAFETY OF FLIGHT
Section 1. Meteorology
Section 2. Altimeter Setting Procedures
Section 3. Wake Turbulence
Section 4. Bird Hazards and Flight over National Refuges, Parks, and Forests
Section 5. Potential Flight Hazards
Section 6. Safety, Accident, and Hazard Reports

CHAPTER 8. MEDICAL FACTS FOR PILOTS
Section 1. Fitness for Flight

CHAPTER 9. AERONAUTICAL CHARTS AND RELATED PUBLICATIONS
Section 1. Types of Charts Available

APPENDICES
Appendix 1. Bird/Other Wildlife Strike Report
Appendix 2. Volcanic Activity Reporting Form (VAR)
Appendix 3. Laser Beam Exposure Questionnaire
Appendix 4. Abbreviations/Acronyms

Pilot/Controller Glossary

Index

---

3)   The *AIM* has a comprehensive and useful index to help you find topics of interest.

c. **FAA Advisory Circular (AC)**

1) The FAA issues advisory circulars (ACs) to provide a systematic means for issuing nonregulatory material of interest to the aviation public.

   a) Unless incorporated into a regulation by reference, the contents of an AC are not binding (i.e., they are only advisory in nature).

   b) An AC is issued to provide guidance and information in its designated subject area or to show a method acceptable to the FAA for complying with a related FAR.

2) ACs are issued in a numbered system of general subject matter areas that correspond with the subject areas in Federal Aviation Regulations (FARs).

   a) The general subject number and the subject areas are as follows:

       00 -- General
       10 -- Procedural Rules
       20 -- Aircraft
       60 -- Airmen
       70 -- Airspace
       90 -- Air Traffic and General Operating Rules
       120 -- Air Carriers, Air Travel Clubs, and Operators for Compensation
              or Hire: Certification and Operations
       140 -- Schools and Other Certificated Agencies
       150 -- Airport Noise Compatibility Planning
       170 -- Navigational Facilities
       180 -- Administrative Regulations
       190 -- Withholding Security Information
       210 -- Flight Information (NOTE: This series is about aeronautical
              charts and does not relate to a FAR part.)

3) If you wish, you may order a free list of the ACs. It is called the *Advisory Circular Checklist*, AC 00-2. See order form on page 149.

d. **Notice to Airmen (NOTAM)**

1) The National Notice to Airmen (NOTAM) System disseminates time-critical aeronautical information, which either is of a temporary nature or is not sufficiently known in advance to permit publication on aeronautical charts or in other operational publications.

   a) NOTAM information is that aeronautical information that could affect the decision to make a flight.

2) NOTAM information is classified into three categories.

   a) **NOTAM (D)**, or distant, includes such information as airport or primary runway closures, changes in the status of navigational aids, ILSs, radar service availability, and other information essential to planned en route, terminal, or landing operations.

      i) This information is disseminated for all navigational facilities that are part of the National Airspace System (NAS), public use airports, seaplane bases, and heliports listed in the *A/FD*.

      ii) NOTAM (D) information is appended to the hourly surface aviation observation (SA) weather reports.

      iii) These NOTAMs will remain available in this manner for the duration of their validity or until published.

b) **NOTAM (L)**, or local, includes such information as taxiway closures, personnel and equipment near or crossing runways, airport rotating beacon outages, and airport lighting that does not affect instrument approach procedure (IAP) criteria (e.g., VASI).

    i) NOTAM (L) information is distributed locally only and is not attached to the hourly SA weather reports.

    ii) A separate file of local NOTAMs is maintained at each FSS for facilities in its area only.

       • NOTAM (L) information for other FSS areas must be specifically requested directly from the FSS that has responsibility for the airport concerned.

    iii) Direct User Access Terminal System (DUATS) vendors are not required to provide NOTAM (L) information.

c) A **Flight Data Center (FDC) NOTAM** is regulatory in nature and includes such information as amendments to published IAPs and other current aeronautical charts. It also advertises temporary flight restrictions caused by such things as natural disasters or large-scale public events that may generate a congestion of air traffic over a site.

    i) FSSs are responsible for maintaining a file of current, unpublished FDC NOTAMs concerning conditions within 400 NM of their facilities.

    ii) FDC information that concerns conditions beyond 400 NM from the FSS or that is already published is given to you only when you request it.

    iii) DUATS vendors will provide FDC NOTAMs only upon site-specific requests using a location identifier.

3) The ***Notices to Airmen Publication (NTAP)*** is issued every 28 days and is an integral part of the NOTAM System. Once a NOTAM is published in the NTAP, the NOTAM is not provided during pilot weather briefings unless specifically requested by the pilot.

a) The *NTAP* consists of two sections.

    i) The first section contains NOTAMs (D) that are expected to remain in effect for an extended period and FDC NOTAMs that are current at the time of publication.

       • Occasionally, some NOTAMs (L) and other unique information are included in this section when they will contribute to flight safety.

    ii) The second section contains special notices that either are too long or concern a wide or unspecified geographic area.

b) The number of the last FDC NOTAM included in the *NTAP* is noted on the first page to assist a pilot in updating the listing with any FDC NOTAMs that may have been issued between the cut-off date and the date the publication is received.

    i) All information contained in the *NTAP* will be carried until the information expires, is canceled, or, in the case of permanent conditions, is published in other publications (e.g., *A/FD*, aeronautical charts, etc.).

c) All new NOTAMs entered, excluding FDC NOTAMs, will be published only if the information is expected to remain in effect for at least 7 days after the effective date of the *NTAP*.

ORDER BLANK (Free Publications)   DATE ___ / ___ / ___

> *For Faster Service, Use a Self-Addressed Mailing Label*
> *When Not Using This Blank. Please Print Or Type All*
> *Information*

Mail to: U.S. Department of Transportation
Subsequent Distribution Office, SVC-121.23
Ardmore East Business Center
3341 Q 75th Avenue
Landover, MD 20785

Help Line: FAX REQUEST TO 301-386-5395 DOT Warehouse
301-322-4961

NUMBER	TITLE	QUANTITY

SVC-121.23
Request Filled By: _____   Date ___ / ___ / ___

1. Out of Stock [reorder in   days]* *   3. Canceled, no replacement

2. Being revised   4. Canceled by _____ [enclosed]

                5. Other: _____

        * * IF YOU DO NOT RECEIVE DESIRED PUBLICATION(S) AFTER YOUR <u>SECOND</u>
    REQUEST, PLEASE CALL FAA'S TOLL-FREE CONSUMER HOTLINE: 1-800-FAA-SURE.

TO COMPLETE ORDER: Enter Name and Address	<u>DO NOT DETACH</u>

NAME

STREET ADDRESS

CITY   STATE   ZIP CODE

e. **Practical Test Standards (PTSs)**

1) FARs specify the areas in which knowledge and skill must be demonstrated by pilot applicants before the issuance of pilot certificates or ratings.

   a) The FARs provide the flexibility to permit the FAA to publish PTSs containing specific tasks in which pilot competency must be demonstrated.

   b) When discussing a PTS task with a student (or your examiner), it is important to cover each element included in the task.

2) PTSs can be purchased from the Government Printing Office or at pilot bookstores (which normally sell commercial reproductions of the government books).

3) The following table is a list of current edition-printings of the Gleim *Flight Maneuvers and Practical Test Prep* books and current PTSs. This table reflects changes through February 2007.

Test	Current Edition-Printing (and date) of Gleim book	Publication Number	Date	No. of Areas of Operation	Tasks
Sport	1-2 (July 2006)	FAA-S-8081-29	December 2004	10	40
Private	4-2 (May 2006)	FAA-S-8081-14A	August 2002	12	47
Instrument	4-1 (October 2004)	FAA-S-8081-4D	April 2004	8	25
Commercial	4-2 (October 2004)	FAA-S-8081-12B	August 2002	11	43
Flight Instructor	4-1 (February 2007)	FAA-S-8081-6C	November 2006	14	72
Flight Instructor - Instrument	N/A	FAA-S-8081-9C	November 2006	10	37
ATP and Type Rating	N/A	FAA-S-8081-5E	August 2006	9	38

f. **Pilot's Operating Handbook (POH)**

1) The FAA requires a Pilot's Operating Handbook (also called an FAA-approved *Airplane Flight Manual*) (FAR 23.1581). POHs are usually 6" × 8" ring notebooks, so pages can be updated, deleted, added, etc. They typically have nine sections:

```
1. General. . . . . . . . . . . . . . . . . . . . . . . . . . . . . . . . . . . . . . . . . Description of the airplane
2. Limitations. . . . . . . . . . . . . . . . . . . . . . . . . . . . . . . . . . . . Description of operating limits
3. Emergency Procedures. . . . . . . . . . . . . . . . . . . . . . . . . . . . What to do in each situation
4. Normal Procedures. . . . . . . . . . . . . . . . . . . . . . . . . . . . . . . . . . . . . . . . . . . Checklists
5. Performance. . . . . . . . . . . . . . . . . . . . . . . . . Graphs and tables of airplane capabilities
6. Weight and Balance. . . . . . . . . . . . . . . . . . . . . . . . Equipment list, airplane empty weight
7. Airplane and Systems Description. . . . . . . . . . . . . . . Description of the airplane's systems
8. Servicing and Maintenance. . . . . . . . . . . . . . . . . . . . . . . Explanation of what and when
9. Supplements. . . . . . . . . . . . . . . . . . . . . . . . . . . . Usually, description of optional equipment
```

2) Most late-model popular airplanes have POHs reprinted as perfect-bound books (called Information Manuals) available at FBOs and aviation bookstores.

END OF TASK

NATIONAL AIRSPACE SYSTEM

II.J. TASK: NATIONAL AIRSPACE SYSTEM

REFERENCES: 14 CFR Part 91; FAA-S-8081-12, FAA-S-8081-14; AIM.

Objective. To determine that the applicant exhibits instructional knowledge of the elements of the national airspace system by describing:

1. Basic VFR Weather Minimums - for all classes of airspace.

2. Airspace classes - the operating rules, pilot certification, and airplane equipment requirements for the following:

   a. Class A.
   b. Class B.
   c. Class C.
   d. Class D.
   e. Class E.
   f. Class G.

3. Special use airspace (SUA).

4. Temporary flight restrictions (TFR).

A. General Information

1. The objective of this task is to determine your instructional knowledge of the elements of the national airspace system.

B.   Task Elements

1.   **Basic VFR Weather Minimums - for all classes of airspace**

a.   The basic VFR weather minimums for all classes of airspace are shown below (FAR 91.155).

Visibility and Cloud Clearance Requirements for VFR

Airspace	Flight Visibility	Distance from Clouds
Class A	Not Applicable	Not applicable
Class B	3 SM	Clear of clouds
Class C	3 SM	500 ft. below 1,000 ft. above 2,000 ft. horiz.
Class D	3 SM	500 ft. below 1,000 ft. above 2,000 ft. horiz.
Class E:		
Less than 10,000 ft. MSL	3 SM	500 ft. below 1,000 ft. above 2,000 ft. horiz.
At or above 10,000 ft. MSL	5 SM	1,000 ft. below 1,000 ft. above 1 SM horiz.

Airspace	Flight Visibility	Distance from Clouds
Class G:		
1,200 ft. or less above the surface (regardless of MSL altitude)		
Day	1 SM	Clear of clouds
Night, except as provided in 1) below	3 SM	500 ft. below 1,000 ft. above 2,000 ft. horiz.
More than 1,200 ft. above the surface but less than 10,000 ft. MSL		
Day	1 SM	500 ft. below 1,000 ft. above 2,000 ft. horiz.
Night	3 SM	500 ft. below 1,000 ft. above 2,000 ft. horiz.
More than 1,200 ft. above the surface and at or above 10,000 ft. MSL	5 SM	1,000 ft. below 1,000 ft. above 1 SM horiz.

1)   An airplane may be operated clear of clouds in Class G airspace at night below 1,200 ft. AGL when the visibility is less than 3 SM but not less than 1 SM in an airport traffic pattern and within 1/2 mi. of the runway.

2)   Except when operating under a special VFR clearance, you may not operate your airplane beneath the ceiling under VFR within the lateral boundaries of the surface areas of Class B, Class C, Class D, or Class E airspace designated for an airport when the ceiling is less than 1,000 ft.

a)   You may not take off, land, or enter the traffic pattern of an airport unless ground visibility is at least 3 SM. If ground visibility is not reported, flight visibility must be at least 3 SM.

3)   Special VFR (SVFR)

a)   **SVFR conditions** means that the current weather conditions are less than that required for basic VFR flight while operating within the lateral boundaries of the Class B, Class C, Class D, or Class E surface area that has been designated for an airport and in which some aircraft are permitted to operate under VFR.

b) SVFR operations may only be conducted

    i) With an ATC clearance (You must request the clearance.)

    ii) Clear of clouds

    iii) With flight visibility of at least 1 SM

c) To take off or land under a SVFR clearance, the ground visibility must be at least 1 SM.

    i) If ground visibility is not reported, then flight visibility must be at least 1 SM.

d) To request a SVFR clearance at night, you must have an instrument rating.

2. **Airspace classes - the operating rules, pilot certification, and airplane equipment requirements for the following:**

Airspace Classification Diagram

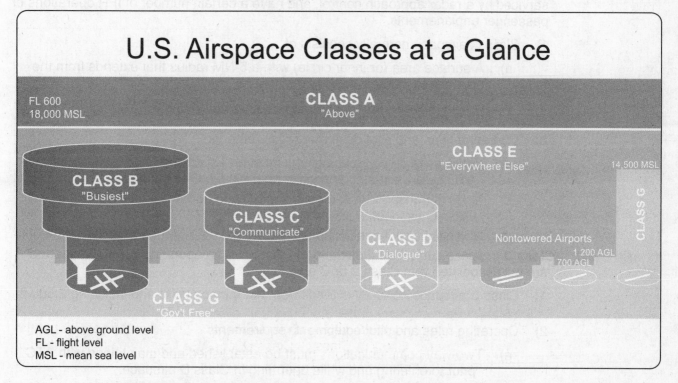

a. **Class A** airspace is generally the airspace from 18,000 ft. MSL up to and including flight level (FL) 600, including the airspace overlying the waters within 12 NM of the coast of the 48 contiguous states and Alaska.

    1) Operating rules and pilot/equipment requirements

        a) An IFR clearance to enter and operate within Class A airspace is mandatory. Thus, you must be instrument-rated to act as PIC of an airplane in Class A airspace.

        b) Two-way radio communication, appropriate navigational capability, and a Mode C transponder are required.

b. **Class B** airspace is generally the airspace from the surface to 10,000 ft. MSL surrounding the nation's busiest airports in terms of IFR operations or passenger enplanements (e.g., Atlanta, Chicago).

1) The configuration of each Class B airspace area is individually tailored and consists of a surface area and two or more layers.

2) Operating rules and pilot/equipment requirements for VFR operations

    a) An ATC clearance is required prior to operating within Class B airspace.

    b) Two-way radio communication capability is required.

    c) Mode C transponder is required within and above the lateral limits of Class B airspace and within 30 NM of the primary airport.

    d) The PIC must be at least a private pilot or a student or recreational pilot who is under the supervision of a CFI.

c. **Class C** airspace surrounds those airports that have an operational control tower, are serviced by a radar approach control, and have a certain number of IFR operations or passenger enplanements.

1) Class C airspace normally consists of

    a) A surface area (or inner circle) with a 5-NM radius that extends from the surface to 4,000 ft. AGL

    b) A shelf area (or outer circle) with a 10-NM radius that extends from 1,200 ft. to 4,000 ft. AGL

2) Operating rules and pilot/equipment requirements

    a) Two-way radio communications must be established and maintained with ATC before entering and while operating in Class C airspace.

    b) Mode C transponder is required within and above the lateral limits of Class C airspace.

    c) No specific pilot certification is required.

d. **Class D** airspace surrounds those airports that have an operating control tower and are not associated with Class B or C airspace.

1) Class D airspace normally extends from the surface up to and including 2,500 ft. AGL.

2) Operating rules and pilot/equipment requirements

    a) Two-way communications must be established and maintained with ATC prior to entering and while operating in Class D airspace.

    b) No specific pilot certification is required.

e. **Class E** airspace is any controlled airspace that is not Class A, B, C, or D airspace.

1) Except for 18,000 ft. MSL (the floor of Class A airspace), Class E airspace has no defined vertical limit, but rather it extends upward from either the surface or a designated altitude to the overlying or adjacent controlled airspace.

2) There are no specific pilot certification or equipment requirements to operate under VFR in Class E airspace.

f. **Class G** airspace is that airspace that has not been designated as Class A, Class B, Class C, Class D, or Class E airspace (i.e., it is uncontrolled airspace).

    1) No specific pilot certification or airplane equipment is required in Class G airspace.

    NOTE: While generally there is no equipment required to operate VFR in Class E or Class G airspace, there are some airports with an operational control tower (i.e., a controlled airport) within the surface area of Class E or Class G airspace. In these circumstances, you must establish and maintain two-way radio communication with the control tower if you plan to operate to, from, or through an area within 4 NM from the airport, from the surface up to and including 2,500 ft. AGL.

3. **Special use airspace (SUA)**

    a. Special use airspace

        1) **Prohibited areas** -- airspace within which flight is prohibited. Such areas are established for security or other reasons of national welfare.

        2) **Restricted areas** -- airspace within which flight, while not wholly prohibited, is subject to restrictions. Restricted areas denote the existence of unusual, often invisible hazards to aircraft, such as artillery firing, aerial gunnery, or guided missiles.

        3) **Warning areas** -- airspace of defined dimensions, extending from 3 NM outward from the coast of the U.S., that contains activity that may be hazardous to nonparticipating aircraft. The purpose of a warning area is to warn nonparticipating pilots of the potential danger (such as those in restricted areas).

            a) A warning area may be located over domestic or international waters or both.

        4) **Military operations areas (MOA)** -- airspace established to separate certain military training activities from IFR traffic.

            a) VFR aircraft should operate with caution while in an active MOA.

        5) **Alert areas** -- depicted on aeronautical charts to inform nonparticipating pilots of areas that may contain a high volume of pilot training or an unusual type of aerial activity.

        6) **Controlled firing areas** -- areas containing activities which, if not conducted in a controlled environment, could be hazardous to nonparticipating aircraft.

            a) The activities are suspended immediately when spotter aircraft, radar, or ground lookout positions indicate an aircraft might be approaching the area.

    b. Other airspace areas

        1) **Airport advisory areas** encompass the areas within 10 SM of airports that have no operating control towers but where FSSs are located. At such locations, the FSS provides advisory service to arriving and departing aircraft. Participation in the Local Airport Advisory (LAA) program is recommended but not required.

        2) **Military training routes (MTRs)** are developed for use by the military for the purpose of conducting low-altitude (below 10,000 ft. MSL), high-speed training (more than 250 kt.).

        3) Temporary flight restrictions (FAR 91.137) may be put into effect in the vicinity of any incident or event which by its nature may generate such a high degree of public interest that hazardous congestion of air traffic is likely.

        4) Flight limitations in the proximity of space flight operations (FAR 91.143) are designated in a NOTAM.

5) Flight restrictions in the proximity of Presidential and other parties (FAR 91.141) are put into effect by a regulatory NOTAM to establish flight restrictions.

6) Tabulations of parachute jump areas in the U.S. are contained in the *A/FD*.

7) **VFR flyway** is a general flight path not defined as a specific course but used by pilots planning flights into, out of, through, or near complex terminal airspace to avoid Class B airspace.

    a) VFR flyways are depicted on the reverse side of some of the VFR terminal area charts.

    b) An ATC clearance is not required to fly these routes since they are not in Class B airspace.

8) **VFR corridor** is airspace through Class B airspace, with defined vertical and lateral boundaries, in which aircraft may operate without an ATC clearance or communication with ATC. A VFR corridor is, in effect, a hole through the Class B airspace.

9) **Class B airspace VFR transition route** is a specific flight course depicted on a VFR terminal area chart for transiting a specific Class B airspace.

    a) These routes include specific ATC-assigned altitudes, and you must obtain an ATC clearance prior to entering the Class B airspace.

    b) On initial contact, you should inform ATC of your position, altitude, route name desired, and direction of flight.

        i) After a clearance is received, you must fly the route as depicted and, most importantly, follow ATC instructions.

10) **Terminal area VFR route** is a specific flight course for optional use by pilots to avoid Class B, Class C, and Class D airspace areas while operating in complex terminal airspace (e.g., Los Angeles).

    a) An ATC clearance is not required to fly these routes.

11) **National security areas (NSA)** -- airspace established at locations where there is a requirement for increased security and safety of ground facilities.

    a) Pilots are requested to voluntarily avoid flying through the depicted NSA.

    b) When necessary, flight in a NSA may be prohibited, and this prohibition will be disseminated by NOTAM.

12) **Terminal Radar Service Area (TRSA)**

    a) TRSAs are not controlled airspace from a regulatory standpoint (i.e., they do not fit into any of the airspace classes) because TRSAs were never subject to the rule-making process.

        i) Thus, TRSAs are not contained in FAR Part 71 nor are there any TRSA operating rules in FAR Part 91.

        ii) TRSAs are areas where participating pilots can receive additional radar services, known as TRSA Service.

    b) The primary airport(s) within the TRSA are Class D airspace.

        i) The remaining portion of the TRSA normally overlies Class E airspace beginning at 700 or 1,200 ft. AGL.

    c) Pilots operating under VFR are encouraged to participate in the TRSA Service. However, participation is voluntary.

    d) TRSAs are depicted on sectional charts with a solid black line and with altitudes for each segment expressed in hundreds of feet MSL.

        i) The Class D portion is depicted with a blue segmented line.

4. **Temporary Flight Restrictions (TFRs)**

    a. The FAA may issue a Notice to Airmen (NOTAM) to establish temporary flight restrictions

        1) To protect persons and property from a hazard associated with an incident on the surface

        2) To provide a safe environment for the operation of disaster relief aircraft

        3) To be observed in airspace above events generating a high degree of public interest

    b. When a NOTAM is issued to protect persons and property from a hazard associated with an incident on the surface, you may not operate your airplane in the area unless it is directed by an official in charge of on-scene emergency activities.

    c. When a NOTAM is issued to provide a safe environment for the operation of disaster relief aircraft, you may not operate your airplane in that area unless one of the following conditions is met:

        1) Your airplane is involved in relief activity and directed by an official in charge on the scene.

        2) Your airplane is carrying law enforcement officials.

        3) The operation is conducted directly to or from an airport in the area or is necessitated because VFR flight is impracticable, notice is given to the proper authority for receiving disaster relief advisories, relief activities are not hampered, and the flight is not solely for observation of the disaster.

        4) Your airplane is carrying properly accredited news representatives, a proper flight plan is filed, and the flight is above the altitude used by relief aircraft.

    d. When a NOTAM is issued to be observed in airspace above events generating a high degree of public interest, you may not operate your airplane in the area unless one of the following conditions is met:

        1) The operation is conducted directly to or from an airport in the area or is necessitated because VFR flight is impracticable.

        2) Your airplane is carrying incident or event personnel or law enforcement officials.

        3) Your airplane is carrying properly accredited news representatives, a proper flight plan is filed, and the flight is above the altitude used by relief aircraft.

    e. Flight plans filed and notice given must include the following:

        1) Aircraft identification, type, and color
        2) Radio frequencies to be used
        3) Times of entry and exit from the area
        4) Name of news organization and purpose of flight
        5) Any other information requested by ATC

END OF TASK

NAVIGATION SYSTEMS AND RADAR SERVICES

II.K. TASK: NAVIGATION SYSTEMS AND RADAR SERVICES

   REFERENCES: FAA-H-8083-3, FAA-H-8083-15; FAA-S-8081-12, FAA-S-8081-14; AIM.

Objective. To determine that the applicant exhibits instructional knowledge of the elements related to navigation systems and radar services by describing:

1.   One ground-based navigational system (VOR/VORTAC, NDB, DME, and LORAN).

2.   Satellite-based navigation system.

3.   Radar service and procedures.

4.   Global positioning systems (GPS).

A.   General Information

   1.   The objective of this task is to determine your instructional knowledge of the elements related to navigation systems and radar service.

   2.   Additional reading: See *Pilot Handbook* for the following:

      a.   Modules 3.19-3.21 of Chapter 3, Airports, Air Traffic Control, and Airspace, for a five-page discussion on ATC radar, transponder operation, and radar services available to VFR aircraft

      b.   Chapter 10, Navigation Systems, for a 26-page discussion on various navigation systems, such as VOR, ADF, LORAN, and GPS

B.   Task Elements

   1.   **Ground-based navigational systems (VOR/VORTAC, NDB, DME, and LORAN)**

      a.   VOR/VORTAC principles and equipment

         1)   The VHF Omnidirectional Range (VOR) transmits two signals – an omnidirectional signal and a rotating signal.

            a)   The omnidirectional signal pulsates 30 times per second, and the rotating signal rotates at a rate of 30 revolutions per second.

            b)   The omnidirectional signal is timed so that it pulsates at the same instant that the rotating signal passes through magnetic north.

            c)   The VOR receiver in your airplane times the interval between reception of the omnidirectional pulse and the rotating signal.

               i)   The receiver converts this time interval into degrees as your magnetic bearing to or from the station.

         2)   The VOR can be thought of as projecting 360 signals (or radials) out in all directions from the station.

            a)   From a top view, the radials can be visualized as being similar to spokes radiating from the hub of a wheel.

            b)   VOR radials are referenced to magnetic north.

               i)   Thus, a radial is defined as a line of magnetic bearing extending FROM the VOR.

            c)   To aid in orientation, a compass rose referenced to magnetic north is superimposed on aeronautical charts at the station location.

         3)   The VOR ground station is a small, low building topped with a flat white disk upon which are located the antennas and a fiberglass antenna shelter. It has the appearance of an inverted ice cream cone about 30 ft. in height.

            a)   VOR ground stations transmit within a VHF band of 108.00 to 117.95 MHz.

               i)   Since VHF signals are subject to line-of-sight restrictions, the reception range varies proportionally to the altitude of the receiving equipment.

4) The VOR equipment in your airplane includes an antenna, a receiver with a tuning device, and a VOR navigation instrument.

    a) VOR signals from the ground station are received through the antenna and sent to the receiver, which interprets and separates the navigation information. Then this information is displayed on the navigation instrument.

b. NDB principles and equipment

1) There are two parts to the nondirectional beacon (NDB) system.

    a) The ground-based antenna that emits radio signals is called a nondirectional beacon (NDB).

    b) The cockpit navigation radio or automatic direction finder (ADF) receives radio signals in the low-to-medium frequency bands of 190 kHz to 1750 kHz.

2) The equipment in the airplane includes two antennas, a receiver with a tuning device, and a navigational display.

    a) The two antennas are the loop antenna and the sense antenna.

        i) A loop antenna is used as the directional antenna.

            • A loop antenna determines the direction in which the signal is the strongest, but it cannot determine whether the station is in front of or behind the airplane (known as loop ambiguity).

        ii) A nondirectional sense antenna allows the ADF to solve the problem of loop ambiguity and thus determine the direction of the signal.

c. DME principles and equipment

1) VORTAC and VOR/DME ground stations provide distance information to those airplanes equipped with distance-measuring equipment (DME).

    a) DME operates on a UHF band of 962 to 1213 MHz, and, like the VOR signal, it is subject to line-of-sight restrictions.

    b) When using the DME, the same VOR/DME or VORTAC frequency can be tuned in. The DME will then be tuned to the correct UHF band. This is called a **paired frequency**.

2) In the operation of DME, your airplane first transmits a signal (interrogation) to the ground station. The ground station (transponder) then transmits a signal back to your airplane.

    a) The DME in your airplane records the round-trip time of this signal exchange. From this, it can compute

        i) Distance (NM) to the station
        ii) Groundspeed (kt.) relative to the station
        iii) Time (min.) to the station at the current groundspeed

3) The mileage readout is the direct distance from the airplane to the DME ground facility. This is commonly referred to as **slant-range** distance.

    a) The difference between a measured distance on the surface and the DME slant-range distance is known as slant-range error.

d. LORAN principles and equipment

1) Long Range Navigation (LORAN) is a system of 27 transmitters that emit signals in the low frequency range.

    a) LORAN was originally used for navigation and nonprecision approaches.

    b) The advent of GPS has led to the gradual phase-out of the LORAN system.

2) The LORAN system consists of a signal processor, navigation computer, control/display, and antenna. Each must be working for the system to function properly.

    a) Navigation information is given based on the time it took a signal to go from a transmitter to the signal processor. The navigation computer is able to determine position to a destination, cross track error, ground speed, and estimated time of arrival based on the time it took the signals to arrive.

2. Satellite-based navigation system

a. Satellite-based navigation systems include

    1) Global Positioning System (GPS)

        a) GPS is covered in detail Element 4 of this task.

    2) Wide Area Augmentation System (WAAS)

    3) Local Area Augmentation System (LAAS)

b. Wide Area Augmentation System (WAAS)

    1) WAAS is a Satellite Based Augmentation System (SBAS) that improves GPS signals to a level at which they may be used for precision approaches.

    2) Worst-case WAAS accuracy is approximately 25 feet 95% of the time.

    3) Like conventional GPS, WAAS includes a ground segment, a space segment, and a user segment.

    4) WAAS is expected to eventually allow precision approaches nationwide at a much lower cost than traditional precision approaches.

c. Local Area Augmentation System (LAAS)

    1) LAAS is a Ground Based Augmentation System (GBAS) that improves GPS signals to a level at which they may be used for precision approaches.

    2) LAAS functions much the same as WAAS, but puts more reliance on ground stations for signal correction and improvement.

    3) LAAS is considered to be less cost effective than WAAS.

    4) LAAS is expected to be capable of handling Category III instrument approaches.

3. Radar services and procedures

a. ATC radar facilities provide a variety of services to participating VFR aircraft on a workload-permitting basis.

    1) To participate, you must be able to communicate with ATC, be within radar coverage, and be radar identified by the controller.

    2) Among the services provided are

        a) VFR radar traffic advisory service (commonly known as flight following)

        b) Terminal radar programs

        c) Radar assistance to lost aircraft

b. The ATC system is based on a hierarchy of controlling agencies. The order is

    1) Ground control.

    2) Clearance delivery.

    3) Tower control.

    4) Approach control.

    5) Center control (ARTCC).

4. **Global positioning system (GPS)**

  a. GPS is a satellite-based radio navigational, positioning, and time transfer system operated by the Department of Defense (DOD).

   1) GPS provides highly accurate position and velocity information and precise time on a continuous global basis to an unlimited number of users and is not affected by weather.

  b. A GPS receiver measures distance from a satellite using the travel time of a radio signal.

   1) Each satellite transmits a specific code called a **course/acquisition (CA) code**, which contains information on the satellite's position, the GPS system time, and the health and accuracy of the transmitted data.

    a) Knowing the speed at which the CA code travels (approximately 186,000 miles per second) and the exact broadcast time, the distance traveled by the signal can be computed from the arrival time.

     i) GPS satellites have very accurate atomic clocks in order to calculate signal travel time.

   2) The GPS receiver matches each satellite's CA code with an identical copy of the code contained in the receiver's database.

    a) By shifting its copy of the satellite's code in a matching process and by comparing this shift with its internal clock, the receiver can calculate how long it took the signal to travel from the satellite to the receiver.

  c. In addition to knowing the distance to a satellite, a GPS receiver needs to know the satellite's exact position in space, which is known as its ephemeris.

   1) Each satellite transmits information about its exact orbital location.
   2) The GPS receiver uses the information to establish the location of the satellite.

  d. Using the calculated distance and position information supplied by the satellite, the GPS receiver mathematically determines its position by triangulation.

   1) The GPS receiver needs at least four satellites to yield a three-dimensional position (latitude, longitude, and altitude) and time solution.

   2) The GPS receiver computes navigational values, such as distance and bearing to a waypoint, groundspeed, etc., by using the airplane's known position and referencing that to the receiver's database.

  e. The GPS constellation of 24 satellites is designed so that a minimum of five are always observable by a user anywhere on earth.

  f. The GPS receiver verifies the usability of the signals received from the satellites through **receiver autonomous integrity monitoring (RAIM)** to determine if a satellite is providing corrupted information.

END OF TASK

LOGBOOK ENTRIES AND CERTIFICATE ENDORSEMENTS

II.L. TASK: LOGBOOK ENTRIES AND CERTIFICATE ENDORSEMENTS

REFERENCES: 14 CFR Part 61; AC 61-65.

Objective. To determine that the applicant exhibits instructional knowledge of the elements related to logbook entries and certificate endorsements by describing:

1. Required logbook entries for instruction given.

2. Required student pilot certificate endorsements, including appropriate logbook entries.

3. Preparation of a recommendation for a pilot practical test, including appropriate logbook entry for

   a. Initial pilot certification.
   b. Additional pilot certification.
   c. Additional aircraft qualification.

4. Required endorsement of a pilot logbook for the satisfactory completion of the required FAA flight review.

5. Required flight instructor records.

A.   General Information

   1.   The objective of this task is to determine your instructional knowledge of the elements related to logbook entries and certificate endorsements.

B.   Task Elements

   1.   **Required logbook entries for instruction given**

      a.   FAR 61.189, Flight Instructor Records, requires you to sign the logbook of each person to whom you give flight or ground training and specify in that book the amount of instruction time and the date on which it was given. Normally, the following data are recorded:

         1)   Date
         2)   Airplane make and model
         3)   Airplane identification number
         4)   Total time of flight
         5)   Place or points of departure and arrival
         6)   Type of pilot experience or training
         7)   Conditions of flight

      b.   **Recreational pilots** are required to have the following logbook endorsements, as appropriate.

         1)   Endorsement for recreational pilot to act as PIC within 50 NM of the airport where instruction was received: FAR 61.101(b)

            I certify that (First name, MI, Last name) has received the required training of Sec. 61.101(b). I have determined he/she is competent to operate at the (name of airport).

            _____   _____   _____   _____
            Date   *Signature*   *CFI No.*   *Expiration Date*

            a)   If a recreational pilot wants to establish a new 50-NM radius around another airport, additional ground and flight training and a new endorsement will be required.

2) Endorsement for recreational pilot with less than 400 flight hours and with no logged PIC time within the preceding 180 days: FAR 61.101(f)

I certify that (First name, MI, Last name) has received the required 180-day recurrent training of Sec. 61.101(f) in a (make and model aircraft). I have determined him/her proficient to act as PIC of that aircraft.

_____ _____ _____ _____
Date   Signature   CFI No.   Expiration Date

3) Endorsement for a recreational pilot to act as PIC on a flight that exceeds 50 NM from the departure airport: FAR 61.101(c)

I certify that (First name, MI, Last name) has received the required cross-country training of Sec. 61.101(c). I have determined that he/she is proficient in cross-country flying of Subpart E of Part 61.

_____ _____ _____ _____
Date   Signature   CFI No.   Expiration Date

4) Endorsement for a recreational pilot to conduct solo flights for the purpose of obtaining an additional certificate or rating while under the supervision of an authorized flight instructor: FAR 61.101(i)

I certify that (First name, MI, Last name) has received the required training of Sec. 61.87 in a (make and model aircraft). I have determined that he/she is prepared to conduct a solo flight on (date) under the following conditions: (List all conditions which require endorsement, e.g., flight which requires communication with ATC, flight in an aircraft for which the pilot does not hold a category/class rating, etc.).

_____ _____ _____ _____
Date   Signature   CFI No.   Expiration Date

c. The following are logbook endorsements for specialized instruction.

1) Endorsement for a pilot to act as PIC in a complex airplane (an airplane that has a retractable landing gear, flaps, and a controllable pitch propeller): FAR 61.31(e)

I certify that (First name, MI, Last name), (pilot certificate) (certificate number) has received the required training of Sec. 61.31(e) in a (make and model of complex airplane). I have determined that he/she is proficient in the operation and systems of a complex airplane.

_____ _____ _____ _____
Date   Signature   CFI No.   Expiration Date

2) Endorsement for a pilot to act as PIC in a high-performance airplane (an airplane with an engine of more than 200 horsepower): FAR 61.31(f)

I certify that (First name, MI, Last name), (pilot certificate) (certificate number) has received the required training of Sec. 61.31(f) in a (make and model of high-performance airplane). I have determined that he/she is proficient in the operation and systems of a high-performance airplane.

_____ _____ _____ _____
Date   Signature   CFI No.   Expiration Date

3) Endorsement to act as PIC in a pressurized aircraft capable of high-altitude operations: FAR 61.31(g)

I certify that (First name, MI, Last name), (pilot certificate) (certificate number) has received the required training of Sec. 61.31(g) in a (make and model of pressurized aircraft). I have determined that he/she is proficient in the operation and systems of a pressurized aircraft.

———————————————————   ————   —————————
Date   Signature   CFI No.   Expiration Date

4) Endorsement for a pilot to act as PIC in an airplane for which the FAA requires type-specific training: FAR 61.31(i)

I certify that I have given Mr./Ms. _____ type-specific training in a (make and model aircraft) and find him/her proficient in the operation of the aircraft and its systems.

———————————————————   ————   —————————
Date   Signature   CFI No.   Expiration Date

5) Endorsement for a pilot to act as PIC in a tailwheel airplane: FAR 61.31(i)

I certify that (First name, MI, Last name), (pilot certificate) (certificate number) has received the required training of Sec. 61.31(i) in a (make and model of tailwheel airplane). I have determined that he/she is proficient in the operation of a tailwheel airplane.

———————————————————   ————   —————————
Date   Signature   CFI No.   Expiration Date

6) Endorsement for a pilot who does not hold an appropriate category/class rating to act as PIC of an aircraft in solo operations: FAR 61.31(d)(3)

I certify that (First name, MI, Last name) has received the training as required by Sec. 61.31(d)(3) to serve as a PIC in a (category and class of aircraft). I have determined that he/she is prepared to serve as PIC in that (make and model of aircraft).

———————————————————   ————   —————————
Date   Signature   CFI No.   Expiration Date

   a) You may want to stipulate additional conditions in the above endorsement.

7) Endorsement for completing a phase of the WINGS Program: FAR 61.56(e)

I certify that (First name, MI, Last name), (pilot certificate) (certificate number) has satisfactorily completed Phase No. ____ of a WINGS program on (date).

———————————————————   ————   —————————
Date   Signature   CFI No.   Expiration Date

2. **Required student pilot certificate endorsements, including appropriate logbook entries**

    a. Student pilot certificates provide for CFI endorsements on the back for solo flight and solo cross-country flight.

        1) The student pilot certificate must be endorsed prior to the student's first solo flight in each make and model of airplane.

        2) The student pilot certificate must be endorsed prior to the student's first solo cross-country flight.

            a) This endorsement is for aircraft category (e.g., airplane), not make and model of aircraft (e.g., Cessna 152).

        3) Shown below is an example student pilot certificate endorsed.

Front

UNITED STATES OF AMERICA
Department of Transportation **EE-**5342031
Federal Aviation Administration

**MEDICAL CERTIFICATE   3rd   CLASS
AND STUDENT PILOT CERTIFICATE**

This certifies that *(Full name and address):*

Richmond, Kane Everett
7771 Coral Way
N. Ft. Myers, Fl. 33903

Date of Birth	Ht.	Wt.	Hair	Eyes	Sex
1/30/67	5'9"	140	Brn.	Brn.	M

has met the medical standards prescribed in Part 67, Federal Aviation Regulations, for this class of Medical Certificate.

Limitations

None

Date of Examination	Examiner's Serial No.
1/26/07	11967-1

Examiner
Signature   *E. W. Williams, II, D.O.*
Typed Name   E.W. Williams II, D.O.

AIRMAN'S SIGNATURE
Kane Everett Richmond

FAA Form 8420-2 (7-92) Supersedes Previous Edition

Back

PASSENGER-CARRYING PROHIBITED

CONDITIONS OF ISSUE: This certificate shall be in the personal possession of the airman at all times while exercising the privileges of his or her airman certificate. The issuance of a medical certificate by an Aviation Medical Examiner may be reversed by the FAA within 60 days. Section 61.19 of the Federal Aviation Regulations (FAR) sets forth the duration of a student pilot certificate. Unless otherwise limited, the duration of a medical certificate is set forth in §61.23 of the FAR. The holder of this certificate is governed by the provisions of FAR §§ 61.53, 63.19, and 65.49(d) relating to physical deficiency (14 CFR Parts 61, 63, and 65).

CERTIFICATED INSTRUCTOR'S ENDORSEMENT FOR STUDENT PILOTS

I certify that the holder of this certificate has met the requirements of the regulations and is competent for the following:

	DATE	MAKE AND MODEL OF AIRCRAFT	INSTRUCTOR'S SIGNATURE	No.	Exp. Date
				INSTRUCTOR'S CERT.	
A. To Solo The Following Aircraft	6-3-00	C-152	*Tracey Lin Law*	264750091	7/09
		AIRCRAFT CATEGORY			
B. To Make Solo Cross-Country Flights		Airplane			
		Glider			
		Rotorcraft			

b. Student pilot logbooks must be endorsed for solo flight. The following are examples of the required endorsements.

1) Endorsement for presolo aeronautical knowledge: FAR 61.87(b)

I certify that (First name, MI, Last name) has satisfactorily completed the presolo knowledge exam of Sec. 61.87(b) for the (make and model aircraft).

Date	*Signature*	*CFI No.*	*Expiration Date*

2) Endorsement for presolo flight training: FAR 61.87(c)

I certify that (First name, MI, Last name) has received the required presolo training in a (make and model of aircraft). I have determined that he/she has demonstrated the proficiency of Sec. 61.87(d) and is proficient to make solo flights in (make and model of aircraft).

Date	*Signature*	*CFI No.*	*Expiration Date*

3) Endorsement for solo flight (each additional 90-day period): FAR 61.87(n)

I certify that (First name, MI, Last name) has received the required training to qualify for solo flying. I have determined that he/she meets the applicable requirements of Sec. 61.87(n) and is proficient to make solo flights in (make and model).

Date	*Signature*	*CFI No.*	*Expiration Date*

4) Endorsement for solo landings and takeoffs at another airport within 25 NM: FAR 61.93(b)(1)

I certify that (First name, MI, Last name) has received the required training of Sec. 61.93(b)(1). I have determined that he/she is proficient to practice solo takeoffs and landings at (airport name). The takeoffs and landings at (airport name) are subject to the following conditions: (List any applicable conditions or limitations).

Date	*Signature*	*CFI No.*	*Expiration Date*

5) Endorsement for presolo flight at night: FAR 61.87(c) and (m)

I certify that (First name, MI, Last name) has received the required presolo training in a (make and model aircraft). I have determined that he/she has demonstrated the proficiency of Sec. 61.87(m) and is proficient to make solo flights at night in a (make and model of aircraft).

Date	*Signature*	*CFI No.*	*Expiration Date*

    c.   Student pilot logbooks must be endorsed for solo cross-country flight. The following are examples of these required endorsements.

        1)   Endorsement for solo cross-country flight training: FAR 61.93(c)(1)

                I certify that (First name, MI, Last name) has received the required solo cross-country training. I find he/she has met the applicable requirements of Sec. 61.93 and is proficient to make solo cross-country flights in a (make and model aircraft).

                Date   Signature   CFI No.   Expiration Date

        2)   Endorsement for each solo cross-country flight: FAR 61.93(c)(2)

                I have reviewed the cross-country planning of (First name, MI, Last name). I find the planning and preparation to be correct to make the solo flight from (location) to (destination) via (route of flight) with landings at (name the airports) in a (make and model aircraft) on (date). (List any applicable conditions or limitations.)

                Date   Signature   CFI No.   Expiration Date

        3)   Endorsement for repeated solo cross-country flights not more than 50 NM from the point of departure: FAR 61.93(b)(2)

                I certify that (First name, MI, Last name) has received the required training in both directions between and at both (airport names). I have determined that he/she is proficient of Sec. 61.93(b)(2) to conduct repeated solo cross-country flights over that route, subject to the following conditions: (List any applicable conditions or limitations).

                Date   Signature   CFI No.   Expiration Date

    d.   Additionally, student pilot logbooks are required to be endorsed before solo flight in Class B airspace. The following are examples of these required endorsements.

        1)   Endorsement for solo flight in Class B airspace: FAR 61.95(a)

                I certify that (First name, MI, Last name) has received the required training of Sec. 61.95(a). I have determined he/she is proficient to conduct solo flights in (name of Class B) airspace. (List any applicable condition or limitations.)

                Date   Signature   CFI No.   Expiration Date

        2)   Endorsement for solo flight to, from, or at an airport located within Class B airspace: FAR 61.95(a) and 91.131(b)(1)

                I certify that (First name, MI, Last name) has received the required training of Sec. 61.95(a)(1). I have determined that he/she is proficient to conduct solo flight operations at (name of airport). (List any applicable conditions or limitations.)

                Date   Signature   CFI No.   Expiration Date

3. **Preparation of a recommendation for a pilot practical test, including appropriate logbook entries**

    a.   You and your student must complete and sign FAA Form 8710-1, Application Certificate and/or Rating Application, which is given to the FAA inspector/examiner at the practical test. See the three-page presentation of the FAA Form 8710-1 (and instructions) in Part 1, Study Unit 4, Your FAA Practical (Flight) Test, beginning on page 46.

        1)   A practical test, whether or not satisfactorily completed, uses up the instructor's endorsement for that test.

            a)   A new FAA Form 8710-1 is required to be completed for each retest conducted for a certificate or rating.

    b.   While not a part of this element, we have also included the recommended wording to be used as an endorsement attesting to a person's aeronautical knowledge in order to take the pilot knowledge (written) test.

    c.   The endorsement for flight proficiency must be in the applicant's logbook and presented to the FAA inspector/examiner at the practical test.

    d.   The following are examples of endorsements for aeronautical knowledge (i.e., for the pilot knowledge test) and flight proficiency (i.e., for the practical test).

        1)   **Sport pilot**

            a)   Endorsement for aeronautical knowledge test: FAR 61.307(a)

I certify that (First name, MI, Last name) has received the required training of Sec. 61.309. I have trained him/her or reviewed his/her home-study materials and have determined that he/she is prepared for the (name the knowledge test).

Date	Signature	CFI No.	Expiration Date

            b)   Endorsement for the practical test: FAR 61.307(b)

I certify that (First name, MI, Last name) has received the required training of Sec. 61.309 and 61.311. I have determined that he/she is prepared for the (name of practical test). He/She has demonstrated satisfactory knowledge of the subject areas found deficient on his/her knowledge test.

Date	Signature	CFI No.	Expiration Date

        2)   **Recreational pilot**

            a)   Endorsement for aeronautical knowledge test: FAR 61.35(a)(1) and 61.96(b)(3)

I certify that (First name, MI, Last name) has received the required training of Sec. 61.97(b). I have determined that he/she is prepared for the (name the knowledge test).

Date	Signature	CFI No.	Expiration Date

      b)   Endorsement for flight proficiency/practical test: FAR 61.96(b)(5), 61.98(a) and (b), and 61.99

            I certify that (<u>First name, MI, Last name</u>) has received the required training of Sec. 61.98(b) and 61.99. I have determined that he/she is prepared for the (<u>name the practical test</u>). He/She has demonstrated satisfactory knowledge of the subject areas found deficient on his/her knowledge test.

                _____   _____   _____   _____
                Date   *Signature*   *CFI No.*   *Expiration Date*

3) **Private pilot**

      a)   Endorsement for aeronautical knowledge test: FAR 61.35(a)(1), 61.103(d), and 61.105

            I certify that (<u>First name, MI, Last name</u>) has received the required training of Sec. 61.105. I have determined he/she is prepared for the (<u>name the knowledge test</u>).

                _____   _____   _____   _____
                Date   *Signature*   *CFI No.*   *Expiration Date*

      b)   Endorsement for flight proficiency/practical test: FAR 61.103(f), 61.107(b), and 61.109

            I certify that (<u>First name, MI, Last name</u>) has received the required training of Sec. 61.107 and 61.109. I have determined he/she is prepared for the (<u>name the practical test</u>). He/She has demonstrated satisfactory knowledge of the subject areas found deficient on his/her knowledge test.

                _____   _____   _____   _____
                Date   *Signature*   *CFI No.*   *Expiration Date*

4) **Instrument rating**

    NOTE: You must be a CFII to provide the following endorsements.

      a)   Endorsement for aeronautical knowledge test: FAR 61.35(a)(1) and 61.65(a) and (b)

            I certify that (<u>First name, MI, Last name</u>) has received the required training of Sec. 61.65(b). I have determined that he/she is prepared for the (<u>name the knowledge test</u>).

                _____   _____   _____   _____
                Date   *Signature*   *CFI No.*   *Expiration Date*

      b)   Endorsement for flight proficiency/practical test: FAR 61.65(a)(6)

            I certify that (<u>First name, MI, Last name</u>) has received the required training of Sec. 61.65(c) and (d). I have determined he/she is prepared for the Instrument-(<u>Airplane, Helicopter, or Powered-lift</u>) practical test.

                _____   _____   _____   _____
                Date   *Signature*   *CFI No.*   *Expiration Date*

5) **Commercial pilot**

    a) Endorsement for aeronautical knowledge test: FAR 61.35(a)(1) and 61.123(c)

        I certify that (<u>First name, MI, Last name</u>) has received the required training of Sec. 61.125. I have determined that he/she is prepared for the (<u>name the knowledge test</u>).

Date	Signature	CFI No.	Expiration Date

    b) Endorsement for flight proficiency/practical test: FAR 61.123(e) and 61.127

        I certify that (<u>First name, MI, Last name</u>) has received the required training of Sec. 61.127 and 61.129. I have determined he/she is prepared for the (<u>name the practical test</u>).

Date	Signature	CFI No.	Expiration Date

6) **Flight instructor other than flight instructors with a sport pilot rating -- airplane**

    a) Endorsement for fundamentals of instructing: FAR 61.183(d) and 61.185(a)(1)

        I certify that (<u>First name, MI, Last name</u>) has received the required fundamentals of instruction training of Sec. 61.185(a)(1).

Date	Signature	CFI No.	Expiration Date

    b) Endorsement for flight proficiency/practical test: FAR 61.183(g) and 61.187(a) and (b)

        I certify that (<u>First name, MI, Last name</u>) has received the required training of Sec. 61.187(b). I have determined he/she is prepared for the CFI-(<u>aircraft category and class</u>) practical test.

Date	Signature	CFI No.	Expiration Date

    c) Endorsement for spin training: FAR 61.183(i)(1)

        I certify that (<u>First name, MI, Last name</u>) has received the required training of Sec. 61.183(i). I have determined that he/she is competent and proficient in instructional skills for training stall awareness, spin entry, spins, and spin recovery procedures.

Date	Signature	CFI No.	Expiration Date

    NOTE: An instructor endorsement is not required for the FOI or any instructor knowledge test.

d) Endorsement for CFII rating/practical test: FAR 61.183(g) and 61.187(a) and (b)(7)

I certify that (First name, MI, Last name) has received the required CFII training of Sec. 61.187(b)(7). I have determined he/she is prepared for the CFII-(airplane, helicopter, or powered-lift) practical test.

_____   _____   _____   _____
Date   Signature   CFI No.   Expiration Date

7) **Flight instructors with a sport pilot rating**

a) Endorsement for the Fundamentals of Instructing Knowledge Test: FAR 61.405(a)(1)

I certify that (First name, MI, Last name) has received the required training of Sec. 61.405. I trained this applicant or reviewed his/her study materials for the aeronautical areas listed in Sec. 61.407. I have determined that he/she is ready for the Fundamentals of Instructing Test.

_____   _____   _____   _____
Date   Signature   CFI No.   Expiration Date

b) Endorsement for the Flight Instructor Knowledge Test: FAR 61.405(a)(2)

I certify that (First name, MI, Last name) has received the required training of Sec. 61.405. I trained this applicant or reviewed his/her study materials for the aeronautical areas for a sport pilot certificate applicable to the aircraft category and class for which flight instructor privileges are sought. I have determined that he/she is ready for the (name of knowledge test).

_____   _____   _____   _____
Date   Signature   CFI No.   Expiration Date

c) Endorsement for the Practical Test: FAR 61.405(b)(1)(i)

I certify that (First name, MI, Last name) has received the required training of Sec. 61.409. I have determined he/she is prepared for the Sport Pilot Instructor-(aircraft category and class, make and model) practical test.

_____   _____   _____   _____
Date   Signature   CFI No.   Expiration Date

d) Endorsement for spin training: FAR 61.405(b)(1)(ii)

I certify that (First name, MI, Last name) has received the required training of Sec. 61.405(b)(1)(ii). I have determined that he/she is competent and proficient in instructional skills for training stall awareness, spin entry, spins, and spin recovery procedures.

_____   _____   _____   _____
Date   Signature   CFI No.   Expiration Date

8) **Additional aircraft or type ratings**

a) Endorsement for an airman seeking an additional aircraft rating (other than ATP): FAR 61.63(b) or (c)

I certify that (First name, MI, Last name), (pilot certificate) (certificate number) has received the required training for an additional (name the aircraft category/class rating). I have determined that he/she is prepared for the (name the practical test) for the addition of a (name the aircraft category/class rating).

Date	Signature	CFI No.	Expiration Date

b) Endorsement for a type rating only; the pilot already holds the appropriate category or class rating (other than ATP): FAR 61.63(d)(2) and (3)

I certify that (First name, MI, Last name) has received the required training of Sec. 61.63(d)(2) and (3) for an addition of a (name the type rating).

Date	Signature	CFI No.	Expiration Date

c) Endorsement for a type rating concurrently with an additional category or class rating (other than ATP): FAR 61.63(d)(2) and (3)

I certify that (First name, MI, Last name) has received the required training of Sec. 61.63(d)(2) and (3) for an addition of a (name the category/class/type rating). I have determined that he/she is prepared for the (name the practical test) for the addition of a (name the aircraft category/class/type rating).

Date	Signature	CFI No.	Expiration Date

d) Endorsement for a type rating only; the pilot already holds the appropriate category or class rating (at the ATP level): FAR 61.157(b)(2)

I certify that (First name, MI, Last name) has received the required training of Sec. 61.157(b)(1) for an addition of a (name the type rating).

Date	Signature	CFI No.	Expiration Date

e) Endorsement for a type rating concurrently with an additional category or class rating (at the ATP level): FAR 61.157(b)(2)

I certify that (First name, MI, Last name) has received the required training of Sec. 61.157(b)(1) for an addition of a (name the category/class/type rating).

Date	Signature	CFI No.	Expiration Date

9) Endorsement to certify completion of prerequisites for a practical test: FAR
    61.39(a)(6)

    *I have given (<u>First name, MI, Last name</u>) flight instruction in preparation for a
    (<u>type of practical test</u>) practical test within the preceding 60 days and find
    him/her competent to pass the test and (if appropriate) to have satisfactory
    knowledge of the subject areas in which the applicant was shown to be
    deficient by his/her airman written test.*

    <u>Date</u>   <u>Signature</u>   <u>CFI No.</u>   <u>Expiration Date</u>

10) Endorsement for retesting after a failure of a knowledge or practical test: FAR
    61.49

    *I certify that (<u>First name, MI, Last name</u>) has received the additional (flight
    and/or ground) training as required by Sec. 61.49. I have determined that
    he/she is prepared for the (<u>name the knowledge/practical test</u>).*

    <u>Date</u>   <u>Signature</u>   <u>CFI No.</u>   <u>Expiration Date</u>

    NOTE: You may also complete the endorsement in the space provided at the
    bottom of the applicant's computer test report in the case of a failure on a
    knowledge test. You must sign the block provided for the instructor's
    endorsement on the reverse side of FAA Form 8710-1 for each retake of a
    practical test. An applicant may retake either a knowledge or practical test if he
    or she has received additional instruction and an instructor's endorsement.

11) Review of a home study curriculum: FAR 61.35(a)(1)

    *I certify that I have reviewed the home study curriculum of (<u>First name, MI, Last
    name</u>). I have determined he/she is prepared for the (<u>name the knowledge
    test</u>).*

    <u>Date</u>   <u>Signature</u>   <u>CFI No.</u>   <u>Expiration Date</u>

12) Endorsement for a ground instructor who does not meet the recent experience
    requirements: FAR 61.217(b)

    *I certify that (<u>First name, MI, Last name</u>) has demonstrated satisfactory
    proficiency on the appropriate ground instructor knowledge and training
    subjects of Sec. 61.213(a)(3) and (4).*

    <u>Date</u>   <u>Signature</u>   <u>CFI No.</u>   <u>Expiration Date</u>

4. **Required endorsement of a pilot logbook for the satisfactory completion of the required FAA flight review**

   a.   If you conduct a flight review, you must endorse the pilot's logbook after (s)he successfully completes the flight review. An example is shown below.

       1)   Endorsement for completion of flight review: FAR 61.56(a) and (c)

           I certify that (First name, MI, Last name), (pilot certificate) (certificate number) has satisfactorily completed a flight review of Sec. 61.56(a) on (date).

           ——————— ——————————— ———— ——————
           Date   *Signature*   *CFI No.*   *Expiration Date*

   b.   Once you add your instrument rating to your flight instructor certificate (i.e., instrument instructor or CFII), you will be able to perform instrument proficiency checks for instrument-rated pilots.

       1)   Endorsement for completion of an instrument proficiency check: FAR 61.57(d)

           I certify that (First name, MI, Last name), (pilot certificate) (certificate number) has satisfactorily completed the instrument proficiency check of Sec. 61.57(d) in a (list make and model of aircraft) on (date).

           ——————— ——————————— ———— ——————
           Date   *Signature*   *CFI No.*   *Expiration Date*

   c.   If the pilot does not satisfactorily complete the flight review or the instrument proficiency check, you will not make a logbook endorsement.

       1)   You should make a logbook entry for the instruction given.

5. **Required flight instructor records**

   a.   You are required to maintain a record in your flight instructor logbook or in a separate document containing the following:

       1)   The name of each person whose logbook or student pilot certificate you endorsed for solo flight privileges, including the type and date of each endorsement.

       2)   The name of each person you have signed off to take a knowledge or practical test, including the kind of test, the date, and the result of the test.

   b.   These records must be retained for a period of 3 years.

END OF TASK -- END OF STUDY UNIT

STUDY UNIT III
PREFLIGHT PREPARATION

This study unit explains the five tasks (A-E) of Preflight Preparation. These tasks are knowledge only. Your FAA inspector/examiner is required to test you on at least one of the five tasks in this area of operation; therefore, you must be prepared for all five.

CERTIFICATES AND DOCUMENTS

III.A. TASK: CERTIFICATES AND DOCUMENTS

REFERENCES: 14 CFR Parts 43, 61, 91; FAA-H-8083-3, AC 61-23/FAA-H-8083-25; FAA-S-8081-12, FAA-S-8081-14, POH/AFM.

Objective. To determine that the applicant exhibits instructional knowledge of the elements related to certificates and documents by describing:

1. The training requirements for the issuance of a recreational, private, and commercial pilot certificate.

2. The privileges and limitations of pilot certificates and ratings at recreational, private, and commercial levels.

3. Class and duration of medical certificate.

4. Recent pilot flight experience requirements.

5. Required entries in pilot logbook or flight record.

A. General Information

1. The objective of the task is to determine your instructional knowledge of the elements related to certificates and documents.

B.   Task Elements

1.   **Training requirements for the issuance of a recreational, private, and commercial pilot certificate**

a.   Refer to *Pilot Handbook*, Chapter 4, Federal Aviation Regulations. Note that this is a very broad element that covers many FAR sections (most of which are outlined in about 15 pages in *Pilot Handbook*).

Recreational Pilots

61.96	Applicability and Eligibility Requirements: General
61.97	Aeronautical Knowledge
61.98	Flight Proficiency
61.99	Aeronautical Experience
61.100	Pilots Based on Small Islands
61.101	Recreational Pilot Privileges and Limitations

Private Pilots

61.103	Eligibility Requirements: General
61.105	Aeronautical Knowledge
61.107	Flight Proficiency
61.109	Aeronautical Experience
61.110	Night Flying Exceptions
61.111	Cross-Country Flights: Pilots Based on Small Islands
61.113	Private Pilot Privileges and Limitations: Pilot-in-Command
61.117	Private Pilot Privileges and Limitations: Second-in-Command of Aircraft Requiring More than One Required Pilot

Commercial Pilots

61.123	Eligibility Requirements: General
61.125	Aeronautical Knowledge
61.127	Flight Proficiency
61.129	Aeronautical Experience
61.131	Exception to Night Flying Requirements
61.133	Commercial Pilot Privileges and Limitations: General

2.   **Privileges and limitations of pilot certificates and ratings at recreational, private, and commercial levels.**

a.   Recreational pilot privileges and limitations may be found in FAR 61.101.
b.   Private pilot privileges and limitations may be found in FAR 61.113.
c.   Commercial pilot privileges and limitations may be found in FAR 61.133.

3. **Class and duration of medical certificates**

   a. A first-class medical certificate expires at the end of the last day of

      1) The sixth month after the date of examination for operations requiring an airline transport pilot certificate.

      2) The 12th month after the date of examination for operations requiring only a commercial pilot certificate.

      3) The period specified in item c. below, for operations requiring only a private, recreational, or student pilot certificate.

   b. A second-class medical certificate expires at the end of the last day of

      1) The 12th month after the date of examination for operations requiring a commercial pilot certificate.

      2) The period specified in item c. below, for operations requiring only a private, recreational, or student pilot certificate.

   c. A third-class medical certificate for operations requiring a private, recreational, or student pilot certificate expires at the end of the last day of

      1) The 36th month after the day of examination if the person has not reached his/her 40th birthday on or before the date of the examination.

      2) The 24th month after the date of examination if the person has reached his or her 40th birthday on or before the date of examination.

   d. Flight instructors are required to hold at least a third-class medical certificate only if they are a required flight crewmember.

      1) Even though flight instructors are paid pilots, the FAA regards the fee paid by the student as payment for instruction, not for pilot services.

         a) Therefore, a second-class medical certificate is not required.

4. **Recent pilot flight experience requirements**

   a. A pilot must complete a flight review and other operating experience requirements to be able to act as PIC.

      1) For a complete discussion of both, see 61.56 Flight Review and 61.57 Recent Flight Experience: Pilot in Command

5. **Required entries in pilot logbook or flight record**

   a. The only information that is required to be entered into a pilot's logbook is that which is required for a certificate or rating, recency requirements, and the flight review.

   b. For each lesson or flight logged, the following must be entered into the pilot logbook:

      1) General

         a) This includes the date, total flight time or lesson time, and the departure and arrival location.

      2) Type of pilot experience or training

         a) The various types are solo, pilot in command, second in command, flight and ground training received from an instructor, or training received in a flight simulator or flight training device.

      3) Conditions of the flight

         a) This may be day or night, actual instrument, or simulated instrument in flight or in a flight simulator.

END OF TASK

WEATHER INFORMATION

III.B. TASK: WEATHER INFORMATION

REFERENCES: AC 00-6, AC 00-45, AC 61-23/FAA-H-8083-25; FAA-S-8081-12, FAA-S-8081-14.

Objective. To determine that the applicant exhibits instructional knowledge of the elements related to weather information by:

1. Importance of a thorough preflight weather briefing.
2. Various means and sources of obtaining weather information.
3. Use of real time weather reports, forecasts, and charts for developing scenario based training.
4. In-flight weather advisories.
5. Recognition of aviation weather hazards to include wind shear.
6. Factors to be considered in making a "go/no-go" decision.

A. General Information

1. The objective of this task is to determine your instructional knowledge of the elements related to weather information.

2. Gleim's *Aviation Weather and Weather Services* is a 450-page book in outline/illustration format that combines the FAA's *Aviation Weather* (AC 00-6A) and *Aviation Weather Services* (AC 00-45E) and numerous FAA publications into one easy-to-understand book. It is a single reference that covers all aspects of aviation weather, weather reports, and weather forecasts.

   a. We recommend this book as part of every flight instructor's reference library.

3. On July 1, 1996, the U.S. converted to the International Civil Aviation Organization (ICAO) and the World Meteorological Organization (WMO) standard for aviation weather reporting and forecasting.

   a. **METAR elements.** A METAR contains the following sequence of elements in the following order:

      1) Type of report
      2) ICAO station identifier
      3) Date and time of report
      4) Modifier (as required)
      5) Wind
      6) Visibility
      7) Runway visual range (RVR)
      8) Weather
      9) Sky condition
      10) Temperature/dew point
      11) Altimeter
      12) Remarks (RMK)

      NOTE: The elements in the body of a METAR report are separated by a space, except temperature and dew point, which are separated with a solidus, /. When an element does not occur or cannot be observed, that element is omitted from that particular report.

b. **Example of a METAR report**

METAR KGNV 201953Z 24015KT 3/4SM R28/2400FT +TSRA
BKN008 OVC015CB 26/25 A2985 RMK TSB32RAB32

To aid in the discussion, we have divided the report into the 12 elements:

METAR	KGNV	201953Z		24015KT	3/4SM	R28/2400FT	+TSRA
1)	2)	3)	4)	5)	6)	7)	8)

BKN008 OVC015CB	26/25	A2985	RMK TSB32RAB32
9)	10)	11)	12)

1) Aviation routine weather report
2) Gainesville, FL
3) Observation taken on the 20th day at 1953 UTC (or Zulu)
4) Modifier omitted; i.e., not required for this report
5) Wind 240° true at 15 kt.
6) Visibility 3/4 SM
7) Runway 28, runway visual range 2,400 ft.
8) Thunderstorm with heavy rain
9) Ceiling 800 ft. broken, 1,500 ft. overcast, cumulonimbus clouds
10) Temperature 26°C, dew point 25°C
11) Altimeter 29.85
12) Remarks, thunderstorm began at 32 min. past the hour, rain began at 32. min. past the hour.

c. **TAF elements.** A TAF contains the following sequence of element in the following order [items 1)-9)]. Forecast change indicators [items 10)-12)] and probability forecast [item 13)] are used as appropriate.

Communications Header	Forecast of Meteorological Conditions	Time Elements
1) Type of report	5) Wind	10) Temporary (TEMPO)
2) ICAO station identifier	6) Visibility	11) From (FM)
3) Date and time of origin	7) Weather	12) Becoming (BECMG)
4) Valid period date and time	8) Sky condition	13) Probability (PROB)
	9) Wind Shear (optional)	

d. **Example of a TAF**

TAF
KOKC 051130Z 051212 14008KT 5SM BR BKN030 WS018/32030KT
TEMPO 1316 1 1/2SM BR FM1600 16010KT P6SM SKC
BECMG 2224 20013G20KT 4SM SHRA OVC020
PROB40 0006 2SM TSRA OVC008CB=

To aid in the discussion, we have divided the TAF above into the following 13 elements:

TAF	KOKC	051130Z	051212	14008KT	5SM	BR	BKN030	WS018/32030KT
1)	2)	3)	4)	5)	6)	7)	8)	9)

TEMPO 1316 1 1/2SM BR   FM1600 16010KT P6SM SKC
         10)   11)

BECMG 2224 20013G20KT 4SM SHRA OVC020   PROB40 0006 2SM TSRA OVC008CB=
              12)   13)

1) Routine terminal aerodrome forecast
2) Oklahoma City, OK
3) Forecast prepared on the 5th day at 1130 UTC (or Z)
4) Forecast valid from the 5th day at 1200 UTC until 1200 UTC on the 6th
5) Wind 140° true at 8 kt.
6) Visibility 5 statute miles
7) Visibility obscured by mist
8) Ceiling 3,000 ft. broken
9) Low-level wind shear at 1,800 ft., wind 320° true at 30 kt.
10) Temporary (spoken as occasional) visibility 1 1/2 SM in mist between 1300 UTC and 1600 UTC
11) From (or after) 1600 UTC, wind 160° true at 10 kt., visibility more than 6 SM, sky clear
12) Becoming (gradual change) wind 200° true at 13 kt. gusts to 20 kt., visibility 4 SM in moderate rain showers, ceiling 2,000 ft. overcast between 2200 UTC and 2400 UTC
13) Probability (40% chance) between 0000 UTC and 0600 UTC of visibility 2 SM, thunderstorm, moderate rain, ceiling 800 ft. overcast, cumulonimbus clouds. The equal sign (=) indicates the end of the TAF.

B. Task Elements

1. **Importance of a thorough preflight weather briefing**

    a. Obtaining a preflight weather briefing is the first step in determining if the flight can be conducted safely and where and when problems may occur during the proposed flight.

    b. When conducting a flight away from the vicinity of a departure airport, you are required to become familiar with the weather reports and weather forecasts for that flight (FAR 91.103).

    c. If you know what to expect while flying, then any unforecast conditions will alert you to possible weather hazards. Be safe -- obtain a weather briefing.

2. **Various means and sources of obtaining weather information**

   a.   A good weather briefing starts with developing an awareness of the overall weather pattern before attempting to get a detailed weather briefing. The following sources can provide information on the overall weather pattern:

      1)   Telephone information briefing service (TIBS)
      2)   Pilots' automatic weather answering service (PATWAS)
      3)   Transcribed weather broadcast (TWEB)
      4)   Good television weather report, e.g., The Weather Channel
      5)   Numerous weather sites are available on the Internet.

   b.   The next step is to obtain a more detailed weather briefing that is specific for your flight.

      1)   The following sources can provide you with this briefing:

         a)   Flight Service Station (FSS)
         b)   Direct User Access Terminal System (DUATS)
         c)   National Weather Service (NWS)
         d)   Supplemental weather service location (SWSL)

      2)   Both the FSS and DUATS provide the same service and provide you with NOTAM information and filing of flight plans. NWS and SWSLs provide you with weather information only.

   c.   In-flight weather may be obtained from the following sources:

      1)   En Route Flight Advisory Service (EFAS, also known as Flight Watch)
      2)   Hazardous In-Flight Weather Advisory Service (HIWAS)
      3)   TWEB

3. **Use of real time weather reports, forecasts, and charts for developing scenario based training**

   a.   You will need to show how you would instruct your student to obtain, read, and analyze weather reports, forecasts, and charts. To do this, you must have a good understanding of how to use the various reports, forecasts, and charts.

   b.   Weather reports and forecasts

      1)   Aviation routine weather report (METAR)
      2)   Pilot weather report (PIREP)
      3)   Radar weather report (SD)
      4)   Satellite weather pictures
      5)   Aviation area forecast (FA)
      6)   Terminal aerodrome forecast (TAF)
      7)   In-flight advisories (AWW, WST, WS, CWA, WA)
      8)   Winds and temperatures aloft forecast (FD)
      9)   TWEB text products
      10) Center weather service unit products
      11) Hurricane advisory (WH)
      12) Convective outlook (AC)
      13) Severe weather watch bulletin (WW)
      14) Alert messages (AWW)

   c.   Weather charts

      1)   Surface Analysis Chart
      2)   Weather Depiction Chart
      3)   Radar Summary Chart
      4)   Significant Weather Prognostic Charts
      5)   Winds and Temperatures Aloft Chart
      6)   Composite Moisture Stability Chart
      7)   Convective Outlook Chart
      8)   Constant Pressure Analysis Charts
      9)   Volcanic Ash Forecast Transport and Dispersion Chart

4.   **In-flight weather advisories**

   a.   **Pilot weather reports (PIREPs)**

      1)   No more timely or helpful weather observation fills the gaps between reporting stations better than observations and reports made by fellow pilots. Aircraft in flight are the **only** source of directly observed cloud tops, icing, and turbulence.

      2)   A PIREP is usually transmitted as one of a group of PIREPs collected by a state or as a remark appended to the surface aviation weather report.

   b.   **Convective SIGMET (WST)**

      1)   Convective SIGMETs are issued in the contiguous 48 states (i.e., none for Alaska and Hawaii) for any of the following:

         a)   Severe thunderstorm due to

            i)   Surface winds greater than or equal to 50 kt.
            ii)   Hail at the surface greater than or equal to 3/4 in. in diameter
            iii)   Tornadoes

         b)   Embedded thunderstorms

         c)   A line of thunderstorms

         d)   Thunderstorms greater than or equal to intensity level 4 affecting 40% or more of an area of at least 3,000 square mi.

      2)   Any convective SIGMET implies severe or greater turbulence, severe icing, and low-level wind shear (gust fronts, downbursts, microbursts, etc.) and will not be specified in the advisory. A convective SIGMET may be issued for any convective situation that the forecaster feels is hazardous to all categories of aircraft.

   c.   **SIGMET (WS)**

      1)   A SIGMET advises of non-convective weather that is potentially hazardous to all aircraft. In the conterminous U.S., items covered are

         a)   Severe icing not associated with thunderstorms

         b)   Severe or extreme turbulence not associated with thunderstorms

         c)   Dust storms, sandstorms, or volcanic ash lowering surface or in-flight visibilities to below 3 SM

         d)   Volcanic eruption

         e)   Tropical storms or hurricanes

      2)   In Alaska and Hawaii, there are no convective SIGMETs. These criteria are added:

         a)   Tornadoes

         b)   Lines of thunderstorms

         c)   Embedded thunderstorms

         d)   Hail equal to or greater than 3/4 in. in diameter

    d. **AIRMET (WA)**

       1) AIRMETs are advisories issued only to amend the area forecast (FA) concerning weather phenomena that are of operational interest to all aircraft and potentially hazardous to aircraft having limited capability because of lack of equipment, instrumentation, or pilot qualifications.

          a) AIRMETs concern weather of less severity than that covered by convective SIGMETs or SIGMETs.

       2) AIRMETs are valid for 6 hr. and will contain details of conditions when one or more of the following occur or are forecast to occur:

          a) Moderate icing

          b) Moderate turbulence

          c) Sustained surface winds of 30 kt. or more

          d) Ceiling less than 1,000 ft. and/or visibility less than 3 SM affecting over 50% of the area at one time

          e) Extensive mountain obscuration

    e. **Notices to Airmen (NOTAMs)**

       1) NOTAMs are listed here because pilots will obtain NOTAM information from a FSS specialist or DUATS.

       2) See the discussion on the types of NOTAMs and how they are published beginning on page 147 of this book.

5. **Recognition of aviation weather hazards to include wind shear**

    a. Aviation weather hazards include

       1) Thunderstorms

       2) Turbulence

       3) Icing

       4) IFR conditions (hazard to VFR pilots)

    b. These hazards are included in the following aviation weather products:

       1) Area forecast

       2) WST, WS, WA

       3) Hurricane advisory

       4) Significant weather prognostic charts

6.   **Factors to be considered in making a "go/no-go" decision**

a.   In a well-equipped airplane with a proficient pilot flying, any ceiling and visibility within legal weather minimums should be flyable. In a poorly equipped airplane or with a new or rusty pilot, marginal VFR (MVFR) should be avoided.

b.   Another factor to consider is the weather. MVFR in smooth air caused by a stalled front is considerably different from heavy turbulence ahead of a strong front or in a squall line. The following forecast conditions may lead to a no-go decision:

1)   Thunderstorms

2)   Embedded thunderstorms

3)   Lines of thunderstorms

4)   Fast-moving fronts or squall lines

5)   Flights that require you to cross strong or fast-moving fronts

6)   Reported turbulence that is moderate or greater (Remember, moderate turbulence in a Boeing 727 is usually severe in a Cessna 152.)

7)   Icing

8)   Fog (Unlike when in a ceiling, you usually cannot maintain visual references with ground fog. This is especially important if sufficient fuel may be a concern.)

c.   These factors must be considered in relation to the equipment to be flown. Thunderstorms are less of a problem in a radar-equipped airplane. The only way to fly safely is to weigh each factor against the other by using common sense and experience.

d.   Flying is a continuing process of decision-making throughout the whole flight. You must use your certificate to gain experience, but you must also temper the pursuit of experience so you do not go beyond your capabilities or the capabilities of your airplane.

e.   A final factor to be considered in the go/no-go decision is your physical and mental condition. Are you sick, tired, upset, depressed, etc.? These factors greatly affect your ability to handle normal and abnormal problems.

END OF TASK

OPERATION OF SYSTEMS

III.C. TASK: OPERATION OF SYSTEMS

REFERENCES: AC 61-23/FAA-H-8083-25, FAA-H-8083-3; FAA-S-8081-12, FAA-S-8081-14; POH/AFM.

Objective. To determine that the applicant exhibits instructional knowledge of the elements related to the operation of systems, as applicable to the airplane used for the practical test, by describing the following systems:

1. Primary flight controls and trim.
2. Flaps, leading edge devices, and spoilers.
3. Powerplant and propeller.
4. Landing gear.
5. Fuel, oil, and hydraulic.
6. Electrical.
7. Avionics.
8. Pitot static, vacuum/pressure, and associated instruments.
9. Environmental.
10. Deicing and anti-icing.

A. General Information

1. The objective of this task is to determine your instructional knowledge of the elements related to the operation of systems.

a. This task is make- and model-specific, and applies to the complex airplane used on your practical test.

2. For your review, see *Pilot Handbook* for the following:

a. Chapter 1, Airplanes and Aerodynamics, Module 1.4, Flight Controls and Control Surfaces, for a five-page discussion on the primary flight controls, trim devices, flaps, leading edge devices, and spoilers

b. Chapter 2, Airplane Instruments, Engines, and Systems, for a 50-page discussion of the operation of the various airplane instruments, engines, and systems

3. To prepare for this task, systematically study, not read, sections 1, 7, 8, and 9 of your POH:

a. Section 1: General
b. Section 7: Airplane and Systems Descriptions
c. Section 8: Airplane Handling, Service and Maintenance
d. Section 9: Supplement (Optional Systems Description and Operating Procedures)

4. Finally, make a list of the make and model of all avionics equipment in your training airplane. Make yourself conversant with the purpose, operation, and capability of each unit and be able to explain these systems to a student.

B. Task Elements

1. **Primary flight controls and trim**

a. The airplane's attitude is controlled by the deflection of the primary flight controls.

1) The primary flight controls are the rudder, elevator (or stabilator on some airplanes), and ailerons.

b. Trim devices are commonly used to relieve you of the need of maintaining continuous pressure on the primary flight controls.

1) The most common trim devices used on trainer-type airplanes are trim tabs and anti-servo tabs.

2. **Flaps, leading edge devices, and spoilers**

    a. Wing flaps are used on most airplanes. Flaps increase both lift and drag and have three important functions:

        1) First, they permit a slower landing speed, which decreases the required landing distance.

        2) Second, they permit a comparatively steep angle of descent without an increase in speed, making it possible to clear obstacles safely when making a landing approach to a short runway.

        3) Third, they may also be used to shorten the takeoff distance and provide a steeper climb path.

    b. High-lift leading edge devices are applied to the leading edge of the airfoil.

        1) Fixed slots direct airflow to the upper wing surface.

            a) This allows for smooth airflow over the wing at increasing angles of attack and delays the airflow separation.

            b) The wing camber is not increased.

            c) Stalls are delayed to a higher angle of attack.

        2) A slat consists of a leading edge segment that is free to move on tracks.

            a) At low angles of attack, the slat is held flush against the leading edge.

            b) At high angles of attack, either a low pressure area at the wing's leading edge or pilot-operated controls force the slat to move forward.

                i) This opens a slot and allows the air to flow smoothly over the wing's upper surface, delaying the airflow separation.

    c. Spoilers, found only on certain airplane designs and most gliders, are mounted on the upper surface of each wing. Their purpose is to "spoil" or disrupt the smooth flow of air over the wing to reduce the wing's lifting force. It is a means of increasing the rate of descent without increasing the airplane's speed.

3. **Powerplant and propeller**

    a. An airplane's engine is commonly referred to as the powerplant. Not only does the engine provide power to propel the airplane, but it also powers the units that operate a majority of the airplane's systems.

    b. You should be able to explain your airplane's powerplant, including

        1) The operation of the engine
        2) Engine type and horsepower
        3) Ignition system
        4) Induction system
        5) Cooling system
        6) Controls
        7) Fire detection

    c. The airplane propeller consists of two or more blades and a central hub to which the blades are attached. Each blade of an airplane propeller is essentially a rotating wing which produces forces that create the thrust to pull or push the airplane through the air.

        1) Most light, trainer-type airplanes have a **fixed-pitch propeller**; i.e., the pitch of the propeller blades is fixed by the manufacturer and cannot be changed.

        2) Most complex airplanes have a **constant-speed propeller** system; i.e., the pitch of the propeller blades can be changed in flight by the pilot.

    d. The power needed to rotate the propeller blades is furnished by the engine. The engine rotates the airfoils of the blades through the air at high speeds, and the propeller transforms the rotary power of the engine into forward thrust.

4. **Landing gear**

   a. The landing gear system supports the airplane during the takeoff run, landing, taxiing, and when parked. The landing gear can be fixed or retractable and must be capable of steering, braking, and absorbing shock.

     1) Most light trainer-type airplanes are equipped with fixed landing gear.
     2) Most high-performance airplanes are equipped with retractable landing gear.

5. **Fuel, oil, and hydraulic systems**

   a. The fuel system stores fuel and transfers it to the airplane engine.

   b. The oil system provides a means of storing and circulating oil throughout the internal components of a reciprocating engine.

     1) Each engine is equipped with an oil pressure gauge and an oil temperature gauge to be monitored to determine that the oil system is functioning properly.

   c. Most small airplanes have an independent hydraulic brake system powered by master cylinders in each main landing gear wheel, similar to those in your car.

     1) An airplane with retractable landing gear may use hydraulic fluid in the operation of the landing gear.

   d. You should be able to explain the fuel, oil and hydraulic systems for your airplane, including

     1) Approved fuel grade(s) and quantity (usable and nonusable)
     2) Oil grade and quantity (minimum and maximum operating levels)
     3) Hydraulic systems (i.e., brakes, landing gear, etc.)

6. **Electrical system**

   a. Electrical energy is required to operate the starter, navigation and communication radios, lights, and other airplane equipment.

   b. You should be able to explain the electrical system for your airplane, including

     1) Battery location, voltage, and capacity (i.e., amperage)
     2) Electrical system and alternator (or generator) voltage and capacity

       a) Advantages and disadvantages of an alternator and a generator
     3) Circuit breakers and fuses -- location and purpose
     4) Ammeter indications

7. The **avionics systems** is all your airplane's aviation electronic equipment.

   a. Be able to explain and instruct how all your communication and navigation systems operate.

   b. Make a list of the make, model, type of radio, and related equipment in your airplane. As appropriate, consult and study their instruction manuals.

8. **Pitot-static/vacuum system and associated flight instruments**

   a. The pitot-static system provides the source for the operation of the

     1) Altimeter
     2) Vertical speed indicator
     3) Airspeed indicator

   b. The vacuum/pressure system provides the source for the operation of the following gyroscopic flight instruments:

     1) Heading indicator
     2) Attitude indicator

   c. While not normally part of the vacuum/pressure system, the turn coordinator is a gyroscopic flight instrument but is normally powered by the electrical system.

9. **Environmental system**

   a. The environmental system includes

      1) Heater and defroster
      2) Cooling and ventilation
      3) Pressurization system
      4) Oxygen system

   b. Be able to explain the system(s) in your airplane.

10. **Deicing and anti-icing systems**

    a. Deicing systems (e.g., carburetor heat, boots) remove ice after it has formed on a surface.

    b. Anti-icing systems (e.g., pitot heat, alcohol used on propellers and windshields) prevent the formation of ice on a surface.

END OF TASK

PERFORMANCE AND LIMITATIONS

III.D. TASK: PERFORMANCE AND LIMITATIONS

　　　　REFERENCES: FAA-H-8083-3, AC 61-23/FAA-H-8083-25, AC 61-84; FAA-S-8081-12,
　　　　　　　　　　　　FAA-S-8081-14; POH/AFM.

Objective. To determine that the applicant exhibits instructional knowledge of the elements related to performance and limitations by describing:

1. Determination of weight and balance condition.

2. Use of performance charts, tables, and other data in determining performance in various phases of flight.

3. Effects of exceeding airplane limitations.

4. Effects of atmospheric conditions on performance.

5. Factors to be considered in determining that the required performance is within the airplane's capabilities.

A. General Information

　　1. The objective of this task is to determine your instructional knowledge of the elements related to performance and limitations.

　　　　a. You must be able to describe the procedure for determining performance and limitations as you would teach it to your students.

　　2. This task is make- and model-specific and applies to the most complex airplane used on your practical test. This task covers sections 2, 5, and 6 of your POH.

　　　　a. Section 2: Limitations
　　　　b. Section 5: Performance
　　　　c. Section 6: Weight and Balance/Equipment List

　　3. Airplane performance can be defined as the ability to operate or function, i.e., the ability of an airplane to accomplish certain things that make it useful for certain purposes.

　　　　a. The various items of airplane performance result from the combination of airplane and powerplant characteristics.

　　　　　　1) The aerodynamic characteristics and weight of the airplane generally define the power and thrust requirements at various conditions of flight.

　　　　　　2) Powerplant characteristics generally define the power and thrust available at various conditions of flight.

　　　　　　3) The matching of these characteristics is done by the manufacturer.

B. Task Elements

1. **Determination of weight and balance condition**

    a. Section 6, Weight and Balance/Equipment List, in the airplane's POH will contain the data needed to determine the weight and balance condition. The weight and balance condition should be determined for takeoff, en route, and landing.

    b. For a three-page discussion on the effect of weight and balance on performance and how to determine the weight and balance condition, see Task II.H., Weight and Balance, beginning on page 135.

2. **Use of performance charts, tables, and other data in determining performance in various phases of flight**

    a. Performance charts, tables, and/or data used to calculate the airplane's performance is located in Section 5, Performance, of the airplane's POH.

        1) You must be able to explain the use of each performance chart, table, and other data (i.e., how and why) in the airplane's POH.

    b. As a review of the use of performance charts, see the 13-page discussion of the following charts in Chapter 5, Airplane Performance and Weight and Balance, in *Pilot Handbook*.

        1) Density altitude chart
        2) Takeoff distance chart
        3) Rate-of-climb chart
        4) Cruise and range performance chart
        5) Glide performance chart
        6) Crosswind component chart
        7) Landing distance chart

3. **Effects of exceeding airplane limitations**

    a. Operating limitations are found in Section 2, Limitations, of the airplane's POH. These limits establish the boundaries (i.e., flight envelope) in which the airplane must be operated.

        1) You should be able to explain to your student the adverse effects (i.e., what, why, and how) of exceeding an airplane's limitations. These may include

            a) Attempting a takeoff or landing without a long enough runway

            b) Not having enough fuel to make your airport of intended landing, while cruising at a high power setting

            c) Exceeding the airplane's structural or aerodynamic limits by being over gross weight and/or outside center of gravity limits

4. **Effects of atmospheric conditions on performance**

    a. Air density is perhaps the single most important factor affecting airplane performance. Density is defined as mass or weight per unit of volume. The general rule is as air density decreases, so does airplane performance.

        1) Temperature, altitude, barometric pressure, and humidity all affect air density. The density of the air DECREASES

            a) As air temperature INCREASES
            b) As altitude INCREASES
            c) As barometric pressure DECREASES
            d) As humidity INCREASES

2) The engine produces power in proportion to density of the air.

    a) As air density decreases, the power output of the engine decreases.

3) The propeller produces thrust in proportion to the mass of air being accelerated through the rotating blades.

    a) As air density decreases, propeller efficiency decreases.

4) The wings produce lift as a result of the air passing over and under them.

    a) As air density decreases, the lift efficiency of the wing decreases.

b. At power settings of less than 75% or at density altitudes above 5,000 ft., it is essential that normally aspirated engines be leaned for maximum power on takeoff, unless equipped with an automatic altitude mixture control.

1) The excessively rich mixture adds another detriment to overall performance.

2) Turbocharged engines need not be leaned for takeoff in high density altitude conditions because they are capable of producing manifold pressure equal to or higher than sea level pressure.

3) At airports of higher elevations, such as those in the western U.S., high temperatures sometimes have such an effect on density altitude that safe operations may be impossible.

    a) Even at lower elevations with excessively high temperature or humidity, airplane performance can become marginal, and it may be necessary to reduce the airplane's weight for safe operations.

5. **Factors to be considered in determining that the required performance is within the airplane's capabilities**

a. As a flight instructor, you must display sound judgment when determining whether the required performance is within your airplane's and your own capabilities and operating limitations.

1) Remember that the performance charts in your POH do not make allowance for pilot proficiency or mechanical deterioration of the airplane.

2) You can determine your airplane's performance in all phases of flight if you follow and use the performance charts in your POH.

b. As a CFI, you must instruct your student in the use of performance charts and then develop his/her judgment in determining whether or not the flight can be safely conducted.

END OF TASK

AIRWORTHINESS REQUIREMENTS

III.E. TASK: AIRWORTHINESS REQUIREMENTS

REFERENCES: 14 CFR Parts 39, 43; FAA-S-8081-12, FAA-S-8081-14; POH/AFM.

Objective. To determine that the applicant exhibits instructional knowledge of the elements related to required airworthiness by explaining:

1. Required instruments and equipment for day/night VFR.

2. Procedures and limitations for determining airworthiness of the airplane with inoperative instruments and equipment with and without a minimum equipment list (MEL).

3. Requirements and procedures for obtaining a special flight permit.

4. Airworthiness directives, compliance records, maintenance/inspection requirements, and appropriate records.

5. Procedures for deferring maintenance on aircraft without an approved MEL.

A. General Information

    1. The objective of this task is to determine your knowledge of aircraft airworthiness requirements.

    2. Additional Reading: See Chapter 4, Federal Aviation Regulations, in *Pilot Handbook*, for the following:

        a. FAR 91.213, Inoperative Instruments and Equipment, for a detailed discussion of operating an airplane with or without an approved minimum equipment list (MEL).

B. Task Objectives

    1. **Explain required instruments and equipment for day/night VFR**

        a. You may not operate a powered civil aircraft with a standard category U.S. airworthiness certificate without the specified operable instruments and equipment.

        b. Required equipment: VFR - day

            1) Airspeed indicator

            2) Altimeter

            3) Magnetic direction indicator (compass)

            4) Tachometer for each engine

            5) Oil pressure gauge for each engine using a pressure system

            6) Temperature gauge for each liquid-cooled engine

            7) Oil temperature gauge for each air-cooled engine

            8) Manifold pressure gauge for each altitude engine

            9) Fuel gauge indicating the quantity of fuel in each tank

            10) Landing gear position indicator, if the aircraft has a retractable landing gear

            11) For small airplanes certificated after March 11, 1996, an approved anticollision light system

            12) Approved flotation gear for each occupant and one pyrotechnic signaling device if the aircraft is operated for hire over water beyond power-off gliding distance from shore

            13) Approved safety belt with approved metal-to-metal latching device for each occupant who is 2 yr. of age or older

            14) For small civil airplanes manufactured after July 18, 1978, an approved shoulder harness for each front seat

15) An emergency locator transmitter (ELT), if required by FAR 91.207
16) For normal, utility, and acrobatic category airplanes with a seating configuration, excluding pilot seats, of nine or less, manufactured after December 12, 1986, a shoulder harness for each seat in the airplane

c. Required equipment: VFR - night

1) All equipment listed in b. on the previous page
2) Approved position (navigation) lights
3) Approved aviation red or white anticollision light system on all U.S.-registered civil aircraft
4) If the aircraft is operated for hire, one electric landing light
5) An adequate source of electricity for all electrical and radio equipment
6) A set of spare fuses or three spare fuses for each kind required which are accessible to the pilot in flight

2. **Explain procedures and limitations for determining the airworthiness of an airplane having inoperative instruments and equipment, both with and without an MEL.**

a. Except as provided in FAR 91.213, you may not take off in an airplane with any inoperative instruments or equipment installed, i.e., in an airplane that is not airworthy.

b. FAR 91.213 describes the acceptable methods for the operation of an airplane with certain inoperative instruments and equipment that are not essential for safe flight. These acceptable methods of operation are

1) Operation with an approved minimum equipment list (MEL)

a) An MEL is a specific inoperative equipment document for a particular make and model aircraft by serial and registration number.

b) An MEL is designed to provide owners/operators with the authority to operate an aircraft with certain items or components inoperative, provided the FAA finds an acceptable level of safety maintained by

i) Appropriate operations limitations
ii) A transfer of the function to another operating component
iii) Reference to other instruments or components providing the required information

2) Operation without an MEL (probably the way your airplane is operated)

a) You may take off in an aircraft with inoperative instruments and equipment without an approved MEL provided the inoperative instruments and equipment are not

i) Part of the VFR-day type certification instruments and equipment under which the aircraft was type certificated
ii) Indicated as required on the aircraft's equipment list or on the Kinds of Operations Equipment List for the kind of flight operation being conducted
iii) Required by any FAR
iv) Required by an airworthiness directive

b) The inoperative instruments or equipment must be

i) Removed from the airplane with the cockpit control placarded and the maintenance properly recorded, or
ii) Deactivated and placarded "inoperative."

       c)   A determination must be made by a certificated and appropriately rated pilot or an appropriately certificated mechanic that the inoperative instrument or equipment does not constitute a hazard to the aircraft.

       d)   By following these procedures, the aircraft is considered to be in a properly altered condition acceptable to the FAA.

3.   **Explain requirements and procedures for obtaining a special flight permit**

    a.   When a special flight permit is required

       1)   A special flight permit may be issued to an airplane with inoperable instruments or equipment under FAR Part 21, Certification Procedures for Products and Parts.

          a)   This can be done despite any provisions listed in FAR 91.213.

       2)   Special flight permits may be issued for an airplane that does not currently meet applicable airworthiness requirements but is capable of safe flight in order for the pilot to fly the airplane to a base where repairs, alterations, or maintenance can be performed or to a point of storage (FAR 21.197).

    b.   Procedures for obtaining a special flight permit

       1)   To obtain a special flight permit, you must submit a written request to the nearest FSDO indicating

          a)   The purpose of the flight

          b)   The proposed itinerary

          c)   The crew required to operate the airplane (e.g., pilot, co-pilot)

          d)   The ways, if any, the airplane does not comply with the applicable airworthiness requirements

          e)   Any restriction that you consider is necessary for safe operation of your airplane

          f)   Any other information considered necessary by the FAA for the purpose of prescribing operating limitations

4.   **Explain airworthiness directives, compliance records, maintenance/inspection requirements, and appropriate records.**

    a.   Airworthiness directives

       1)   Airworthiness directives (ADs) are issued by the FAA to require correction of unsafe conditions found in an airplane, an airplane engine, a propeller, or an appliance when such conditions exist and are likely to exist or develop in other products of the same design.

          a)   ADs may be divided into two categories:

             i)   Those of an emergency nature requiring immediate compliance

             ii)   Those of a less urgent nature requiring compliance within a relatively longer period of time

          b)   ADs are regulatory (i.e., issued under FAR Part 39, Airworthiness Directives) and must be complied with unless a specific exemption is granted.

b. Compliance records

1) FAR 91.417, Maintenance Records, requires that a record be maintained that shows the current status of applicable ADs, including the method of compliance, the AD number, the revision date, and the signature and certificate number of the repair station or mechanic who performed the work.

a) If the AD involves recurring action (e.g., an inspection every 50 hr.), a record must be kept of the time and date when the next action is required.

c. Maintenance/inspection requirements

1) The maintenance requirements on aircraft that are used in commercial operations (i.e., flight training, charter, etc.) are more stringent than on non-commercial Part 91, which requires a maintenance inspection only on an annual basis.

a) All aircraft must undergo an annual inspection by a certificated airframe and powerplant (A&P) mechanic who also possesses an Inspection Authorization (IA).

b) Aircraft used to carry people for hire or to provide flight instruction for hire must undergo an annual or 100-hr. inspection within the preceding 100 hr. of flight time. The 100-hr. interval may be exceeded by no more than 10 hr. to facilitate transport of the aircraft to a maintenance location where the inspection can be performed.

i) However, if the 100-hr. inspection is overflown, the next inspection will be due after 100 hr. of flight time less the amount overflown.

ii) EXAMPLE: If the check is performed at the 105-hr. point, the next 100-hr. check is due at the end of 95 hr., not 100 hr.; thus, it would be due at the 200-hr. point.

c) Based on the specific make and model aircraft, further checks beyond the 100-hr. check may be necessary to comply with the FARs. This additional maintenance may be required at the 50-, 150-, or 250-hr. point.

d) You may not use an ATC transponder unless it has been tested and inspected within the preceding 24 calendar months.

e) The emergency locator transmitter (ELT) battery must be replaced after half its useful life has expired (as established by the transmitter manufacturer), or after 1 hr. of cumulative use.

i) The ELT must be inspected every 12 calendar months for

• Proper installation
• Battery corrosion
• Operation of the controls and crash sensor
• Sufficient signal radiated from its antenna

    d.   Appropriate record keeping

        1)   Examine the engine logbooks and the airframe logbook of your training airplane (presumably the one you will use for your practical test), and ask your instructor for assistance as appropriate. Locate and paperclip the most recent signoff for

            a)   Annual inspection

            b)   100-hr. inspection

            c)   Transponder and static system inspection

            d)   ELT inspection and battery expiration (The expiration date is on the outside of the ELT and in the airframe logbook.)

        2)   The owner or operator is primarily responsible for maintaining an airplane in an airworthy condition and for ensuring compliance with all pertinent ADs.

            a)   The term operator includes the PIC. Thus, as PIC, you are responsible for ensuring that the airplane is maintained in an airworthy condition (e.g., 100-hr. and/or annual inspections) and that there is compliance with all ADs.

    e.   An easy way to remember the required documents is by using the memory aid **ARROW**

> **A** irworthiness certificate
> **R** egistration
> **R** adio station license
> **O** perating limitations
> **W** eight and balance

> NOTE: A radio station license is required only if the airplane is flown outside of U.S. airspace (i.e., to another country). Additionally, on these flights you are required to have a restricted radiotelephone operator permit. These are requirements of the Federal Communications Commission (FCC), not FAA requirements.

5.   **Explain procedures for deferring maintenance on aircraft without an approved MEL.**

    a.   You may take off in an aircraft with inoperative instruments and equipment without an approved MEL provided the inoperative instruments and equipment are not

        1)   Part of the VFR-day type certification instruments and equipment under which the aircraft was type certificated

        2)   Indicated as required on the aircraft's equipment list or on the Kinds of Operations Equipment List for the kind of flight operation being conducted.

        3)   Required by any FAR

        4)   Required by an airworthiness directive

    b.   The inoperative instruments or equipment must be

        1)   Removed from the airplane with the cockpit control placarded and the maintenance properly recorded, or

        2)   Deactivated and placarded "inoperative."

    c.   A determination must be made by a certificated and appropriately rated pilot or an appropriately certificated mechanic that the inoperative instrument or equipment does not constitute a hazard to the aircraft.

    d.   By following these procedures, the aircraft is considered to be in a properly altered condition acceptable to the FAA.

END OF TASK -- END OF STUDY UNIT

STUDY UNIT IV
PREFLIGHT LESSON ON A MANEUVER
TO BE PERFORMED IN FLIGHT

This study unit explains the task of preparing and presenting a preflight lesson on a selected maneuver. The maneuver will be one of the 39 maneuvers set forth in Study Units VII through XIV.

Your discussions, explanations, and descriptions should follow the recommended teaching procedures and techniques outlined in Study Unit I, Fundamentals of Instructing (pages 65 to 93), which is a synthesis of the FAA's *Aviation Instructor's Handbook* (FAA-H-8083-9).

Your FAA inspector/examiner will select one maneuver (task) from the tasks in Study Units VII through XIV. You will then be required to **prepare and present a preflight lesson on the selected maneuver as the lesson would be taught to your student**. This maneuver will probably be emphasized during the flight portion of your practical test.

NOTE: There is only one task in this area of operation, but it can be applied to any one of the 39 tasks in Study Units VII through XIV, which are listed on the next page. We recommend that you bring a completed Flight Maneuver Analysis Sheet for each of these 39 maneuvers to your FAA Practical Test to provide notes for your preflight lesson. Flight Maneuver Analysis Sheets are explained at the end of this study unit.

FLIGHT INSTRUCTOR PTS FLIGHT MANEUVERS

SUMMARY OF TASKS

Areas of Operation	No. of Tasks
VII. Takeoffs, Landings, and Go-Arounds	9
VIII. Fundamentals of Flight	4
IX. Performance Maneuvers	4
X. Ground Reference Maneuvers	4
XI. Slow Flight, Stalls, and Spins	8
XII. Basic Instrument Maneuvers	5
XIII. Emergency Operations	4
XIV. Postflight Procedures	1
TOTALS	39

LISTING OF TASKS

VII. TAKEOFFS, LANDINGS, AND GO-AROUNDS
- A. Normal and Crosswind Takeoff and Climb
- B. Short-Field Takeoff and Maximum Performance Climb
- C. Soft-Field Takeoff and Climb
- F. Normal and Crosswind Approach and Landing
- G. Slip to a Landing
- H. Go-Around/Rejected Landing
- I. Short-Field Approach and Landing
- J. Soft-Field Approach and Landing
- K. Power-off 180° Accuracy Approach and Landing

VIII. FUNDAMENTALS OF FLIGHT
- A. Straight-and-Level Flight
- B. Level Turns
- C. Straight Climbs and Climbing Turns
- D. Straight Descents and Descending Turns

IX. PERFORMANCE MANEUVERS
- A. Steep Turns
- B. Steep Spirals
- C. Chandelles
- D. Lazy Eights

X. GROUND REFERENCE MANEUVERS
- A. Rectangular Course
- B. S-Turns Across a Road
- C. Turns Around a Point
- D. Eights on Pylons

XI. SLOW FLIGHT, STALLS, AND SPINS
- A. Maneuvering during Slow Flight
- B. Power-On Stalls (Proficiency)
- C. Power-Off Stalls (Proficiency)
- D. Crossed-Control Stalls (Demonstration)
- E. Elevator Trim Stalls (Demonstration)
- F. Secondary Stalls (Demonstration)
- G. Spins
- H. Accelerated Maneuver Stalls (Demonstration)

XII. BASIC INSTRUMENT MANEUVERS
- A. Straight-and-Level Flight
- B. Constant Airspeed Climbs
- C. Constant Airspeed Descents
- D. Turns to Headings
- E. Recovery from Unusual Flight Attitudes

XIII. EMERGENCY OPERATIONS
- A. Emergency Approach and Landing (Simulated)
- B. Systems and Equipment Malfunctions*
- C. Emergency Equipment and Survival Gear*
- D. Emergency Descent

XIV. POSTFLIGHT PROCEDURES
- A. Postflight Procedures

*May not be considered a "maneuver to be performed in flight."

MANEUVER LESSON

IV.A. TASK: MANEUVER LESSON

REFERENCES: FAA-H-8082-3, FAA-H-8083-9, FAA-H-8083-25; FAA-S-8081-12, FAA-S-8081-14; POH/AFM.

Objective. To determine that the applicant exhibits instructional knowledge of the selected maneuver by:

1. Stating the purpose.
2. Giving an accurate, comprehensive oral description, including the elements and common errors.
3. Using instructional aids, as appropriate.
4. Describing the recognition, analysis, and correction of common errors.

A. General Information

1. The objective of this area of operation is to determine your ability to prepare and present a preflight lesson on a selected flight maneuver.

2. To prepare for this area of operation, it is best that you prepare a lesson plan for each flight maneuver in Study Units VII through XIV.

   a. You can tell your FAA inspector/examiner that you have a prepared lesson plan.

   b. Your FAA inspector/examiner can choose to let you use your prepared lesson plan, or (s)he may ask you to develop a lesson plan from scratch.

   c. Preparing lesson plans in advance is a good learning exercise that also provides you with a reference library for future use as a CFI.

3. The FAA recommends a written lesson plan for all flight and ground lessons. The objective is to help CFIs monitor their instruction and students' progress. A sample lesson plan for ground reference maneuvers appears on page 201. The following components are recommended for each lesson plan:

   a. Objective
   b. Elements
   c. Schedule
   d. Equipment
   e. Instructor's actions
   f. Student's actions
   g. Completion standards

4. The lesson plan idea is to control instructional activity by CFI planning, execution, and evaluation. Student participation in lesson planning, execution, and evaluation is essential. Remember that you, as flight instructor, are the facilitator for your student's learning and understanding.

5. After the discussion of the FAA's task elements, there are two additional sections for your study and consideration.

   a. Content of a maneuver lesson
   b. Use of Flight Maneuver Analysis Sheets (FMASs)

B.   Task Objectives

1.   **When you begin your lesson, you should clearly state the objective (purpose) of the maneuver.**

a.   Stating the objective explains to your student the reason for the lesson.

b.   The objective is what you expect your student to know or do at the completion of the lesson.

c.   When, why, and how would the student need to know and use the skills required in this task?

1)   What are the safety implications of these skills? Always emphasize safety during practical tests and as an instructor.

2.   **Give an accurate and comprehensive oral description of the maneuver, including the elements and common errors.**

a.   You need to know at what level your student is so you can use appropriate language.

1)   Your level of detail will be different when teaching a student pilot as opposed to a commercial pilot applicant.

2)   Draw on previous lessons and knowledge to enhance student interest.

b.   A comprehensive description does not mean a long drawn-out discussion.

1)   Keep it centered on the elements of the maneuver, and draw on student participation to determine when your student shows an understanding.

c.   The elements will tell the student what knowledge and skills are required in the maneuver.

1)   They break the maneuver into smaller, more manageable components.

d.   Review the common errors described in the various FAA references, but add one or two of your own and discuss why these happen.

3.   **Use instructional aids to help your student to visually understand the maneuver.**

a.   Use a chalkboard to diagram the segments of a maneuver.

b.   Use a model airplane to help provide a three-dimensional view of the maneuver.

c.   Remember that your aids should be simple and compatible with the learning outcome you want to achieve.

4.   **Describe the recognition, analysis, and correction of common errors of the maneuver.**

a.   When you discuss the common errors, do not just tell your student what they are. Explain how to recognize, analyze (i.e., why they occur), and correct them.

b.   Instructional aids may be useful in visually demonstrating common errors to your student.

5.   Remember to be energetic and positive throughout your instruction period. Ask your FAA inspector/examiner questions, as you would your student.

a.   If you are asked a question by your FAA inspector/examiner and you are not sure of the answer, admit that you do not know and explain how you would locate the correct answer.

1)   Always give your student a commitment as to when you will give him/her the answer, and follow through.

CONTENT OF A MANEUVER LESSON

Sample Lesson Plan for a 90-Minute Instructional Flight Period

LESSON: GROUND REFERENCE MANEUVERS

STUDENT: _____ **DATE:** _____

OBJECTIVE
- To develop the student's skill in planning and following a pattern over the ground, compensating for wind drift at varying angles

CONTENT
- Use of ground references to control path
- Observation and control of wind effect
- Control of airplane attitude, altitude, and heading

SCHEDULE
- Preflight Discussion..:10
- Instructor Demonstrations..:25
- Student Practice...:45
- Postflight Critique...:10

EQUIPMENT
- Chalkboard for preflight discussion
- IFR visor for maneuvers reviewed

INSTRUCTOR'S ACTIONS
- *PREFLIGHT* -- Discuss lesson objective.
  Diagram S-turns, turns around a point, and rectangular course on chalkboard.
- *INFLIGHT* -- Demonstrate elements.
  Demonstrate following a road, S-turns, turns around a point, and rectangular course. Coach student practice.
- *POSTFLIGHT* -- Critique student performance and make study assignment.

STUDENT'S ACTIONS
- *PREFLIGHT* -- Discuss lesson objective and resolve questions.
- *INFLIGHT* -- Review previous maneuvers, including power-off stalls and maneuvering during slow flight.
  Perform each new maneuver as directed.
- *POSTFLIGHT* -- Ask pertinent questions.

COMPLETION STANDARDS
- Student should demonstrate competency in maintaining orientation, airspeed within 10 kt., altitude within 100 ft., and headings within 10°, and in making proper correction for wind drift.

A. The FAA's written lesson plan as illustrated above implies that this lesson plan is implemented within a 90-min. period.

   1. A lesson plan should include home study and preparation by your student so (s)he can learn and understand all of the concepts underlying each flight maneuver.

   2. A blank lesson plan is provided on page 563 so that you may make copies for your use.

B. Explain to your FAA inspector/examiner that the presentation you are giving in a preflight briefing would follow your study assignment of the flight maneuver in your previous postflight briefing.

   1. A part of that assignment would be for your student to have completed a Flight Maneuver Analysis Sheet (discussed on the next page).

   2. Note the FAA's sample lesson plan above and the time allocations of 10 min., 25 min., 45 min., and 10 min.

     a. A student needs much more than 10 min. to study, learn, and understand a flight maneuver; therefore, home study is imperative.

     b. Point this out to your FAA inspector/examiner.

      c.   Show your FAA inspector/examiner your FMAS as you would like your students to have completed it.

          1)   Presumably you will have 39 completed FMASs, i.e., one for each of the flight maneuver tasks.

C.   Follow up on the use of lesson plans: Note that the objectives, elements, and completion standards change from maneuver to maneuver.

    1.   The schedule, equipment, instructor's actions, and student's actions do NOT change appreciably.

    2.   Thus, the lesson plan is more generic to maneuvers than the FMAS, which is very specific to maneuvers.

      a.   The FMAS should be completed by the student at home and discussed/critiqued by the flight instructor during the preflight discussion.

FLIGHT MANEUVER ANALYSIS SHEET (FMAS)

A.   Purpose: FMASs are to facilitate the study, learning, and understanding of flight maneuvers before they are executed.

    1.   FMASs promote conceptual knowledge that supports and enhances motor skills.

    2.   FMASs must be completed in the student's own words -- NOT copied word-for-word from another source. Students should

      a.   Read, study, learn, and understand
      b.   Explain the maneuver on the FMAS

    3.   Prior to flight, the FMAS should be critiqued by the CFI.

B.   The FAA emphasizes purpose, description, instructional aids, and common errors in its practical test standards.

    1.   The "objectives/purpose" is at the top of the FMAS.

    2.   Description includes

      a.   Flight path
      b.   Power settings
      c.   Altitude
      d.   Airspeed
      e.   Control forces
      f.   Timing
      g.   Completion standards
      h.   Emphasis on visual scanning and clearing turns

    3.   Instructional aids: This FMAS is a great example!

      a.   Chalkboards and model airplanes are other examples.
      b.   Reference book figures, diagrams, and illustrations are also good.

    4.   Common errors are at the bottom of the reverse side of the FMAS.

      a.   You may wish to photocopy the FMAS on only one side and thus make the FMAS two pages in length.
      b.   You may also wish to photocopy the FMAS onto legal paper to provide additional space at the bottom for explanations and notes.
      c.   Make 39 copies of the FMAS for your personal use in studying, learning, and understanding these 39 flight maneuvers.

> Note: This is a sample for steep turns in a C-152. Please photocopy FMASs from pages 14-15 for use with your students.

CFI _____

Student _____

Date _____

GLEIM'S
FLIGHT MANEUVER ANALYSIS SHEET

1. **MANEUVER**   *Steep Turns*

2. **OBJECTIVES/PURPOSE**   *To develop smoothness, coordination, orientation, division of attention, and control techniques in the performance of steep turns.*

3. **FLIGHT PATH (visual maneuvers)**

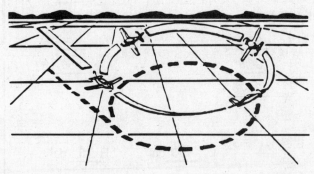

4. **POWER SETTINGS**   5. **ALT**   6. **A/S**

MP	RPM		SEGMENT OF MANEUVER	ALT	A/S
	2300	a.	*Entry*	*2,000*	*95*
	2350	b.	*Maneuver*	*2,000*	*95*
	2300	c.	*Recovery*	*2,000*	*95*

Pencil in expected indication on each of 6 flight instruments on reverse side.

7. **CONTROL FORCES**

a. *Entry -- coordinated aileron and rudder to a 45° bank. As bank steepens beyond 30°, add back elevator pressure as necessary to maintain altitude, trim.*

b. *Maneuver -- aileron pressure as necessary to maintain a 45° bank. Elevator pressure as necessary to maintain altitude.*

c. *Recovery -- lead rollout by 22° of heading. Use coordinated aileron and rudder to wings level, and release back elevator pressure (or forward pressure to counteract trim) to maintain level flight.*

8. **TIME(S), TIMING** _____

9. **TRAFFIC CONSIDERATIONS**   **CLEARING TURNS REQUIRED** *Yes*

10. **COMPLETION STANDARDS/ATC CONSIDERATIONS**   *Entry speed should be at V_A or the manufacturer's recommended entry speed. Roll into a coordinated 360° turn; maintain a 45° bank, ±5°; and roll out on the entry heading, ±10°. Maintain entry altitude, ±100 ft., and airspeed, ±10 kt.*

AIRPLANE MAKE/MODEL   *Cessna 152*

WEIGHT			**AIRSPEEDS**	
Gross	1633		V_{SO}	35
Empty	1136		V_{S1}	40
Pilot/Pasngrs	340		V_X	54
Baggage	10		V_Y	67
Fuel (gal × 6)	147		V_A	103
			V_{NO}	111
			V_{NE}	149
			V_{FE}	85

45° Bank

CENTER OF GRAVITY

Fore Limit	32.45"	V_{LO}	--
Aft Limit	36.5"	V_R	50
Current CG	34.78"		

PRIMARY vs. SUPPORTING INSTRUMENTS

(IFR maneuvers) -- instruments: AI, ASI, ALT, TC, HI, VSI; RPM and/or MP
(most relevant to instrument instruction)

	PITCH	BANK	POWER
ENTRY			
primary	___	___	___
supporting	___	___	___
ESTABLISHED			
primary	___	___	___
supporting	___	___	___

FUEL

Capacity	L _13_ gal	R _13_ gal
Current Estimate	L _13_ gal	R _13_ gal
Endurance (Hr.)	3.7	
Fuel-Flow -- Cruise (GPH)	6.5	

PERFORMANCE DATA

	Airspeed	Power* MP	RPM
Takeoff Rotation	50	--	FULL
Climbout	67	--	FULL
Cruise Climb	75	--	FULL
Cruise Level	90 - 95	--	2300
Cruise Descent	100 - 105	--	2100
Approach**	--	--	--
Approach to Land (Visual)	55	--	1500
Landing Flare	50–	--	IDLE

* If you do not have a constant-speed propeller, ignore manifold pressure (MP).
**Approach speed is for holding and performing instrument approaches.

COMMON ERRORS

1. *Improper pitch, bank, and power coordination during entry and rollout.*

2. *Uncoordinated use of flight controls.*   3. *Inappropriate control applications.*

4. *Improper techniques in correcting altitude deviations.*   5. *Loss of orientation.*

6. *Excessive deviation from desired heading during rollout.*

STUDY UNIT V
PREFLIGHT PROCEDURES

This study unit explains the five tasks (A-D and G) of Preflight Procedures. These tasks include both knowledge and skill. Your FAA inspector/examiner is required to test you on at least one of the five tasks in this area of operation; therefore, you must be prepared for all five.

PREFLIGHT INSPECTION

V.A. TASK: PREFLIGHT INSPECTION

REFERENCES: FAA-H-8083-3; FAA-S-8081-12, FAA-S-8081-14; POH/AFM.

Objective. To determine that the applicant:

1. Exhibits instructional knowledge of the elements of a preflight inspection, as applicable to the airplane used for the practical test, by describing--

    a. Reasons for the preflight inspection, items that should be inspected, and how defects are detected.

    b. Importance of using the appropriate checklist.

    c. How to determine fuel and oil quantity and contamination.

    d. Detection of fuel, oil, and hydraulic leaks.

    e. Inspection of the oxygen system, including supply and proper operation (if applicable).

    f. Inspection of the flight controls.

    g. Detection of visible structural damage.

    h. Removal of tie-downs, control locks, and wheel chocks.

    i. Removal of ice and frost.

    j. Importance of the proper loading and securing of baggage, cargo, and equipment.

    k. Use of sound judgment in determining whether the airplane is airworthy and in condition for safe flight.

2. Exhibits instructional knowledge of common errors related to a preflight inspection by describing--

    a. Failure to use or the improper use of checklist.

    b. Hazards which may result from allowing distractions to interrupt a visual inspection.

    c. Inability to recognize discrepancies to determine airworthiness.

    d. Failure to assure servicing with the proper fuel and oil.

    e. Failure to ensure proper loading and securing of baggage, cargo, and equipment.

3. Demonstrates and simultaneously explains a preflight inspection from an instructional standpoint.

A. General Information

1. The objective of this task is to determine your instructional knowledge of the elements and common errors related to a visual inspection. You will also need to demonstrate and simultaneously explain a visual inspection from an instructional standpoint.

2. You, as a pilot in command, are responsible for determining whether your airplane is airworthy and safe to fly. FAR 91.7 states, "The pilot in command is responsible for determining whether that aircraft is in condition for safe flight."

B. Task Objectives

1. **Exhibit your instructional knowledge by explaining the following elements of a preflight inspection, as applicable to the airplane used for the practical test.** This is make and model specific; study your POH.

    a. **Reasons for the preflight inspection, items that should be inspected, and how defects are detected**

        1) The objective of the preflight (visual) inspection is to ensure that your airplane has no obvious problems prior to taking off. The preflight is carried out in a systematic walk around the airplane and begins in the cockpit.

            a) Make sure all necessary documents, maps, safety equipment, etc., are aboard.

            b) Check to ensure all inspections are current (e.g., 100-hr., annual, transponder).

            c) Make sure your airplane has the required equipment for the flight you are about to take, e.g., Mode C transponder for an operation in Class B or Class C airspace.

2) Next, inspect items outside of the airplane to determine that the airplane is in condition for safe flight.

3) Defects can be detected only if you follow the checklist and actually look for something wrong in each item you check for on the airplane.

4) Your student will perform the preflight the same way you do as his/her CFI. Ensure that you instruct, demonstrate, and perform a thorough preflight inspection.

b. **Importance of using the appropriate checklist**

1) Each airplane has a specific list of preflight procedures recommended by the airplane manufacturer, which are found in Section 4, Normal Procedures, of the POH.

a) The written checklist is a systematic set of procedures.

2) As you walk around your airplane, you will always be looking at some item on the airplane. Thus, your inspection will most likely be more detailed than the checklist in the POH.

a) Always have your checklist in hand to be used as a reference to ensure that all items have been checked. If you become distracted during the preflight inspection, you should use the checklist to determine the last item to be checked.

c. **How to determine fuel and oil quantity and contamination**

1) You should first check the fuel gauges for indication of fuel quantity. Next, visually check the fuel quantity in the fuel tanks.

a) These two indications should match.

b) If the fuel tank is not full or at a position on a tab installed by the manufacturer, you may not know exactly how much fuel is in the tank.

i) At this time, you will need to use some type of calibrated fuel quantity dipstick to determine the fuel quantity.

2) Usually the oil is stored in a sump at the bottom of the engine crankcase. An opening to the oil sump is provided through which oil may be added, and a dipstick is provided to measure the oil level.

a) Your POH will specify the quantity of oil needed for safe operation.

b) Always make certain that the oil filler cap and the oil dipstick are secure after adding and/or checking the oil level. If these are not properly secured, oil loss may occur.

c) Use only the type and grade of oil recommended by the engine manufacturer, or its equivalent. Never use any oil additive that has not been recommended by the engine manufacturer or authorized by the FAA.

i) The type and grade of oil to use can be found in the POH or on placards on or near the oil filler cap.

d) The wrong type of oil or an insufficient oil supply may interfere with any or all of the basic oil functions and can cause serious engine damage and/or an engine failure during flight.

3) Fuel or oil contamination can be due to improper fuel or oil grade, water, or other contaminants.

4) Fuel grade

    a) Refer to your POH for the manufacturer's recommendation regarding the minimum grade. Dyes are added by the refinery to help you identify the various grades of aviation fuel.

        i) 80 is red.

        ii) 100LL is blue.

        iii) 100 is green.

        iv) Jet fuel is clear.

    b) Every aircraft engine has been designed to use a specific grade of aviation fuel for satisfactory performance.

        i) When you are faced with a shortage of the correct grade of fuel, always use the alternate fuel grade specified by the manufacturer or the next higher grade.

    c) DO NOT USE AUTOMOTIVE FUEL unless an FAA supplemental type certificate (STC) has been obtained for your airplane that approves auto gas use.

5) Fuel contamination safeguards

    a) Always assume that the fuel in your airplane is contaminated. A transparent container should be used to collect a generous fuel sample from each sump drainage valve at the lowest point of each tank and from other parts of the fuel system.

    b) Water, the most common fuel contaminant, is usually caused by condensation inside the tank.

        i) Since water is heavier than the fuel, it will be located at the lowest levels in the fuel system.

        ii) If water is found in the first sample (i.e., clear liquid or bubbles at the bottom of the container), drain further samples until no water appears.

    c) Also check for other contaminants, e.g., dirt, sand, rust.

        i) Keep draining until no trace of the contaminant appears.

        ii) A preventive measure is to avoid refueling from cans and drums, which may introduce fuel contaminants, such as dirt or other impurities.

    d) Wait at least 15 min. after your airplane has been refueled before you take a fuel sample.

        i) This will allow time for any contaminants to settle to the bottom of the tank.

6) Oil contamination is more difficult to detect than fuel contamination.

    a) When checking the oil level, you should look for any discoloration of the oil, e.g., a milky color.

        i) Discoloration is an indication of contamination, and the oil and filter must be changed.

    b) Using the proper grade of oil will prevent contamination caused by mixing different grades of oil.

    c) Ensuring that the oil and filter are changed when scheduled will help prevent an oil system failure caused by an excessive build-up of particles in the oil.

d. **Detection of fuel, oil, and hydraulic leaks**

   1) Check to see that there are no oil puddles or other leakages (fuel, hydraulic) under the plane, inside the engine cowling, or on the wheel struts.

      a) If the airplane is equipped with a constant speed propeller, you must check for oil leakage around the spinner and on the propeller blades.

         i) Remember to stay clear of the propeller and its arc while preflighting because the engine may start if the propeller is moved. Treat the area as a danger zone.

   2) If in doubt, seek assistance from a qualified mechanic to look at any leakage. Know the cause and make the necessary repairs before flying.

e. **Inspection of the oxygen system, including supply and proper operation (if applicable)**

   1) If your airplane is equipped with an oxygen system, follow the preflight procedures in the POH.

   2) Check the oxygen pressure gauge for the pressure reading.

      a) The POH will have graphs or tables to determine the amount of oxygen in the system and the duration of the supply depending on the number of people and the flight altitude.

   3) You should ensure that the mask(s) fit(s) properly and that the system is operating properly.

   4) Do not handle oxygen equipment with greasy hands or permit an accumulation of oil or grease near the system.

f. **Inspection of the flight controls**

   1) Visually inspect the flight control surfaces (ailerons, elevators, rudder) to ensure that they move smoothly and freely for their entire movement span.

      a) They also must be securely attached, with no missing, loose, or broken nuts, bolts, or rivets.

   2) Inspect any mass balance weights on control surfaces (designed to keep the control surface's center of gravity forward of the hinge so as to preclude possible shudder).

   3) Check to see that the control yoke moves in the proper direction as the control surfaces move.

   4) Place the flaps in the full down position to examine the attaching bolts and the entire flap surface.

      a) Ensure that the flaps operate correctly with the flap control and lock into position.

g. **Detection of visible structural damage**

   1) Check for dents, cracks, or tears (cloth cover) on all surfaces of the airplane. These can disrupt the smooth airflow and change your airplane's performance.

      a) Surface deformities can lead to, or may be caused from, structural weakness and/or failures due to the stress that is put on the airplane during flight.

         i) These deformities result from bent or broken underlying structure.

      b) One method of checking the wings on a cloth-covered airplane is to grasp the wing spars at the wing tip and gently push down and pull up.

         i) Any damage may be evident by sound and/or wrinkling of the skin.

2) Inspect the propeller for nicks and/or cracks. A small nick not properly repaired can become a stress point where a crack could develop and the blade could break.

3) If you have any doubts, seek assistance from a qualified mechanic.

h. **Removal of tie-downs, control locks, and wheel chocks**

1) Tie-downs are chains or ropes used to secure the airplane to the ground from three locations on the airplane: usually, the midpoint of each wing and the tail.

a) Tie-down hooks or eyelets are provided at these locations on most airplanes.

2) Control locks keep the control surfaces stationary so they do not move back and forth in the wind. Three methods are used for this purpose:

a) A control or "gust" lock is a pin that prevents the control yoke from turning or moving in or out.

b) The control yoke can be secured tightly with a seatbelt.

c) In older airplanes, sometimes the aileron, elevator, and rudder are clamped to adjacent stationary surfaces so they cannot move.

3) Chocks are normally blocks of wood placed both in front of and behind a tire to keep the airplane from rolling.

i. **Removal of ice and frost**

1) Frost, ice, frozen rain, or snow may accumulate on parked airplanes. All of these should be removed before takeoff.

a) Ice is removed by parking the airplane in a heated hangar or spraying deicing compounds on the airplane.

b) Frost should also be removed from all airfoils before flight. Even small amounts can disrupt the airflow, increase stall speed, and reduce lift.

j. **Importance of the proper loading and securing of baggage, cargo, and equipment**

1) Secure all baggage, cargo, and equipment during the preflight inspection. Make sure everything is in its place and secure.

a) You do not want items flying around the cockpit if you encounter turbulence.

2) Cargo and baggage should be secured to prevent movement that could damage the airplane and/or cause a shift in the airplane's CG.

a) An item of cargo is not more secure because it is heavy; it is more dangerous because it moves with greater force.

k. **Use of sound judgment in determining whether the airplane is airworthy and in condition for safe flight**

1) During your preflight inspection of your airplane, you must note any discrepancies and make sound judgments on the airworthiness of your airplane.

a) As pilot in command, you are responsible for determining that the airplane is airworthy.

b) If you have any doubt, you should ask someone with more experience and/or knowledge.

c) Do not attempt a flight unless you are completely satisfied that the airplane is safe and airworthy.

2) After you have completed the preflight inspection, take a step back and look at your entire airplane.

a) During your inspection, you were looking at individual items for airworthiness. Now you should look at the airplane as a whole and ask, "Is this airplane safe to fly?"

2. **Exhibit your instructional knowledge by explaining the following common errors related to a visual inspection.**

a. **Failure to use or the improper use of the checklist**

1) Checklists are guides for ensuring that all necessary items are checked in a logical sequence.

2) You must not get the idea that the list is merely a crutch for poor memory.

b. **Hazards which may result from allowing distractions to interrupt a visual inspection**

1) Being distracted could lead to you missing items on the checklist or not recognizing a discrepancy.

a) You must keep your thoughts on the preflight inspection.

2) If you are distracted, either start at the beginning of the preflight inspection or repeat the preceding two or three items.

c. **Inability to recognize discrepancies to determine airworthiness**

1) You must understand what you are looking at during the preflight inspection.

2) Look for smaller items, such as missing screws, drips of oil, etc.

3) As a CFI, you should quiz your student by having him/her explain what to check for on the preflight inspection.

d. **Failure to assure servicing with the proper fuel and oil**

1) It is easy to determine whether the correct grade of fuel has been used. Even if you are present during fueling, you should be in the habit of draining a sample of fuel from the airplane to check for the proper grade and for any contamination.

2) Oil is not color-coded for identification. You will need to check the proper grade before you or any line personnel add oil to the airplane.

e. **Failure to ensure proper loading and securing of baggage, cargo, and equipment**

1) Improperly secured cargo, baggage, and equipment can be a distraction during normal flight and may become projectiles during turbulence or in the event of a forced landing.

2) Items that move in the aircraft also change the CG.

a) If the CG shifts beyond the forward or aft limit, the airplane may become difficult or impossible to control.

3. **Demonstrate and simultaneously explain a preflight inspection from an instructional standpoint.**

    a. Your knowledge, understanding, and demonstration of the preflight inspection must be to a commercial pilot level.

        1) Since there are no definitive standards in either the private or commercial PTSs, your level of performance will be determined by your FAA inspector/examiner through the exercise of subjective judgment.

    b. You will be evaluated on your ability to simultaneously explain and demonstrate the key elements of the preflight inspection.

    c. Point out any deficiency in proper preflight procedures to your FAA inspector/examiner as an example of an error, and explain how it should be corrected (including how and why it occurred).

4. This task does not require you to analyze and correct common errors of a visual inspection, but you can and will (as a CFI) critique your student's preflight inspection.

    a. If your FAA inspector/examiner has any deficiency in proper preflight inspection procedures, point out the deficiency to him/her as an example of a student error, and explain how it should be corrected (including how and why it occurred).

END OF TASK

SINGLE-PILOT RESOURCE MANAGEMENT

V.B. TASK: SINGLE-PILOT RESOURCE MANAGEMENT

REFERENCES: FAA-H-8083-3; FAA-S-8081-12, FAA-S-8081-14; POH/AFM.

Objective. To determine that the applicant:

1. Exhibits instructional knowledge of the elements of single-pilot resource management by describing --

    a. Proper arranging and securing of essential materials and equipment in the cockpit.

    b. Proper use and/or adjustment of cockpit items such as safety belts, shoulder harnesses, rudder pedals, and seats.

    c. Occupant briefing on emergency procedures and use of safety belts.

    d. Proper utilization of all resources required to operate a flight safely, dispatchers, weather briefers, maintenance personnel, and air traffic control.

2. Exhibits instructional knowledge of common errors related to single-pilot resource management by describing --

    a. Failure to place and secure essential materials and equipment for easy access during flight.

    b. Failure to properly adjust cockpit items, such as safety belts, shoulder harnesses, rudder pedals, and seats.

    c. Failure to provide proper adjustment of equipment and controls.

    d. Failure to provide occupant briefing on emergency procedures and use of safety belts.

    e. Failure to utilize all resources required to operate a flight safely.

3. Demonstrates and simultaneously explains single-pilot resource management from an instructional standpoint.

A. General Information

1. The objective of this task is to determine your instructional knowledge of the elements and common errors related to single-pilot resource management. You will also need to demonstrate and simultaneously explain single-pilot resource management from an instructional standpoint.

    a. This task involves both maintaining an organized cockpit and understanding the aeronautical decision-making process.

2. Utilizing all appropriate checklists is an element of single-pilot resource management listed in both the private and commercial PTSs, but not here.

    a. You and your student should use the appropriate checklist for a specific phase of flight while on the ground or in the air (e.g., before starting engine, during climb, before landing, etc.).

        1) You must set the example for your student.

    b. We emphasize (as does the FAA) the appropriate use of checklists throughout this book.

        1) A checklist provides a listing of "actions" and/or "confirmations." For example, you either "open cowl flaps" or confirm that cowl flaps are open.

        2) If the desired condition is not available, you have to decide whether to accept the situation or take action. For example, if your engine oil temperature is indicating a higher than normal temperature while en route, you may continue your flight or attempt to divert for a landing, depending upon the level of overheating and relative changes in the temperature.

        3) Each item on the checklist requires evaluation and possible action:

            a) Is the situation safe?

            b) If not, what action is required?

            c) Is the overall airplane/environment safe when you take all factors into account?

      4)   There are different types of checklists:

          a)   "Read and do" -- e.g., before-takeoff checklist

          b)   "Do and read" -- e.g., in reacting to emergencies, doing everything that comes to mind and then confirming or researching in the POH

      5)   In other words, checklists are not an end in and of themselves. Checklists are a means of flying safely. Generally, they are to be used as specified in the POH to accomplish safe flight.

   c.   ALL CHECKLISTS should be read aloud at all times.

      1)   Call out each item on the checklist as you undertake the action or make the necessary observation.

   d.   When using a checklist, you (and your student) must consider proper scanning vigilance and division of attention at all times.

3.   Single-pilot resource management is more than just maintaining an organized and neat cockpit. Single-pilot resource management is a process that combines you, your airplane, and the environment for safer and more efficient operations.

   a.   Some of the elements of cockpit management include

      1)   Communication -- the exchange of information with ATC, FSS personnel, maintenance personnel, and other pilots

          a)   To be effective, you must develop good speaking and listening skills.

      2)   Decision making and problem solving -- the manner in which you respond to problems that you encounter from preflight preparation to your postflight procedures

      3)   Situational awareness -- your knowledge of how you, your airplane, and the environment are interacting. This is a continuous process throughout your flight.

          a)   As you increase your situational awareness, you will become a safer pilot by being able to identify clues that signify a problem prior to an impending accident or incident.

      4)   Standardization -- your use of standardized checklists and procedures

          a)   Checklist discipline will help you because you will develop a habit of reading a checklist item and then performing the task.

          b)   Procedural learning is learning a standardized procedure pattern while using the checklist as a backup, as you may do in the first few steps of an emergency.

      5)   Leader/follower -- Below are the desirable characteristics of both.

          a)   A leader will manage those resources that contribute to a safe flight, e.g., ensuring the proper quantity and grade of fuel.

          b)   A good follower will ask for help at the first indication of trouble.

      6)   Psychological factors -- your attitude, personality, and motivation in the decision-making process

          a)   Hazardous attitudes include anti-authority, impulsive, invulnerable, macho, and resigned.

          b)   Personality is the way you cope with problems.

          c)   Your motivation to achieve a goal can be internal (you are attracted to the goal) or external (an outside force is driving you to perform).

      7)   Planning ahead -- anticipation of and preparation for future situations

          a)   Always think and stay ahead of what needs to be done at any specific time.

        b)   Always picture your location and your heading with respect to nearby NAVAIDs, airports, and other geographical fixes.

        c)   Confirm your present position and anticipate future positions with as many NAVAIDs as possible, i.e., use them all.

    8)   Stress management -- the manner in which you manage the stress in your life will follow you into the cockpit. Stress is your reaction to a perceived threat (real or not) to your body's equilibrium.

        a)   Learn to reduce the stress in your life or to cope with it better.

B.   Task Objectives

1.   **Exhibit your instructional knowledge by explaining the following elements of single-pilot resource management.**

    a.   **Proper arranging and securing of essential materials and equipment in the cockpit**

        1)   On every flight, you should be in the habit of organizing and neatly arranging your materials and equipment in an efficient manner that makes them readily available.

        2)   Be in the habit of "good housekeeping."

           a)   A disorganized cockpit will complicate even the simplest of flights.

        3)   Organization will contribute to safe and efficient flying.

        4)   The cockpit and/or cabin should be checked for loose articles or cargo that may be tossed about if turbulence is encountered and must be secured.

    b.   **Proper use and/or adjustment of cockpit items, such as safety belts, shoulder harnesses, rudder pedals, and seats**

        1)   Adjust and lock your seat, rudder pedals, safety belts, and shoulder harness, if adjustments can be made, to a safe position that assures full control movement.

        2)   On each flight, ensure that you and your student are seated in the same position.

           a)   If the seat is adjustable, it should be moved so that your knees are slightly bent with the balls of your feet placed on the rudder pedals.

               i)   This position will allow full movement of the pedals.

           b)   If the seat is not adjustable, cushions should be used to provide proper seating.

               i)   If this adjustment is not made, comfort and ease of control movements may be sacrificed.

        3)   You and your student must be able to see both inside and outside the cockpit without straining.

           a)   Poor vision will not only cause apprehension and confusion, but will also present a hindrance to the control of the airplane.

        4)   If the seat is adjustable, it is important to ensure that you and your student check that the seat is locked in position.

           a)   Accidents caused by the pilot losing control of the airplane have occurred as a result of the seat moving.

        5)   When your student is positioned correctly for vision and operation of the pedals, (s)he needs to check that the control yoke has freedom of full movement.

           a)   Full movement includes full forward, backward, and side movement.

6) Once you and your student are comfortably seated, the safety belt and shoulder harness, if installed, should immediately be fastened and adjusted to a comfortably snug fit.

   a) Safety belts and shoulder harnesses should be properly fastened even if the engine is only to be run up momentarily.

   b) You and your student are required to wear your safety belts and shoulder harnesses, if equipped, during taxi, takeoff, and landing.

   c) While at your crewmember station, you are required to keep your safety belt fastened.

c. **Occupant briefing on emergency procedures and use of safety belts**

1) You (i.e., the pilot in command) are required to brief each passenger on how to fasten and unfasten the safety belt and, if installed, the shoulder harness (FAR 91.107).

   a) You cannot taxi, take off, or land before notifying each passenger to fasten his/her safety belt and shoulder harness, if installed, and ensuring that (s)he has done so.

2) At this time, you need to brief your passengers on the emergency procedures of the airplane that are relevant to them.

   a) Inform them of what they should do before and after an off-airport landing is made.

   b) You can determine if a passenger is competent to assist you in reading an emergency checklist. Assistance would allow you to perform the tasks as they are read item by item.

   c) Ensure that each passenger can open all exit doors and unfasten safety belts.

3) Instruct your student (FAA inspector/examiner) on how to give a passenger briefing and about the items that must be included in that briefing.

d. **Proper utilization of all resources required to operate a flight safely, dispatchers, weather briefers, maintenance personnel, and air traffic control**

1) To make informed decisions during flight operations, you must be aware of the resources found both inside and outside the cockpit.

   a) In addition to identifying resources, you must also evaluate whether you have time to use a particular resource and the impact that its use will have upon the safety of flight.

   b) EXAMPLE: The assistance of ATC may be useful if you are lost. However, in an emergency situation when action needs to be taken quickly, time may not be available to contact ATC before actions must be taken to address the situation.

2) Be prepared to explain to your FAA Inspector or DPE how you utilized all available resources throughout your preflight planning to make a go/no-go decision. In addition, be prepared to explain the resources available to you during the flight to aid in your decision-making process.

2. **Exhibit your instructional knowledge by explaining the following common errors related to single-pilot resource management:**

a. **Failure to place and secure essential materials and equipment for easy access during flight**

1) Do not use the top of the instrument panel as a storage area.

2) You need to demonstrate your methods of maintaining an organized cockpit and stress the safety factors of being organized.

  b. **Failure to properly adjust cockpit items, such as safety belts, shoulder harnesses, rudder pedals, and seats**

    1) The proper adjustment of safety belts, shoulder harnesses, rudder pedals, and seats is important to comfort and safety.

      a) Complete these tasks on the ground because performing them in flight can present a serious hazard.

  c. **Failure to provide proper adjustment of equipment and controls**

    1) Help your student determine the proper seat and control adjustment, and stress to him/her to always use those positions when flying.

      a) Ensure proper control movement after the seat is adjusted for vision and operation of controls.

  d. **Failure to provide occupant briefing on emergency procedures and use of safety belts**

    1) The pilot is required to provide a proper safety briefing to passengers before every flight.

      a) Your passengers should fully understand how to use the safety belts and what to do in the event of an emergency.

    2) Your passengers will be more at ease and better prepared after a complete preflight briefing.

  e. **Failure to utilize all resources required to operate a flight safely**

    1) Teach your students to utilize all available resources during preflight planning and during the flight.

      a) It may be necessary to give examples showing how each resource is properly utilized. For example, you may consult with an A&P to determine the effects of an inoperative piece of equipment.

      b) Be sure that you teach your students to use all available resources. In the example above, you would also check the limitations section of the flight manual and the FARs to be sure that the aircraft can be flown legally.

3. **Demonstrate and simultaneously explain single-pilot resource management from an instructional standpoint.**

  a. Your knowledge, understanding, and demonstration of single-pilot resource management procedures must be to a commercial pilot level.

    1) Since there are no definitive standards in either the private or commercial PTSs, your level of performance will be determined by your FAA inspector/examiner through the exercise of subjective judgment.

  b. You will be evaluated on your ability to simultaneously explain and demonstrate the key elements of single-pilot resource management.

  c. Point out any deficiency in proper single-pilot resource management procedures to your FAA inspector/examiner as an example of an error, and explain how it should be corrected (including how and why it occurred).

4. This task does not require you to analyze and correct common errors of single-pilot resource management, but you can and will (as a CFI) critique your student's single-pilot resource management procedures.

  a. If your FAA inspector/examiner has any deficiency in proper single-pilot resource management procedures, point out the deficiency to him/her as an example of a student error, and explain how it should be corrected (including how and why it occurred).

END OF TASK

ENGINE STARTING

V.C. TASK: ENGINE STARTING

REFERENCES: AC 61-23/FAA-H-8083-25, AC 91-13, AC 91-55; FAA-H-8083-3; FAA-S-8081-12, FAA-S-8081-14; POH/AFM.

Objective. To determine that the applicant:

1. Exhibits instructional knowledge of the elements of engine starting, as appropriate to the airplane used for the practical test, by describing --

    a. Safety precautions related to starting.

    b. Use of external power.

    c. Effect of atmospheric conditions on starting.

    d. Importance of following the appropriate checklist.

    e. Adjustment of engine controls during start.

    f. Prevention of airplane movement during and after start.

    g. Safety procedures for hand propping an aircraft.

    h. Carburetor fire hazard.

2. Exhibits instructional knowledge of common errors related to engine starting by describing --

    a. Failure to properly use the appropriate checklist.

    b. Failure to use safety precautions related to starting.

    c. Improper adjustment of engine control during start.

    d. Failure to assure proper clearance of the propeller.

3. Demonstrates and simultaneously explains engine starting from an instructional standpoint.

A. General Information

   1. The objective of this task is to determine your instructional knowledge of the elements and common errors related to engine starting. You will also need to demonstrate and simultaneously explain engine starting from an instructional standpoint.

B. Task Objectives

   1. **Exhibit your instructional knowledge by explaining the following elements of engine starting, as appropriate to the airplane used for the practical test.**

      a. **Safety precautions related to starting**

        1) Position the airplane properly.

          a) Always start the engine with enough room in front of the airplane so you can turn off the engine if the brakes fail.

          b) Also, do not start the engine with the tail of the airplane pointed toward an open hangar door, parked cars, or a group of bystanders (i.e., think about direction of prop blast).

            i) It is a violation of FAR 91.13 to operate an airplane on any part of the surface of an airport in a careless or reckless manner that endangers the life or property of another.

          c) Be cautious of loose debris, e.g., rocks or dirt, that can become projectiles when you start the engine.

        2) Set the brakes.

          a) Some airplanes have a parking brake that should be set in the manner prescribed in the POH.

          b) In airplanes without a parking brake, you must ensure that the airplane's brakes are set, normally by applying appropriate pressure on the toe (or pedal) brakes.

    c)   Before starting the engine, remember to position the airplane to avoid creating a hazard.

        i)   If for some reason the brakes are not set properly and the airplane moves forward when the engine is started, you must have an area in which you can stop the airplane by engine shut-down.

   3)   Clear the area.

    a)   Determine that the area around the airplane is clear by observing the area and shouting, "Clear prop!" out your open window, before cranking the engine.

        i)   Allow a few seconds for a response if someone is nearby or under the airplane.

b. **Use of external power**

   1)   Some airplanes are equipped with an external power receptacle.

    a)   An external receptacle allows the connection of an external power source (battery) to the airplane's electrical system without accessing the battery in the airplane.

    b)   Also, an external battery can be connected directly to the airplane's battery to provide power to the starter.

    c)   Read the POH for the correct procedures.

c. **Effect of atmospheric conditions on starting**

   1)   During cold weather, the oil in the airplane's engine becomes very thick. There are several methods to assist in starting a cold engine. Check the POH for the recommended procedure.

    a)   One method is pulling the propeller through (turning it) several times to loosen the oil.

        i)   This method saves battery energy, which is already low due to the low temperature.

        ii)   When performing this procedure, ensure that the ignition/magneto switch is off, the throttle is closed, the mixture is in the lean/idle cut-off position, nobody is standing in or near the propeller arc, the parking brake is on, and the airplane is chocked and/or tied down.

        iii)   A loose or broken groundwire on either magneto could cause the engine to fire or backfire.

    b)   Cold weather starting can be made easier by preheating the engine.

        i)   Many FBOs in cold weather locations offer this service.

        ii)   Small, portable heaters are available that can blow hot air into the engine to warm it.

        iii)   Preheating is generally required when outside air temperatures are below 0°F and is recommended by most engine manufacturers when the temperature is below 20°F.

     c) To start a cold engine, prime it with fuel first.

        i) In carburetor engines, the primer is a small manual or electric pump that draws fuel from the tanks and vaporizes it directly into one or two of the cylinders through small fuel lines.

- Continuous priming may be required to keep the engine running until sufficient engine heat is generated to vaporize the fuel.

        ii) In most fuel-injected engines, priming is accomplished by turning the electric fuel pump ON and moving the mixture control to full RICH until an indication is noted on the fuel flow meter. The engine is now primed.

- Move the mixture control to idle cut-off to complete the priming process.

     d) After a cold engine has been started, it should be idled at low RPMs for 2 to 5 min. to allow the oil to warm and begin circulating throughout the system.

2) During hot weather and/or with a hot engine, the cylinders tend to become overloaded with fuel, possibly leading to a flooded engine situation.

     a) Follow the appropriate checklist for either a HOT or FLOODED engine in the POH.

        i) Flooded engine normally requires you to have the mixture in the lean position and the throttle full open.

- Taking these steps helps clear the cylinders of the excess fuel and allows the engine to start.

        ii) As the engine starts, ensure that you retard the throttle and move the mixture to rich.

     b) Fuel-injected engines have a disadvantage because they may have difficulty starting in hot weather and/or when the engine is hot, due mainly to fuel vaporization.

        i) When the engine is shut down, the air temperature inside the engine cowling increases. As it increases, the fuel in the injection system vaporizes, creating a **vapor lock** that prevents the engine from starting.

- In most airplanes, the electric fuel pump is used to move fuel into the distribution lines, thus cooling them and removing the fuel vapor.

        ii) On very hot days, you may continue to have vaporization problems after the engine is started.

- These problems are caused by insufficient cooling from the propeller slipstream.
- Monitor the fuel-flow gauge for a fluctuation, and use the electric fuel pump to purge the system.

d. **Importance of following the appropriate checklist**

1) It is vital that you and your student be in the habit of using the appropriate checklist for every operation in flying.

     a) The use of checklists ensures that every item is completed and checked.

2) You must use the checklist in the POH for the before-starting and the starting procedures.

e. **Adjustment of engine controls during start**

1) Adjust the engine controls.

   a) While activating the starter and during ground operations while the engine is running, one hand should be kept on the throttle at all times.

      i) During starting, this allows you to advance the throttle if the engine falters or to prevent excessive RPM just after starting.

   b) When starting a hot or flooded engine and some fuel-injected engines, the starting procedure may be to set the throttle and then engage the starter while moving the mixture control from the idle-cutoff position to the rich position.

      i) Once the engine engages, your hand should move to the throttle control.

2) Avoid excess engine RPM and temperatures.

   a) You must monitor the engine instruments during ground operations.

      i) The POH will have the recommended RPM and temperature ranges for the warm-up and other ground operations.

      ii) If the engine temperature begins to rise and the airplane has adjustable cowl flaps, check that they are open.

   b) Follow the checklist in the POH if the engine temperature begins to rise above the normal operating range.

3) Check the engine instruments after engine start.

   a) As soon as the engine is started and operating, you should check the oil pressure gauge. If it does not rise to the normal operating range in about 30 sec. in summer or 60 sec. in winter, the engine may not be receiving proper lubrication and should be shut down immediately.

   b) Check all other engine instruments to ensure that they are also operating within the normal limits as prescribed in the POH.

f. **Prevention of airplane movement during and after start**

1) The brakes need to be set during the starting procedure.

   a) Some airplanes have a parking brake, which should be set in the manner prescribed in the POH.

   b) In airplanes without a parking brake, ensure that the airplane's brakes are set, normally by applying appropriate pressure on the toe (or pedal) brakes.

   c) Before starting the engine, remember to position the airplane to avoid creating a hazard.

      i) If for some reason the brakes are not set properly and the airplane moves forward when the engine is started, you must have an area in which the airplane can be stopped by engine shut-down.

2) The airplane must be prevented from moving after you start the engine. This is done with the brakes.

   a) You must look outside the airplane to ensure that you are not moving. Be aware of what is happening around you.

g. **Safety procedures for hand propping an aircraft.**

1) Even though most airplanes are equipped with electric starters, you should know the procedures and dangers involved in hand propping so you can properly instruct your student.

a) It is recommended that a competent pilot or a qualified person thoroughly familiar with the operation of all the controls be seated at the controls in the cockpit and that the person turning the propeller be thoroughly familiar with this technique.

b) The traditional approach with the person "pulling through" the propeller in front of the propeller follows:

i) To start the plane, the propeller is rotated in the clockwise direction (as seen from the cockpit).

ii) Never lean into the propeller as you pull it. You must not be in a position where you would fall forward if your feet slip. You should have one foot forward and one foot back. As you pull down, you should shift your weight to your rear foot and back away.

iii) You should NOT wrap your fingers around the propeller. You should have your fingers just over the trailing edge. This is to prevent injury if the engine misfires (or backfires) and turns backward.

- Before pulling the propeller through slowly so the engine sucks gas into the cylinders, the person pulling the propeller should shout, "Brakes on, mag (magneto, i.e., ignition) off." The pilot inside the plane should confirm by repeating, "Brakes on, mag off." But the person on the propeller should always assume the mag switch is on. That is, (s)he should always be in a safe position relative to the propeller and keep the area clear.
- Before moving the propeller, the pilot outside should check the brakes by pushing on the innermost portions of the propeller blades to see that the airplane cannot be pushed backward.
- After the propeller is pulled through a few times, the propeller should be positioned with the left side (when facing the airplane from the front) at the 10 to 11 o'clock position. This will facilitate spinning the propeller by pulling down against the engine's compression.
- When the person on the propeller is ready to attempt starting and the propeller (left blade when facing the nose of the plane) is at 10 to 11 o'clock, the person on the propeller should shout, "Brakes on, throttle cracked, mag on." The pilot in the plane should make the required adjustments and repeat, "Brakes on, throttle cracked, mag on."
- Note that some pilots prefer to use the term "contact" rather than "mag on."
- After checking to see that the brakes are on (by pushing the propeller), the person on the propeller should then spin the propeller as hard as possible by pulling down on the left blade. (S)he should stand on firm ground and close enough to the propeller to be able to step away easily. As the propeller is pulled through, the person should step back away from the propeller to avoid being hit as the engine starts.

  c) If you, as the pilot, are starting the engine yourself, you should stand behind the propeller on the right side when facing the direction the airplane is facing. This position will permit you to place one foot in front of a main tire in addition to having the brakes set full on or having the airplane tied down (both if possible). Note that this is an extremely dangerous procedure. USE EXTREME CARE.

   i) Some pilots also advocate this approach when there is a pilot in the cockpit at the controls.

 h. **Carburetor fire hazard**

  1) In carburetor-equipped airplanes, there is some danger of fire in the carburetor on engine start.

   a) Carburetor fires can be the result of over-priming before engine start.

    i) This may be the result of excess priming with a primer or opening the throttle beyond a safe point at engine start.

    ii) Excess fuel in the carburetor may ignite when the starter is engaged.

    iii) There is typically an audible backfire.

  2) To extinguish the fire, continue engaging the starter while simultaneously ensuring the primer is locked, closing the throttle, and leaning the mixture.

   a) The fire should be starved and the engine will turn over normally.

2. **Exhibit your instructional knowledge by explaining the following common errors related to engine starting.**

 a. **Failure to properly use the appropriate checklist**

  1) You must instill in your students the habit of properly using the correct checklist for engine starting.

  2) The use of checklists ensures that every item is completed and checked in a logical order.

 b. **Failure to use safety precautions related to starting**

  1) Remember to position the airplane away from any person or property.

   a) Any debris close to the airplane will be blown behind you during the start.

  2) Visually and verbally clear the area before starting.

 c. **Improper adjustment of engine control during start**

  1) Constantly monitor the engine instruments while the engine is operating.

 d. **Failure to assure proper clearance of the propeller**

  1) During the visual inspection, the propeller path should be checked for debris or obstructions, especially on the ground.

  2) Before starting, assure that no person or object will be struck by the propeller.

3. **Demonstrate and simultaneously explain engine starting from an instructional standpoint.**

    a. Your knowledge, understanding, and demonstration of engine starting must be to a commercial pilot level.

        1) Since there are no definitive standards in either the private or commercial PTSs, your level of performance will be determined by your FAA inspector/examiner through the exercise of subjective judgment.

    b. You will be evaluated on your ability to explain and demonstrate simultaneously the key elements of engine starting.

    c. Point out any deficiency in your engine starting procedures to your FAA inspector/examiner as an example of an error, and explain how it should be corrected (including how and why it occurred).

4. This task does not require you to analyze and correct common errors of engine starting, but you can and will (as a CFI) critique your student's engine starting procedures.

    a. If your FAA inspector/examiner does not perform the engine starting procedures properly, point out any deficiency to him/her as an example of a student error, and explain how it should be corrected (including how and why it occurred).

END OF TASK

TAXIING

V.D. TASK: TAXIING

> REFERENCES: FAA-H-8083-3; FAA-S-8081-12, FAA-S-8081-14; POH/AFM.

Objective. To determine that the applicant:

1. Exhibits instructional knowledge of the elements of taxiing by describing --

   a. Proper brake check and correct use of brakes.

   b. Compliance with airport/taxiway surface marking, signals, and ATC clearances or instructions.

   c. How to control direction and speed.

   d. Flight control positioning for various wind conditions.

   e. Procedures used to avoid other aircraft and hazards.

   f. Avoiding runway incursions.

2. Exhibits instructional knowledge of common errors related to taxiing by describing --

   a. Improper use of brakes.

   b. Improper positioning of the flight controls for various wind conditions.

   c. Hazards of taxiing too fast.

   d. Failure to comply with airport/taxiway surface marking, signals, and ATC clearances or instructions.

3. Demonstrates and simultaneously explains taxiing from an instructional standpoint.

4. Analyzes and corrects simulated common errors related to taxiing.

A. General Information

1. The objective of this task is to determine your instructional knowledge of the elements and common errors related to taxiing. You will also need to demonstrate and simultaneously explain taxiing from an instructional standpoint.

2. See Chapter 3, Airports, Air Traffic Control, and Airspace, in *Pilot Handbook*, for the following:

   a. Module 3.1, Runway and Taxiway Markings for a 15-page discussion on taxiway, holding position, and other markings and airport signs.

   b. Module 3.12, Ground Control, for a one-page discussion on ATC clearances while taxiing.

3. You must instruct your student to use and follow the taxi checklist, if any, that is in the POH. It may be combined as "after-engine start/taxiing." It is a good habit to include the following in the checklist before starting to taxi.

   a. Set the heading indicator to the magnetic compass.

      1) While taxiing, check that the heading indicator moves freely and indicates known headings, e.g., runway direction.

   b. Set the attitude indicator. It may not have had enough time for the gyro to stabilize, but within 5 min. it should be stable with a level attitude.

      1) While you are taxiing, the attitude indicator should not indicate a bank.

   c. Set the altimeter to the altimeter setting, if available. If not, set it to the airport elevation.

   d. If the airplane has a clock, set it to the correct time.

   e. By setting the flight instruments before taxiing, you and your student will have a base on which to make a determination of their proper operation.

B.   Task Objectives

    1.   **Exhibit your instructional knowledge by explaining the following elements of taxiing.**

        a.   **Proper brake check and correct use of brakes**

            1)   To perform a brake check on the airplane, begin moving the airplane forward by gradually adding power (push the throttle forward slowly) to increase the engine RPM.

                a)   Reduce the power to idle as soon as the airplane begins rolling, and immediately apply the brakes to stop the forward motion of the airplane.

                b)   If there is any question about the operation of the brakes, shut down the engine immediately and have the brakes checked.

            2)   During your demonstration of taxiing, emphasize to your student the use of proper braking technique.

                a)   Do not "ride" the brakes. Riding the brakes means using the brakes constantly, usually while excessive power is used.

                b)   Develop the skill of taxiing as if the brakes were inoperative. Use the brakes only when a reduction in engine speed is not sufficient.

                c)   Airplane brakes are generally toe brakes, i.e., on the top of the rudder pedals. Thus, when pressure is applied to the top of a rudder pedal, that brake is activated (i.e., like a brake in a car).

                    i)   Normally (i.e., when not applying the brakes) the foot should be positioned on the rudder pedal so that the ball of the foot is at the bottom of the pedal and the heel of the foot rests on the floorboard. This foot position minimizes the possibility of riding the brakes.

                d)   Some older airplanes have heel brakes, which are activated by separate small brakes just in front and to the inside of the rudder pedals. These, as the description implies, are activated by the heel of the foot.

            3)   Always emphasize the use of both brakes together with the same pressure. The ONLY time differential braking should be used for directional purposes is for sharp turns at VERY LOW speeds, i.e., barely moving. At all other times, apply the brakes evenly, and use the airplane's nosewheel or tailwheel (steerable with the rudder pedals) for directional control.

            4)   Completion standards for this element (commercial and private PTSs)

                a)   ***Perform a brake check immediately after the airplane begins moving.***
                b)   ***Control direction and speed without excessive use of brakes.***

        b.   **Compliance with airport/taxiway surface marking, signals, and ATC clearances or instructions**

            1)   While demonstrating taxiing procedures, explain the meaning of, and comply with, airport markings, signals, and ATC clearances.

                a)   As a CFI, quiz your student to ensure that (s)he knows the meanings of airport markings, signs, signals, and ATC clearances.

        c.   **How to control direction and speed**

            1)   Taxiing is the controlled movement of the airplane under its own power while on the ground.

            2)   The brakes are used primarily to stop the airplane at a desired point, to slow the airplane, or to aid in making a sharp controlled turn.

                a)   Whenever used, they must be applied smoothly, evenly, and cautiously at all times.

3) More engine power may be required to start moving the airplane forward, or to start or stop a turn, than is required to keep it moving in any given direction.

   a) When extra power is used, the throttle should immediately be retarded once the airplane begins moving to prevent accelerating too rapidly.

4) Usually when operating on a soft or muddy field, you must maintain the taxi speed or power slightly above that required under normal field operations; otherwise, the airplane may come to a stop.

   a) Full power may be required to get the airplane moving, causing mud or stones to be picked up by the propeller and resulting in damage.

   b) The use of additional power during taxiing will result in more slipstream acting on the rudder, thus providing better control.

5) Taxiing nosewheel airplanes

   a) Taxiing an airplane equipped with a nosewheel is relatively simple. Nosewheel airplanes generally have better ground-handling characteristics (relative to tailwheel airplanes). The nosewheel is usually connected to the rudder pedals by a mechanical linkage.

   b) When starting to taxi, the airplane should always be allowed to roll forward slowly so the nosewheel turns straight ahead in order to avoid turning into an adjacent airplane or nearby obstruction.

   c) All turns conducted with a nosewheel airplane are started using the rudder pedals.

      i) Power may be applied after entering the turn to counteract the increase in friction during the turn.

      ii) If it is necessary to tighten the turn after full rudder pedal deflection has been reached, the inside brake may be used as needed to aid in turning the airplane.

   d) When stopping the airplane, always stop with the nosewheel straight in order to relieve any strain on the nose gear and to make it easier to start moving again.

      i) In particular, you should straighten the nosewheel when positioning the airplane for the before-takeoff (pretakeoff) checklist during which you run up (operate at relatively high RPM) the airplane's engine.

6) Taxiing tailwheel airplanes

   a) Taxiing a tailwheel-type airplane is usually more difficult than taxiing nosewheel-equipped airplanes because the tailwheel provides less directional control than a nosewheel. Also, tailwheel airplanes tend to turn so the nose of the aircraft points itself into the wind (this is referred to as weathervaning).

      i) The tendency for tailwheel airplanes to weathervane is greatest in a crosswind situation.

      ii) Generally, brakes play a much larger role in taxiing tailwheel-equipped airplanes.

b)   Since a tailwheel-type airplane rests on its tailwheel as well as on the main landing wheels, it assumes a nose-high attitude when on the ground. In most cases, this attitude causes the engine to restrict the pilot's forward vision.

    i)   It may be necessary to weave the airplane right and left while taxiing to see and avoid collision with any objects or hazardous surface conditions in front of the nose.

    ii)   The weave, zigzag, or short S-turns must be done slowly, smoothly, and cautiously.

7)   There is no set rule for a safe taxiing speed. What is safe under some conditions may be hazardous under others.

a)   The primary requirement is safe, positive control -- the ability to stop or turn where and when desired.

b)   Normally, the speed should be at a rate at which movement of the airplane is dependent on the throttle, that is, slow enough so that when the throttle is closed, the airplane can be stopped promptly.

8)   Very sharp turns or attempting to turn at too great a speed must be avoided as both tend to exert excessive pressure on the landing gear, and such turns are difficult to control once started.

d.   **Flight control positioning for various wind conditions**

1)   The wind is a very important consideration when operating an airplane on the ground (especially tailwheel airplanes). The objective is to keep the airplane firmly on the ground, i.e., not let the wind blow the airplane around.

a)   If a wind from the side gets under the wing, it can lift the wing up and even blow the airplane over sideways. A wind from the rear can get under the tail of the airplane and blow the airplane over to the front.

b)   Caution is recommended. Avoid sudden bursts of power and sudden braking.

2) When taxiing in windy conditions, the control surfaces should be positioned as shown in the following diagram.

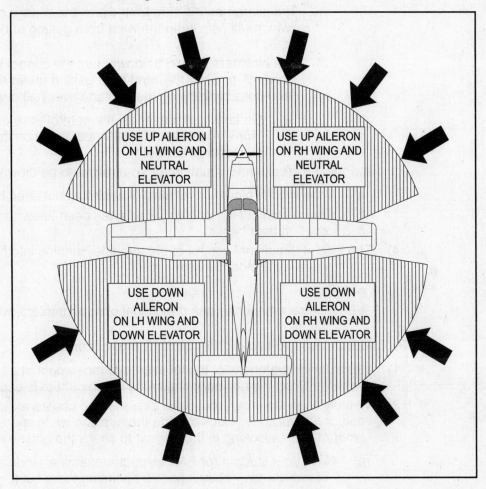

USE UP AILERON ON LH WING AND NEUTRAL ELEVATOR

USE UP AILERON ON RH WING AND NEUTRAL ELEVATOR

USE DOWN AILERON ON LH WING AND DOWN ELEVATOR

USE DOWN AILERON ON RH WING AND DOWN ELEVATOR

a) When the wind is from any forward direction, the control yoke should be turned or pushed fully toward the wind.

   i) The aileron on the side from which the wind is coming will be up, and the wind flowing over the wing will hold the wing down (rather than lifting the wing, which would permit the wind to get under the wing and possibly blow the airplane over on its back).

   ii) The elevators should be in a neutral position, i.e., the control yoke held neither forward nor back, to permit the nosewheel to carry its normal weight and be used for directional control.

      • On tailwheel airplanes, the elevators should be up, i.e., control yoke or stick pulled back, to keep the tail firmly down so the tailwheel can provide directional control.

      b)   When the wind is from any rearwind direction, the control yoke should be turned or pushed fully away from the wind.

          i)   The aileron on the side from which the wind is coming will be down, which will help keep the wind from getting under the wing and lifting it.

          ii)   The elevators should be down, i.e., the control yoke pushed full forward, to deter the wind from getting under the tail, raising the tail, and possibly blowing the airplane over (tail over front).

              •   On tailwheel airplanes, the control yoke or stick is also held full forward to keep the tailwheel firmly on the ground for directional control.

3)   It would be an extreme situation for an airplane to be blown (flipped) on its back.

      a)   If the wind is blowing that hard, you should not even be in the airplane.

      b)   Also, watch out for jet blast, which has been known to blow small airplanes over on their backs.

4)   The loss of directional control of the airplane, running into something, or running off the runway is most likely due to

      a)   Panic

      b)   Incorrect response to a directional control problem, which aggravates the situation

e.   **Procedures used to avoid other aircraft and hazards**

1)   Maintaining awareness of the location and movement of all other aircraft and vehicles along the taxi path and in the traffic pattern is essential to safety.

2)   Visually scan the area around the airplane and constantly look for other traffic and/or obstructions. Look outside the airplane while spending a minimum amount of time looking in the cockpit to check the engine and flight instruments.

      a)   Have your student (or FAA inspector/examiner) indicate his/her awareness of traffic and/or obstructions by pointing them out to you.

3)   Monitor the appropriate radio frequency (i.e., ground control or the CTAF) for traffic and possible conflicts.

4)   Apply right-of-way rules and maintain adequate spacing behind other aircraft.

      a)   Generally, the right-of-way rules apply as they do while in the air. That is, approaching head-on, alter your course to the right; yield to an airplane on the right.

          i)   Ground control (at an airport with an operating control tower) may instruct one aircraft to stop or yield to another.

          ii)   If in doubt, always yield to other aircraft. Be safe.

      b)   Avoid being too close to another airplane's prop or jet wash, which could cause you to lose control of the airplane. Maintain a safe separation.

5)   Avoid creating hazards to other people or property while taxiing.

      a)   Be sure that the airplane's wings will clear all other airplanes or obstructions.

          i)   If in doubt, stop.

      b)   Avoid prop-washing people, aircraft, or vehicles while taxiing.

          i)   FAR 91.13 prohibits a pilot from operating an airplane in a careless or reckless manner that endangers the life or property of another.

          ii)   Be polite when operating around people and/or property.

f.   Avoiding runway incursions

1)   The potential for runway incidents and accidents can be reduced through adequate planning, coordination, and communication. The following are some practices to help prevent a runway incursion:

   a)   Read back all runway/taxiway crossing and/or hold instructions.

   b)   Review airport layouts prior to taxi, before landing if able, and while taxiing as needed.

   c)   Review NOTAMs for up-to-date information.

   d)   Be familiar with airport signs and markings.

   e)   Request progressive taxi from ATC when unsure of taxi route.

   f)   Check for traffic before crossing any runway or taxiway or entering a runway or taxiway.

   g)   Make sure aircraft position and taxi lights are on whenever aircraft is moving.

   h)   When landing, clear the active runway in a timely fashion.

   i)   Use proper phraseology and good radio discipline at all times.

   j)   Write down complex taxi instructions.

2.   **Exhibit your instructional knowledge by explaining the following common errors related to taxiing.**

a.   Improper use of brakes

1)   The most common error is the tendency to ride the brakes while taxiing.

   a)   Correct this error by using the throttle to slow the airplane down, and use the brakes to stop the airplane completely.

b.   Improper positioning of the flight controls for various wind conditions

1)   Always know the direction of the wind in relation to the airplane. Use all available means to determine direction, such as wind sock or ground control.

2)   Picture the wind relative to the airplane at any given time by means of the heading indicator.

   a)   EXAMPLE: If the airplane is heading 090° and the wind is from 240°, use the heading indicator to determine that the wind is a right-quartering tailwind.

3)   Coach your student on the proper position and, after having explained the reason on a previous flight, ask him/her to explain to you why the controls are positioned for the wind.

c.   Hazards of taxiing too fast

1)   This error occurs from the improper use of the throttle and sometimes results from feeling rushed to get to the run-up area.

2)   Taxi slowly in the ramp area and at a speed at which you can stop or turn where and when you desire.

   a)   Normally it should be at such a speed that, when the throttle is closed, the airplane can be stopped promptly.

    d.   **Failure to comply with airport/taxiway surface marking, signals, and ATC clearances or instructions**

        1)   Before starting to taxi at a controlled airport, ask yourself whether the taxi instructions make sense and whether you understand the clearance.

            a)   Contact ground control for clarification.

        2)   Before starting to taxi at a controlled airport, ask your student to repeat to you the taxi instructions. If you hear something wrong, take the following steps:

            a)   Question your student.
            b)   Instruct him/her to contact ground control for clarification.

        3)   While taxiing, ask your student to identify markings and signals.

            a)   When approaching critical areas, such as hold lines or intersections, ask your student if (s)he is cleared to cross or if (s)he must hold.

3.   **Demonstrate and simultaneously explain taxiing from an instructional standpoint.**

    a.   Your knowledge, understanding, and demonstration of taxiing must be to a commercial pilot level.

        1)   Since there are no definitive standards in either the private or commercial PTSs, your level of performance will be determined by your FAA inspector/examiner through the exercise of subjective judgment.

    b.   You will be evaluated on your ability to simultaneously explain and demonstrate the key elements of taxiing.

    c.   Point out any deficiency in taxiing to your FAA inspector/examiner as an example of an error, and explain how it should be corrected (including how and why it occurred).

4.   **Analyze and correct simulated common errors related to taxiing.**

    a.   As your FAA inspector/examiner is taxiing the airplane, you are required to recognize, analyze, and correct any error.

    b.   Explain what the error is, and then explain what (s)he should do to correct the error.

    c.   If your FAA inspector/examiner does not perform proper taxiing procedures, point out the deficiency to him/her as an example of a student error, and explain how it should be corrected (including how and why it occurred).

END OF TASK

BEFORE TAKEOFF CHECK

V.G. TASK: BEFORE TAKEOFF CHECK

REFERENCES: AC 60-14; FAA-H-8083-3; FAA-S-8081-12, FAA-S-8081-14; POH/AFM.

Objective. To determine that the applicant:

1. Exhibits instructional knowledge of the elements of the before takeoff check by describing --

   a. Positioning the airplane with consideration for other aircraft, surface conditions, and wind.

   b. Division of attention inside and outside the cockpit.

   c. Importance of following the checklist and responding to each checklist item.

   d. Reasons for ensuring suitable engine temperatures and pressures for run-up and takeoff.

   e. Method used to determine that airplane is in a safe operating condition.

   f. Importance of reviewing takeoff performance airspeeds, expected takeoff distances, and emergency procedures.

   g. Method used for ensuring that the takeoff area or path is free of hazards.

   h. Method of avoiding runway incursions and ensures no conflict with traffic prior to taxiing into takeoff position.

2. Exhibits instructional knowledge of common errors related to the before takeoff check by describing --

   a. Failure to properly use the appropriate checklist.

   b. Improper positioning of the airplane.

   c. Improper acceptance of marginal engine performance.

   d. An improper check of flight controls.

   e. Hazards of failure to review takeoff and emergency procedures.

   f. Failure to avoid runway incursions and ensure no conflict with traffic prior to taxiing into takeoff position.

3. Demonstrates and simultaneously explains a before takeoff check from an instructional standpoint.

4. Analyzes and corrects simulated common errors related to a before takeoff check.

A. General Information

   1. The objective of this task is to determine your instructional knowledge of the elements and common errors related to the pretakeoff check. You will also need to demonstrate and simultaneously explain a pretakeoff check from an instructional standpoint.

   2. The before takeoff (pretakeoff) check is the systematic procedure for making a last-minute check of the engine, controls, systems, instruments, and radio prior to flight.

      a. Normally, it is performed after taxiing to a position near the takeoff end of the runway.

      b. Taxiing to that position usually allows sufficient time for the engine to warm up to at least minimum operating temperatures and ensures adequate lubrication of the internal moving parts of the engine before operating the engine at high power settings.

      c. Some people differentiate between a ground checklist and a pretakeoff list. We have combined both, as in this task. Both are done on a taxiway near the end of the active runway. The objective is to ascertain that all systems, instruments, etc., are working properly and are ready for flight.

   3. Your POH will explain the proper operating limitations while you are performing your before takeoff check.

      a. Any deviation from these normal operating limits means that there is a possible malfunction and you should return to the ramp to determine the cause.

B. Task Objectives

1. **Exhibit your instructional knowledge by explaining the following elements of the before takeoff check.**

    a. **Positioning the airplane with consideration for other aircraft, surface conditions, and wind**

       1) As you taxi into the designated run-up area, turn the airplane somewhat diagonal so you will not prop blast any aircraft behind you.

       2) The FAA recommends that the airplane be positioned into the wind, as nearly as possible, to obtain more accurate operating indications and to minimize engine overheating when the engine is run up.

          a) In older airplanes with radial engines, the rule was to turn into the wind to provide as much cooling as possible for the engine. Generally, cooling is not a problem for most modern airplane engines.

          b) If the airplane has problems with overheating, position the airplane into the wind to obtain the maximum cooling effect possible.

       3) The airplane should be positioned on a firm surface, smooth turf, or paved surface that is free of debris.

          a) Otherwise, the propeller will pick up pebbles, dirt, mud, sand, or other loose particles and hurl them backward, not only damaging the tail of the airplane, but often inflicting damage to the propeller itself.

       4) Straighten the airplane's nosewheel before stopping, as the magneto check requires an engine run-up which puts considerable stress on the airplane's nosewheel (which is better absorbed with the nosewheel straight).

       5) Position the airplane in a direction so that, in the event of brake failure, the airplane will not run into another aircraft, a ditch, a sign, etc.

    b. **Division of attention inside and outside the cockpit** (especially during the engine run-up)

       1) If the parking brake slips or if the application of the toe brakes is inadequate for the amount of power applied, the airplane could move forward unnoticed if your and your student's attention is fixed inside the airplane.

    c. **Importance of following the checklist and responding to each checklist item**

       1) Follow the appropriate checklist, item by item.

       2) Be critical of the airplane's performance, and determine whether the airplane meets the performance guidelines in the POH.

          a) Do not accept any unacceptable levels of airplane performance.

       3) As you read each checklist item aloud, demonstrate to your student by touching the indicated control and switching or adjusting it to the prescribed position.

          a) If the control is an instrument gauge, the instrument reading should be said aloud and pointed at.

          b) Instruct your student that (s)he should not just perform the procedures but rather should interpret them to you.

d. **Reasons for assuring suitable engine temperatures and pressures for run-up and takeoff**

1) Scan all the airplane's engine instruments (including temperatures and pressures) periodically from engine start on to ensure they are suitable for engine run-up and takeoff. Do not let the airplane's engine overheat.

2) Most of the engine warm-up will have been conducted during taxi.

3) If equipped, a cylinder head temperature gauge is useful to determine adequate operating temperature.

    a) Without this type of instrument, the takeoff can be made when the throttle can be advanced to full power without the engine faltering.

4) In extremely cold weather, the cylinder heads may overheat before an indication is shown on the oil temperature gauge.

    a) This overheating may occur because the oil heats more slowly in cold weather.

5) Operating too cold an engine can be just as hazardous. Even after winter preheating, you may have to run the airplane's engine with the cowl flap (if equipped) closed to attain sufficient temperature for takeoff.

e. **Method used to determine that airplane is in a safe operating condition**

1) You, as the pilot in command, are responsible for determining whether the airplane is in condition for safe flight (FAR 91.7). Remember that everything on your checklist is very important to ensure that the airplane is safe for flight.

2) Stop at each discrepancy and note its effect(s). How is any problem covered by another instrument, piece of equipment, pilot workload, etc.? Relate problems to FARs.

3) Exercise sound judgment in determining that the airplane is safe for flight.

    a) If you have any doubts, return to the ramp for further investigation.

f. **Importance of reviewing takeoff performance airspeeds, expected takeoff distances, and emergency procedures**

1) Review the V_R, V_X, V_Y, and other takeoff performance airspeeds for the airplane.

    a) As you reach these airspeeds, plan to call them out loud.

2) From your preflight planning, you have already determined the expected takeoff distance for the conditions.

    a) Confirm that the runway and wind conditions are adequate to meet performance expectations.

3) Takeoff emergency procedures are set forth in Section 3, Emergency Procedures, of the POH. Prepare ahead for all contingencies. Be prepared at all times to execute an emergency landing if you lose an engine. Remember, **maintain airspeed** so you control your situation rather than enter a stall/spin.

    a) The most common emergency on takeoff is the loss of engine power during the takeoff roll or during the takeoff climb.

        i) If engine power is lost during the takeoff roll, pull the throttle to idle, apply the brakes, and slow the airplane to a stop.

        ii) If you are just lifting off the runway and you lose your engine power, try to land the airplane on the remaining runway. Leave it in the flair attitude, which it is already in. It will settle back down to the ground; i.e., land it like a normal landing.

            • It is very important not to lower the nose because you do not want to come down on the nosewheel.

        iii) If engine power is lost any time during the climbout, a general rule is that, if the airplane is above 500 to 1,000 ft. AGL, you may have enough altitude to turn back and land on the runway from which you have just taken off. This decision must be based on distance from airport, wind condition, obstacles, etc.

            • Watch your airspeed! Avoiding a stall is the most important consideration. Remember that the control yoke should be forward (nose down) for more airspeed.

        iv) If the airplane is below 500 ft. AGL, do not try to turn back. If you turn back, you will probably either stall or hit the ground before you get back to the runway.

            • The best thing to do is to land the airplane straight ahead. Land in a clear area, if possible.

            • If you have no option but to go into trees, slow the airplane to just above the stall speed (as close to the treetops as possible) to strike the trees with the slowest forward speed possible.

4) You should review your departure procedure before you depart the run-up area.

    a) Know your initial direction of flight after takeoff.

    b) At a controlled airport, ATC will issue you a clearance on how to depart the traffic pattern.

    c) At an uncontrolled airport, you should depart the traffic pattern by continuing straight out or exiting with a 45° left turn (right turn if the runway has a right-hand traffic pattern) beyond the departure end of the runway, after reaching traffic pattern altitude.

g. **Method used for assuring that the takeoff area or path is free of hazards**

1) Prior to taxiing onto the runway, visually check that the takeoff area or path is clear of other aircraft, vehicles, persons, livestock, wildlife (including birds), etc.

2) At controlled airports, this is a function of ATC, but you must also check for hazards.

3) At uncontrolled airports, you should announce your intentions on the CTAF and, if possible, make a 360° turn on the taxiway in the direction of the runway traffic pattern to look for other aircraft.

h. **Method of avoiding runway incursions and ensures no conflict with traffic prior to taxiing into takeoff position**

   1) Traffic separation is a function (not a responsibility) of ATC at controlled airports.
   2) Constant vigilance and monitoring of the CTAF is the process at uncontrolled airports.
   3) Whether at a controlled or an uncontrolled airport, you must make the final judgment in assuring adequate clearance from other traffic.

      a) If any doubt exists, wait for the traffic to clear. Be safe, and do not put yourself in a dangerous position.

   4) As a CFI, you will need to develop this judgment with your student.

2. **Exhibit your instructional knowledge by explaining the following common errors related to the before takeoff (pretakeoff) check.**

   a. **Failure to properly use the appropriate checklist**

      1) You must be in the habit of properly using the appropriate checklist.
      2) The use of checklists ensures that every item is completed and checked in a logical order.

   b. **Improper positioning of the airplane**

      1) Position the airplane so you will not prop blast any aircraft behind you.
      2) The FAA recommends that the airplane be positioned into the wind as nearly as possible.
      3) The airplane should be on a surface that is firm and free of debris.

   c. **Improper acceptance of marginal engine performance**

      1) A pilot may feel that (s)he has to complete a flight at a certain time and therefore may accept marginal engine performance.

         a) Be safe. Marginal engine performance is not acceptable and may lead to a hazardous condition.

   d. **An improper check of flight controls**

      1) The flight controls should be visually checked for proper positioning and movement.
      2) The control yoke should move freely in the full range of positions.
      3) Call aloud the proper position and visually check it.

   e. **Hazards of failure to review takeoff and emergency procedures**

      1) Before taxiing onto the runway, review the critical airspeeds used for takeoff, the takeoff distance required, and takeoff emergency procedures.
      2) You will then be thinking about this review during the takeoff roll. It helps prepare you for any type of emergency that may occur.

   f. **Failure to avoid runway incursions and ensure no conflict with traffic prior to taxiing into takeoff position**

      1) You, the pilot in command, are responsible for collision avoidance.

         a) ATC is not responsible but works with pilots to maintain separation.

      2) Other airplanes are not the only hazards you must look for. Vehicles, people, and livestock could be in a hazardous position during the takeoff.
      3) As a CFI, assist in the development of your student's awareness of what is happening around him/her on the airport.

3. **Demonstrate and simultaneously explain a before takeoff check from an instructional standpoint.**

    a.   Your knowledge, understanding, and demonstration of the before takeoff check must be to a commercial pilot level.

        1)   Since there are no definitive standards in either the private or commercial PTSs, your level of performance will be determined by your FAA inspector/examiner through the exercise of subjective judgment.

    b.   You will be evaluated on your ability to explain and demonstrate simultaneously the key elements of a pretakeoff check.

    c.   Point out any deficiency in the before takeoff check to your FAA inspector/examiner as an example of an error, and explain how it should be corrected (including how and why it occurred).

4. **Analyze and correct simulated common errors related to a before takeoff check.**

    a.   As your FAA inspector/examiner is conducting a pretakeoff check, you must be able to recognize, analyze, and correct any error.

    b.   Explain what the error is, and then explain what (s)he should do to correct the error.

    c.   If your FAA inspector/examiner does not perform the before takeoff check properly, point out any deficiency to him/her as an example of a student error, and explain how it should be corrected (including how and why it occurred).

END OF TASK -- END OF STUDY UNIT

STUDY UNIT VI
AIRPORT OPERATIONS

This study unit explains the three tasks (A-C) of Airport Operations. These tasks include both knowledge and skill. Your FAA inspector/examiner is required to test you on at least one of the three tasks in this area of operation; therefore, you must be prepared for all three.

This study unit covers tasks that you use every time you go flying. We will assume that you have knowledge of the FAA regulatory and suggested procedures for these tasks. If you need to update yourself on these procedures, see the detailed discussion of radio communication procedures, ATC light signals, airport traffic patterns, and airport/runway markings and lighting in Chapter 3, Airports, Air Traffic Control, and Airspace, in *Pilot Handbook*.

RADIO COMMUNICATIONS AND ATC LIGHT SIGNALS

> **VI.A. TASK: RADIO COMMUNICATIONS AND ATC LIGHT SIGNALS**
>
>   REFERENCES: AC 61-23/FAA-H-8083-25; FAA-H-8083-3; FAA-S-8081-12, FAA-S-8081-14;
>   POH/AFM.
>
> **Objective.**   To determine that the applicant:
>
> 1. Exhibits instructional knowledge of the elements of radio communications and ATC light signals by describing --
>
>   a. Selection and use of appropriate radio frequencies.
>
>   b. Recommended procedure and phraseology for radio communications.
>
>   c. Receipt, acknowledgement of, and compliance with, ATC clearances and instructions.
>
>   d. Interpretation of, and compliance with, ATC light signals.
>
> 2. Exhibits instructional knowledge of common errors related to radio communications and ATC light signals by describing --
>
>   a. Use of improper frequencies.
>
>   b. Improper procedure and phraseology when using radio communications.
>
>   c. Failure to acknowledge, or properly comply with, ATC clearances and instructions.
>
>   d. Failure to understand, or to properly comply with, ATC light signals.
>
> 3. Demonstrates and simultaneously explains radio communication procedures from an instructional standpoint.
>
> 4. Analyzes and corrects simulated common errors related to radio communications and ATC light signals.

A. General Information

   1. The objective of this task is to determine your instructional knowledge of the elements and common errors related to radio communications and ATC light signals.

   2. See Chapter 3, Airports, Air Traffic Control, and Airspace, in *Pilot Handbook*, for a six-page discussion on radio communications and phraseology and communications at uncontrolled and controlled airports.

B. Task Objectives

   1. **Exhibit your instructional knowledge by explaining the following elements of radio communications and ATC light signals.**

      a. **Selection and use of appropriate radio frequencies**

         1) You should always continue to work to make your radio technique as professional as possible. Selecting the appropriate frequency is obviously essential.

         2) Your preflight planning should include looking up the frequencies of all facilities that you might use and/or need during your flight.

            a) This information can be obtained from a current *A/FD*, sectional charts, etc.

            b) Write this information on your navigation log, or organize it so you can locate it easily in the cockpit.

         3) You may still have to look up frequencies while you are flying.

         4) Always plan ahead as to frequencies needed.

            a) Listen to hand-offs by your controller to airplanes ahead of you.

            b) Look up frequencies before you need them.

b. **Recommended procedure and phraseology for radio communications**

1) Radio communications is a very important task of flying, especially when you are working with ATC. The single most important concept in radio communication is understanding.

   a) Using standard phraseology enhances safety and is a mark of professionalism in a pilot.

      i) Jargon, chatter, and "CB" slang have no place in aviation radio communications.

2) In virtually all situations, radio broadcasts can be thought of as

   a) Whom you are calling
   b) Who you are
   c) Where you are
   d) What you want to do

3) As an instructor, you are responsible for developing your student's ability to communicate on the radio effectively.

   a) Your student will use the same procedures and phraseology that you use. Thus, you must ensure that you are using proper procedures and phraseology.

c. **Receipt, acknowledgment of, and compliance with, ATC clearances and instructions**

1) Make sure your radios, speakers, and/or headset are in good working order so you can plainly hear radio communications. Acknowledge all ATC clearances by repeating key points, e.g., "Taxi to (or across) Runway 10," "Position and hold," "Clear for takeoff Runway 24," or "Left downwind 6," followed by your call sign.

   a) You must read back all runway hold short instructions.

   b) Always repeat altitudes and headings.

   c) Do not hesitate with "Say again" if your clearance was blocked or you did not hear or understand it.

   d) As appropriate, ask for amplification or clarification, e.g., ask for **progressives** if you need taxi routing instructions.

2) FAR 91.123 states that once you, as pilot in command, obtain a clearance from ATC, you may not deviate from that clearance except in an emergency.

   a) You have the responsibility for the safe operation of your airplane.

   b) If you cannot accept a clearance from ATC (e.g., flying into clouds), inform ATC of the reason you cannot accept and obtain a new clearance.

3) FAR 91.3 states that you, the pilot in command, are directly responsible for, and the final authority as to, the operation of your airplane.

   a) As a safe and competent pilot, you should obtain clarification on any clearance that you do not understand or feel would put you in a bad situation.

d.   **Interpretation of, and compliance with, ATC light signals**

1)   ATC light signals have the meaning shown in the following table:

Light Signal	On the Ground	In the Air
Steady green	Cleared for takeoff	Cleared to land
Flashing green	Cleared for taxi	Return for landing *(to be followed by steady green at proper time)*
Steady red	Stop	Give way to other aircraft and continue circling
Flashing red	Taxi clear of landing area (runway) in use	Airport unsafe -- do not land
Flashing white	Return to starting point on airport	Not applicable
Alternating red and green	General warning signal -- exercise extreme caution	General warning signal -- exercise extreme caution

2)   You can operate at an airport with an operating control tower within Class D, Class E, or Class G airspace without an operating two-way radio if the weather conditions are at or above 1,000 ft. ceiling and/or 3 SM visibility, visual contact with the tower is maintained, and you receive appropriate ATC clearances (i.e., ATC light signals).

a)   Arriving aircraft

   i)   If you receive no response to your transmission inbound, you may have a radio failure.

   ii)   If you are receiving tower transmissions, but none are directed toward you, you should suspect a transmitter failure.

   - Determine the direction and flow of traffic, enter the traffic pattern, and look for light signals.
   - During daylight, acknowledge tower transmissions or light signals by rocking your wings. At night, acknowledge by blinking the landing or navigation lights.
   - After landing, telephone the tower to advise them of the situation.

   iii)   If you are receiving no transmissions on tower or ATIS frequency, suspect a receiver failure.

   - Transmit to the tower in the blind your position, situation, and intention to land.
   - Determine the flow of traffic, enter the pattern, and wait for light signals.
   - Acknowledge signals as described above and by transmitting in the blind.
   - After landing, telephone the tower to advise them of the situation.

        b)   Departing aircraft

           i)   If you experience radio failure prior to leaving the parking area, make every effort to have the equipment repaired before the flight.

           ii)   If you are unable to have the malfunction repaired, call the tower by telephone and request authorization to depart without two-way radio communications.

- If tower authorization is granted, you will be given departure information and requested to monitor the tower frequency or watch for light signals, as appropriate.
- During daylight, acknowledge tower transmissions or light signals by promptly executing action authorized by light signals.
  - When in the air, rock your wings.
  - At night, acknowledge by blinking the landing or navigation lights.

           iii)   If your radio malfunctions after departing the parking area (ramp), watch the tower for light signals or monitor the appropriate (ground or tower) frequency. However, you should return to the ramp.

    3)   Practice light signals. Periodically ask ground control to direct some practice light signals toward your student. Also, periodically review the meaning of the light signals. Most people have a terrible time remembering them. It helps to study the signals in the order that you would expect to receive them. You may want to reproduce the light gun signals chart and place it on your kneeboard permanently.

2.   **Exhibit your instructional knowledge by explaining the following common errors related to radio communications and ATC light signals.**

    a.   **Use of improper frequencies**

        1)   This error is caused by inadequate planning, misreading the frequency on the chart or flight log, or mistuning the frequency on the radio.

        2)   Double-check and read aloud the frequency numbers that are to be set in the radio.

        3)   Monitor the frequency before transmitting. Often you can confirm the correct frequency by listening to other transmissions.

    b.   **Improper procedure and phraseology when using radio communications**

        1)   Think about what you are going to say before you transmit.

        2)   Be sensitive to the controller's workload and tailoring your broadcasts to match. Often pilots are taught correct phraseology only and never taught how to abbreviate transmissions on busy ATC frequencies.

    c.   **Failure to acknowledge, or properly comply with, ATC clearances and instructions**

        1)   This error normally occurs because you did not hear or understand the message.

        2)   Developing your ability to divide your attention will help you not to miss ATC messages.

        3)   Ask ATC to repeat its message or ask for clarification. Do not assume what ATC meant or instructed.

    d.   **Failure to understand, or to properly comply with, ATC light signals**

        1)   Periodically review the different light gun signals and their meanings.

        2)   Using practice light signals, if you operate where you can ask ground control to direct some toward you, will help you learn ATC light signals.

        3)   Reviewing and practicing (if possible) will help you understand and comply with ATC light signals.

3.   **Demonstrate and simultaneously explain radio communication procedures from an instructional standpoint.**

    a.   You should be able to explain and demonstrate radio communication.

    b.   You might also ask the control tower to flash you a green or red signal to demonstrate light signals and then explain their meaning and how they are used.

    c.   Point out any deficiency you have while operating the radio or discussing ATC light signals to your FAA inspector/examiner as an example of an error, and explain how it should be corrected (including how and why it occurred).

4.   **Analyze and correct simulated common errors related to radio communications and ATC light signals.**

    a.   Point out to your FAA inspector/examiner any deficiency (s)he has while operating the radio or discussing ATC light signals as an example of a student error, and explain how it should be corrected (including how and why it occurred).

END OF TASK

TRAFFIC PATTERNS

VI.B. TASK: TRAFFIC PATTERNS
REFERENCES: AC 61-23/FAA-H-8083-25; FAA-H-8083-3; AC 90-42, AC 90-66; FAA-S-8081-12,
FAA-S-8081-14; AIM.

Objective.   To determine that the applicant:

1. Exhibits instructional knowledge of the elements of traffic patterns by describing --

   a. Operations at airport with and without operating control towers.

   b. Adherence to traffic pattern procedures, instructions, and rules.

   c. How to maintain proper spacing from other traffic.

   d. How to maintain the desired ground track.

   e. Wind shear and wake turbulence avoidance procedures.

   f. Orientation with the runway in use.

   g. How to establish a final approach at an appropriate distance from the runway or landing area.

   h. Use of checklist.

2. Exhibits instructional knowledge of common errors related to a traffic patterns by describing --

   a. Failure to comply with traffic pattern instructions, procedures, and rules.

   b. Improper correction for wind drift.

   c. Inadequate spacing from other traffic.

   d. Poor altitude or airspeed control.

3. Demonstrates and simultaneously explains traffic patterns from an instructional standpoint.

4. Analyzes and corrects simulated common errors related to traffic patterns.

A. General Information

   1. The objective of this task is to determine your instructional knowledge of the elements and common errors related to traffic patterns. You will also need to demonstrate and simultaneously explain traffic patterns from an instructional standpoint.

   2. See Chapter 3, Airports, Air Traffic Control, and Airspace, in *Pilot Handbook*, for

      a. A two-page discussion on airport traffic patterns
      b. A seven-page discussion on wake turbulence

   3. Safety first! Commit to it and practice it. Always look for traffic and talk about it, and help your student to develop this habit. Ask your FAA inspector/examiner to watch for traffic.

B. Task Objectives

   1. **Exhibit your instructional knowledge by explaining the following elements of traffic patterns.**

      a. **Operations at airports with and without operating control towers**

         1) At an airport with an operating control tower, the controller will direct when and where you should enter the traffic pattern.

            a) Once in the pattern, the controller may request that you perform some maneuvers for better traffic spacing, including

               i) Shortening or extending the downwind leg,
               ii) Increasing or decreasing your speed, or
               iii) Performing a 360° turn or S-turns to provide spacing ahead of you.

2)   To enter the traffic pattern at an airport without an operating control tower, inbound pilots are expected to observe other aircraft already in the pattern and to conform to the traffic pattern in use.

    a)   If no other aircraft are in the pattern, traffic and wind indicators on the ground must be checked to determine which runway and traffic pattern direction should be used.

        i)   Overfly the airport at least 500 to 1,000 ft. above the traffic pattern altitude.

        ii)   After the proper traffic pattern direction has been determined, proceed to a point well clear of the pattern before descending to the pattern altitude.

    b)   When approaching an airport for landing, enter the traffic pattern at a 45° angle to the downwind leg at the midpoint of the runway.

        i)   Always be at the proper traffic pattern altitude before entering the pattern.

3)   Departing a traffic pattern

    a)   At controlled airports, ATC will generally approve the most expedient turnout for the direction of flight.

    b)   At uncontrolled airports, depart straight out or with a 45° turn in the direction of the traffic pattern after reaching pattern altitude.

b.   **Adherence to traffic pattern procedures, instructions, and rules**

1)   Established airport traffic patterns assure that air traffic flows into and out of an airport in an orderly manner. Airport traffic patterns establish

    a)   The direction and placement of the pattern

        i)   At uncontrolled airports, left traffic is required, unless indicated by the traffic pattern indicators in the segmented circle.

            • The *A/FD* will indicate right traffic when applicable.

        ii)   At controlled airports, the direction of the traffic pattern will be specified by ATC.

    b)   The altitude at which the pattern is to be flown

        i)   The normal traffic pattern altitude for small airplanes is 1,000 ft. AGL, unless otherwise specified in the *A/FD*.

        ii)   The traffic pattern altitude for turbine-powered and large airplanes is 1,500 ft. AGL, and their pattern will normally be wider than those of single-engine airplanes.

    c)   The procedures for entering and departing the pattern

        i)   At uncontrolled airports, the FAA recommends entering the pattern at a 45° angle abeam the midpoint of the runway on the downwind leg.

            • When departing the traffic pattern, airplanes should continue straight out or exit with a 45° left turn (right turn for right traffic pattern) beyond the departure end of the runway after reaching pattern altitude.

        ii)   At controlled airports, ATC will specify pattern entry and departure procedures.

2) There is a basic rectangular airport traffic pattern which you should use unless modified by ATC or by approved visual markings at the airport. Thus, all you need to know is

   a) The basic rectangular traffic pattern

   b) Visual markings and typical ATC clearances which modify the basic rectangular pattern

   c) Reasons for modifying the basic pattern

3) The traffic pattern

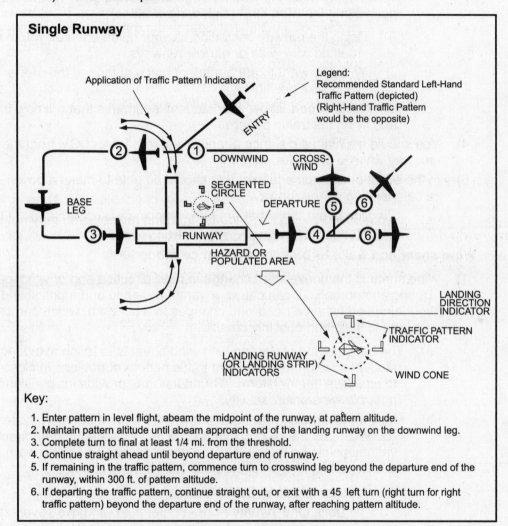

Single Runway

Application of Traffic Pattern Indicators

Legend:
Recommended Standard Left-Hand Traffic Pattern (depicted)
(Right-Hand Traffic Pattern would be the opposite)

ENTRY

② DOWNWIND ① CROSS-WIND

BASE LEG

SEGMENTED CIRCLE

DEPARTURE ⑤ ⑥

③ RUNWAY ④ ⑥

HAZARD OR POPULATED AREA

LANDING DIRECTION INDICATOR

TRAFFIC PATTERN INDICATOR

LANDING RUNWAY (OR LANDING STRIP) INDICATORS

WIND CONE

Key:

1. Enter pattern in level flight, abeam the midpoint of the runway, at pattern altitude.
2. Maintain pattern altitude until abeam approach end of the landing runway on the downwind leg.
3. Complete turn to final at least 1/4 mi. from the threshold.
4. Continue straight ahead until beyond departure end of runway.
5. If remaining in the traffic pattern, commence turn to crosswind leg beyond the departure end of the runway, within 300 ft. of pattern altitude.
6. If departing the traffic pattern, continue straight out, or exit with a 45 left turn (right turn for right traffic pattern) beyond the departure end of the runway, after reaching pattern altitude.

4) You must comply with appropriate FARs (i.e., speed, weather minimums, operating rules), ATC instructions (if appropriate), and any local airport rules or procedures.

c. **How to maintain proper spacing from other traffic**

1) As you fly in the traffic pattern, you must observe other traffic and maintain separation, especially when smaller airplanes may have relatively slower approach speeds than your airplane.

   a) Faster aircraft typically fly a wider pattern than slower aircraft.

2) At an airport with an operating control tower, the controller may instruct you to adjust your traffic pattern to provide separation.

3) Remember, whether you are at a controlled or an uncontrolled airport, you are responsible for seeing and avoiding other aircraft.

d.   **How to maintain the desired ground track**

1)   The procedures used to correct for wind drift are explained in Task X.A., Rectangular Course, beginning on page 378.

2)   The maximum bank angle that should be used in the traffic pattern is 30°.

3)   You will need to help your student to learn how far from the runway to position the airplane.

a)   Proper positioning can be done by placing the runway centerline at a specified point on the leading edge of the wing (low-wing airplane) or a point along the wing strut (high-wing airplane).

i)   Using the runway centerline as your guide enables you to use this method with wide or narrow runways.

ii)   You may want to put a mark or piece of tape at the proper wing or wing strut position.

b)   At the local airport, show your student landmarks that will help in entering and flying the traffic pattern.

4)   You should maintain a distance of approximately 1/2 to 1 SM from the landing runway while in the traffic pattern.

5)   In the event of an engine failure, you should be able to make a power-off glide to a safe landing on the runway.

a)   If you extend your pattern too far from the runway, you may not be able to glide to the runway, especially in strong winds.

e.   **Wind shear and wake turbulence avoidance procedures**

1)   Wind shear is the unexpected change in wind direction and/or windspeed. During an approach, it can cause severe turbulence and a possible decrease to your airspeed (when a headwind changes to a tailwind), which can cause your airplane to stall (and possibly crash).

a)   The best method of dealing with wind shear is to teach avoidance. Your student should never conduct traffic pattern operations in close proximity to an active thunderstorm. Thunderstorms provide visible signs of possible wind-shear activity.

b)   Many large airfields now have a low-level wind shear alert system (LLWAS). By measuring differences in windspeed and/or direction at various points on the airfield, the controller will be able to warn arriving and departing aircraft of the possibility of wind shear.

i)   An example of an LLWAS alert:

Delta One Twenty Four - center field wind two seven zero at one zero - south boundary wind one four zero at three zero.

ii)   Elsewhere, pilot reports (PIREPs) from airplanes preceding you on the approach can be very informational.

c) If you are conducting an approach with possible wind shear or a thunderstorm nearby, you should consider

    i) Using more power during the approach

    ii) Flying the approach at a faster airspeed (general rule: adding 1/2 the gust factor to your airspeed)

    iii) Staying as high as feasible on the approach until it is necessary to descend for a safe landing

    iv) Initiating a go-around at the first sign of a change in airspeed or an unexpected pitch change. The most important factor is to go to full power and get the airplane climbing.

       • Many accidents caused by wind shear are due to a severe downdraft (or a rapid change from headwind to tailwind) punching the aircraft into the ground. In extreme cases, even the power of an airliner is unable to counteract the descent.

2) In addition to the wingtip vortices, wake turbulence also includes rotor wash (from helicopters) and jet engine exhaust (and propeller) wash.

    a) Helicopters

      i) In a slow hover taxi or a stationary hover near the surface, helicopter main rotor(s) generate(s) downwash producing high-velocity outwash vortices to a distance approximately three times the diameter of the rotor.

         • When rotor downwash hits the surface, the vortex circulation is outward, upward, around, and away from the main rotors in all directions.

         • You should avoid operating your airplane within three rotor diameters of any helicopter in a slow hover taxi or stationary hover.

      ii) In forward flight, departing or landing helicopters produce a pair of strong, high-speed trailing vortices similar to wingtip vortices of larger fixed-wing aircraft.

         • You should use caution when operating behind or crossing behind landing and departing helicopters.

    b) Jet engine exhaust

      i) During ground operations, jet engine blast (thrust stream turbulence) can cause damage and upsets if encountered at close range.

         • Thus, you need to maintain an adequate separation during ground operations.

f. **Orientation with the runway in use**

1) While conducting airport traffic pattern operations, you must remain oriented with the active runway.

2) Know which runway is in use, and plan to enter properly and remain in the correct traffic pattern.

3) When approaching an airport, visualize your position from the airport and the relative direction of the runway. Use the airplane's heading indicator to assist you.

4) Runway incursion is a concern at airports with parallel or intersecting runways in use.

a) Runway incursion avoidance is accomplished by flying the correct traffic pattern for the runway you are to use.

b) Confirm runway number with heading indicator during all traffic pattern legs.

g. **How to establish a final approach at an appropriate distance from the runway or landing area**

1) The descent from traffic pattern altitude to the landing normally begins when the airplane is abeam the touchdown point.

a) The descent may be delayed if you need to extend the downwind leg due to other traffic.

2) Begin the turn to the base leg when the touchdown point is approximately 45° behind the wing.

a) If there is a strong wind, the base leg should be closer to the runway than normal.

i) This is because the decrease in groundspeed on final will allow a steeper approach glide slope.

3) The turn to final should be no closer than 1/4 SM, at an altitude appropriate for the selected glide slope.

a) A general rule: Turn to final at approximately 3/4 SM at 500 ft. AGL.

4) The placement of the base leg, the rate of descent, and how to adjust the pattern when following other aircraft can be very confusing to a student pilot.

a) You must help develop this judgment in your student by explaining clearly the "why" of how one arrives at the appropriate position on final.

h. **Use of checklist**

1) Prior to or as you enter the airport traffic pattern (usually on the downwind leg), conduct a before-landing checklist to be sure that you and your airplane are ready to land.

2) Many pilots use **GUMPS**:

> **G**as
> **U**ndercarriage
> **M**ixture
> **P**rops
> **S**afety belts and shoulder harnesses

2. **Exhibit your instructional knowledge by explaining the following common errors related to traffic patterns.**

   a. **Failure to comply with traffic pattern instructions, procedures, and rules**

      1) Noncompliance with ATC instructions may be caused by not understanding or hearing radio communications.

         a) You must learn to divide your attention while in the traffic pattern between flying, collision avoidance, performing checklists, and radio communications.

      2) As a CFI, you should ask your student questions about his/her plan of action in the traffic pattern to ensure (s)he understands the instructions, procedures, and rules.

   b. **Improper correction for wind drift**

      1) Remember that a traffic pattern is no more than a rectangular course and should be performed in the same manner.

   c. **Inadequate spacing from other traffic**

      1) This error occurs when you turn onto a traffic pattern leg too soon or you are flying an airplane that is faster than the one you are following.

      2) Help your student in developing judgment on the amount of separation needed.

   d. **Poor altitude or airspeed control**

      1) Know the airspeeds at various points in the traffic pattern.
      2) Check the flight and engine instruments.

3. **Demonstrate and simultaneously explain traffic patterns from an instructional standpoint.**

   a. Your demonstration must be to commercial pilot skill level.

      1) Completion standards for this maneuver (commercial and private PTSs)

         a) *Maintain and hold the traffic pattern altitude, ±100 ft., and the appropriate airspeed, ±10 kt.*

   b. You will be evaluated on your ability to demonstrate and explain simultaneously the key elements of traffic patterns.

   c. If you do not perform the maneuver to FAA completion standards, point out the deficiency to your FAA inspector/examiner as an example of an error, and explain how it should be corrected (including how and why it occurred).

4. **Analyze and correct simulated common errors related to traffic patterns.**

   a. You will be evaluated on your ability to recognize, analyze, and then provide a critique of the error(s) your FAA inspector/examiner made in performing traffic patterns.

   b. Explain how and why it was wrong and what (s)he should do to correct the error(s).

   c. If your FAA inspector/examiner does not perform the maneuver to FAA completion standards, point out the deficiency to him/her as an example of a student error, and explain how it should be corrected (including how and why it occurred).

END OF TASK

AIRPORT, RUNWAY AND TAXIWAY SIGNS, MARKINGS, AND LIGHTING

VI.C. TASK: AIRPORT, RUNWAY AND TAXIWAY SIGNS, MARKINGS, AND LIGHTING

REFERENCES: AC 61-23/FAA-H-8083-25; FAA-S-8081-12, FAA-S-8081-14; FAA/ASY-20 95/001; AIM.

Objective. To determine that the applicant:

1. Exhibits instructional knowledge of the elements of airport, runway and taxiway signs, markings, and lighting by describing --

   a. Identification and proper interpretation of airport, runway and taxiway signs and markings with emphasis on runway incursion avoidance.

   b. Identification and proper interpretation of airport, runway and taxiway lighting with emphasis on runway incursion avoidance.

2. Exhibits instructional knowledge of common errors related to airport, runway and taxiway signs, markings, and lighting by describing --

   a. Failure to comply with airport, runway and taxiway signs and markings.

   b. Failure to comply with airport, runway and taxiway lighting.

   c. Failure to use proper runway incursion avoidance procedures.

3. Demonstrates and simultaneously explains airport, runway and taxiway signs, markings, and lighting from an instructional standpoint.

4. Analyzes and corrects simulated common errors related to airport, runway and taxiway signs, markings, and lighting.

A. General Information

   1. The objective of this task is to determine your instructional knowledge of the elements of airport and runway markings and lighting.

      a. This is an instructional knowledge only task.

   2. The FAA has established standard airport and runway markings. Since most airports are marked in this manner, it is important for you to know and understand these markings.

      a. This same standardization is also found in airport lighting and other airport visual aids.

   3. You need to be able to identify, interpret, and explain the various runway and taxiway markings and airport lighting since you may be flying in and out of various airports.

   4. See Chapter 3, Airports, Air Traffic Control, and Airspace, in *Pilot Handbook*, for a 21-page discussion on airport, runway and taxiway signs, markings; airport lighting; visual glideslope indicators; and wind and landing direction indicators and segmented circles.

B. Task Objectives

   1. **Exhibit your instructional knowledge of airport, runway and taxiway signs, markings, and lighting by describing the following:**

      a. **Identification and proper interpretation of airport, runway and taxiway signs and markings with emphasis on runway incursion avoidance.**

         1) Runway markings include

            a) Runway designators
            b) Runway centerline marking
            c) Runway aiming point marking
            d) Runway touchdown zone markers
            e) Runway side stripes
            f) Runway shoulder markings
            g) Runway threshold markings
            h) Runway threshold bar
            i) Demarcation bar
            j) Chevrons

2) Taxiway markings include

a) Taxiway centerline
b) Taxiway edge markings
c) Taxiway shoulder markings
d) Runway holding position markings
e) Holding position markings for instrument landing system
f) Holding position markings for taxiway/taxiway intersections
g) Surface painted taxiway direction signs
h) Surface painted location signs
i) Geographic position markings
j) Surface painted holding position signs

3) Other airport markings include

a) Vehicle roadway markings
b) VOR receiver checkpoint markings
c) Non-movement area boundary markings
d) Marking and lighting of permanently closed runways and taxiways
e) Temporarily closed runway and taxiway markings
f) Helicopter landing areas

4) There are six types of airport signs:

a) Mandatory instruction signs
b) Location signs
c) Direction signs
d) Destination signs
e) Information signs
f) Runway distance remaining signs

b. **Identification and proper interpretation of airport, runway and taxiway lighting with emphasis on runway incursion avoidance.**

1) Airport lighting includes

a) Approach light system
b) Runway lights/runway edge lights
c) Touchdown zone lighting
d) Runway centerline lighting
e) Threshold lights
f) Runway end identifier lights
g) Various types of visual approach slope indicator lights
h) Airport rotating beacon

2. **Exhibit your instructional knowledge by explaining the following common errors related to airport, runway and taxiway signs, markings, and lighting.**

a. **Failure to comply with airport, runway and taxiway signs and markings**

1) Make sure you know the meaning and purpose of every sign and marking.

a) If you are unsure of the meaning or purpose of a sign or marking, stop the airplane and ask ATC for clarification or find the meaning before proceeding.

b. **Failure to comply with airport, runway and taxiway lighting**

1) Because of reduced visibility, it is often hard to find your way on the ground at an unfamiliar airport at night.

a) Make sure you know the meaning of all airport lights before proceeding.

    c.   **Failure to use proper runway incursion avoidance procedures**

        1)   There is no good excuse for a runway incursion.

        2)   At an airport with an operating control tower, do not cross the hold-short lines until you have been cleared for takeoff or cleared to taxi onto or across the runway.

            a)   At an airport without an operating control tower, make sure you look both ways and announce your intentions before taxiing onto the runway.

            b)   Never taxi on the runway when there is another airplane on final approach.

3.   **Demonstrate and simultaneously explain airport, runway and taxiway signs, markings, and lighting from an instructional standpoint.**

    a.   As you are conducting ground operations, you should identify the signs, markings, and lighting you encounter and explain their meanings.

4.   **Analyze and correct simulated common errors related to airport, runway and taxiway signs, markings, and lighting.**

    a.   You, as CFI, will critique your student's ability to identify and interpret these markings and lights.

    b.   If your FAA inspector/examiner has any deficiency in his/her explanation and/or interpretation of airport and runway markings and lighting, explain them to him/her as an example of a student error, and provide the correct interpretation (including how and why it occurred).

END OF TASK -- END OF STUDY UNIT

STUDY UNIT VII
TAKEOFFS, LANDINGS, AND GO-AROUNDS

NOTE: Tasks D and E are seaplane tasks.

This study unit explains the nine tasks (A-C, F-K) of Takeoffs, Landings, and Go-Arounds. These tasks include both knowledge and skill. Your FAA inspector/examiner is required to test you on at least two takeoff and two landing tasks; therefore, you must be prepared for all nine.

This study unit explains and describes the factors involved and the pilot techniques required for safely taking the airplane off the ground, departing the takeoff area, approaching the landing area, and landing the airplane under conditions requiring maximum aircraft performance as well as normal circumstances.

Though the takeoff and climb is one continuous process, it can be divided into phases.

1. The **takeoff roll** is that portion of the maneuver during which your airplane is accelerated to an airspeed that provides sufficient lift for it to become airborne.

2. The **liftoff**, or rotation, is the act of becoming airborne as a result of the wings lifting the airplane off the ground or your rotating the nose up, increasing the angle of attack to start a climb.

3. The **initial climb** begins when your airplane leaves the ground and a pitch attitude is established to climb away from the takeoff area. Normally, it is considered complete when your airplane has reached a safe maneuvering altitude or an en route climb has been established.

Like the takeoff and climb, the approach and landing can be divided into phases.

1. The **base leg** is that portion of the traffic pattern during which you must accurately judge the distance in which your airplane must descend to the landing point.

2. The **final approach** is the last part of the traffic pattern during which your airplane is aligned with the landing runway and a straight-line descent is made to the point of touchdown. The descent rate (descent angle) is governed by your airplane's height and distance from the intended touchdown point and by the airplane's groundspeed.

3. The **roundout**, or **flare**, is that part of the final approach during which your airplane makes a transition from the approach attitude to the touchdown or landing attitude.

4. The **touchdown** is the actual contact or touching of the main wheels of your airplane on the landing surface, as the full weight of the airplane is being transferred from the wings to the wheels.

5. The **after-landing roll**, or **rollout**, is the forward roll of your airplane on the landing surface after touchdown while the airplane's momentum decelerates to a normal taxi speed or a stop.

NORMAL AND CROSSWIND TAKEOFF AND CLIMB

VII.A. TASK: NORMAL AND CROSSWIND TAKEOFF AND CLIMB

REFERENCES: FAA-H-8083; FAA-S-8081-12, FAA-S-8081-14; POH/AFM.

Objective. To determine that the applicant:

1. Exhibits instructional knowledge of the elements of a normal and crosswind takeoff and climb by describing --

   a. Procedures before taxiing onto the runway to ensure runway incursion avoidance.

   b. Normal and crosswind takeoff and lift-off procedures.

   d. Proper climb attitude, power setting, and airspeed (V_Y).

   e. Proper use of checklist.

2. Exhibits instructional knowledge of common errors related to a normal and crosswind takeoff and climb by describing --

   a. Improper runway incursion avoidance procedures.

   b. Improper use of controls during a normal or crosswind takeoff.

   c. Inappropriate lift-off procedures.

   d. Improper climb attitude, power setting, and airspeed (V_Y).

   e. Improper use of checklist.

3. Demonstrates and simultaneously explains a normal or a crosswind takeoff and climb from an instructional standpoint.

4. Analyzes and corrects simulated common errors related to a normal or a crosswind takeoff and climb.

NOTE: Element **1.c.** (omitted) is a seaplane element.

A. General Information

1. The objective of this task is to determine your instructional knowledge of the elements and common errors related to a normal and crosswind takeoff and climb. You will also be required to demonstrate and simultaneously explain this task to your FAA inspector/examiner.

2. Normal takeoff and climb is one in which your airplane is headed directly into the wind or the wind is very light, and the takeoff surface is firm, with no obstructions along the takeoff path, and is of sufficient length to permit your airplane to accelerate gradually to normal climbing speed.

   a. A crosswind takeoff and climb is one in which your airplane is NOT headed directly into the wind.

3. Section 4, Normal Procedures, in your POH will provide you with the proper airspeeds, e.g., V_R, V_Y, and also the proper configuration.

   a. Best rate of climb (V_Y) is the speed that will produce the greatest gain in altitude for a given unit of time. V_Y gradually decreases as the density altitude increases.

B. Task Objectives

1. **Exhibit your instructional knowledge by explaining the following elements of a normal and crosswind takeoff and climb.**

   a. **Procedures before taxiing onto the runway to ensure runway incursion avoidance**

      1) Always make sure to visually clear the area before taxiing onto the runway.

         a) This applies to airports with and without an operating control tower.

2) If you are at an airport with an operating control tower, do not cross the hold-short lines until you are given a clearance to take off.

   a) At an airport without an operating control tower, announce your intentions before taxiing onto the runway.

   b) Never taxi onto the runway when there is another airplane on final approach.

b. **Normal and crosswind takeoff and lift-off procedures**

1) For a normal takeoff, the ailerons should be neutral at the start of the takeoff.

   a) For a crosswind takeoff, the ailerons should be FULLY deflected at the start of the takeoff.

   b) The aileron should be up on the upwind side of the airplane (i.e., the control yoke turned toward the wind).

   c) This aileron position will impose a downward force on the upwind wing to counteract the lifting force of the crosswind and prevent that wing from rising prematurely.

2) Check that the trim is in the proper takeoff setting.

   a) The proper setting may be shown by a mark on the trim indicator that identifies the takeoff position or by a setting in degrees that is specified in the POH.

   b) In a tailwheel airplane, the correct trim setting will allow the airplane to attain the correct takeoff attitude of its own accord.

3) Recheck that the mixture is set in accordance with the POH and the propeller is at high RPM (constant speed propeller).

   a) Power should be added smoothly to allow for a controllable transition to flying airspeed.

      i) Applying power too quickly can cause engine surging, backfiring, and a possible overboost situation (turbocharged engines). These conditions cause unnecessary engine wear as well as possible failure.

      ii) Applying power too slowly wastes runway length.

   b) Use the power setting that is recommended in the POH.

   c) Engine instruments must be monitored during the entire maneuver.

      i) Listen for any indication of power loss or engine roughness.

      ii) Monitoring enables you to notice immediately any malfunctions or indication of insufficient power or other potential problems. Do not commit to liftoff unless all engine indications are normal.

4) Rudder pressure must be promptly and smoothly applied to counteract yawing forces (from wind and/or torque), so the airplane continues straight down the center of the runway.

   a) During a crosswind takeoff roll, you will normally apply downwind rudder pressure since on the ground the airplane (especially tailwheel-type) will tend to weathervane into the wind.

   b) When takeoff power is applied, the elements of torque, which yaws the airplane to the left, may be sufficient to counteract the weathervaning tendency caused by a right crosswind.

      i) On the other hand, a left crosswind may aggravate the left-turning tendency of the airplane.

   c) Use whatever rudder pressure is required to keep the airplane straight down the centerline of the runway.

5) During crosswind takeoffs, the aileron deflection into the wind should be decreased as the airspeed increases.

a) As the forward speed of the airplane increases and the ailerons become effective in controlling the airplane about its longitudinal axis, the holding of full aileron into the wind should be reduced.

b) You will feel increasing pressure on the controls as the ailerons become more effective.

i) The objective is to release enough pressure to keep the wings level.

ii) Some aileron pressure will need to be maintained throughout the takeoff roll to prevent the crosswind from raising the upwind wing.

6) As the airplane accelerates, check the airspeed indicator to ensure that the needle is moving and operating properly.

a) Call out the airspeed as the airplane accelerates to V_R, e.g., "40, 50, 60."

b) If the airspeed indicator is not moving, abort the takeoff, and return to the ramp to determine the problem.

7) The best takeoff attitude requires only minimal pitch adjustments just after liftoff to establish the best rate of climb airspeed, V_Y. The airplane should be allowed to fly off the ground in its normal takeoff (i.e., best rate of climb) attitude, if possible.

a) Your airplane's V_R _____.

8) If your POH does not recommend a V_R, use the following procedure from the FAA's *Flight Training Handbook* (AC 61-21).

a) When all the flight controls become effective during the takeoff roll in a nosewheel-type airplane, back elevator pressure should be gradually applied to raise the nosewheel slightly off the runway, thus establishing the liftoff attitude. This is referred to as rotating.

i) In tailwheel-type airplanes, the tail should first be allowed to rise off the ground slightly to permit the airplane to accelerate more rapidly.

b) At this point, the position of the nose in relation to the horizon should be noted, then elevator pressure applied as necessary to hold this attitude.

i) On both types of airplanes, the wings must be kept level by applying aileron pressure as necessary.

9) As a CFI, explain to your student that forcing the airplane into the air by applying excessive back pressure would only result in an excessively high pitch attitude and may delay the takeoff.

a) Excessive and rapid changes in pitch attitude result in proportionate changes in the effects of torque, thus making the airplane more difficult to control.

b) If your student forces the airplane to leave the ground before adequate speed is attained, the wing's angle of attack may be excessive, causing the airplane to settle back onto the runway or stall.

c) Also, jerking the airplane off the ground reduces passenger comfort.

10) If not enough back pressure is held to maintain the correct takeoff attitude or if the nose is allowed to lower excessively, the airplane may settle back to the runway. This occurs because the angle of attack is decreased and lift is diminished to the point where it will not support the airplane.

11) Many high-performance airplanes require conscious rearward elevator pressure at V_R to establish the liftoff.

    a) Without this conscious control pressure, the airplane may start to wheelbarrow (i.e., the main wheels break ground before the nose wheel).

    b) Note that, in general, high-performance airplanes have heavier control pressures and require more deliberate application of control movements.

12) During takeoffs in a strong, gusty wind, increase V_R to provide an additional margin of safety in the event of sudden changes in wind direction immediately after liftoff.

d. **Proper climb attitude, power setting, and airspeed (V_Y)**

1) At liftoff, the airplane should be at the approximate pitch attitude to establish V_Y. Make small pitch corrections to maintain V_Y, and trim the airplane as necessary.

    a) Your airplane's V_Y _____.

2) Landing gear retraction (if retractable) is normally started when you can no longer land on the remaining runway and a positive rate of climb is established on the V_{SI}.

    a) Before retracting the landing gear, apply the brakes momentarily to stop the rotation of the wheels to avoid excessive vibration on the gear mechanism.

        i) Centrifugal force caused by the rapidly rotating wheels expands the diameter of the tires, and if mud or other debris has accumulated in the wheel wells, the rotating wheels may rub as they enter.

    b) An airplane with retractable landing gear may have a V_Y for both gear up and gear down.

        i) Your airplane's V_Y (gear down) _____
               V_Y (gear up) _____

    c) Make any necessary pitch adjustment to maintain the proper V_Y.

    d) Follow the gear retraction procedure in your POH.

        i) Normally the landing gear is retracted before the flaps.

3) The wing flaps are normally retracted (if extended) after the surrounding terrain and obstacles have been cleared.

    a) Retract the flaps smoothly, and make the needed pitch adjustments to maintain V_Y.

4) After the recommended climbing airspeed (V_Y) has been well established and a safe maneuvering altitude has been reached (normally 500 to 1,000 ft. AGL), the power (manifold pressure and/or RPM) should be adjusted (if appropriate) to the recommended climb setting and the pitch attitude adjusted for cruise climb airspeed.

    a) Cruise climb offers the advantages of higher airspeed for increased engine cooling, higher groundspeed, better visibility ahead of the airplane, and greater passenger comfort.

    b) Most small trainer-type airplane manufacturers recommend maintaining maximum power until reaching your selected cruising altitude.

    c) Follow the procedures in your POH.

5) You must comply with any established noise abatement procedure.

    a) A noise abatement policy is developed by the airport authority or city and is a local ordinance. Thus, a pilot can be cited by the city for violation of the policy.

    b) The *A/FD* will list that an airport has a noise abatement procedure in effect under "airport remarks."

        i) Other pilot guides may contain more detailed information on an airport's noise abatement procedures.

    c) The key to complying with noise abatement is to put as much distance as possible between the airplane and the ground, as quickly as possible.

        i) Use the longest runway available.

        ii) Rotate at V_R and climb out at V_X or V_Y, as recommended in the POH.

        iii) Reduce power to climb power, and transition to a cruise climb as appropriate.

           • The reduction to climb power will reduce the noise of the engine, and the transition to cruise climb airspeed will reduce the time the airplane is over the noise monitors and noise sensitive areas.

6) Completion standards for this element

    a) Commercial PTS

        i) ***Maintain takeoff power and V_Y, ±5 kt. during the climb.***

        ii) ***Retract the landing gear and flaps after a positive rate of climb is established.***

    b) Private PTS

        i) Maintain takeoff power and V_Y, +10/–5 kt. during the climb.

        ii) Retract the landing gear, if appropriate, and flaps after a positive rate of climb is established.

7) In a crosswind takeoff as the nosewheel or tailwheel rises off the runway, holding the aileron control into the wind should result in the downwind wing rising and the downwind main wheel lifting off the runway first, with the remainder of the takeoff roll being made on the other main wheel (i.e., on the side from which the wind is coming).

    a) This result is preferable to side skipping (which would occur if you did not turn the control yoke into the wind and use opposite rudder).

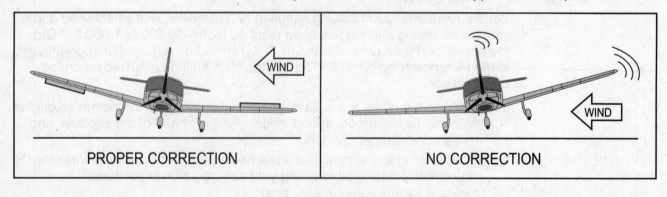

| PROPER CORRECTION | NO CORRECTION |

b)   If a significant crosswind exists, the main wheels should be held on the ground slightly longer than in a normal takeoff so that a smooth but very definite liftoff can be made.

   i)   Accomplish this by slightly less back pressure on the control yoke as you near V_R.

   ii)   This procedure will allow the airplane to leave the ground under more positive control so that it will definitely remain airborne while the proper amount of drift correction is established.

   iii)   More importantly, it will avoid imposing excessive side loads on the landing gear and prevent possible damage that would result from the airplane settling back to the runway while drifting (due to the crosswind).

c)   As both main wheels leave the runway and ground friction no longer resists drifting, the airplane will be slowly carried sideways with the wind unless you maintain adequate drift correction.

8)   In the initial crosswind climb, the airplane will be slipping (upwind wing down to prevent drift and opposite rudder to align your flight path with the runway) into the wind sufficiently to counteract the drifting effect of the wind and to increase stability during the transition to flight.

a)   After your airplane is safely off the runway and a positive rate of climb is established, the airplane should be headed toward the wind to establish just enough crab to counteract the wind, and then the wings should be rolled level. The climb while in this crab should be continued so as to follow a ground track aligned with the runway centerline.

b)   Center the ball in the inclinometer with proper rudder pressure throughout the climb.

   i)   An uncoordinated flight condition will reduce the airplane's climb performance.

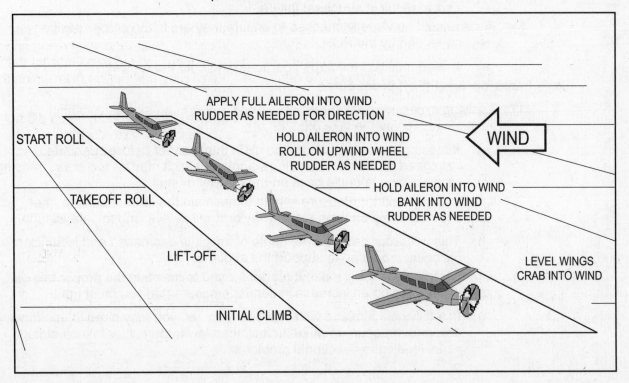

9) Maintain a straight track over the extended runway centerline until a turn is required.

    a) In a crosswind condition, after you leave the initial side slip for liftoff and enter the crab for climbout, the crab should be maintained as needed to continue along the extended runway centerline until a turn on course or the crosswind leg is initiated.

    b) It is important to remain aligned with the runway to avoid the hazards of drifting into obstacles or the path of another aircraft which may be taking off from a parallel runway.

e. **Proper use of checklist**

1) Complete the takeoff and climb checklist from Section 4, Normal Procedures, of the POH.

2. **Exhibit your instructional knowledge by explaining the following common errors related to a normal and a crosswind takeoff and climb.**

a. **Improper runway incursion avoidance procedures**

1) Do not taxi onto the runway until you are cleared to take off or to position and hold (at an airport with an operating control tower) and you are sure the area is clear by checking visually (at any airport).

b. **Improper use of controls during a normal or crosswind takeoff**

1) If a crosswind is present, FULL aileron should be held into the wind initially.

2) As the forward speed of the airplane increases and the ailerons become more effective, the mechanical holding of full aileron should be reduced.

3) Some aileron pressure must be maintained to keep the upwind wing from rising.

4) If the upwind wing rises, a "skipping" action may develop.

    a) The crosswind tends to move the airplane sideways.

    b) This side skipping imposes severe side stresses on the landing gear and could result in structural failure.

5) Flaps should be visually checked to ensure they are in the proper position recommended by the POH.

    a) If used, position the flaps prior to taxiing onto the runway.

c. **Inappropriate lift-off procedures**

1) Applying excessive back pressure will only result in an excessively high pitch attitude and delay the take off.

    a) If the airplane is forced into the air in this manner before adequate airspeed is attained, the wing's angle of attack may be too great, causing the airplane to settle back on the runway or stall.

2) If not enough elevator pressure is held to maintain the correct attitude, the airplane may settle back on the runway and will delay climb to safe altitude.

    a) This will occur because the angle of attack is decreased and insufficient lift is being produced to support the airplane.

3) Improper trim setting will make it harder for you to maintain the proper takeoff attitude by causing an increase in control pressure that you must hold.

    a) In a tailwheel airplane with improper trim set, you may need to use forward elevator pressure to raise the tail, then lower the tail for takeoff attitude, thus leading to directional problems.

    d. **Improper climb attitude, power setting and airspeed (V_Y)**

      1) Use your POH checklists to determine the proper climb configuration and airspeed.

      2) Maintain airspeed by making small pitch changes by outside visual references; then cross-check with the airspeed indicator.

    e. **Improper use of checklist**

      1) Use the appropriate checklist whenever necessary.

3. **Demonstrate and simultaneously explain a normal or a crosswind takeoff and climb from an instructional standpoint.**

    a. The demonstration must be to commercial pilot skill level.

    b. You will be evaluated on your ability to explain and demonstrate simultaneously the key elements of a normal or a crosswind takeoff and climb.

    c. If you do not perform the maneuver to FAA completion standards, point out the deficiency to your FAA inspector/examiner as an example of an error, and explain how it should be corrected (including how and why it occurred).

4. **Analyze and correct simulated common errors related to a normal or a crosswind takeoff and climb.**

    a. As your FAA inspector/examiner is performing a normal or a crosswind takeoff and climb, you must be able to recognize, analyze, and correct any error.

    b. Explain what the error is, and then explain what (s)he should do to correct the error.

    c. If your FAA inspector/examiner does not perform the maneuver to FAA completion standards, point out the deficiency to him/her as an example of a student error, and explain how it should be corrected (including how and why it occurred).

END OF TASK

SHORT-FIELD TAKEOFF AND MAXIMUM PERFORMANCE CLIMB

VII.B. TASK: SHORT-FIELD TAKEOFF AND MAXIMUM PERFORMANCE CLIMB

REFERENCES: FAA-H-8083-3; FAA-S-8081-12, FAA-S-8081-14; POH/AFM.

Objective. To determine that the applicant:

1. Exhibits instructional knowledge of the elements of a short-field takeoff and climb by describing --

   a. Procedures before taxiing onto the runway to ensure runway incursion avoidance.

   b. Short-field takeoff and lift-off procedures.

   c. Initial climb attitude and airspeed (V_X) until obstacle is cleared (50 feet/16 meters AGL).

   d. Proper use of checklist.

2. Exhibits instructional knowledge of common errors related to a short-field takeoff and climb by describing --

   a. Improper runway incursion avoidance procedures.

   b. Improper use of controls during a short-field takeoff.

   c. Improper lift-off procedures.

   d. Improper initial climb attitude, power setting, and airspeed (V_X) to clear obstacle.

   e. Improper use of checklist.

3. Demonstrates and simultaneously explains a short-field takeoff and climb from an instructional standpoint.

4. Analyzes and corrects simulated common errors related to a short-field takeoff and climb.

A. General Information

1. The objective of this task is to determine your instructional knowledge of the elements and common errors related to a short-field takeoff and climb. You will also be required to demonstrate and simultaneously explain this task to your FAA inspector/examiner.

2. When taking off from a field where the available runway is short and/or where obstacles must be cleared, you must operate your airplane to its maximum capability.

   a. Positive and accurate control of your airplane attitude and airspeed is required to obtain the shortest ground roll and the steepest angle of climb.

3. You should review Section 4, Normal Procedures, in the POH for the proper airspeeds, e.g., V_R, V_X, V_Y, and also the proper flap setting.

   a. Best angle of climb, V_X, is the speed which will result in the greatest gain of altitude for a given distance over the ground. This speed increases slowly as higher density altitudes are encountered.

      1) In some airplanes, a deviation of 5 kt. from V_X can result in a significant reduction in climb performance.

   b. Always climb above the altitude of obstacles (usually powerlines or treelines) before accelerating to V_Y. This acceleration involves a reduction in pitch.

4. Consult Section 5, Performance, of the POH before attempting a short-field takeoff, specifically taking into account the existing temperature, barometric pressure, field length, type of runway surface, and airplane operating condition.

   a. Since the performance charts assume good pilot technique, you should consider short-field takeoff proficiency.

   b. Recognize that, in some situations, you should NOT attempt to take off because the margin of safety is too small. You may have to

      1) Remove fuel, people, baggage.
      2) Wait for different wind and/or temperature conditions.
      3) Have the airplane moved to a safer takeoff location.

B.   Task Objectives

1.   **Exhibit your instructional knowledge by explaining the following elements of a short-field takeoff and climb.**

a.   **Procedures before taxiing onto the runway to ensure runway incursion avoidance**

1)   Before taxiing onto the runway, you must make sure that you have sufficient time to execute the takeoff before any aircraft in the traffic pattern turns onto final approach.

a)   Check that the runway is clear of other aircraft, vehicles, persons, or other hazards.

b)   This precaution should be observed at both controlled and uncontrolled airports.

b.   **Short-field takeoff and lift-off procedures**

1)   You will need to demonstrate that you should taxi to the very beginning of the takeoff runway (i.e., threshold) and come to a complete stop with the airplane aligned with the runway centerline.

a)   Taxiing to the beginning of the runway may require a back taxi. Announce your intentions on the CTAF or request permission from the tower, as appropriate.

b)   At this position, you should study the runway and related ground reference points (e.g., buildings, trees, obstacles).

i)   Your observations will provide a frame of reference for directional control during takeoff.

2)   Determine the crosswind condition and apply appropriate aileron deflection (i.e., toward the wind).

a)   Ailerons should be FULLY deflected at the start of the takeoff roll.

b)   The aileron should be up on the upwind side of the airplane (i.e., the control yoke turned toward the wind).

c)   This aileron position will impose a downward force on the upwind wing to counteract the lifting force of the crosswind and prevent that wing from rising.

3)   The elevator should be held neutral with the proper trim setting.

a)   The proper trim setting may be found in the POH and identified by a mark on the trim indicator or by a number of degrees.

4)   Recheck that the mixture is set in accordance with the POH and the propeller is at high RPM (constant speed propeller).

5)   Apply brakes and add takeoff (i.e., full, or as recommended in the POH) power; then release the brakes smoothly. Confirm that the engine is developing takeoff power under prevailing conditions before releasing the brakes.

a)   Applying power too quickly can cause engine surging, backfiring, and a possible overboost situation (turbocharged engines). These conditions cause unnecessary engine wear as well as possible failure.

b)   Engine instruments must be monitored during the entire maneuver.

i)   Monitoring enables you to notice immediately any malfunctions or indication of insufficient power or other potential problems. Do not hesitate to abort the take-off if all engine indications are not normal.

ii)   Listen for any indication of power loss or engine roughness.

    6)   Rudder pressure must be promptly and smoothly applied to counteract yawing forces, so the airplane continues straight down the center of the runway.

        a)   Maintain directional control of the airplane on the surface as discussed in item 4) on page 257 of Task VII.A., Normal and Crosswind Takeoff and Climb.

    7)   During crosswind takeoffs, the aileron deflection into the wind should be decreased as the airspeed increases.

        a)   Some aileron pressure must be maintained to keep the upwind wing from rising, i.e., keep the wings level.

    8)   At the recommended rotation speed (V_R) specified in the POH, you should smoothly raise the nose of the airplane to the attitude that will deliver the best angle of climb airspeed, V_X.

        a)   If no rotation airspeed is recommended, accelerate to V_X minus 5 kt. and rotate to V_X attitude.

        b)   DO NOT attempt to raise the nose until V_R because this will create unnecessary drag and will prolong the takeoff roll.

            i)   In a nosewheel-type airplane, keep the elevator in a neutral position to maintain a low drag attitude until rotation.

            ii)   In a tailwheel-type airplane, the tail should be allowed to rise off the ground slightly and then be held in this tail-low flight attitude until the proper liftoff or rotation airspeed is attained.

        c)   Your airplane's V_R _____.

        d)   Some high-performance airplanes require conscious rearward elevator pressure at V_R to establish the liftoff.

            i)   Without this conscious control pressure, the airplane may start to wheelbarrow (i.e., the main wheels break ground before the nose wheel).

            ii)   Note that, in general, high-performance airplanes have heavier control pressures and require more deliberate application of control movements.

  c.   **Initial climb attitude and airspeed (V_X) until the obstacle is cleared (50 ft. AGL)**

    1)   While practicing your short-field takeoff and climb, you should learn the pitch attitude required to maintain V_X.

        a)   Observe the position of the airplane's nose on the horizon.
        b)   Note the position of the aircraft bar on the attitude indicator.

    2)   Rotate to this predetermined pitch angle as soon as you reach V_R. As you climb out at V_X, you should maintain visual references, but occasionally glance at the attitude indicator and airspeed indicator to check the pitch angle and airspeed.

        a)   Your airplane's V_X _____.

    3)   If not enough back pressure is held to maintain the correct takeoff attitude or if the nose is allowed to lower excessively, the airplane may settle back to the runway. This occurs because the angle of attack is decreased and lift is diminished to the point where it will not support the airplane.

4) Retract the landing gear, if retractable, and flaps after a positive rate of climb has been established.

    a) Before retracting the landing gear, apply the brakes momentarily to stop the rotation of the wheels to avoid excessive vibration on the gear mechanism.

        i) Centrifugal force caused by the rapidly rotating wheels expands the diameter of the tires, and if mud or other debris has accumulated in the wheel wells, the rotating wheels may rub as they enter.

    b) When to retract the landing gear varies among manufacturers. Thus, it is important that you know what procedure is prescribed by your airplane's POH.

        i) Some recommend gear retraction after a positive rate of climb has been established while others recommend gear retraction only after the obstacles have been cleared.

        ii) Normally, the landing gear will be retracted before the flaps.

    c) Make necessary pitch adjustments to maintain the appropriate airspeed.

    d) Flaps are normally retracted when you are clear of any obstacle(s), the best rate-of-climb (V_Y) pitch attitude has been established, and airspeed is stable.

        i) Raise the flaps in increments (if appropriate) to avoid sudden loss of lift and settling of the airplane.

5) After establishing V_Y, completing gear and flap retraction, and reaching a safe maneuvering altitude (normally 500 to 1,000 ft. AGL), reduce the power to the normal cruise climb setting and adjust pitch for cruise climb airspeed.

    a) Most trainer-type airplane manufacturers recommend maintaining maximum power to your selected cruising altitude.

    b) Use the power setting recommended in your POH.

6) Comply with any noise abatement procedures.

    a) For more information on noise abatement procedures, see Task VII.A., Normal and Crosswind Takeoff and Climb, item 5) on page 260.

7) Completion standards for this element

    a) Commercial PTS

        i) ***Establish a pitch attitude that will maintain the recommended obstacle clearance airspeed, or V_X, +5/–0 kt., until the obstacle is cleared, or until the airplane is 50 ft. above the surface.***

        ii) ***After clearing the obstacle, establish the pitch attitude for V_Y, accelerate to, and maintain V_Y, ±5 kt. during the climb.***

        iii) ***Retract the landing gear, if appropriate, and flaps after clear of any obstacles, or as recommended by the manufacturer.***

    b) Private PTS

        i) Establish a pitch attitude that will maintain the recommended obstacle clearance airspeed, or V_X, +10/– 5kt., until the obstacle is cleared, or until the airplane is 50 ft. above the surface.

        ii) After clearing the obstacle, establish the pitch attitude for V_Y, accelerate to V_Y, and maintain V_Y, +10/– 5kt., during the climb.

        iii) Retract the landing gear, if appropriate, and flaps after clear of any obstacles or as recommended by the manufacturer.

d. **Proper use of checklist**

1) Follow the short-field takeoff checklist in the POH.

2) Complete the checklist for climb to ensure the airplane is in the proper configuration for the continued climb to the desired cruising altitude.

2. **Exhibit your instructional knowledge by explaining the following common errors related to a short-field takeoff and climb.**

a. **Improper runway incursion avoidance procedures**

1) Do not taxi onto the runway until you are cleared to take off or to position and hold (at an airport with an operating control tower) and you are sure the area is clear by checking visually (at any airport).

b. **Improper use of controls during a short-field takeoff**

1) If a crosswind is present, FULL aileron should be held into the wind initially to prevent the crosswind from raising the upwind wing.

   a) This aileron pressure should be gradually decreased as airspeed increases.

2) Flaps should be visually checked to ensure that they are in the proper position recommended by the POH.

   a) The short-field takeoff performance chart in your POH will also list the flap setting used to attain the chart performance.

   b) Position the flaps prior to taxiing onto the active runway.

c. **Improper lift-off procedures**

1) Applying excessive back pressure will only result in an excessively high pitch attitude and delay the takeoff.

   a) If you attempt to force the airplane into the air in this manner before adequate airspeed is attained, the wing's angle of attack may be too great, causing the airplane to settle back on the runway or stall.

2) If not enough back pressure is held to maintain the correct V_x attitude, the airplane may settle back on the runway.

   a) This will occur because the angle of attack is decreased and insufficient lift is being produced to support the airplane.

3) Improper trim setting will make it harder for you to maintain the proper takeoff attitude by causing an increase in control pressure that you must hold.

d. **Improper initial climb attitude, power setting, and airspeed (V_x) to clear obstacle**

1) Use your POH checklists to determine the proper climb configuration and airspeed.

2) Wing flaps and landing gear should not be retracted until you are clear of the obstacle.

3) Maintain V_x until the obstacle is cleared; then accelerate and maintain VY.

4) Maintain pitch attitude by outside references; then cross-check the airspeed and attitude indicators.

e. **Improper use of checklist**

1) Use the appropriate checklist whenever necessary.

3. **Demonstrate and simultaneously explain a short-field takeoff and climb from an instructional standpoint.**

   a. The demonstration must be to commercial pilot skill level.

   b. You will be evaluated on your ability to explain and demonstrate simultaneously the key elements of a short-field takeoff and climb.

   c. If you do not perform the maneuver to FAA completion standards, point out the deficiency to your FAA inspector/examiner as an example of an error, and explain how it should be corrected (including how and why it occurred).

4. **Analyze and correct simulated common errors related to a short-field takeoff and climb.**

   a. As your FAA inspector/examiner is performing a short-field takeoff and climb, you must be able to recognize, analyze, and correct any error.

   b. Explain what the error is, and then explain what (s)he should do to correct the error.

   c. If your FAA inspector/examiner does not perform the maneuver to FAA completion standards, point out the deficiency to him/her as an example of a student error, and explain how it should be corrected (including how and why it occurred).

END OF TASK

SOFT-FIELD TAKEOFF AND CLIMB

VII.C. TASK: SOFT-FIELD TAKEOFF AND CLIMB

REFERENCES: FAA-H-8083-3; FAA-S-8081-12, FAA-S-8081-14; POH/AFM.

Objective. To determine that the applicant:

1. Exhibits instructional knowledge of the elements of a soft-field takeoff and climb by describing --

   a. Procedures before taxiing onto the runway or takeoff area to ensure runway incursion avoidance.

   b. Soft-field takeoff and lift-off procedures.

   c. Initial climb attitude and airspeed, depending on if an obstacle is present.

   d. Proper use of checklist.

2. Exhibits instructional knowledge of common errors related to a soft-field takeoff and climb by describing --

   a. Improper runway incursion avoidance procedures.

   b. Improper use of controls during a soft-field takeoff.

   c. Improper lift-off procedures.

   d. Improper climb attitude, power setting, and airspeed (V_Y) or (V_X).

   e. Improper use of checklist.

3. Demonstrates and simultaneously explains a soft-field takeoff and climb from an instructional standpoint.

4. Analyzes and corrects simulated common errors related to a soft-field takeoff and climb.

A. General Information

1. The objective of this task is to determine your instructional knowledge of the elements and common errors related to a soft-field takeoff and climb. You will also be required to demonstrate and simultaneously explain this task to your FAA inspector/examiner.

2. Before landing at a soft (unpaved) field, determine your capability to take off in your airplane from that field. Also, consider the possibility of damage and extra wear on your airplane. You may decide to wait until the takeoff surface conditions improve.

   a. If the need arises to make a soft-field departure, consult the recommendations provided by the manufacturer in your airplane's POH.

      1) As a CFI, have your student practice and perfect soft-field takeoffs.

3. If the airplane is parked on a soft surface, there is a possibility that other airplanes or the wind may have blown unwanted debris onto your airplane. Such materials, when trapped in the control surfaces, may jam the controls or limit their travel, which can cause disaster.

   a. Soft fields are often remote fields. Birds and animals can seek refuge or build nests (even overnight) under the cowling, in landing gear wheel wells, and elsewhere.

   b. Also, be cautious of possible vandalism of your airplane at remote airfields.

4. Inspect your taxi route and your takeoff runway. Normally, you should walk the entire route carefully.

   a. Note wet or soft spots and mark them as necessary (use pieces of cloth or paper tied to objects, e.g., fence posts, or anchor them to the ground at the side of the runway with stakes, sticks, etc.).

   b. Determine and mark your takeoff abort point -- exactly where you will cut power if not airborne.

      1) 75% of V_R by the halfway point on the takeoff surface is a general rule.

5. If the airplane wheels have settled into the ground, move the airplane forward before getting into the cockpit.

    a. Use leverage of the wing by holding the wingtip and rocking the wingtip back and forth.

    b. Be careful not to stress the nose wheel with side loads (have someone lift the nose or push down on the tail).

    c. Use help as available.

6. The goals of a soft-field takeoff are

    a. To get the airplane airborne as soon as possible

    b. To transfer as much weight as possible to the wings to minimize wheel friction with the soft surface.

7. In a nose-wheel airplane, enough back elevator pressure should be applied to establish a positive angle of attack.

    a. Back elevator pressure and the resulting increased angle of attack reduce the weight supported by the nosewheel.

    b. The nose-high attitude will allow the weight to transfer from the wheels to the wings as lift is developed.

8. In a tailwheel airplane, the tailwheel should be raised barely off the soft runway surface.

    a. Thus, tailwheel drag on the soft surface is eliminated.

    b. The angle of attack produced in this attitude is still high enough to allow the airplane to leave the ground at the earliest opportunity and transfer weight from the main wheels to the wings.

B. Task Objectives

1. **Exhibit your instructional knowledge by explaining the following elements of a soft-field takeoff and climb.**

    a. **Procedures before taxiing onto the runway to ensure runway incursion avoidance**

        1) Always make sure to visually clear the area before taxiing onto the runway.

            a) This applies to both airports with and without an operating control tower.

        2) If flying at an airport with an operating control tower, don't cross the hold-short lines until you are given a clearance to take off or to position and hold.

            a) At an airport without an operating control tower, announce your intentions before taking the runway.

            b) Never taxi onto the runway when there is another airplane on final approach.

    b. **Soft-field takeoff and lift-off procedures**

        1) Determine the crosswind condition and apply appropriate aileron deflection (i.e., toward the wind).

            a) Ailerons should be FULLY deflected at the start of the takeoff.

            b) The aileron should be up on the upwind side of the airplane (i.e., the control yoke turned toward the wind).

            c) This aileron position will impose a downward force on the upwind wing to counteract the lifting force of the crosswind and prevent that wing from rising.

2) In a nosewheel airplane, enough back elevator pressure should be applied to establish a positive angle of attack.

    a) In a tailwheel-type airplane, the control stick should be slightly forward to allow the tailwheel to become airborne as quickly as possible.

3) Trim should be in the takeoff position as prescribed in the POH.

    a) The takeoff position may be identified by a mark on the trim indicator or by a setting in number of degrees.

4) Recheck that the mixture is set in accordance with the POH and the propeller is at high RPM (constant speed propeller).

    a) Power should be applied smoothly and as rapidly as possible to allow for a controllable transition to flying airspeed.

        i) Applying power too quickly can cause engine surging, backfiring, and a possible overboost situation (turbocharged engines). These conditions cause unnecessary engine wear as well as possible failure.

        ii) Applying power too slowly wastes runway length.

    b) Engine instruments must be monitored during the entire maneuver.

        i) Monitoring enables you to notice immediately any malfunction or indication of insufficient power or other potential problems.

        ii) You should listen for any indication of power loss or engine roughness.

        iii) Do not hesitate to abort the takeoff if there is any indication of trouble.

    c) You should check the airspeed indicator for movement as the airplane accelerates.

        i) If the airspeed indicator does not move, abort the takeoff.

5) Rudder pressure must be promptly and smoothly applied to counteract yawing forces (from torque), so the airplane continues straight down the center of the runway.

    a) Throughout the takeoff roll, adjust control pressures to maintain the runway centerline.

6) When the airplane is held at a nose-high attitude throughout the takeoff, the wings will, as speed increases and lift develops, progressively relieve the wheels of more and more of the airplane's weight, thus reducing the drag on the wheels caused by surface irregularities or adhesion.

    a) If the pitch attitude is accurately maintained during the takeoff roll, the airplane should become airborne at an airspeed slower than a safe climb speed because of the action of ground effect.

        i) As the airplane picks up speed, back pressure must be reduced somewhat to avoid an excessive angle of attack, while weight is transferred to the wings and liftoff into ground effect is accomplished as quickly as possible.

c. **Initial climb attitude and airspeed, depending on if an obstacle is present**

1) After the airplane becomes airborne, lower the nose of the airplane very gently with the wheels just clear of the surface to allow the airplane to accelerate in ground effect to V_Y.

a) Failure to level off (i.e., maintain constant altitude) means the airplane will climb out of ground effect at too slow an airspeed, and the increase in drag can reduce the lift sufficiently to cause the airplane to settle back onto the takeoff surface.

NOTE: While not an element of this task, if obstacles are present at the departure end of the takeoff area, you should accelerate to climb at V_X and climb at V_X until cleared of the obstacle(s).

2) Ground effect is due to the interference of the ground surface with the airflow patterns about the airplane in flight.

a) The vertical component of the airflow around the wing is restricted, which alters the wing's upwash, downwash, and wingtip vortices.

b) The reduction of the wingtip vortices alters the spanwise lift distribution and reduces the induced angle of attack and induced drag.

i) Thus, the wing will require a lower angle of attack in ground effect to produce the same lift coefficient or, if a constant angle of attack is maintained, an increase in lift coefficient will result.

c) Since induced drag is reduced due to ground effect, required thrust at low airspeeds is reduced.

d) In order for ground effect to be of significant magnitude, the wing must be at a height one-half of its span or less.

e) Due to the reduced drag in ground effect, the airplane may seem capable of takeoff well below the recommended speed.

i) As the airplane climbs out of ground effect with a deficiency of speed, the greater induced drag may result in very marginal climb performance.

f) In some cases (e.g., high density altitude), the airplane may become airborne and attempt to climb out of ground effect with a deficiency of airspeed and then settle back to the runway.

3) Once you have accelerated to V_Y in ground effect, you should establish the pitch attitude for V_Y.

4) Completion standards for this element

a) Commercial PTS

i) *Lift off at the lowest possible airspeed and remain in ground effect while accelerating to V_X or V_Y, as appropriate.*

ii) *Establish a pitch attitude for V_X or V_Y, as appropriate, and maintain the selected airspeed ±5 kt. during the climb.*

iii) *Retract the landing gear, if appropriate, and flaps after clear of any obstacles, or as recommended by the manufacturer.*

b) Private PTS

i) Lift off at the lowest possible airspeed and remain in ground effect while accelerating to V_X or V_Y, as appropriate.

ii) Establish a pitch attitude for V_X or V_Y, as appropriate, and maintain the selected airspeed +10/−5 kt., during the climb.

iii) Retract the landing gear, if appropriate, and flaps after clear of any obstacles or as recommended by the manufacturer.

d. **Proper use of checklist**

1) Locate and study the soft-field checklist and any amplified procedures in Section 4, Normal Procedures, of the POH.

2) Complete the checklist for the climb to ensure that the airplane is in the proper configuration for the continued climb to cruising altitude.

3) Follow the checklist(s) in the POH.

2. **Exhibit your instructional knowledge by explaining common errors related to a soft-field takeoff and climb.**

a. **Improper runway incursion avoidance procedures**

1) Always make sure to visually clear the area before taxiing onto the runway.

a) This applies to airports with and without an operating control tower.

2) If you are at an airport with an operating control tower, do not cross the hold-short lines until you are given a clearance to take off or to taxi onto or across the runway.

a) At an airport without an operating control tower, announce your intentions before taking the runway.

b) Never taxi onto the runway when there is another airplane on final approach.

b. **Improper use of controls during a soft-field takeoff**

1) The control yoke should be held in the full back position and turned into the crosswind (if appropriate).

a) Gradually decrease aileron pressure into the wind as the airspeed increases.

2) If wing flaps are recommended by the POH, they should be lowered prior to taxiing onto the takeoff surface.

c. **Improper lift-off procedures**

1) During the takeoff roll, excessive back elevator pressure may cause the angle of attack to exceed that required for a climb, which would generate more drag.

a) In a nosewheel-type airplane, excessive back elevator pressure may also cause the tail of your airplane to drag on the ground.

2) You must slowly lower the nose of the airplane after liftoff to allow the airplane to accelerate in ground effect.

a) If done too quickly, the airplane will settle back onto the takeoff surface.

3) Attempting to climb without the proper airspeed may cause the airplane to settle back onto the runway due to the increase in drag.

d. **Improper climb attitude, power setting, and airspeed (V_Y) or (V_X)**

1) Follow the procedures in the POH.

2) You must learn to fly the airplane by the numbers and teach your student to fly in the same manner. Failure to do so means reduced performance, which may be disastrous on a short soft-field takeoff, especially if there is an obstacle to be cleared.

e. **Improper use of checklist**

1) Use the appropriate checklist whenever necessary.

END OF TASK

NORMAL AND CROSSWIND APPROACH AND LANDING

VII.F. TASK: NORMAL AND CROSSWIND APPROACH AND LANDING

REFERENCES: FAA-H-8083-3; FAA-S-8081-12, FAA-S-8081-14; POH/AFM.

Objective. To determine that the applicant:

1. Exhibits instructional knowledge of the elements of a normal and a crosswind approach and landing by describing --

   a. How to determine landing performance and limitations.

   b. Configuration, power, and trim.

   c. Obstructions and other hazards which should be considered.

   d. A stabilized approach at the recommended airspeed to the selected touchdown area.

   e. Course of action if selected touchdown area is going to be missed.

   f. Coordination of flight controls.

   g. A precise ground track.

   h. Wind shear and wake turbulence avoidance procedures.

   i. Most suitable crosswind procedure.

   j. Timing, judgment, and control procedure during roundout and touchdown.

   k. Directional control after touchdown.

   l. Use of brakes.

   m. Use of checklist.

2. Exhibits instructional knowledge of common errors related to a normal and a crosswind approach and landing by describing --

   a. Improper use of landing performance data and limitations.

   b. Failure to establish approach and landing configuration at appropriate time or in proper sequence.

   c. Failure to establish and maintain a stabilized approach.

   d. Inappropriate removal of hand from throttle.

   e. Improper procedure during roundout and touchdown.

   f. Poor directional control after touchdown.

   g. Improper use of brakes.

3. Demonstrates and simultaneously explains a normal or a crosswind approach and landing from an instructional standpoint.

4. Analyzes and corrects simulated common errors related to a normal or crosswind approach and landing.

A. General Information

1. The objective of this task is to determine your instructional knowledge of the elements and common errors related to a normal and a crosswind approach and landing. You will be required to demonstrate and simultaneously explain this task from an instructional standpoint.

2. A normal approach and landing is one in which engine power is available, the wind is light or the final approach is made directly into the wind, the final approach path has no obstacles, and the landing surface is firm and of ample length to bring the airplane to a stop gradually.

   a. A crosswind approach and landing involves the same basic principles as a normal approach and landing except the wind is blowing across rather than parallel to the final approach path.

      1) Virtually every landing will require at least some slight crosswind correction.

3. See *Pilot Handbook* for the following:

   a. Module 3.7, Wake Turbulence, for a seven-page discussion on wake turbulence and avoidance procedures.

   b. Module 5.9, Crosswind Performance, for a one-page discussion on determining the crosswind component.

   c. Module 5.10, Landing Performance, for a four-page discussion on determining landing performance (distance).

B.   Task Objectives

1.   **Exhibit your instructional knowledge by explaining the following elements of a normal and a crosswind approach and landing.**

a.   **How to determine landing performance and limitations**

1)   Landing performance (e.g., distance and crosswind component) is determined by using the appropriate chart in Section 5, Performance, of the POH.

a)   Landing distance is a required preflight action.

2)   Landing limitations (e.g., maximum weight or crosswind limitation) are located in Section 2, Limitations, of the POH.

b.   **Configuration, power, and trim**

1)   The airspeeds and configuration for your airplane during a normal and crosswind approach and landing are found in Section 4, Normal Procedures, of the POH.

2)   Properly configuring your airplane throughout the various approach segments will assist you in flying a stabilized approach.

a)   On the downwind leg, you should complete the before-landing checklist in your POH, which includes gear extension (if retractable).

i)   You should reduce the power and hold altitude constant, and when the airspeed slows below the maximum flap extended speed (V_{FE}), you should partially lower the flaps and begin your descent.

- In your airplane, V_{FE} _____.

ii)   You should plan to be in this configuration prior to beginning your descent from traffic pattern altitude.

- Normally the descent begins when you are abeam your selected touchdown point.

b)   On the base leg, the flaps may be extended further, but full flaps are not recommended.

c)   Once aligned with the runway centerline on the final approach, you should make the final flap selection. This is normally full flaps.

i)   In turbulent air or strong gusty winds, you may elect not to use full flaps. This will allow you to maintain control more easily at a higher approach speed.

- With less than full flaps, your airplane will be in a higher nose-up attitude.

3)   The approach and landing configuration means that the gear is down (if retractable), wing flaps are extended, and you are maintaining a reduced power setting.

4)   During the approach to a landing, power is considerably lower-than-cruise setting, and the airplane is flying at a relatively slower airspeed. Thus, trim must be used throughout the maneuver to compensate for the change in aerodynamic forces.

c.   **Obstructions and other hazards which should be considered**

1)   You should consider the wind conditions and obstacles when planning your approach.

a)   A strong headwind on final will cause you to position the base leg closer to the approach end of the runway than you would if the wind were light.

i)   Another method is based on the concept of always flying the same familiar glide path to any airport. Thus, in a strong headwind, you would use power and a lower rate of descent on final.

b) Obstacles along the final approach path (e.g., trees, towers, construction equipment) will cause you to select an aim point sufficiently beyond the obstacles for safety.

c) Be aware of traffic, both in the air and on the ground.

    i) Look out for vehicles and/or people on or near the runway.

2) The presence of strong, gusting winds or turbulent air may require you to increase your airspeed on final approach. This provides for more positive control of your airplane.

a) The gust factor, the difference between the steady state wind and the maximum gust, should be factored into your final approach airspeed in some form.

    i) It should also be added to your various approach segment airspeeds for downwind, base, and final.

b) One recommended technique is to use the normal approach speed plus one-half the gust factor.

    i) EXAMPLE: If the normal approach speed is 70 kt. and the wind gusts increase 20 kt., an airspeed of 80 kt. is appropriate.

    ii) Some pilots add all of the steady wind and one-half the gust, or all of the gust and no steady wind.

c) Remember, your airspeed and whatever gust factor you select to add to your final approach speed should be flown only after all maneuvering has been completed and your airplane has been lined up on the final approach.

d) When using a higher-than-normal approach speed, it may be expedient to use less than full flaps on landing.

e) Follow the recommended procedures in your POH.

3) After considering the conditions, you should select a touchdown point that is beyond the runway's landing threshold but well within the first one-third portion of the runway.

4) Once you have selected your touchdown point, you need to select your aim point. The aim point will be the point at the end of your selected glide path, not your touchdown point. Thus, your aim point will be short of your touchdown point.

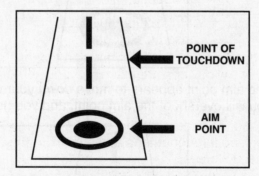

d. **A stabilized approach at the recommended airspeed to the selected touchdown area**

1) The term **stabilized approach** means that the airplane is in a position where minimum input of all controls will result in a safe landing.

a) Excessive control input at any juncture could be an indication of improper planning.

2) The objective of a good final approach is to descend at an angle and airspeed that will permit your airplane to reach the desired touchdown point at an airspeed that will result in a minimum of floating just before touchdown.

a) A fundamental key to flying a stabilized approach is the interrelationship of pitch and power.

i) This interrelationship means that any changes to one element in the approach equation (e.g., airspeed, attitude) must be compensated for by adjustments in the other.

b) Power should be adjusted as necessary to control the airspeed, and the pitch attitude adjusted SIMULTANEOUSLY to control the descent angle or to attain the desired altitudes along the approach path.

i) By lowering the nose of your airplane and reducing power to keep your approach airspeed constant, you can descend at a higher rate to correct for being too high in the approach.

c) The important point is never to let your airspeed drop below your approach speed and never to let your airplane sink below the selected glide path.

3) When you are established on final, you should use pitch to fly your airplane to the aim point.

a) If the aim point has no apparent movement in your windshield, then you are on a constant glide path to the aim point. No pitch correction is needed.

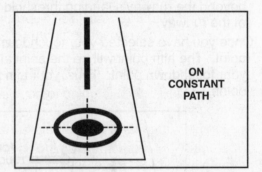

ON CONSTANT PATH

b) If the aim point appears to move down your windshield or toward you, then you will overshoot the aim point, and you need to pitch down.

OVERSHOOT

i) As you pitch down, reduce power to maintain approach speed.

c) If the aim point appears to move up your windshield or away from you, then you will undershoot the aim point, and you need to pitch up.

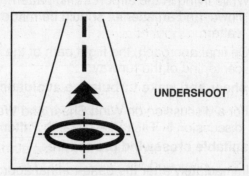

UNDERSHOOT

i) As you pitch up, increase power to maintain approach speed.

4) On final approach, you should use the airspeed in your POH. In the absence of the manufacturer's recommended airspeed, a speed equal to $1.3\ V_{S0}$ should be used.

a) EXAMPLE: If V_{S0} in your airplane is 60 kt., the airspeed on final approach should be 78 kt. (1.3×60).

b) In your airplane, final approach speed (POH) _____, or $1.3\ V_{S0}$ _____.

c) Make necessary adjustments to that speed if you are in turbulent air or strong, gusty winds.

5) Standards for this element

a) Commercial pilot

i) *Maintain a stabilized approach and the recommended airspeed, or in its absence, not more than $1.3\ V_{SO}$, ± 5 kt. with wind gust factor applied.*

b) Private pilot

i) Maintain a stabilized approach and the recommended airspeed, or in its absence, not more than $1.3\ V_{SO}$, +10/–5 kt. with wind gust factor applied.

e. **Course of action if selected touchdown area is going to be missed**

1) If the touchdown point is going to be missed, the pilot should perform a go-around.

2) You should state this intention before the landing is started.

3) While your FAA inspector/examiner is flying, do not allow him/her to complete a landing that should have been a go-around.

f. **Coordination of flight controls**

1) The flight controls should always be used in a proper and coordinated manner, e.g., turning or performing a slip.

2) When turning, use no more than a 30° bank angle.

3) The most dangerous situation is one in which a pilot is turning from base to final and attempts to increase the rate of turn by use of the rudder only.

a) This improper use of the rudder is the beginning of an improper cross-control situation which could lead to a stall.

b) See the four-page discussion of a crossed-control stall in Task XI.D., Crossed-Control Stalls, beginning on page 434.

g.   **A precise ground track**

1)   While flying in the airport traffic pattern, a good rectangular course should be flown, and an attempt should be made not to cut the corners; i.e., fly a square pattern.

2)   On final approach, the flight path of the airplane must track on the extended centerline of the runway.

h.   **Wind shear and wake turbulence avoidance procedures**

1)   For a discussion on Wind Shear and Wake Turbulence, see the two-page discussion in Task VI.B., Traffic Patterns, beginning on page 248.

i.   **Most suitable crosswind procedure**

1)   Immediately after the base-to-final approach turn is completed, the longitudinal axis of your airplane should be aligned with the centerline of the runway so that drift (if any) will be recognized immediately.

2)   On a crosswind approach, there are two usual methods of maintaining the proper ground track on final approach. These are the crab method and the side-slip (wing-low) method.

3)   The crab method is used first by establishing a heading (crab) toward the wind with the wings level so that your airplane's ground track remains aligned with the centerline of the runway.

   a)   This heading is maintained until just prior to touchdown, when the longitudinal axis of the airplane must be quickly aligned with the runway.

      i)   A high degree of judgment and timing is required in removing the crab immediately prior to touchdown.

   b)   This method is best to use while on a long final approach until you are on a short final, when you should change to the wing-low method.

   c)   Maintaining a crab as long as possible increases passenger comfort.

4)   The wing-low method is recommended in most cases since it will compensate for a crosswind at any angle, but more importantly, it will enable you to simultaneously keep your airplane's ground track and longitudinal axis aligned with the runway centerline throughout the approach and landing.

   a)   To use this method, align your airplane's heading with the centerline of the runway, note the rate and direction of drift, and then promptly apply drift correction by lowering the upwind wing.

      i)   The amount the wing must be lowered depends on the rate of drift.

   b)   When you lower the wing, the airplane will tend to turn in that direction. Thus, it is necessary to simultaneously apply sufficient opposite rudder pressure to prevent the turn and keep the airplane's longitudinal axis aligned with the runway.

      i)   Drift is controlled with aileron, and the heading with rudder.

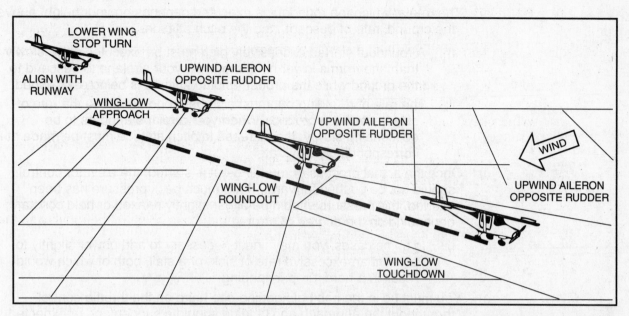

c) Your airplane will now be slipping into the wind just enough that both the resultant flight path and the ground track are aligned with the runway.

d) In a very strong crosswind, the required bank may be so steep that full opposite rudder will not prevent a turn. The wind is too strong to land safely on that particular runway with those wind conditions.

    i) Since the airplane's capabilities would be exceeded, it is imperative that the landing be made on a more favorable runway either at that airport or at an alternate airport.

j. **Timing, judgment, and control procedure during roundout and touchdown**

1) The roundout (flare) is a slow, smooth transition from a normal approach attitude to a landing attitude. When your airplane, in a normal descent, approaches what appears to be about 10 to 20 ft. above the ground, the roundout should be started and, once started, should be a continuous process until the airplane touches down.

a) To start the roundout, reduce power to idle and gradually apply back elevator pressure to increase the pitch attitude and angle of attack slowly. This will cause your airplane's nose to rise gradually toward the desired landing attitude.

    i) The angle of attack should be increased at a rate that will allow your airplane to continue settling slowly as forward speed decreases.

b) When the angle of attack is increased, the lift is momentarily increased, thereby decreasing the rate of descent.

    i) Since power is normally reduced to idle during the roundout, the airspeed will gradually decrease. Decreasing airspeed, in turn, causes lift to decrease again, which must be controlled by raising the nose and further increasing the angle of attack.

    ii) During the roundout, the airspeed is being decreased to touchdown speed while the lift is being controlled so your airplane will settle gently onto the runway.

        c)   The rate at which the roundout is executed depends on your height above the ground, rate of descent, and the pitch attitude.

            i)   A roundout started excessively high must be executed more slowly than one from a lower height to allow your airplane to descend to the ground while the proper landing attitude is being established.

            ii)   The rate of rounding out must also be proportionate to the rate of closure with the ground. When your airplane appears to be descending slowly, the increase in pitch attitude must be made at a correspondingly slow rate.

        d)   Once the actual process of rounding out is started, the elevator control should not be pushed forward. If too much back pressure has been exerted, this pressure should be either slightly relaxed or held constant, depending on the degree of error.

            i)   In some cases, you may find it necessary to add power slightly to prevent an excessive rate of sink, or a stall, both of which would result in a hard, drop-in landing.

        e)   You must be in the habit of keeping one hand on the throttle control throughout the approach and landing should a sudden and unexpected hazardous situation require an immediate application of power.

    2)   The touchdown is the gentle settling of your airplane onto the runway. The touchdown should be made so that your airplane will touch down on the main gear at approximately stalling speed.

        a)   As your airplane settles, the proper landing attitude must be attained by application of whatever back elevator pressure is necessary.

        b)   It seems contradictory that the way to make a good landing is to try to hold your airplane's wheels a few inches off the ground as long as possible with the elevator.

            i)   Normally, when the wheels are about 2 or 3 ft. off the ground, the airplane will still be settling too fast for a gentle touchdown. Thus, this descent must be retarded by further back pressure on the elevators.

            ii)   Since your airplane is already close to its stalling speed and is settling, this added back pressure will only slow up the settling instead of stopping it. At the same time, it will result in your airplane's touching the ground in the proper nose-high landing attitude.

3) During a normal landing, a nosewheel-type airplane should contact the ground in a tail-low attitude, with the main wheels touching down first so that no weight is on the nose wheel.

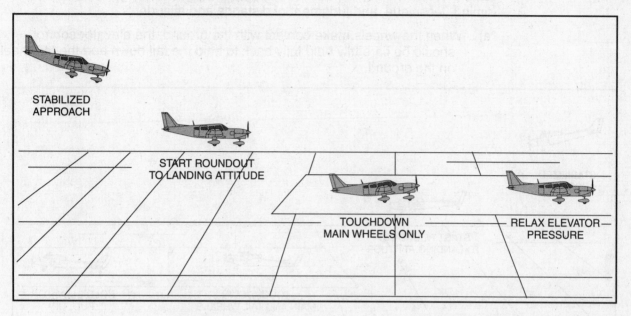

STABILIZED APPROACH

START ROUNDOUT TO LANDING ATTITUDE

TOUCHDOWN MAIN WHEELS ONLY

RELAX ELEVATOR PRESSURE

a) After the main wheels make initial contact with the ground, back pressure on the elevator control should be held to maintain a positive angle of attack for aerodynamic braking and to hold the nosewheel off the ground until the airplane decelerates, thus protecting the relatively fragile forward landing gear.

b) As the airplane's momentum decreases, back pressure should be sufficient to hold the nosewheel off the ground until just prior to lack of elevator control effectiveness.

  i) This will prevent oversensitive steering and ground directional control problems, such as swerving.

  ii) At the same time, it will allow the full weight of the airplane to rest on the main gear for better braking action.

4) During a normal landing in a tailwheel-type airplane, the roundout and touchdown should be timed so that the wheels of the main landing gear and tailwheel touch down simultaneously (i.e., a 3-point landing). This requires fine timing, technique, and judgment of distance and altitude.

    a) When the wheels make contact with the ground, the elevator control should be carefully held fully back to hold the tail down and the tailwheel on the ground.

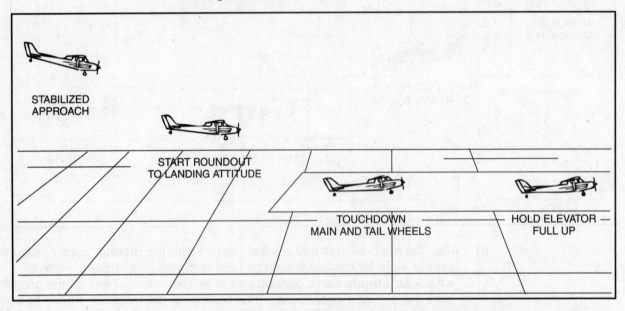

STABILIZED APPROACH

START ROUNDOUT TO LANDING ATTITUDE

TOUCHDOWN MAIN AND TAIL WHEELS

HOLD ELEVATOR FULL UP

        i) For the airplane equipped with a steerable tailwheel, holding the tailwheel on the ground provides more positive directional control and prevents any tendency for the airplane to nose over.

        ii) If the tailwheel is not on the ground, easing back on the elevator control may cause the airplane to become airborne again because the change in attitude will increase the angle of attack and produce enough lift for the airplane to fly.

5) The roundout during a crosswind approach landing can be made as in a normal landing approach, but the application of a crosswind correction must be continued as necessary to prevent drifting.

    a) Since the airspeed decreases as the roundout progresses, the flight controls gradually become less effective. Thus, the crosswind correction being held would become inadequate.

        i) It is therefore necessary to increase the deflection of the rudder and ailerons gradually to maintain the proper amount of drift correction.

    b) Do not level the wings. Keep the upwind wing down throughout the crosswind roundout.

        i) If the wings are leveled, your airplane will begin drifting and the touchdown will occur while drifting, which imposes severe side stresses (loads) on the landing gear.

6) During a crosswind touchdown, you must make prompt adjustments in the crosswind correction to assure that your airplane does not drift as it touches down.

   a) The crosswind correction should be maintained throughout the roundout, and the touchdown made on the upwind main wheel.

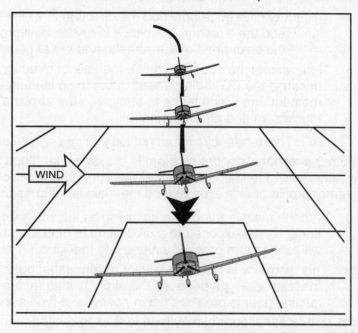

   i) As the forward momentum decreases after initial contact, the weight of the airplane will cause the downwind main wheel to settle gradually onto the runway.

   b) In those airplanes having nosewheel steering interconnected with the rudder, the nosewheel may not be aligned with the runway as the wheels touch down because opposite rudder is being held in the crosswind correction.

   i) This is the case in airplanes which have no centering cam built into the nose gear strut to keep the nosewheel straight until the strut is compressed.

   ii) To prevent swerving in the direction the nosewheel is offset, the corrective rudder pressure must be promptly relaxed just as the nosewheel touches down.

7) Completion standards for this element

   a) ***Touch down smoothly at the approximate stalling speed at, or within 200 ft. beyond, a specified point with no drift, and with your airplane's longitudinal axis aligned with and over the centerline.***

   i) Private pilot applicants must touch down at, or within 400 ft. beyond, a specified point.

k. **Directional control after touchdown**

1) The landing process must never be considered complete until your airplane decelerates to normal taxi speed during the landing roll or has been brought to a complete stop when clear of the runway.

   a) Accidents have occurred as the result of pilots abandoning their vigilance and positive control after getting the airplane on the ground.

2)   You must be alert for directional control problems immediately upon and after touchdown due to the ground friction on the wheels. The friction creates a pivot point on which a moment arm can act.

   a)   This is especially true in tailwheel-type airplanes because, unlike nosewheel-type airplanes, the CG is behind the main wheels.

   i)   Any difference between the direction in which the airplane is traveling and the direction in which it is headed will produce a moment about the pivot point of the wheels, and the airplane will tend to swerve.

   b)   Nosewheel-type airplanes make the task of directional control easier because the CG, being ahead of the main landing wheels, presents a moment arm which tends to straighten the airplane's path during the touchdown and after-landing roll.

   i)   This tendency should not lull you into a false sense of security.

3)   Another directional control problem in crosswind landings is due to the weathervaning tendency of your airplane. Characteristically, an airplane has a greater profile or side area behind the main landing gear than forward of it.

   a)   With the main landing wheels acting as a pivot point and the greater surface area exposed to a crosswind behind the pivot point, the airplane will tend to turn or weathervane into the wind.

   b)   This tendency is characteristic of all airplanes, but it is more prevalent in the tailwheel type because the airplane's surface area behind the main landing gear is greater than in nosewheel-type airplanes.

4)   Loss of directional control may lead to an aggravated, uncontrolled, tight turn on the ground (i.e., a ground loop).

   a)   The combination of centrifugal force acting on the CG and ground friction on the main wheels resisting it during the ground loop may cause the airplane to tip, or lean, enough for the outside wingtip to contact the ground.

   i)   This may impose a great enough sideward force to collapse the landing gear.

   b)   Tailwheel-type airplanes are most susceptible to ground loops late in the after-landing roll because rudder effectiveness decreases with the decreasing airflow along the rudder surface as the airplane slows.

5)   The ailerons serve the same purpose on the ground as they do in the air; they change the lift and drag components of the wings.

   a)   While your airplane is decelerating during the after-landing roll, more and more aileron must be applied to keep the upwind wing from rising.

   b)   Since your airplane is slowing down and there is less airflow around the ailerons, they become less effective. At the same time, the relative wind the upwind wing.

   i)   Consequently, when the airplane is coming to a full stop, the aileron control must be held FULLY toward the wind.

l. **Use of brakes**

1) If available runway permits, the speed of the airplane should be allowed to dissipate in a normal manner by the friction and drag of the wheels on the ground.

a) Brakes may be used if needed to slow the airplane. This is normally done near the end of the after-landing roll to ensure the airplane is moving slowly enough to exit the runway in a controlled manner.

2) Use equal pressure on brakes to avoid loss of directional control, i.e., ground looping.

a) Start with a light pressure and gradually increase as the elevator control becomes ineffective.

m. **Use of checklist**

1) The before-landing checklist should be completed on the downwind leg.
2) Use the checklist in your POH.

2. **Exhibit your instructional knowledge by explaining the following common errors related to a normal and a crosswind approach and landing.**

a. **Improper use of landing performance data and limitations**

1) This error may be due to a lack of understanding of the performance charts in the POH.

2) Use the POH to determine the appropriate airspeeds for a normal and crosswind approach and landing.

3) In gusty or strong crosswinds, use the crosswind component chart to determine that the crosswind component does not exceed the airplane's crosswind limitations.

4) You must use the POH to determine performance data and limitations and should not attempt to do better than the data.

b. **Failure to establish approach and landing configuration at appropriate time or in proper sequence**

1) You must use the before-landing checklist in the POH to ensure that you follow the proper sequence in establishing the correct approach and landing configuration for the airplane.

2) You should initially start the checklist at midpoint on the downwind leg with the power reduction beginning once the airplane is abeam of the intended point of landing.

a) By the time the airplane is on the final approach and aligned with the runway centerline, you should confirm that the airplane is in the final landing configuration by completing the checklist once again.

c. **Failure to establish and maintain a stabilized approach**

1) Once on final and aligned with the runway, you should make small adjustments to pitch and power to establish the correct descent angle (i.e., glide path) and airspeed.

a) Remember that you must make simultaneous adjustments to both pitch and power.

b) Large adjustments will make the approach a roller coaster ride.

2)   Low final approach

    a)   When the base leg is too low, insufficient power is used, landing flaps are extended prematurely, or the velocity of the wind is misjudged, sufficient altitude may be lost to cause the airplane to be well below the proper glide path.

        i)   In this situation, your student would have to apply considerable power to fly the airplane (at an excessively low altitude) up to the runway threshold.

    b)   When it is realized that the runway will not be reached unless appropriate action is taken, immediately apply power to maintain the airspeed while the pitch attitude is raised to increase lift and stop the descent.

        i)   When the proper approach path has been intercepted, the correct approach attitude should be reestablished and the power reduced again.

        ii)   Do NOT increase the pitch attitude without increasing the power, since the airplane will decelerate rapidly. It may approach the critical angle of attack and stall.

    c)   If there is any doubt about the successful outcome of the approach, execute an immediate go-around.

3)   Slow final approach

    a)   When the airplane is flown at too slow an airspeed on the final approach, your student's judgment of the rate of descent and the height of roundout may be defective.

    b)   During an excessively slow approach, the wing is operating near the critical angle of attack and, depending on the pitch attitude changes and control usage, the airplane may stall or sink rapidly, hitting the ground hard.

    c)   When your student notices a slow-speed approach, (s)he should apply power to accelerate the airplane and increase the lift to reduce the sink rate and to prevent a stall.

        i)   Do NOT increase the angle of attack because this may cause a stall.

        ii)   Applying power should be done at a high enough altitude to reestablish the correct approach speed and attitude.

    d)   If too slow and/or too low, execute an immediate go-around.

d.   **Inappropriate removal of hand from throttle**

1)   One hand should always remain on the control yoke or stick.

2)   The other hand should remain on the throttle unless operating the microphone or making an adjustment, such as trim or flaps.

    a)   Once you are on short final, your hand should remain on the throttle even if ATC gives you instruction (e.g., cleared to land).

        i)   Your first priority is to fly your airplane and avoid doing tasks which may distract you from maintaining control.

        ii)   Fly first; talk later.

3)   It is important that your student be in the habit of keeping one hand on the throttle in case a sudden and unexpected hazardous situation should require an immediate application of power.

e. **Improper procedure during roundout and touchdown**

1) High roundout

   a) This error occurs when the roundout has been made too rapidly and the airplane is flying level too high above the runway.

      i) If you continue the roundout, you will increase the wings' angle of attack to the critical angle while reducing the airspeed. Thus, you will stall your airplane and drop hard onto the runway.

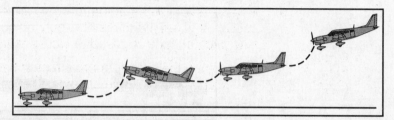

   b) To correct this, the pitch attitude should be held constant until the airplane decelerates enough to again start descending. Then the roundout can be continued to establish the proper landing attitude.

      i) Use this technique only when there is an adequate amount of airspeed. It may be necessary to add a small amount of power to prevent the airspeed from decreasing excessively and to avoid losing lift too rapidly.

   c) Although back pressure on the elevator control may be relaxed slightly, the nose should not be lowered any perceptible amount to make the airplane descend when relatively close to the runway.

      i) The momentary decrease in lift resulting from lowering the nose (i.e., decreasing angle of attack) may be so great that a nosewheel-type airplane may contact the ground with the nosewheel, which may then collapse.

      ii) Instruct your student to execute a go-around any time it appears that the nose should be lowered significantly.

2) Late or rapid roundout

   a) Starting the roundout too late or pulling the elevator control back too rapidly to prevent the airplane from touching down prematurely can impose a heavy load factor on the wing and cause an accelerated stall.

      i) This dangerous situation may cause the airplane to land extremely hard on the main landing gear and then bounce back into the air.

         • As the airplane contacts the ground, the tail will be forced down very rapidly by the back pressure on the elevator and by inertia acting downward on the tail.

   b) Recovery requires prompt and positive application of power prior to occurrence of the stall.

      i) Recovery may be followed by a normal landing, if sufficient runway is available; otherwise, execute an immediate go-around.

3) Floating during roundout

a) This error is caused by using excessive speed on the final approach. Before touchdown can be made, your airplane may be well past the desired landing point, and the available runway may be insufficient.

b) When diving an airplane on final approach to land at the proper point, there will be an appreciable increase in airspeed. Consequently, the proper touchdown attitude cannot be established without producing an excessive angle of attack and lift, usually causing the airplane to gain altitude.

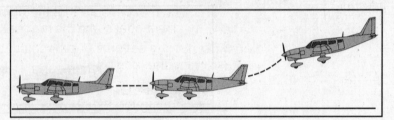

c) Failure to anticipate ground effect may also result in floating.

d) The recovery will depend on the amount of floating, the effect of a crosswind (if any), and the amount of runway remaining.

i) You must smoothly and gradually adjust the pitch attitude as your airplane decelerates to touchdown speed and starts to settle so that the proper landing attitude is attained at the moment of touchdown.

• The slightest error in judgment will result in either ballooning or bouncing.

ii) If there is limited runway remaining or the airplane drifts sideways, you should immediately execute a go-around.

4) Ballooning during roundout

a) If your student misjudges the rate of descent during a landing and thinks the airplane is sinking faster than it should, (s)he may tend to increase the pitch attitude and angle of attack too rapidly.

i) As a result, not only does the descent stop, but the airplane actually starts to climb (i.e., balloon).

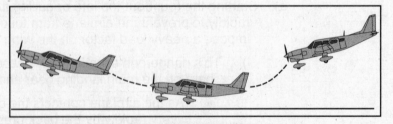

ii) Ballooning can be dangerous because the height above the ground is increasing and the airplane may rapidly approach a stall.

• The altitude gained in each instance will depend on the airspeed or the rapidity with which the pitch attitude is increased.

b) When ballooning is slight, a constant landing attitude may be held and the airplane allowed to settle onto the runway.

    i) You must be extremely cautious of ballooning when there is a crosswind present because the crosswind correction may be inadvertently released or it may become inadequate.

    ii) Due to the lower airspeed after ballooning, the crosswind affects the airplane more. Consequently, the wing will have to be lowered even further to compensate for the increased drift.

       • You must ensure that the upwind wing is down and that directional control is maintained with the rudder.

c) Depending on the severity of ballooning, the use of power may be helpful in cushioning the landing.

    i) By adding power, you can increase thrust to keep the airspeed from decelerating too rapidly and the wings from suddenly losing lift, but the throttle must be closed immediately after touchdown.

    ii) Remember that left-turning tendencies will have been created as power was applied; thus it will be necessary to use rudder pressure to counteract this effect.

d) When ballooning is excessive, instruct your student that it is best to execute a go-around IMMEDIATELY. Do NOT attempt to salvage the landing. Power must be applied before the airplane enters a stall.

5) Bouncing during touchdown

a) When the airplane contacts the ground with a sharp impact as the result of an improper attitude or an excessive rate of sink, it tends to bounce back into the air.

    i) Though the airplane's tires and shock struts provide some springing action, the airplane does not bounce as does a rubber ball.

    ii) The airplane rebounds into the air because the wing's angle of attack was abruptly increased, producing a sudden addition of lift.

       • The change in angle of attack is the result of inertia instantly forcing the airplane's tail downward when the main wheels contact the ground sharply.

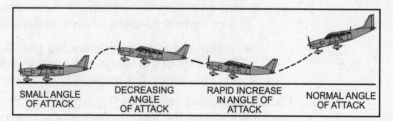

SMALL ANGLE OF ATTACK   DECREASING ANGLE OF ATTACK   RAPID INCREASE IN ANGLE OF ATTACK   NORMAL ANGLE OF ATTACK

    iii) The severity of the bounce depends on the airspeed at the moment of contact and the degree to which the angle of attack, or pitch attitude, was increased.

b) The corrective action for a bounce is the same as for ballooning and similarly depends on the severity.

   i) When the bounce is very slight and there is no extreme change in the airplane's pitch attitude, a follow-up landing may be executed by applying sufficient power to cushion the subsequent touchdown and by smoothly adjusting the pitch to the proper touchdown attitude.

   ii) Remember that addition of power requires immediate right rudder to preclude a yaw to the left. A yaw to the left means uncoordinated flight controls, which may result in the left wing stalling first and dropping to strike the ground.

c) Extreme caution and alertness must be exercised, especially when there is a crosswind. The crosswind correction will normally be released by inexperienced pilots when the airplane bounces.

   i) When one main wheel of the airplane strikes the runway, the other wheel will touch down immediately afterwards, and the wings will become level.

   ii) Then, with no crosswind correction as the airplane bounces, the wind will cause the airplane to roll with the wind, thus exposing even more surface to the crosswind and drifting the airplane more rapidly.

   iii) Remember, the upwind wing will have to be lowered even further to compensate for the increased drift due to the slower airspeed.

d) When a bounce is severe, it is best to execute a go-around immediately.

6) Hard landing

a) When the airplane contacts the ground during a landing, its vertical speed is instantly reduced to zero. Unless provision is made to slow this vertical speed and cushion the impact of touchdown, the force of contact with the ground may be so great as to cause structural damage to the airplane.

b) The purpose of pneumatic tires, rubber or oleo shock absorbers, and other such devices is, in part, to cushion the impact and to increase the time in which the airplane's vertical descent is stopped.

   i) The importance of this cushion may be understood from the computation that a 6-in. free fall on landing is roughly equal to a 340-fpm descent.

   ii) Within a fraction of a second, the airplane must be slowed from this rate of vertical descent to zero, without damage.

   • During this time, the landing gear together with some aid from the lift of the wings must supply whatever force is needed to counteract the force of the airplane's inertia and weight.

c) The lift decreases rapidly as the airplane's forward speed is decreased, and the force on the landing gear increases as the shock struts and tires are compressed by the impact of touchdown.

   i) When the descent stops, the lift will practically be zero, leaving the landing gear alone to carry both the airplane's weight and inertial forces.

   ii) The load imposed at the instant of touchdown may easily be three or four times the actual weight of the airplane, depending on the severity of contact.

7) Touchdown in a drift or crab

a) If your student has not taken adequate corrective action to avoid drift during a crosswind landing, the main wheels' tire treads offer resistance to the airplane's sideward movement in respect to the ground. Consequently, any sideward velocity of the airplane is abruptly decelerated, as shown in the figure below.

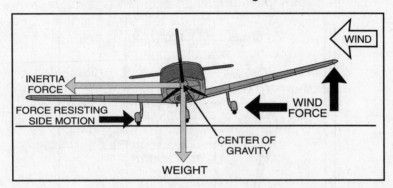

i) This resistance creates a moment around the main wheel when it contacts the ground, tending to roll the airplane.

ii) If the windward landing gear is raised by the action of this moment, all of the weight and shock of landing will be borne by one main wheel, possibly causing structural damage.

b) It is vital to prevent drift and keep the longitudinal axis of the airplane aligned with the runway during the roundout and touchdown.

c) There are three factors that will cause the longitudinal axis and the direction of motion to be misaligned during touchdown.

i) Drifting
ii) Crabbing
iii) A combination of both

f. **Poor directional control after touchdown**

1) Ground loop

a) A ground loop is an uncontrolled turn during ground operation that may occur while taxiing or taking off, but especially during the after-landing roll.

i) It is not always caused by drift or weathervaning, although one of these may cause the initial swerve. Other reasons may include careless use of the rudder, an uneven ground surface, or a soft spot that retards one main wheel of the airplane.

b)   Due to the characteristics of an airplane equipped with a tailwheel, the forces that cause a ground loop increase as the swerve increases.

   i)   The initial swerve develops centrifugal force which, acting at the CG (located behind the main wheels), swerves the airplane even more.

   ii) If allowed to develop, the centrifugal force produced may become great enough to tip the airplane until one wing strikes the ground.

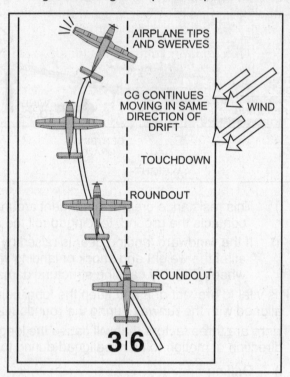

c)   A nosewheel-type airplane is less prone to ground loop. Since the CG is located forward of the main landing gear, any time a swerve develops, centrifugal force acting on the CG will tend to stop the swerving action.

d)   If the airplane touches down while drifting or in a crab, instruct your student to apply aileron toward the high wing and stop the swerve with the rudder.

e)   Brakes should be used to correct for turns or swerves only when the rudder is inadequate. You must exercise caution when applying corrective brake action because it is very easy to over-control and aggravate the situation.

   i)   If brakes are used, sufficient brake should be applied on the low-wing (outside of the turn) to stop the swerve.

   ii) When the wings are approximately level, the new direction must be maintained until the airplane has slowed to taxi speed or has stopped.

2) Wing rising after touchdown

    a) When landing in a crosswind, there may be instances when a wing will rise during the after-landing roll. This may occur whether or not there is a loss of directional control, depending on the amount of crosswind and the degree of corrective action.

    b) Anytime an airplane is rolling on the ground in a crosswind condition, the upwind wing is receiving a greater force from the wind than the downwind wing. This causes a lift differential.

        i) Also, the wind striking the fuselage on the upwind side may further raise the wing by tending to tip or roll the fuselage.

    c) When the effects of these two factors are great enough, one wing may rise even though directional control is maintained. If no correction is applied, it is possible that a wing will rise sufficiently to cause the other one to strike the ground.

    d) In the event a wing starts to rise during the after-landing roll, instruct your student immediately to apply more aileron pressure toward the high wing and continue to maintain forward direction with the rudder pedals.

        i) The sooner the aileron control is applied, the more effective it will be.

        ii) The further a wing is allowed to rise before taking corrective action, the more airplane surface is exposed to the force of the crosswind, thus reducing the effectiveness of the aileron.

        iii) Also, when one wing rises or is lighter than the other, the wheel on that side does not grip as well, so it will slip and lose contact with the ground.

g. **Improper use of brakes**

1) Never apply the brakes until the airplane is firmly on the runway under complete control.

2) Use the minimum amount of braking required, and let the airplane slow by the friction and drag of the wheels on the ground, if runway length permits.

3) Use equal pressure on both brakes to prevent swerving and/or loss of directional control.

3. **Demonstrate and simultaneously explain a normal or a crosswind approach and landing from an instructional standpoint.**

a. Your demonstration must be to commercial pilot skill level.

b. You will be evaluated on your ability to explain and demonstrate simultaneously the key elements of this task.

c. If you do not perform the maneuver to FAA completion standards, point out the deficiency to your FAA inspector/examiner as an example of an error, and explain how it should be corrected (including how and why it occurred).

4. **Analyze and correct simulated common errors related to a normal or crosswind approach and landing.**

a. As your FAA inspector/examiner is performing this task, you must be able to recognize, analyze, and correct any error.

b. Explain what the error is, and then explain what (s)he should do to correct the error.

c. If your FAA inspector/examiner does not perform the maneuver to FAA completion standards, point out the deficiency to him/her as an example of a student error, and explain how it should be corrected (including how and why it occurred).

END OF TASK

SLIP TO A LANDING

VII.G. TASK: SLIP TO A LANDING

 REFERENCES: FAA-H-8083-3; FAA-S-8081-14, POH/AFM.

Objective. To determine that the applicant:

1. Exhibits instructional knowledge of the elements of a slip (forward and side) to a landing by describing --

 a. Configuration, power, and trim.

 b. Obstructions and other hazards which should be considered.

 c. A stabilized slip at the appropriate airspeed to the selected touchdown area.

 d. Possible airspeed indication errors.

 e. Proper application of flight controls.

 f. A precise ground track.

 g. Wind shear and wake turbulence avoidance procedures.

 h. Timing, judgment, and control procedure during transition from slip to touchdown.

 i. Directional control after touchdown.

 j. Use of brakes.

 k. Use of checklist.

2. Exhibits instructional knowledge of common errors related to a slip (forward and side) to a landing by describing --

 a. Improper use of landing performance data and limitations.

 b. Failure to establish approach and landing configuration at appropriate time or in proper sequence.

 c. Failure to maintain a stabilized slip.

 d. Inappropriate removal of hand from throttle.

 e. Improper procedure during transition from the slip to the touchdown.

 f. Poor directional control after touchdown.

 g. Improper use of brakes.

3. Demonstrates and simultaneously explains a forward or sideslip to a landing from an instructional standpoint.

4. Analyzes and corrects simulated common errors related to a forward or sideslip to a landing.

A. General Information

 1. The objective of this task is to determine your instructional knowledge of the elements and common errors related to a slip to a landing. You will be required to demonstrate and simultaneously explain this task from an instructional standpoint.

 a. The discussion for this task will focus primarily on forward slips.
 b. Sideslips are used only when making a crosswind landing.

 1) For a complete discussion of crosswind landing techniques, see Task VII.F., Normal and Crosswind Approach and Landing, beginning on page 275.

 2. The primary purpose of forward slips is to dissipate altitude without increasing the airplane's speed, particularly in airplanes not equipped with flaps.

 a. There are many circumstances requiring the use of forward slips, such as during flap failure in a landing approach over obstacles and in the execution of forced landings, when it is always wise to allow an extra margin of altitude for safety in the original estimate of the approach.

 b. Altitude is lost in a slip by increasing drag caused by the airflow striking the wing-low side of the airplane. The L/D ratio decreases, which causes the rate of descent to increase.

 3. The primary purpose of a sideslip is to allow the airplane to touch down with zero drift (sideward motion relative to the runway) and with the airplane's longitudinal axis aligned with the direction of movement (i.e., with the nose pointing straight down the runway) when a crosswind condition exists.

4. While the performance of a slip to a landing is required only on the Private Pilot PTS, it is a good maneuver to review with all pilots.

    a. Sideslips will be covered as a natural byproduct of performing crosswind landings.

5. See Module 3.7, Wake Turbulence, in *Pilot Handbook* for a seven-page discussion on wake turbulence and avoidance procedures.

B. Task Objectives

1. **Exhibit your instructional knowledge by explaining the following elements of a slip (forward or side) to a landing.**

    a. **Configuration, power, and trim**

        1) The forward slip is a descent with one wing lowered and the airplane's longitudinal axis at an angle to the flight path. The flight path remains the same as before the slip was begun.

            a) If there is any crosswind, the slip will be more effective and easier to recover if made toward the wind.

            b) Slipping should be done with the engine idling. There is little logic in slipping to lose altitude if the power is still being used.

        2) The sideslip is a descent (usually followed by a touchdown) with one wing lowered and the airplane's longitudinal axis aligned with the flight path.

            a) See the discussion of Task VII.F., Normal and Crosswind Approach and Landing, beginning on page 276.

        3) The use of slips has definite limitations.

            a) Some pilots may try to lose altitude by using a violent forward slip rather than by smooth maneuvering, exercising good judgment, and using only a slight or moderate slip.

                i) In emergency landings, this erratic practice will invariably lead to trouble since excess speed may prevent the airplane from touching down anywhere near the proper point, and very often may result in its overshooting the entire field.

            b) A successful sideslip is not possible in all wind conditions.

                i) At some wind speeds and angles, there may not be sufficient rudder authority to align the longitudinal axis with the runway.

        4) Check your POH for any limitations on the use of a forward slip.

            a) Some airplanes may be prohibited from performing a forward slip with the wing flaps extended.

            b) Other airplanes may have a time limitation (i.e., slip can be used no more than 30 sec.).

        5) Trim should be adjusted as necessary.

    b. **Obstructions and other hazards which should be considered**

        1) Depending on the wind conditions, landing surface (i.e., hard surface or soft surface), and any obstructions, you should select the touchdown point as you would for a normal, soft, or short field.

            a) Remember to select an aim point, which will be the point at the end of your selected glide path.

            b) See Task VII.F., Normal and Crosswind Approach and Landing, beginning on page 277, for a discussion of selecting and using an aim point.

2) Be aware of traffic, both in the air and on the ground.

a) Watch for vehicles and/or people on or near the runway.

c. **A stabilized slip at the appropriate airspeed to the selected touchdown area**

1) You should only establish a forward slip once you are assured that you can safely land in the desired area.

a) You will need to establish your airplane higher on final, since the slip will result in a steeper than normal descent.

2) Apply carburetor heat (if applicable), and reduce power to idle.

a) Extend the flaps, unless your POH prohibits slips with flaps extended.
b) Establish a pitch attitude that will maintain a normal final approach speed.

3) Completion standard for this element (private pilot only)

a) Establish the slipping attitude at the point from which a landing can be made using the recommended approach and landing configuration and airspeed, and adjust pitch attitude and power as required.

d. **Possible airspeed indication errors**

1) Because of the location of the pitot tube and static vent(s) in some airplanes, the airspeed indicator may have a considerable degree of error when the airplane is in a slip.

a) If your airplane has only one static vent (normally on the left side of the fuselage), airspeed indications may be in error during a slip.

i) In a slip to the left (i.e., left-wing down), ram air pressure will be entering the static tube, and the pressure in the airspeed indicator will be higher than normal.

• Thus, the airspeed indication will be lower than the actual speed.

ii) In a slip to the right, a low pressure area tends to form on the left side of the airplane, thus lowering the pressure in the static vent and the airspeed indicator.

• Thus, the airspeed indication will be higher than the actual speed.

b) If your airplane has a static vent on each side of the fuselage, these errors tend to cancel each other out.

2) You must recognize a properly performed slip by the attitude of your airplane, the sound of the airflow, and the feel of the flight controls.

e.   **Proper application of flight controls**

1)   Assuming that your airplane is originally in straight flight, you should lower the wing on the side toward which a forward slip is to be made by use of the ailerons (wing down into crosswind, if one exists).

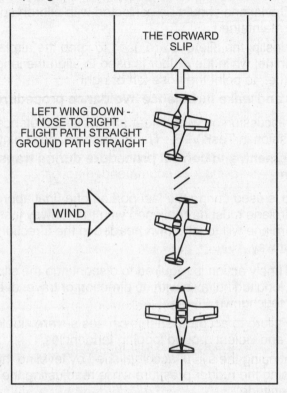

THE FORWARD
SLIP

LEFT WING DOWN -
NOSE TO RIGHT -
FLIGHT PATH STRAIGHT
GROUND PATH STRAIGHT

WIND

a)   Simultaneously, your airplane's nose must be yawed in the opposite direction with the rudder so that the airplane's longitudinal axis is at an angle to its flight path.

i)   If rudder application is delayed, the airplane will turn in the direction of the lowered wing.

ii)   The rudder should be fully deflected and kept that way during the forward slip because the rudder is less effective than the ailerons.

•   The ailerons will be used to maintain a precise ground track.

2)   The nose of the airplane should also be raised as necessary to prevent the airspeed from increasing.

a)   Remember, if your airspeed indicator is subject to errors in slips, you should understand those errors.

i)   Maintain the proper pitch attitude by sight and feel.

b)   Remember to use whatever elevator pressure (forward or back) is necessary to maintain the proper pitch attitude.

3)   Note that a sideslip is aerodynamically identical to a forward slip; the difference between these maneuvers lies in their respective purposes.

a)   A forward slip is used to steepen the descent angle without excessively increasing the airspeed, while a sideslip is used for making a crosswind landing.

b)   One aerodynamic difference is that a forward slip is usually performed with full rudder deflection, while a sideslip normally requires only partial rudder deflection (i.e., enough to keep the nose straight).

f. **A precise ground track**

1) In a forward slip, the degree to which the nose is yawed in the opposite direction from the bank should be such that the airplane maintains a precise ground track over the extended centerline of the runway.

   a) Deviations should be corrected with aileron control while full rudder is maintained.

2) In a sideslip, the ailerons are used to keep the airplane over the runway centerline, while the rudder is used to align the longitudinal axis with the flight path (i.e., to point the nose left or right).

g. **Wind shear and wake turbulence avoidance procedures**

1) For a discussion on wind shear and wake turbulence, see the two-page discussion in Task VI.B., Traffic Patterns, beginning on page 248.

h. **Timing, judgment, and control procedure during transition from slip to touchdown**

1) If a slip is used during the last portion of a final approach, the longitudinal axis of the airplane must be realigned with the runway just prior to touchdown so that the airplane will touch down headed in the direction in which it is moving over the runway.

   a) Timely action is required to discontinue the slip and realign the airplane's longitudinal axis with its direction of travel over the ground before touchdown.

   b) Failure to accomplish this causes severe sideloads on the landing gear and violent ground looping tendencies.

2) Discontinuing the slip is accomplished by leveling the wings and simultaneously releasing the rudder pressure while readjusting the pitch attitude to the normal glide attitude.

   a) If the pressure on the rudder is released abruptly, the nose will swing too quickly into line, and the airplane will tend to acquire excess speed.

      i) Also, momentum may carry the nose of the airplane past straight ahead. Recovery should be smooth.

3) Use the same technique during the roundout and touchdown as discussed in Task VII.F., Normal and Crosswind Approach and Landing, beginning on page 275.

4) Completion standard for this element (private pilot only)

   a) Touch down smoothly at approximate stalling speed, at or within 400 ft. beyond a specified point, with no side drift, and with your airplane's longitudinal axis aligned with and over the runway centerline.

   b) Your performance is expected to be "more precise" than that required for a private pilot.

      i) This more precise performance must be determined by your FAA inspector/examiner through the exercise of subjective judgment.

i. **Directional control after touchdown**

1) Maintain directional control after touchdown as described in Task VII.F., Normal and Crosswind Approach and Landing, beginning on page 285.

j.   **Use of brakes**

1)   If available runway permits, the speed of the airplane should be allowed to dissipate in a normal manner by the friction and drag of the wheels on the ground.

  a)   Brakes may be used if needed to slow the airplane. This is normally done near the end of the after-landing roll to ensure the airplane is moving slowly enough to exit the runway in a controlled manner.

2)   Use equal pressure on the brakes when using them to avoid loss of directional control, i.e., ground looping.

k.   **Use of checklist**

1)   The before-landing checklist should be completed on the downwind leg.

  a)   Remember that, if your airplane has an operating limitation on the use of flaps during a slip, you must comply with the limitation.

2)   After your airplane is clear of the runway, you should stop and complete the after-landing checklist.

2.   **Exhibit your instructional knowledge by explaining the following common errors related to a slip (forward and side) to a landing.**

a.   **Improper use of landing performance data and limitations**

1)   Use your POH to determine the appropriate airspeeds for a normal and crosswind approach and landing.

2)   In gusty and/or strong crosswinds, use the crosswind component chart to determine that you are not exceeding your airplane's crosswind limitations.

3)   Use your POH to determine data and limitations, and do not attempt to do better than the data.

4)   Do not exceed any limitations listed in the POH on the use of a slip.

b.   **Failure to establish approach and landing configuration at appropriate time or in proper sequence**

1)   Use the before-landing checklist in your POH to ensure that you follow the proper sequence in establishing the correct approach and landing configuration for your airplane.

2)   You should be in your final landing configuration, if possible, before entering the slip.

c.   **Failure to maintain a stabilized slip**

1)   Once the decision is made to use a slip, you must use the proper application of flight controls and power to establish the slip.

2)   Stabilize the slip as soon as possible. Avoid large corrections, as they will prevent you from maintaining a stabilized slip.

d.   **Inappropriate removal of hand from throttle**

1)   One hand should remain on the control yoke at all times.

2)   The other hand should remain on the throttle unless operating the microphone or making an adjustment, such as trim or flaps.

  a)   Once you are on short final, your hand should remain on the throttle, even if ATC gives you instruction (e.g., cleared to land).

    i)   Your first priority is to fly your airplane and avoid doing tasks which may distract you from maintaining control.

    ii)   Fly first; talk later.

      3) It is important that you be in the habit of keeping one hand on the throttle in case a sudden and unexpected hazardous situation should require an immediate application of power.

   e. **Improper procedure during transition from the slip to the touchdown**

      1) When recovering from a forward slip, you should smoothly straighten the nose with the rudder and use ailerons as necessary to correct for any crosswind.

      2) If you release the pressure on the rudder too abruptly, the nose will swing too quickly into line, and your airplane's airspeed will increase.

         a) An abrupt release may also cause the nose to swing past straight ahead.

      3) Failure to realign the airplane's longitudinal axis with the runway centerline will cause severe sideloads on the landing gear.

   f. **Poor directional control after touchdown**

      1) Use rudder to steer your airplane on the runway, and increase aileron deflection into the wind as airspeed increases.

      2) See the discussion of common errors of Task VII.F., Normal and Crosswind Approach and Landing, beginning on page 293, concerning ground loops and other directional control problems after touchdown.

   g. **Improper use of brakes**

      1) Use the minimum amount of braking required, and let the airplane slow by the friction and drag of the wheels on the ground, if runway length permits.

      2) Never attempt to apply brakes until the airplane is firmly on the runway under complete control.

      3) Use equal pressure on both brakes to prevent swerving and/or loss of directional control.

3. **Demonstrate and simultaneously explain a forward or sideslip to a landing from an instructional standpoint.**

   a. Your demonstration must be to commercial pilot skill level.

   b. You will be evaluated on your ability to explain and demonstrate simultaneously the key elements of this task.

   c. If you do not perform the maneuver to FAA completion standards, point out the deficiency to your FAA inspector/examiner as an example of an error, and explain how it should be corrected (including how and why it occurred).

4. **Analyze and correct simulated common errors related to a forward or sideslip to a landing.**

   a. As your FAA inspector/examiner is performing this task, you must be able to recognize, analyze, and correct any error.

   b. Explain what the error is, and then explain what (s)he should do to correct the error.

   c. If your FAA inspector/examiner does not perform the maneuver to FAA completion standards, point out the deficiency to him/her as an example of a student error, and explain how it should be corrected (including how and why it occurred).

END OF TASK

GO-AROUND/REJECTED LANDING

VII.H. TASK: GO-AROUND/REJECTED LANDING

REFERENCES: FAA-H-8083-3; FAA-S-8081-12, FAA-S-8081-14; POH/AFM.

Objective. To determine that the applicant:

1. Exhibits instructional knowledge of the elements of a go-around/rejected landing by describing --

   a. Situations where a go-around/rejected landing is necessary.

   b. Importance of making a prompt decision.

   c. Importance of applying takeoff power immediately after the go-around/rejected landing decision is made.

   d. Importance of establishing proper pitch attitude.

   e. Wing flaps retraction.

   f. Use of trim.

   g. Landing gear retraction.

   h. Proper climb speed.

   i. Proper track and obstruction clearance.

   j. Use of checklist.

2. Exhibits instructional knowledge of common errors related to a go-around/rejected by describing --

   a. Failure to recognize a situation where a go-around/rejected landing is necessary.

   b. Hazards of delaying a decision to go-around/rejected landing.

   c. Improper power application.

   d. Failure to control pitch attitude.

   e. Failure to compensate for torque effect.

   f. Improper trim procedure.

   g. Failure to maintain recommended airspeeds.

   h. Improper wing flaps or landing gear retraction procedure.

   i. Failure to maintain proper track during climb-out.

   j. Failure to remain well clear of obstructions and other traffic.

3. Demonstrates and simultaneously explains a go-around/rejected landing from an instructional standpoint.

4. Analyzes and corrects simulated common errors related to a go-around/rejected landing.

A. General Information

   1. The objective of this task is to determine your instructional knowledge of the elements and common errors related to a go-around/rejected landing. You will be required to demonstrate and simultaneously explain this task from an instructional standpoint.

   2. For safety reasons, it may be necessary for you to discontinue your approach and attempt another approach under more favorable conditions.

      a. This procedure is called a go-around from a rejected (balked) landing.

      b. We will use the term "go-around" to refer to this maneuver in the following discussion.

   3. Regardless of the airplane's height above the ground, a safe go-around can be accomplished if

      a. An early decision is made.

      b. A sound plan is followed.

      c. The procedure is performed properly.

   4. Instruct your student to consult the POH for the proper procedure to follow for the airplane.

      a. You should think about the possibility of having to do a go-around during each approach and landing.

B.   Task Objectives

1.   **Exhibit your instructional knowledge by explaining the following elements of a go-around/rejected landing.**

   a.   **Situations where a go-around/rejected landing is necessary**

   1)   Occasionally it may be advisable for safety reasons to discontinue the landing approach and make another approach under more favorable conditions. Unfavorable conditions may include

   a)   Extremely low base-to-final turn

   b)   Too high or too low final approach

   c)   The unexpected appearance of hazards on the runway, e.g., another airplane failing to clear the runway on time

   d)   Wake turbulence from a preceding aircraft

   e)   Wind shear encounter

   f)   Overtaking another airplane on final approach

   g)   ATC instructions to "go around"

   b.   **Importance of making a prompt decision**

   1)   The need to discontinue a landing may arise at any point in the landing process, but the most critical go-around is one started when very close to the ground. A timely decision must be made.

   a)   The earlier you recognize a dangerous situation, the sooner you can decide to reject the landing and start the go-around, and the safer this maneuver will be.

   b)   Never wait until the last possible moment to make a decision.

   c)   As an instructor, you must help develop your student's ability to make good and timely decisions.

   2)   Official reports concerning go-around accidents frequently cite "pilot indecision" as a cause. This happens when a pilot fixates on trying to make a bad landing good, resulting in a late decision to go around.

   a)   This is natural, since the purpose of an approach is a landing.

   b)   Delays in deciding what to do cost valuable runway stopping distance. They also cause loss of valuable altitude as the approach continues.

   c)   If there is any question about making a safe touchdown and rollout, execute a go-around immediately.

   3)   Once you decide to go around, stick to it! Too many airplanes have been lost because a pilot has changed his/her mind and tried to land after all.

   c.   **Importance of applying takeoff power immediately after the go-around/rejected landing decision is made**

   1)   When the decision is made to discontinue an approach and perform a go-around, takeoff power should be applied immediately.

   a)   Adjust carburetor heat to OFF (cold) position, if appropriate.

   b)   Check that the airplane's propeller control is set at maximum (high) RPM and the mixture is full rich or appropriately leaned for high-density altitude airport operations. Checking for the appropriate settings should have been accomplished during the before-landing checklist.

   i)   Avoid overboosting a turbocharged engine.

d. **Importance of establishing proper pitch attitude**

1) As the takeoff power is being applied, you should establish the pitch attitude to maintain V_Y so as to slow or stop the descent.

    a) You may want to establish a pitch attitude for V_X initially, if you need to clear any obstacles.

2) During the initial part of an extremely low go-around, your airplane may "mush" onto the runway and bounce. This situation is not particularly dangerous if the airplane is kept straight and a constant, safe pitch attitude is maintained.

    a) Your airplane will be approaching safe flying speed rapidly, and the advanced power will cushion any secondary touchdown.

3) Establish a climb pitch attitude by use of outside visual references. From your training, you should have a knowledge of the visual clues to attain the climb attitude.

e. **Wing flaps retraction**

1) Immediately after applying power and raising the nose, you should partially retract or place the wing flaps in the takeoff position, as stated in the POH. Use caution in retracting the flaps.

    a) It will probably be wise to retract the flaps intermittently in small increments to allow time for the airplane to accelerate progressively as the flaps are being raised.

    b) A sudden and complete retraction of the flaps at a very low airspeed could cause a loss of lift, resulting in your airplane's settling onto the ground

2) Note that in the commercial and private PTS, the FAA uses the term "approach setting" which may be confusing. Some airplanes have a three-position flap selector, e.g., up, approach, and down, which is why this term was used.

    a) In most training airplanes, your approach will be using full flaps, and during a go-around, your POH will specify how much flaps should be initially retracted.

f. **Use of trim**

1) When takeoff power is applied in the go-around, you must cope with undesirable pitch and yaw.

2) Since you have trimmed your airplane for the approach (i.e., nose-up trim), the nose may rise sharply and veer to the left.

    a) Proper elevator pressure must be applied to maintain a safe climbing pitch attitude.

    b) Right rudder pressure must be increased to counteract torque, or P-factor, and to keep the nose straight.

3) You must use whatever control pressure is required to maintain the proper pitch attitude and to keep your airplane straight. Considerable pressure may be required.

4) While holding your airplane straight and in a safe climbing attitude, you should retrim your airplane to relieve any heavy control pressures.

    a) Since the airspeed will build up rapidly with the application of takeoff power and the controls will become more effective, this initial trim is to relieve the heavy pressures until a more precise trim can be made for the lighter pressures.

     5)   If the pitch attitude is increased excessively in an effort to prevent your airplane from mushing onto the runway, the airplane may stall.

         a)   A stall is especially likely if no trim correction is made and the flaps remain fully extended.

g.   **Landing gear retraction**

     1)   Retract the landing gear, if so equipped, after establishing a positive rate of climb as indicated on the vertical speed indicator.

         a)   Unless otherwise noted in the POH, the flaps are normally retracted (at least partially) before retracting the landing gear.

         b)   On most airplanes, full flaps create more drag than the landing gear.

         c)   In case the airplane should inadvertently touch down as the go-around is initiated, it is desirable to have the gear in the down-and-locked position.

     2)   Never attempt to retract the landing gear until after a rough trim adjustment is accomplished and a positive rate of climb is established.

h.   **Proper climb speed**

     1)   After you have established a positive rate of climb and have retracted the landing gear, adjust the pitch attitude and trim the airplane to climb at V_Y.

     2)   Completion standards for this element

         a)   Commercial PTS

             i)   ***Apply takeoff power immediately, transition to the climb pitch attitude for V_Y, and maintain V_Y, ±5 kt.; retract flaps as appropriate.***

         b)   Private PTS

             i)   Apply takeoff power immediately, transition to the climb pitch attitude for V_Y, and maintain V_Y, +10/-5 kt.; retract flaps as appropriate.

         c)   Both the Private and Commercial PTSs specify landing gear retraction after a positive rate of climb is established.

     3)   From this point on, the procedure is identical to that of a normal climb after takeoff.

i.   **Proper track and obstruction clearance**

     1)   Maintain a ground track parallel to the runway's centerline and at a position that permits you to see the runway.

         a)   Your student needs to maintain visual contact to avoid another dangerous situation if the go-around was made due to an airplane on the runway, especially if that airplane is taking off.

     2)   Maintain an awareness of obstacles, especially if you initiated the go-around at a low altitude.

         a)   You may have to climb at V_X until you have cleared any obstacle along your flight path.

     3)   Now that you have your airplane under control, you can communicate with the tower or the appropriate ground station to advise that you are going around.

j.   **Use of checklist**

     1)   Consult the POH for the proper procedure to follow for your airplane.

     2)   A go-around checklist is an excellent example of a checklist that you will "do and then review." When you execute a go-around, you will do it from memory and then review your checklist after you have initiated and stabilized your go-around.

2. **Exhibit your instructional knowledge by explaining the following common errors related to a go-around/rejected landing.**

   a. **Failure to recognize a situation where a go-around/rejected landing is necessary**

      1) As an instructor, you must help your student learn to make good judgments and decisions.

      2) When there is any doubt of the safe outcome of a landing, the go-around should be initiated immediately.

      3) Do not attempt to salvage a possible bad landing.

   b. **Hazards of delaying a decision to go around/rejected landing**

      1) Delay can lead to an accident because the remaining runway may be insufficient for landing or because delay can prevent the clearing of obstacles on the departure end of the runway.

   c. **Improper power application**

      1) Power should be added smoothly and continuously.

      2) Assure that you have maximum power available at all times during the final approach by completing your before-landing checklist.

      3) On turbocharged engines, a power application that is too rapid may result in an overboost condition.

   d. **Failure to control pitch attitude**

      1) During a go-around, you must divide your attention to accomplish the procedure and maintain control of the airplane.

      2) You must help your student recognize the visual clues as to V_X and V_Y pitch attitudes, and then cross-check with the airspeed indicator.

   e. **Failure to compensate for torque effect**

      1) In a high-power, low airspeed configuration, right rudder pressure must be increased to counteract torque and to keep the airplane's nose straight.

         a) Center the ball in the inclinometer.

   f. **Improper trim procedure**

      1) Initial trim is important to relieve the heavy control pressures.

      2) Since your airplane may be in a nose-up trim configuration, the application of full power may cause the nose to rise sharply.

         a) A considerable amount of forward elevator pressure will be required to maintain the proper pitch attitude and to prevent a stall/spin situation. The use of trim will decrease the pressure you will have to hold.

   g. **Failure to maintain recommended airspeeds**

      1) This error will reduce the climb performance of the airplane and may create unsafe conditions due to obstructions or, if too slow, a stall/spin situation.

   h. **Improper wing flaps or landing gear retraction procedure**

      1) Follow the procedures in the POH.

      2) On most airplanes, the flaps create more drag than the landing gear; thus you should raise (at least partially) the flaps before the landing gear, if retractable.

      3) Retract the landing gear only after a positive rate of climb is established, as indicated on the vertical speed indicator.

    i.   **Failure to maintain proper track during climb-out**

        1)   Not maintaining the proper ground track may cause possible conflicts with other traffic and/or obstructions.

        2)   You are expected by other traffic and/or ATC to maintain a ground track parallel to the runway centerline until at the proper position to turn crosswind.

    j.   **Failure to remain well clear of obstructions and other traffic**

        1)   Climb at V_x if necessary to clear any obstructions.

        2)   Maintain visual contact with other traffic, especially if the go-around was due to departing traffic.

3.   **Demonstrate and simultaneously explain a go-around/rejected landing from an instructional standpoint.**

    a.   Your demonstration must be to commercial pilot skill level.

    b.   You will be evaluated on your ability to explain and demonstrate simultaneously the key elements of this task.

    c.   If you do not perform the maneuver to FAA completion standards, point out the deficiency to your FAA inspector/examiner as an example of an error, and explain how it should be corrected (including how and why it occurred).

4.   **Analyze and correct simulated common errors related to a go-around/rejected landing.**

    a.   As your FAA inspector/examiner is performing this task, you must be able to recognize, analyze, and correct any error.

    b.   Explain what the error is, and then explain what (s)he should do to correct the error.

    c.   If your FAA inspector/examiner does not perform the maneuver to FAA completion standards, point out the deficiency to him/her as an example of a student error, and explain how it should be corrected (including how and why it occurred).

END OF TASK

SHORT-FIELD APPROACH AND LANDING

VII.I. TASK: SHORT-FIELD APPROACH AND LANDING

REFERENCES: FAA-H-8083-3; FAA-S-8081-12, FAA-S-8081-14; POH/AFM.

Objective. To determine that the applicant:

1. Exhibits instructional knowledge of the elements of a short-field approach and landing by describing --

   a. How to determine landing performance and limitations.

   b. Configuration and trim.

   c. Proper use of pitch and power to maintain desired approach angle.

   d. Obstructions and other hazards which should be considered.

   e. Effect of wind.

   f. Selection of touchdown and go-around points.

   g. A stabilized approach at the recommended airspeed to the selected touchdown point.

   h. Coordination of flight controls.

   i. A precise ground track.

   j. Timing, judgment, and control procedure during roundout and touchdown.

   k. Directional control after touchdown.

   l. Use of brakes.

   m. Use of checklist.

2. Exhibits instructional knowledge of common errors related to a short-field approach and landing by describing --

   a. Improper use of landing performance data and limitations.

   b. Failure to establish approach and landing configuration at appropriate time or in proper sequence.

   c. Failure to establish and maintain a stabilized approach.

   d. Improper procedure in use of power, wing flaps, and trim.

   e. Inappropriate removal of hand from throttle.

   f. Improper procedure during roundout and touchdown.

   g. Poor directional control after touchdown.

   h. Improper use of brakes.

3. Demonstrates and simultaneously explains a short-field approach and landing from an instructional standpoint.

4. Analyzes and corrects simulated common errors related to a short-field approach and landing.

A. General Information

1. The objective of this task is to determine your instructional knowledge of the elements and common errors related to a short-field approach and landing. You will be required to demonstrate and simultaneously explain this task from an instructional standpoint.

2. This maximum performance operation requires the use of procedures and techniques for the approach and landing at fields that have a relatively short landing area and/or where an approach must be made over obstacles that limit the available landing area. You must consider the landing surface and wind conditions.

   a. This task is considered a critical maximum performance operation because it requires you to fly your airplane at one of its critical performance capabilities while close to the ground in order to land safely in a confined area.

3. To land within a short field or a confined area, you must have precise, positive control of your airplane's rate of descent and airspeed to produce an approach that will clear any obstacles, result in little or no floating during the roundout, and permit your airplane to be stopped in the shortest possible distance.

4. You must know, understand, and respect both your own and your airplane's limitations.

   a. Think ahead. Do not attempt to land on a short field from which a takeoff is beyond your capability or that of your airplane.

B.   Task Objectives

1.   **Exhibit your instructional knowledge by explaining the following elements of a short-field approach and landing.**

   a.   **How to determine landing performance and limitations**

   1)   Section 5, Performance, of the POH has performance charts on landing distances required to clear a 50-ft. obstacle under the conditions specified on the chart. During your preflight preparation, you need to ensure that you can land in a confined area or short field before attempting to do so.

   a)   Runway surface (paved, grass) and conditions (dry, wet) will affect the landing roll distance.

   2)   Section 4, Normal Procedures, of the POH will describe the recommended procedures to use in a short-field approach and landing.

   3)   Landing limitations (e.g., weight, crosswind, etc.) are listed in Section 2, Limitations, of the POH.

   b.   **Configuration and trim**

   1)   Follow the procedures in the POH to establish the proper short-field approach and landing configuration.

   2)   Use the trim to relieve the control pressures.

   a)   This will result in a nose-up trim.

   c.   **Proper use of pitch and power to maintain desired approach angle**

   1)   After the landing gear (if retractable) and full flaps have been extended, you should simultaneously adjust the pitch attitude and power to establish and maintain the proper descent angle and airspeed.

   2)   Since short-field approaches are power-on approaches, the pitch attitude is adjusted as necessary to establish and maintain the desired rate or angle of descent, and power is adjusted to maintain the desired airspeed.

   a)   However, a coordinated combination of both pitch and power adjustments is required.

   b)   When these adjustments are done properly and the final approach is stabilized, very little change in your airplane's pitch attitude and power will be necessary to make corrections in the angle of descent and airspeed.

   3)   If it appears that the obstacle clearance is excessive and touchdown will occur well beyond the desired spot, leaving insufficient room to stop, power may be reduced while lowering the pitch attitude to increase the rate of descent while maintaining the proper airspeed

   4)   If it appears that the descent angle will not ensure safe clearance of obstacles, power should be increased while simultaneously raising the pitch attitude to decrease the rate of descent and maintain the proper airspeed.

   5)   The final approach is normally started from an altitude of at least 500 ft. higher than the touchdown area, when you are approximately 3/4 to 1 SM from the runway threshold.

   a)   The steeper descent angle means more altitude for a longer period of time, which can be converted to airspeed if needed by lowering the nose. This angle is good for safety because it prevents an approach that is simultaneously too low and too slow.

d. **Obstructions and other hazards which should be considered**

    1) You must know the type of obstructions or hazards that you will need to clear on the approach.

        a) A tree or a row of trees and the height of the tree(s)

        b) Power lines, towers, or other man-made obstructions

    2) The height of obstructions will dictate how steep the approach will have to be.

e. **Effect of Wind**

    1) With a headwind, the descent angle to the runway will be steeper than in a calm wind condition due to the decrease in groundspeed. To correct for this, use more power and a lower rate of descent to maintain the glide path.

        a) The effect will be to shorten the landing distance.

    2) With a tailwind, the groundspeed is higher, which means the touchdown is at a higher speed, significantly lengthening the landing distance.

f. **Selection of touchdown and go-around points**

    1) Select a touchdown aim point that allows you to clear any obstacles and land with the greatest amount of runway available.

        a) Your descent angle (glide path) may be steeper than the one used on a normal approach.

            i) This steeper descent angle helps you pick a touchdown aim point closer to the base of any obstacle, which means a shorter landing distance.

        b) Another method is to select an aim point that will allow you to safely clear the obstacle and land on the remaining runway.

            i) This will allow you to fly your familiar glide path.

    2) You should also select points along the approach path at which you will decide between continuing the approach or executing a go-around.

        a) A go-around may be necessary if you are too low, too high, too slow, too fast, and/or not stabilized on the final approach.

    3) Once you have selected a touchdown point, select an aim point. Remember, your aim point will be the point at the end of your selected glide path, not your touchdown point. Thus, your aim point will be short of your touchdown point.

        a) See Task VII.F., Normal and Crosswind Approach and Landing, beginning on page 277, for a discussion on the use of the aim point.

g. **A stabilized approach at the recommended airspeed to the selected touchdown point**

    1) A stabilized approach and controlled rate of descent can be accomplished only by making minor adjustments to pitch and power while on final approach.

        a) To do this, you must maintain your selected glide path and airspeed.

    2) An excessive amount of airspeed may result in touchdown too far from the runway threshold or an after-landing roll that exceeds the available landing area.

    3) Care must be taken to avoid an excessively low airspeed.

        a) If the speed is allowed to become too slow, an increase in pitch and application of full power may only result in a further rate of descent.

            i) This occurs when the angle of attack is so great and creates so much drag that the maximum available power is insufficient to overcome it.

4)   Completion standards for this element

   a)   Commercial PTS

      i)   ***Maintain a stabilized approach and the recommended approach airspeed (or in its absence, not more than 1.3 V_{so}), ±5 kt., with wind gust factor applied.***

   b)   Private PTS is the same except

      i)   Maintain airspeed +10/–5 kt.

h.   **Coordination of flight controls**

1)   The flight controls should always be used in a proper and coordinated manner, e.g., turning or performing a slip.

2)   When turning, use no more than a 30° bank angle.

3)   The most dangerous situation is when a pilot is turning from base to final and attempts to increase the rate of turn by use of the rudder only.

   a)   This improper use of the rudder is the beginning of an improper cross-control situation, which could lead to a stall.

   b)   See the four-page discussion of a crossed-control stall in Task XI.D., Crossed-Control Stalls, beginning on page 434.

i.   **A precise ground track**

1)   Use the crosswind and directional control techniques described in Task VII.F., Normal and Crosswind Approach and Landing, beginning on page 275.

j.   **Timing, judgment, and control procedure during roundout and touchdown**

1)   Since the final approach over obstacles is made at a steep approach angle and close to the stalling speed, the initiation of the roundout (flare) must be judged accurately to avoid flying into the ground or stalling prematurely and sinking rapidly.

   a)   Smoothly close the throttle during the roundout.

2)   Touchdown should occur at the minimum controllable airspeed at a pitch attitude that will produce a power-off stall.

   a)   Upon touchdown, nosewheel-type airplanes should be held in this positive pitch attitude as long as the elevator/stabilator remains effective. Tailwheel-type airplanes should be firmly held in a three-point attitude to provide aerodynamic braking.

   b)   A lack of floating during the roundout, with sufficient control to touch down properly, is one verification that the approach speed was correct.

3)   Use the proper crosswind technique to ensure your airplane's longitudinal axis is aligned with and over the runway centerline.

    4) Completion standards for this element

     a) Commercial PTS

      i) ***Touch down at or within 100 ft. beyond a specified point, with minimum float, with no side drift, and with the airplane's longitudinal axis aligned with and over the runway centerline.***

     b) Private PTS is the same except

      i) Touch down at or within 200 ft. beyond a specified point.

  k. **Directional control after touchdown**

    1) Use the directional control techniques described in Task VII.F., Normal and Crosswind Approach and Landing, beginning on page 285.

  l. **Use of brakes**

    1) Braking can begin aerodynamically by maintaining the landing attitude after touchdown. Once you are sure that the main gear wheels are solidly in ground contact, begin braking while holding back elevator pressure.

    2) Airplanes with larger flap surfaces may benefit more from leaving the flaps down for drag braking, whereas smaller flaps may be retracted through the rollout to increase wheel contact with the ground and main wheel braking effectiveness.

    3) Follow the procedures in the POH.

  m. **Use of checklist**

    1) The before-landing checklist should be completed on the downwind leg.

    2) After your airplane is clear of the runway, you should stop and complete the after-landing checklist.

2. **Exhibit your instructional knowledge by explaining the following common errors related to a short-field approach and landing.**

  a. **Improper use of landing performance data and limitations**

    1) Use your POH to determine the appropriate airspeeds for a short-field approach and landing.

    2) In gusty and/or strong crosswinds, use the crosswind component chart to determine that you are not exceeding your airplane's crosswind limitations.

    3) Use your POH to determine minimum landing distances, and do not attempt to do better than the data.

    4) The most common error, as well as the easiest to avoid, is to attempt a landing that is beyond the capabilities of your airplane and/or your flying skills. You need to remember that the distance needed for a safe landing is normally less than is needed for a safe takeoff. Plan ahead!

  b. **Failure to establish approach and landing configuration at appropriate time or in proper sequence**

    1) Use the before-landing checklist in your POH to ensure that you follow the proper sequence in establishing the correct approach and landing configuration for your airplane.

    2) You should initially start the checklist at midpoint on the downwind leg with the power reduction beginning once you are abeam of your intended point of landing.

     a) By the time you turn on final and align your airplane with the runway centerline, you should be in the final landing configuration. Confirm this by completing your checklist once again.

c. **Failure to establish and maintain a stabilized approach**

1) Once you are on final and aligned with the runway centerline, you should make small adjustments to pitch and power to establish the correct descent angle (i.e., glide path) and airspeed.

a) Remember, you must make simultaneous adjustments to both pitch and power.

b) Large adjustments will result in a roller coaster ride.

2) A common error in thinking is the idea that a low and slow approach will ensure the shortest landing. This technique exposes you to many dangers, from stalling the aircraft to hitting terrain or trees short of the runway.

a) All POHs will provide a reference speed that will ensure adequate controllability along with short-field performance. Never attempt landing at a speed below the reference speed. You can get behind the power curve, resulting in a high sink rate and possibly an accident.

d. **Improper procedure in use of power, wing flaps, and trim**

1) Use pitch and power adjustments simultaneously to maintain the proper descent angle and airspeed.

2) Wing flaps should be used in accordance with the POH.

3) Trim to relieve control pressures to help in stabilizing the final approach.

e. **Inappropriate removal of hand from throttle**

1) One hand should remain on the control yoke at all times.

2) The other hand should remain on the throttle unless operating the microphone or making an adjustment, such as trim or flaps.

a) Once you are on short final, your hand should remain on the throttle, even if ATC gives you instruction (e.g., cleared to land).

i) Your first priority is to fly your airplane and avoid doing tasks which may distract you from maintaining control.

ii) Fly first; talk later.

3) It is important that your student be in the habit of keeping one hand on the throttle in case a sudden and unexpected hazardous situation should require an immediate application of power.

f. **Improper procedure during roundout and touchdown**

1) Do not attempt to hold the airplane off the ground.

2) See Task VII.F., Normal and Crosswind Approach and Landing, beginning on page 289, for a detailed discussion of general landing errors.

3) Remember that you have limited runway, so when in doubt, go around.

g. **Poor directional control after touchdown**

1) Use rudder to steer the airplane on the runway and increase aileron deflection into the wind as airspeed decreases.

2) See Task VII.F., Normal and Crosswind Approach and Landing, beginning on page 293, for a detailed discussion on ground loops and other directional control problems after touchdown.

h. **Improper use of brakes**

1) Instruct your student that (s)he should never apply the brakes until the airplane is firmly on the runway under complete control.

2) Use equal pressure on both brakes to prevent swerving and/or loss of directional control.

3) Follow the braking procedures described in the POH.

3. **Demonstrate and simultaneously explain a short-field approach and landing from an instructional standpoint.**

    a. Your demonstration must be to commercial pilot skill level.

    b. You will be evaluated on your ability to explain and demonstrate simultaneously the key elements of this task.

    c. If you do not perform the maneuver to FAA completion standards, point out the deficiency to your FAA inspector/examiner as an example of an error, and explain how it should be corrected (including how and why it occurred).

4. **Analyze and correct simulated common errors related to a short-field approach and landing.**

    a. As your FAA inspector/examiner is performing this task, you must be able to recognize, analyze, and correct any error.

    b. Explain what the error is, and then explain what (s)he should do to correct the error.

    c. If your FAA inspector/examiner does not perform the maneuver to FAA completion standards, point out the deficiency to him/her as an example of a student error, and explain how it should be corrected (including how and why it occurred).

END OF TASK

SOFT-FIELD APPROACH AND LANDING

VII.J. TASK: SOFT-FIELD APPROACH AND LANDING

REFERENCES: FAA-H-8083-3; FAA-S-8081-12, FAA-S-8081-14; POH/AFM.

Objective. To determine that the applicant:

1. Exhibits instructional knowledge of the elements of a soft-field approach and landing by describing --

   a. How to determine landing performance and limitations.

   b. Configuration and trim.

   c. Obstructions and other hazards which should be considered.

   d. Effect of wind and landing surface.

   e. Selection of a touchdown area.

   f. A stabilized approach at the recommended airspeed to the selected touchdown area.

   g. Coordination of flight controls.

   h. A precise ground track.

   i. Timing, judgment, and control technique during roundout and touchdown.

   j. Touchdown in a nose-high pitch attitude at minimum safe airspeed.

   k. Proper use of power.

   l. Directional control after touchdown.

   m. Use of checklist.

2. Exhibits instructional knowledge of common errors related to a soft-field approach and landing by describing --

   a. Improper use of landing performance data and limitations.

   b. Failure to establish approach and landing configuration at proper time or in proper sequence.

   c. Failure to establish and maintain a stabilized approach.

   d. Failure to consider the effect of wind and landing surface.

   e. Improper procedure in the use of power, wing flaps, or trim.

   f. Inappropriate removal of hand from throttle.

   g. Improper procedure during roundout and touchdown.

   h. Failure to hold back elevator pressure after touchdown.

   i. Closing the throttle too soon after touchdown.

   j. Poor directional control after touchdown.

   k. Improper use of brakes.

3. Demonstrates and simultaneously explains a soft-field approach and landing from an instructional standpoint.

4. Analyzes and corrects simulated common errors related to a soft-field approach and landing.

A. General Information

    1. The objective of this task is to determine your instructional knowledge of the elements and common errors related to a soft-field approach and landing. You will be required to demonstrate and simultaneously explain this task from an instructional standpoint.

    2. The approach for the soft-field landing is similar to the normal approach used for operating into long, firm landing areas.

        a. The major difference between the two is that during the soft-field landing, the airplane is held 1 to 2 ft. off the surface as long as possible to dissipate the forward speed sufficiently to allow the wheels to touch down gently at minimum speed.

    3. Landing on fields that are rough or have soft surfaces (e.g., snow, mud, sand, or tall grass) requires special techniques.

        a. When landing on such surfaces, you must control your airplane in a manner such that the wings support the weight of the airplane as long as practical.

            1) This minimizes drag and stress put on the landing gear from the rough or soft surfaces.

    4. Follow the procedures prescribed in your POH.

5. In Chapter 5, Airplane Performance and Weight and Balance, of *Pilot Handbook*, see the following:

   a. Module 5.9, Crosswind Performance, for a one-page discussion on determining the crosswind component during a landing.

   b. Module 5.10, Landing Performance, for a four-page discussion on determining landing performance (distance).

B. Task Objectives

   1. **Exhibit your instructional knowledge by explaining the following elements of a soft-field approach and landing.**

      a. **How to determine landing performance and limitations**

         1) Landing performance (e.g., distance and crosswind component) is determined by using the appropriate chart in Section 5, Performance, of the POH.

            a) Landing distance is a required preflight action.

         2) Landing limitations (e.g., maximum weight or crosswind limitation) is located in Section 2, Limitations, of the POH.

      b. **Configuration and trim**

         1) Establish your airplane in the proper soft-field configuration as prescribed in your POH. This configuration is usually similar to that used for a normal approach.

            a) The use of flaps during soft-field landings will aid in touching down at minimum speed and is recommended whenever practical.

               i) In low-wing airplanes, however, the flaps may suffer damage from mud, stones, or slush thrown up by the wheels. In such cases, it may be advisable not to use full flaps.

         2) Trim as necessary.

      c. **Obstructions and other hazards which should be considered**

         1) During your approach, you must look for any hazards or obstructions and then evaluate how they may affect your approach and selection of a suitable touchdown point.

            a) Be aware of traffic, both in the air and on the ground.

            b) Look out for vehicles and/or people on or near the runway.

            c) Check the approach area for any natural or man-made obstacles (e.g., trees, towers, or construction equipment).

            d) Your angle of descent on final approach may need to be steepened if obstacles are present.

      d. **Effect of wind and landing surface**

         1) You must know the wind conditions and the effect they will have upon your airplane's approach and landing performance. The effect of wind on the landing distance may be significant and deserves proper consideration.

            a) A headwind will decrease the landing distance, while a tailwind will greatly increase the landing distance.

            b) This knowledge is important if the landing area is short and/or in a confined area.

2) A soft field is any surface other than a paved one. You must take into account a hard-packed turf or a wet, high grass turf. Know the condition of the landing surface you will be operating into.

   a) If a surface is soft or wet, consider what effect that will have if you perform a crosswind landing, when one main wheel touches down before the other main wheel.

e. **Selection of a touchdown area**

1) After considering the conditions, you should select the most suitable touchdown point.

2) Once you have selected your touchdown point, you need to select your aim point.

   a) See Task VII.F., Normal and Crosswind Approach and Landing, beginning on page 277, for a discussion on the use of the aim point.

f. **A stabilized approach at the recommended airspeed to the selected touchdown area**

1) See Task VII.F., Normal and Crosswind Approach and Landing, beginning on page 278, for a discussion of how to use pitch and power to maintain a stabilized approach.

2) Use the airspeed recommended in Section 4, Normal Procedures, of the POH for a soft-field approach.

   a) If there is no recommended airspeed, use an airspeed that is not more than 1.3 V_{S0}.

3) Completion standards for this element

   a) Commercial PTS

      i) *Maintain a stabilized approach and the recommended airspeed (or in its absence, not more than 1.3 V_{S0}), ±5 kt., with wind gust factor applied.*

   b) Private PTS is the same except

      i) Maintain airspeed +10/–5 kt.

g. **Coordination of flight controls**

1) The flight controls should always be used in a proper and coordinated manner, e.g., turning or performing a slip.

2) When turning, use no more than a 30° bank angle.

3) The most dangerous situation is when a pilot is turning from base to final and attempts to increase the rate of turn by use of the rudder only.

   a) This improper use of rudder is the beginning of an improper crossed-control situation, which could lead to a stall.

   b) See the four-page discussion of a crossed-control stall in Task XI.D., Crossed-Control Stalls, beginning on page 434.

h. **A precise ground track**

1) If a crosswind is present, use the crosswind and directional control techniques described in Task VII.F., Normal and Crosswind Approach and Landing, beginning on page 275.

i. **Timing, judgment, and control technique during roundout and touchdown**

1) Use the same technique during the roundout and touchdown as described in Task VII.F., Normal and Crosswind Approach and Landing, beginning on page 281.

a) The only exception is that you will use partial power during the roundout and touchdown.

2) Do not misjudge the roundout too high, since this may cause the airplane to stall above the surface and drop the airplane in too hard for a soft-surface.

j. **Touchdown in a nose-high pitch attitude at a minimum safe airspeed**

1) Maintain slight power throughout the roundout (flare) to assist in producing as soft a touchdown (i.e., minimum descent rate) as possible.

a) Attempt to hold your airplane about 1 to 2 ft. above the ground as long as possible to allow the touchdown to be made at the slowest possible airspeed with your airplane in a nose-high pitch attitude.

2) In a tailwheel-type airplane, the tailwheel should touch down simultaneously with or just before the main wheels and then should be held down by maintaining firm back elevator pressure throughout the landing roll.

a) This pressure will minimize any tendency for your airplane to nose over and will provide aerodynamic braking.

3) In nosewheel-type airplanes, after the main wheels touch the surface, you should hold sufficient back elevator pressure to keep the nosewheel off the ground until it can no longer aerodynamically be held off the surface.

a) At this time, you should let the nosewheel come down to the ground on its own. Maintain full back elevator pressure at all times on a soft surface.

i) Maintaining slight power during and immediately after touchdown usually will aid in easing the nosewheel down.

NOTE: When teaching a student soft-field landing techniques on a hard surface, most CFIs will have a student hold the nosewheel off the runway until instructed to ease it to the surface. This is an exercise of airplane control by coordinating pitch, power, and rudder.

4) Use the proper crosswind technique to ensure your airplane's longitudinal axis is aligned with and over the runway centerline.

k. **Proper use of power**

1) You must maintain enough speed while taxiing to prevent becoming bogged down on the soft surface.

a) You will often need to increase power after landing on a very soft surface to keep your airplane moving and prevent being stuck.

b) Care must be taken not to taxi excessively fast because, if you taxi onto a very soft area, your airplane may bog down and bend the landing gear and/or nose over.

c) Keep your airplane moving at all times until you are at the point where you will be parking your airplane.

2) Brakes are not needed on a soft surface. Avoid using the brakes because their use may impose a heavy load on the nosegear due to premature or hard contact with the landing surface, causing the nosewheel to dig in.

    a) On a tailwheel-type airplane, the application of brakes may cause the main wheels to dig in, causing the airplane to nose over.

    b) The soft or rough surface itself will normally provide sufficient friction to reduce your airplane's forward speed.

3) Maintain full back elevator pressure and the proper aileron deflection for a crosswind condition while on the ground.

l. **Directional control after touchdown**

1) Use the after-landing directional control techniques described in Task VII.F., Normal and Crosswind Approach and Landing, beginning on page 285.

2) If flaps are used, it is generally inadvisable to retract them during the after landing roll because flap retraction usually is less important than the need for total concentration on maintaining full control of the airplane.

m. **Use of checklist**

1) The before-landing checklist should be completed on the downwind leg of the traffic pattern.

2) On a soft field, the after-landing checklist should normally be accomplished only after you have parked your airplane.

    a) Some items can be done while taxiing (e.g., turning the carburetor heat OFF, if it was used).

    b) You should maintain control of the airplane and, on a soft field, come to a complete stop only at the point at which you are parking your airplane.

2. **Exhibit your instructional knowledge by explaining the following common errors related to a soft-field approach and landing.**

a. **Improper use of landing performance data and limitations**

1) This error may be due to your student's lack of understanding of the performance charts in his/her POH.

2) Use your POH to determine the appropriate airspeeds and performance for a soft-field approach and landing.

3) The most common error, as well as the easiest to avoid, is to attempt a landing that is beyond the capabilities of your airplane and/or your flying skills. Be sure that the surface of the field you plan to use is suitable for landing. Plan ahead!

b. **Failure to establish approach and landing configuration at appropriate time or in proper sequence**

1) Use the before-landing checklist in your POH to ensure that you follow the proper sequence in establishing the correct approach and landing configuration for your airplane.

2) You should initially start the checklist at midpoint on the downwind leg with the power reduction beginning once you are abeam of your intended point of landing.

    a) By the time you turn on final and align your airplane with the runway centerline, you should be in the proper configuration. Confirm this by completing your checklist once again.

c.   **Failure to establish and maintain a stabilized approach**

1)   Once you are on final and aligned with the runway centerline, you should make small adjustments to pitch and power to establish the correct descent angle (i.e., glide path) and airspeed.

a)   Remember, you must make simultaneous adjustments to both pitch and power.

d.   **Failure to consider the effect of wind and landing surface**

1)   Proper planning will ensure knowledge of the landing surface condition, e.g., wet, dry, loose, hard packed.

2)   Understand how the wind affects the landing distance required on a soft field.

e.   **Improper procedure in the use of power, wing flaps, and trim**

1)   Use power pitch adjustments simultaneously to maintain the proper descent angle and airspeed.

2)   Wing flaps should be used in accordance with the POH.

3)   Trim to relieve control pressures to help in stabilizing the final approach.

4)   Remember to maintain slight power throughout the roundout, touchdown, and after-landing roll.

f.   **Inappropriate removal of hand from throttle**

1)   One hand should remain on the control yoke at all times.

2)   The other hand should remain on the throttle unless operating the microphone or making an adjustment, such as trim or flaps.

a)   Your first priority is to fly your airplane and avoid doing tasks that may distract you from maintaining control.

3)   It is important that your student be in the habit of keeping one hand on the throttle in case a sudden and unexpected hazardous situation should require an immediate application of power.

4)   This is particularly true during a soft-field landing in which power is maintained throughout the landing.

g.   **Improper procedure during roundout and touchdown**

1)   Maintain a little power, and hold the airplane off the ground as long as possible.

2)   See Task VII.F., Normal and Crosswind Approach and Landing, beginning on page 289, for a detailed discussion of general landing errors.

3)   Remember, if you have any doubts about the suitability of the field, go around.

h.   **Failure to hold back elevator pressure after touchdown**

1)   In a nosewheel-type airplane, holding back elevator pressure will keep weight off the nosewheel, which otherwise could get bogged down causing the gear to bend and/or nose over the airplane.

2)   In a tailwheel-type airplane, back elevator pressure keeps the tailwheel firmly on the surface to prevent the tendency to nose over.

i.   **Closing the throttle too soon after touchdown**

1)   On a soft field, you must keep the airplane moving at all times.

j.   **Poor directional control after touchdown**

1)   Use rudder to steer your airplane on the landing surface, and increase aileron deflection into the wind as the airspeed decreases.

2)   See Task VII.F., Normal and Crosswind Approach and Landing, beginning on page 293, for a discussion on ground loops and other directional control problems after touchdown.

    k. **Improper use of brakes**

        1) Brakes are not needed on a soft field and should be avoided.

        2) On a very soft surface, you may even need full power to avoid stopping and/or becoming bogged down on the landing surface.

3. **Demonstrate and simultaneously explain a soft-field approach and landing from an instructional standpoint.**

    a. Your demonstration must be to commercial pilot skill level.

    b. You will be evaluated on your ability to explain and demonstrate simultaneously the key elements of this task.

    c. If you do not perform the maneuver to FAA completion standards, point out the deficiency to your FAA inspector/examiner as an example of an error, and explain how it should be corrected (including how and why it occurred).

4. **Analyze and correct simulated common errors related to a soft-field approach and landing.**

    a. As your FAA inspector/examiner is performing this task, you must be able to recognize, analyze, and correct any error.

    b. Explain what the error is, and then explain what (s)he should do to correct the error.

    c. If your FAA inspector/examiner does not perform the maneuver to FAA completion standards, point out the deficiency to him/her as an example of a student error, and explain how it should be corrected (including how and why it occurred).

END OF TASK

POWER-OFF 180° ACCURACY APPROACH AND LANDING

VII.K. TASK: 180° POWER-OFF ACCURACY APPROACH AND LANDING

REFERENCES: FAA-H-8083-3; FAA-S-8081-12.

Objective. To determine that the applicant:

1. Exhibits instructional knowledge of the elements of a 180° power-off accuracy approach and landing by describing --

   a. Configuration and trim.

   b. Effects of wind and selection of a touchdown area.

   c. The key points in the pattern.

   d. A stabilized approach at the recommended airspeed to the selected touchdown area.

   e. Coordination of flight controls.

   f. Timing, judgment, and control procedure during roundout and touchdown.

   g. Directional control after touchdown.

   h. Use of checklist.

2. Exhibits instructional knowledge of common errors related to a 180° power-off accuracy approach and landing by describing --

   a. Failure to establish approach and landing configuration at proper time or in proper sequence.

   b. Failure to identify the key points in the pattern.

   c. Failure to establish and maintain a stabilized approach.

   d. Failure to consider the effect of wind and landing surface.

   e. Improper use of power, wing flaps, and trim.

   f. Improper procedure during roundout and touchdown.

   g. Failure to hold back elevator pressure after touchdown.

   h. Poor directional control after touchdown.

   i. Improper use of brakes.

3. Demonstrates and simultaneously explains a 180° power-off accuracy approach and landing from an instructional standpoint.

4. Analyzes and corrects simulated common errors related to a 180° power-off accuracy approach and landing.

A. General Information

1. The objective of this task is to determine your instructional knowledge of the elements and common errors related to a power-off 180° accuracy approach and landing. You will be required to demonstrate and simultaneously explain this task to your FAA inspector/examiner.

   a. A power-off 180° accuracy approach and landing is an approach and landing made by gliding with the engine idling through a 180° pattern, begun abeam a specified touchdown point on the runway, to a touchdown at or within 200 ft. beyond that point.

      1) The objective of this maneuver is to instill the judgment and procedures necessary for accurately flying the airplane to a safe landing without power.

      2) A practical application of this skill would be in the execution of a simulated or actual forced landing.

B.   Task Objectives

1.   **Exhibit your instructional knowledge by explaining the following elements related to a 180° power-off accuracy approach and landing.**

    a.   **Configuration and trim**

        1)   You should plan to touch down with the airplane in a normal landing configuration; i.e., gear and flaps down.

            a)   Abeam the touchdown point, close the throttle and establish your airplane's best glide speed.

                i)   Be sure to trim for this airspeed.

            b)   Extend the landing gear abeam the touchdown point, just as with any other approach and landing.

            c)   Use flaps as necessary to control the glide path during your approach.

                i)   Remember to retrim the airplane to relieve control pressures after extending flaps.

                ii)   Do not allow the airplane to touch down with less than the manufacturer's minimum recommended flap setting for landing.

                    •   If it is necessary to delay flap extension in order to make your touchdown point, extend the remaining flaps as you cross the runway threshold.

    b.   **Effects of wind and selection of a touchdown area**

        1)   Wind conditions have a major impact on the performance of a 180° accuracy approach and landing. As with any other approach and landing, you should plan this maneuver to land into the wind, but you must also take into account the effects of the wind on your glide path because power is considered to be fixed (i.e., the throttle is closed) and therefore is not available to help control the approach.

            a)   You should begin the maneuver on the downwind leg at your airport's traffic pattern altitude. Abeam the intended point of touchdown, you will close the throttle and establish the manufacturer's recommended best glide speed.

            b)   Plan your base turn according to the existing wind conditions.

                i)   If the wind is strong, you will need to turn the base leg early in order to avoid landing short of the intended touchdown point.

                    •   While heading into a strong wind on final approach, your ground speed will be lower than it would be with a light or calm wind, meaning that you cannot cover as much ground in a power-off glide from a given altitude.

                        ■   Accordingly, you must turn your base leg sooner in order to create a shorter final approach leg so that you are able to reach your intended touchdown point from the altitude available.

    ii)   If the wind is light or calm or if there is a tailwind, you will need to extend the downwind leg and turn the base leg later in order to avoid overshooting the intended touchdown point.

- While heading into a light or calm wind on final or with a tailwind, your ground speed will be higher than it would be with a strong headwind, meaning that you can cover more ground in a power-off glide from a given altitude.
- Accordingly, you must turn your base leg later in order to create a longer final approach leg along which to dissipate altitude.

    iii)   If it is necessary to land with a crosswind, you should select the runway that provides the greatest headwind component (comply with any ATC instructions, however). You must also take into account the effects of the crosswind on your approach.

- If you have a tailwind component on the base leg, the tendency is to be too high on the approach because you will not lose as much altitude during the base leg due to your higher ground speed.
  - You must turn base later or fly a wider pattern (i.e., make the downwind leg farther from the runway) in order to lengthen the base leg.
- If you have a headwind component on the base leg, you will tend to end up too low on the approach because you will lose more altitude during the base leg due to your lower ground speed.
  - You must turn base earlier or fly a smaller pattern (i.e., make the downwind leg closer to the runway) in order to shorten the base leg.

2) You should select a touchdown area that you will be able to safely reach from your altitude under the existing wind conditions without coming uncomfortably close to obstacles or making a landing excessively far down the runway.

3) You should select a touchdown point within the first third of the runway.

    a)   Do not select a touchdown point at the edge of the usable landing area (e.g., on the numbers) because your aiming point will need to be short of the runway if you are to touch down "on the numbers."

       i)   Remember that you will always touch down beyond your aiming point because the airplane floats down the runway during the roundout.

- Your aiming point should never be short of the runway because of the increased risk of landing short of the runway threshold.

       ii)   If obstacle clearance allows, choose a touchdown point that will allow you to use the runway threshold line as an aiming point.

- For most light airplanes, aiming for the threshold line will result in a touchdown 300 to 400 feet down the runway.
- You will become familiar with how far your airplane floats during practice of power-off 180° accuracy approaches and landings.

    b)   Clearly identify the intended touchdown point to your examiner (e.g., the third centerline stripe).

c. **The key points in the pattern**

1) The point abeam the intended touchdown area on the downwind leg is called the "downwind key position."

    a) At this point, you should already have some idea as to the speed and direction of the wind based on any wind correction you have been holding during the downwind leg.

        i) Based on your knowledge of the wind at the downwind key position, you can begin planning the rest of your approach.

2) Immediately after completing the turn to base, you will be in the "base key position" from which you can evaluate your approach.

    a) At this position early in the base leg, you can evaluate how successful your planning has been up to that point and make any necessary corrections early, while you still have time to affect the outcome of the approach.

        i) Observe the intended point of touchdown through the side window of your airplane and look for upward or downward relative motion in the window.

            • Because of your orientation to the touchdown point while on the base leg, there will appear to be sideways relative movement of the touchdown point. This motion can be ignored because it will always be present and provides no useful information.

        ii) If the point has no upward or downward relative motion, you are presently on an acceptable glide path.

            • You should continue with a normal approach and square-off the turn from base to final.

        iii) If the point appears to be moving upward, your present glide path will be too low if a normal approach is made, and you will land short of your intended touchdown point.

            • You should head directly for your intended touchdown point (i.e., "dogleg" the base).

        iv) If the point appears to be moving downward, your present glide path will be too high if a normal approach is made, and you will overshoot your intended touchdown point.

            • You should increase drag by lowering flaps or performing a forward slip, or increase the length of your approach by performing S-turns.

d. **A stabilized approach at the recommended airspeed to the selected touchdown area**

1) Establish a stabilized approach by pitching for the best glide speed, continuously evaluating the effects of the wind on your approach, and making necessary corrections all the way to touchdown.

    a) The relationship between relative movement of the touchdown point and your glide path does not change when you turn final (i.e., if the point is moving down, you are still too high, and vice-versa).

      i)   One thing that you must be careful to avoid when adjusting your glide path is the tendency to wait for very obvious upward or downward movements of the touchdown area before making a correction.

- This technique will result in making large, infrequent corrections that are often excessive (i.e., they are over-corrections which must themselves be corrected for) or made too late to salvage the approach.
- You must be alert to subtle movements of the touchdown area.
  - In order to accurately interpret your glide path, you must "filter out" apparent movements of the touchdown area that are caused by turbulence and look for the general trend "up or down" of the touchdown point.

      ii)   Note that, due to updrafts/downdrafts or gusty conditions, your approach may be high at one instant and low at the next.

- Accordingly, you must constantly make small adjustments to your glide path in order to reach your intended touchdown point.

e. **Coordination of flight controls**

1) The flight controls should always be used in a proper and coordinated manner, e.g., turning or performing a slip.
2) When turning, use no more than a 30° bank angle.
3) The most dangerous situation is when a pilot is turning from base to final and attempts to increase the rate of turn by use of the rudder only.
   a) This improper use of rudder is the beginning of an improper crossed-control situation, which could lead to a stall.
   b) See the four-page discussion of a crossed-control stall in Task XI.D. Crossed-Control Stalls, beginning on page 434.

f. **Timing, judgment, and control procedure during roundout and touchdown**

1) You should make a normal landing in the proper touchdown attitude at or within 200 ft. beyond the touchdown point you have identified to your examiner.
   a) It must be emphasized that, while accurate spot touchdowns are important, a safe and properly-executed approach and landing is essential.
   b) You must never sacrifice a good approach or landing simply to make a touchdown on the desired spot.
      i) Forcing the airplane onto the runway before it is ready to touch down will result in contacting the runway in a flat or nose-low attitude.
         - Wheelbarrowing or porpoising so severe as to cause structural damage may be the result.
      ii) Attempting to stretch the glide by increasing the pitch attitude without adding power may result in a stall at low altitude with insufficient room for recovery.
         - Alternatively, you may experience a hard landing or a bounce so severe as to cause structural damage.

2) Completion standards for this element (commercial pilot only)
   a) ***Touch down in a normal landing attitude at or within 200 ft. beyond a specified touchdown point.***

g. **Directional control after touchdown**

1) Use rudder to control steering on the ground.

2) As the airplane slows down, apply increasing aileron pressure into the wind.

h. **Use of checklist**

1) The before-landing checklist should be completed on the downwind leg before reaching the abeam point.

a) The GUMPS should be completed on both base leg and final.

2) Use the checklist in your POH.

2. **Exhibit your instructional knowledge by explaining the following common errors related to 180° power-off accuracy approach and landings.**

a. **Failure to establish the approach and landing configuration at the proper time or in the proper sequence**

1) Use the before-landing checklist in your POH to ensure that you follow the proper sequence in establishing the correct approach and landing configuration for your airplane.

2) You should initially start the checklist at midpoint on the downwind leg with the throttle being closed once you are abeam your intended point of landing.

a) By the time you turn on final and align your airplane with the runway centerline, you should be in the final landing configuration. Confirm this by completing your checklist once again.

b. **Failure to identify the key points in the pattern**

1) The two "key" points in a power-off 180° accuracy approach and landing are the downwind and base key positions.

a) The downwind key position is on the downwind leg abeam the intended touchdown point.

b) The base key position is on the base leg immediately after completion of the turn from downwind to base.

2) Note that the PTS does not specify that you are required to point out these key positions to your examiner.

a) Your failure to identify these points will be indicated by improper performance of the maneuver; e.g., not closing the throttle at the downwind key position or not recognizing and correcting for an improper glide path at the base key position.

c. **Failure to establish and maintain a stabilized approach**

1) Once you are on final and aligned with the runway centerline, you should make small adjustments to your pitch, configuration (e.g., flaps), and approach path in order to establish the correct descent angle (i.e., glide path) and airspeed.

a) You must make corrections that are small and frequent as you continually evaluate your glide path.

b) Large adjustments will result in a roller coaster ride.

2) Lock in your airspeed and glide path as soon as possible.

a) Never let your airspeed go below the appropriate glide speed.

b) Make necessary corrections in order to stop any upward or downward movement of the touchdown point in the airplane's windscreen.

d. **Failure to consider the effect of wind and landing surface**

1) Proper planning will ensure knowledge of the landing surface condition, e.g., wet, dry, grass, or paved.

a) Because power is theoretically not available to perform a soft-field touchdown, do not attempt a power-off 180° accuracy approach and landing on a surface that is very soft or rough.

2) Understand how the wind affects your approach path and plan accordingly.

e. **Improper use of power, wing flaps, and trim**

1) By definition, a power-off 180° accuracy approach and landing is a power-off maneuver.

a) Accordingly, you must correct for deviations from the desired glide path and airspeed without the use of power.

b) Remember, however, that you must always demonstrate good judgment to your examiner.

i) Do whatever you feel is necessary if the approach does not look like it can be completed safely without the addition of power (i.e., add power or go around).

2) Wing flaps should be used in accordance with your POH to adjust the glide path.

3) Trim to relieve control pressures to help in stabilizing the final approach.

f. **Improper procedure during roundout and touchdown**

1) See Task VII.F., Normal and Crosswind Approach and Landing, beginning on page 289, for a detailed discussion of general landing errors.

2) Remember, if you have any doubts about the outcome of the approach, go around.

g. **Failure to hold back elevator pressure after touchdown**

1) The landing should be made on the main landing gear.

a) Keep pressure off the nosewheel as long as possible.

h. **Poor directional control after touchdown**

1) Use rudder to steer your airplane on the landing surface, and increase aileron deflection into the wind as airspeed decreases.

2) See the discussion of common errors of Task VII.F., Normal and Crosswind Approach and Landing, beginning on page 293, for a discussion on ground loops and other directional control problems after touchdown.

i. **Improper use of brakes**

1) Use the minimum amount of braking required, and let your airplane slow by the friction and drag of the wheels on the ground, if runway length permits.

2) Never attempt to apply brakes until your airplane is firmly on the runway under complete control.

3) Use equal pressure on both brakes to help prevent swerving and/or loss of directional control.

3. **Demonstrate and simultaneously explain a 180° power-off accuracy approach and landing from an instructional standpoint.**

    a. The demonstration must be to commercial pilot skill level.

       1) You will be evaluated on your ability to explain and demonstrate simultaneously the key elements of a power-off 180° accuracy approach and landing.

          a) If you do not perform the maneuver to FAA completion standards, point out the deficiency to your FAA inspector/examiner as an example of an error, and explain how it should be corrected (including how and why it occurred).

4. **Analyze and correct simulated common errors related to a 180° power-off accuracy approach and landing.**

    a. As your FAA inspector/examiner is performing a power-off 180° accuracy approach and landing, you must be able to recognize, analyze, and correct any error.

    b. Explain what the error is, and then explain what (s)he should do to correct the error.

    c. If your FAA inspector/examiner does not perform the maneuver to FAA completion standards, point out the deficiency to him/her as an example of a student error, and explain how it should be corrected (including how and why it occurred).

END OF TASK --- END OF STUDY UNIT

STUDY UNIT VIII
FUNDAMENTALS OF FLIGHT

This study unit explains the four tasks (A-D) of Fundamentals of Flight. These tasks include both knowledge and skill. Your FAA inspector/examiner is required to test you on at least one of the four tasks in this area of operation; therefore, you must be prepared for all four.

During the first few flight lessons, you will introduce your student to, and instruct him/her in, the basic flight maneuvers, i.e., straight-and-level flight, turns, climbs, and descents. While these maneuvers are not specifically listed in the private or commercial PTSs, they are the fundamentals of flying. Every maneuver is either one, or a combination, of the basic flight maneuvers.

Throughout this study unit, we will refer you to Task II.D., Airplane Flight Controls, for a six-page discussion on the effect and use of the flight controls. Additionally, the outline on attitude flying on page 332 is basic to all of the tasks in this study unit.

ATTITUDE FLYING

A. Airplane control is composed of four components: pitch control, bank control, power control, and trim.

1. **Pitch control** is the control of the airplane about its lateral axis (i.e., wingtip to wingtip) by applying elevator pressure to raise or lower the nose, usually in relation to the horizon.

2. **Bank control** is the control of the airplane about its longitudinal axis (i.e., nose to tail) by use of the ailerons to attain the desired angle of bank in relation to the horizon.

3. **Power control** is the control of power or thrust by use of the throttle to establish or maintain a desired airspeed, climb rate, or descent rate in coordination with the attitude changes.

4. **Trim** is used to relieve all possible control pressures held after a desired attitude has been attained.

5. For additional information on the flight controls and control surfaces, there is a three-page discussion/illustration in Task II.D., Airplane Flight Controls, beginning on page 111.

B. The outside references used in controlling the airplane include the airplane's nose and wingtips to show both the airplane's pitch attitude and flight direction and the wings and frame of the windshield to show the angle of bank.

1. The instrument references will be the six basic flight instruments: attitude indicator, heading indicator, altimeter, airspeed indicator, turn coordinator, and vertical speed indicator.

• ASI	• AI	• ALT
• TC	• HI	• VSI

C. The objectives of these basic flight maneuvers are

1. To learn the proper use of the flight controls for maneuvering the airplane

2. To attain the proper attitude in relation to the horizon by use of visual and instrument references

3. To emphasize the importance of dividing your attention and constantly checking all reference points while looking for other traffic

STRAIGHT-AND-LEVEL FLIGHT

VIII.A. TASK: STRAIGHT-AND-LEVEL FLIGHT

REFERENCES: FAA-H-8083-3; FAA-S-8081-14.

Objective. To determine that the applicant:

1. Exhibits instructional knowledge of the elements of a straight-and-level flight by describing --

   a. Effect and use of flight controls.

   b. The Integrated Flight Instruction method.

   c. Outside and instrument references used for pitch, bank, and power control; the cross-check and interpretation of those references; and the control procedure used.

   d. Trim procedure.

   e. Methods that can be used to overcome tenseness and overcontrolling.

2. Exhibits instructional knowledge of common errors related to a straight-and-level flight by describing --

   a. Failure to cross-check and correctly interpret outside and instrument references.

   b. Application of control movements rather than pressures.

   c. Uncoordinated use of flight controls.

   d. Faulty trim procedure.

3. Demonstrates and simultaneously explains a straight-and-level flight from an instructional standpoint.

4. Analyzes and corrects simulated common errors related to straight-and-level flight.

A. General Information

1. The objective of this task is to determine your instructional knowledge of the elements and common errors related to straight-and-level flight. You must be able to demonstrate and simultaneously explain this task to your FAA inspector/examiner from an instructional standpoint.

2. Review the four-page discussion of straight-and-level flight, solely by reference to instruments, in Task XII.A., Straight-and-Level Flight, beginning on page 460.

3. Straight-and-level flight simply means that a constant heading and altitude are maintained.

   a. It is accomplished by making corrections for deviations in direction and altitude from unintentional turns, descents, and climbs.

B. Task Objectives

1. **Exhibit your instructional knowledge by explaining the following elements of straight-and-level flight.**

   a. **Effect and use of flight controls**

      1) See Task II.D., Airplane Flight Controls, beginning on page 111, for a six-page discussion on the effect and use of the flight controls.

   b. **The Integrated Flight Instruction method**

      1) The FAA recommends Integrated Flight Instruction, which means that each flight maneuver (except those requiring ground references) should be learned first by outside visual references and then by instrument references only (i.e., flight instruments).

         a) Thus, instruction in the control of the airplane by outside visual references is **integrated** with instruction in the use of flight instrument indications for the same operations.

      b)   This method will assist your student in developing a habit of monitoring the flight and engine instruments.

           i)   Your student should be able to hold desired altitudes, control airspeed during various phases of flight, and maintain headings.

c.   **Outside and instrument references used for pitch, bank, and power control; the cross-check and interpretation of those references; and the control procedure used**

   1)   The pitch attitude for **level flight** (i.e., constant altitude) is obtained by selecting some portion of the airplane's nose or instrument glare shield as a reference point and then keeping that point in a fixed position relative to the horizon.

      a)   That position should be cross-checked occasionally against the altimeter to determine whether or not the pitch attitude is correct.

           i)   If altitude is being lost or gained, the pitch attitude should be readjusted in relation to the horizon, and then the altimeter should be checked to determine if altitude is being maintained.

      b)   The application of forward or back elevator pressure is used to control this attitude.

           i)   The term "increase the pitch attitude" implies raising the nose in relation to the horizon by pulling back on the control yoke.

           ii)   The term "decreasing the pitch" means lowering the nose by pushing forward on the control yoke.

      c)   The pitch information obtained from the attitude indicator will also show the position of the nose relative to the horizon.

   2)   To achieve **straight flight** (i.e., constant heading), you should select two or more outside visual reference points directly ahead of the airplane (e.g., roads, section lines, towns, lakes, etc.) to form an imaginary line and then keep the airplane headed along that line.

      a)   While using these references, you should occasionally check the heading indicator (HI) to determine that the airplane is maintaining a constant heading.

      b)   Both wingtips should be equidistant above or below the horizon (depending on whether your airplane is a high-wing or low-wing type). Any necessary adjustment should be made with the ailerons to return to a wings level flight attitude.

           i)   Observing the wingtips helps to divert your attention from the airplane's nose and expands the radius of your visual scan, which assists you in collision avoidance.

      c)   The attitude indicator (AI) should be checked for small bank angles, and the heading indicator (HI) checked to note deviations from the desired direction.

   3)   The airspeed will remain constant in straight-and-level flight with a constant power setting.

      a)   Significant changes in airspeed (e.g., power changes) will, of course, require considerable changes in pitch attitude to maintain altitude.

      b)   Pronounced changes in pitch attitude will also be necessary as the flaps and landing gear (if retractable) are operated.

d. **Trim procedure**

1) Straight-and-level flight requires almost no application of control pressure if the airplane is properly trimmed and the air is smooth.

2) Trim the airplane so it will fly straight and level without constant assistance.

   a) This is called "hands-off flight."

   b) The trim controls, when correctly used, are aids to smooth and precise flying.

   c) Improper trim technique usually results in flying that is physically tiring, particularly in prolonged straight-and-level flight.

3) The airplane should be trimmed by first applying control pressure to establish the desired attitude and then adjusting the trim so that the airplane will maintain that attitude without control pressure in hands-off flight.

e. **Methods that can be used to overcome tenseness and overcontrolling**

1) As an instructor, you can help your student to overcome tenseness and overcontrolling.

   a) Suggest that your student use only fingertip control to avoid gripping the control yoke too tightly.

2) Make the flight a pleasurable experience for your student, and make him/her feel capable.

3) Demonstrate the maneuver to your student, and have your student put his/her hand on the control yoke and feet on the rudder pedals so (s)he can feel the control pressures you are using to maintain straight-and-level flight.

   a) During this time, instruct your student on the interpretation of outside references.

   b) Gradually release your pressure on the controls, and inform your student that (s)he is doing a great job of flying the airplane.

4) This method is one way of demonstrating to your student that comfortable pressure on the controls and little control movement are needed to maintain straight-and-level flight.

5) Help build confidence in your student to overcome his/her natural tenseness about these new experiences.

2. **Exhibit your instructional knowledge by explaining the following common errors related to straight-and-level flight.**

a. **Failure to cross-check and correctly interpret outside and instrument references**

1) This error is indicated by your student's having trouble maintaining a constant altitude, heading, or both.

2) During your demonstration, you must have the airplane in straight-and-level flight and explain what to look for as outside references and then how the instrument indications provide the same information.

   a) Then put these together, and have your student verbally interpret both outside and instrument references.

b. **Application of control movements rather than pressures**

1) Applying control movements tends to jerk the airplane, while applying pressure will cause the airplane to react smoothly.

   a) Teach your student the importance of applying pressure to the controls.

c. **Uncoordinated use of flight controls**

    1) This error normally occurs when your student attempts to correct a heading error by yawing the airplane with the rudder instead of by using coordinated aileron and rudder to turn the airplane.

    2) Maintain a coordinated flight condition (i.e., keep the ball centered) throughout this maneuver.

d. **Faulty trim procedure**

    1) Trim should be used, not to substitute for control with the control yoke and rudder, but to relieve pressures already held to stabilize attitude.

    2) Use trim frequently and in small amounts.

    3) Improper adjustment of seat or rudder pedals for comfortable positioning of legs and feet may contribute to trim errors

        a) Tension in the ankles makes it difficult to relax rudder pressures

3. **Demonstrate and simultaneously explain straight-and-level flight from an instructional standpoint**

a. Your demonstration must be to commercial pilot skill level

b. You will be evaluated on your ability to explain and demonstrate simultaneously the key elements of straight-and-level flight

c. Straight-and-level flight is not a separate task in either the private or commercial PTSs

    1) Your performance of straight-and-level flight should be "precise.

        a) This precise performance must be determined by your FAA inspector/examiner through the exercise of subjective judgment

d. If you do not perform the maneuver well, point out the deficiency to your FAA inspector/examiner as an example of an error, and explain how it should be corrected (including how and why it occurred)

4. **Analyze and correct simulated common errors related to straight-and-level flight**

a. As your FAA inspector/examiner is performing this task, you must be able to recognize, analyze, and correct any error.

b. Explain what the error is, and then explain what (s)he should do to correct the error.

END OF TASK

LEVEL TURNS

VIII.B. TASK: LEVEL TURNS

REFERENCES: FAA-H-8083-3; FAA-S-8081-14.

Objective. To determine that the applicant:

1. Exhibits instructional knowledge of the elements of level turns by describing --

   a. Effect and use of flight controls.

   b. The Integrated Flight Instruction method.

   c. Outside and instrument references used for pitch, bank, and power control; the cross-check and interpretation of those references; and the control procedure used.

   d. Trim procedure.

   e. Methods that can be used to overcome tenseness and overcontrolling.

2. Exhibits instructional knowledge of common errors related to level turns by describing --

   a. Failure to cross-check and correctly interpret outside and instrument references.

   b. Application of control movements rather than pressures.

   c. Uncoordinated use of flight controls.

   d. Faulty altitude and bank control.

3. Demonstrates and simultaneously explains level turns from an instructional standpoint.

4. Analyzes and corrects simulated common errors related to level turns.

A. General Information

1. The objective of this task is to determine your instructional knowledge of the elements and common errors related to level turns. You must be able to demonstrate and simultaneously explain this task to your FAA inspector/examiner from an instructional standpoint.

2. A turn is a basic flight maneuver used to change from, or return to, a desired heading. This maneuver involves the coordinated use of the ailerons, rudder, and elevator.

3. Explain to your student that the terms shallow, medium, or steep turns are used to indicate the approximate bank angle to use.

   a. EXAMPLE: A shallow turn uses 20° of bank, a medium turn uses 30° of bank, and a steep turn uses 45° of bank.

4. When introducing the level turn to a student, you should begin with shallow to medium banked turns.

5. See Module 1.8, How Airplanes Turn, in *Pilot Handbook* for a two-page discussion on the forces acting on the airplane during turning flight.

B. Task Objectives

1. **Exhibit your instructional knowledge by explaining the following elements of level turns.**

   a. **Effect and use of flight controls**

      1) See Task II.D., Airplane Flight Controls, beginning on page 111, for a six-page discussion on the effect and use of the flight controls.

   b. **The integrated flight instruction method**

      1) The FAA recommends integrated flight instruction, which means that each flight maneuver (except those requiring ground references) should be learned first by outside visual references and then by instrument references only (i.e., flight instruments).

         a) Thus, instruction in the control of the airplane by outside visual references is **integrated** with instruction in the use of flight instrument indications for the same operations.

b) This method will assist your student in developing a habit of monitoring the flight and engine instruments.

    i) Your student should be able to hold desired altitudes, control airspeed during various phases of flight, and maintain headings.

c. **Outside and instrument references used for pitch, bank, and power control; the cross-check and interpretation of those references; and the control procedure used**

1) Before you start the turn, look in the direction you will be turning to ensure the area is clear of other traffic.

2) To enter a turn, you should simultaneously move the control yoke (i.e., apply aileron control pressure) and rudder pressure in the desired direction.

    a) The speed (or rate) at which your airplane rolls into a bank depends on the rate and amount of control pressure you apply.

        i) The amount of bank depends on how long you keep the ailerons deflected.

    b) Rudder pressure must be enough to keep the ball of the inclinometer (part of the turn coordinator) centered.

        i) If the ball is not centered, step on the ball to recenter.

        ii) EXAMPLE: If the ball is to the right, apply right rudder pressure (i.e., step on the ball) to recenter.

3) The best outside reference for establishing the degree of bank is the angle made by the top of the engine cowling or the instrument panel with respect to the horizon.

    a) Since on most light airplanes the engine cowling is fairly flat, its horizontal angle to the horizon will give some indication of the approximate degree of bank.

    b) Your posture while seated in the airplane is very important in all maneuvers, particularly during turns, since that will affect the alignment of outside visual references.

        i) At first, your student may want to lean away from the turn in an attempt to remain upright in relation to the ground instead of rolling with the airplane.

        ii) This tendency must be corrected from the outset if your student is to learn how to use the visual references properly.

    c) In an airplane with side-by-side seating, your student will be seated in the left seat, and you will be seated in the right seat. Since the seats are to either side of the centerline of the airplane, the apparent position of the airplane's nose in relation to the horizon will be different in turns to the left than it is in turns to the right while you are maintaining the same altitude.

        i) For your student (in the left seat), you must explain that

            • In a turn to the left, the nose may appear level or slightly high.
            • In a turn to the right, the nose will appear to be low.

        ii) For you (in the right seat), just the opposite will be true.

            • In a turn to the left, the nose will appear to be low.
            • In a turn to the right, the nose may appear level or slightly high.

4) Information obtained from the attitude indicator (AI) will show the angle of the wings in relation to the horizon. Use this information to learn to judge the degree of bank based on outside references.

5) In a bank, the total lift consists of both horizontal lift (to turn the airplane) and vertical lift (counteracting weight/gravity).

a) Given the same amount of total lift, there is less vertical lift in a bank than in straight-and-level flight.

b) To maintain altitude, the vertical lift must remain equal to weight. Thus, total lift must be increased.

i) Increase total lift by applying enough back elevator pressure (i.e., increasing the angle of attack) to maintain altitude.

ii) This increase in pitch will cause a slight decrease in airspeed. In a medium banked turn, this slight decrease in airspeed is acceptable and will be regained once the wings are level, so no increase in power is required.

6) As the desired angle of bank is established, aileron and rudder pressures should be released. The release of these pressures will stop the bank from increasing since the aileron control surfaces will be neutral in their streamlined position.

a) The back elevator pressure should not be released but should be held constant or sometimes increased to maintain a constant altitude.

b) Throughout the turn, you should cross-check the references and occasionally include the altimeter to determine whether the pitch attitude is correct.

c) If gaining or losing altitude, adjust the pitch attitude in relation to the horizon, and then recheck the altimeter and vertical speed indicator to determine if altitude is now being maintained.

7) The rollout from a turn is similar to the roll-in except that control pressures are used in the opposite direction. Aileron and rudder pressures are applied in the direction of the rollout or toward the high wing.

a) Lead your rollout by an amount equal to one-half your bank angle.

i) If you are using a 30° bank, begin your rollout approximately 15° before your desired heading.

b) As the angle of bank decreases, the elevator pressure should be released smoothly as necessary to maintain altitude. Remember, when the airplane is no longer banking, the vertical component of lift increases.

c) Since the airplane will continue turning as long as there is any bank, the rollout must be started before reaching the desired heading.

i) The time to begin rollout in order to lead the heading will depend on the rate of turn and the rate at which the rollout will be made.

d) As the wings become level, the control pressures should be gradually and smoothly released so that the controls are neutralized as the airplane resumes straight-and-level flight.

e) As the rollout is completed, attention should be given to outside visual references as well as to the attitude indicator and heading indicator to determine that the wings are leveled precisely and the turn stopped.

d. **Trim procedure**

1) Trim may be used during turns to relieve the back elevator pressure you are holding to maintain altitude.

2) The trim controls, when correctly used, are aids to smooth and precise flying.

3) Improper trim technique usually results in flying that is physically tiring for your student, especially in prolonged or steep turns.

4) The airplane should be trimmed by first applying control pressure to establish the desired attitude and then adjusting the trim so that the airplane will maintain that attitude without control pressure.

5) When rolling out on the desired heading, trim must be adjusted for straight-and-level flight.

e. **Methods that can be used to overcome tenseness and overcontrolling**

1) As an instructor, you can help your student to overcome tenseness and overcontrolling.

a) Suggest that your student use only fingertip control to avoid gripping the control yoke too tightly.

2) Make the flight a pleasurable experience for your student, and make him/her feel capable.

3) Demonstrate the maneuver to your student, and have your student put his/her hand on the control yoke and feet on the rudder pedals so (s)he can feel the control pressures you are using to make a level turn.

a) During this time, instruct your student on the interpretation of outside references.

b) Gradually release your pressure on the controls, and inform your student that (s)he is doing a great job of flying the airplane.

4) This method is one way of demonstrating to your student that comfortable pressure on the controls and little control movement are needed to make a level turn.

5) Help build confidence in your student to overcome his/her natural tenseness about these new experiences.

2. **Exhibit your instructional knowledge by explaining the following common errors related to level turns.**

a. **Failure to cross-check and correctly interpret outside and instrument references**

1) This error is indicated by your student's having trouble maintaining a constant altitude and a constant bank angle and rolling out on a specified heading.

2) During your demonstration, you must show your student the visual and instrument references to use.

a) Then put these together, and have your student verbally interpret both outside and instrument references.

b. **Application of control movements rather than pressures**

1) Applying control movements tends to jerk the airplane, while applying pressure will cause the airplane to react smoothly.

a) Teach your student the importance of applying pressure to the controls.

    c. **Uncoordinated use of flight controls**

        1) This error normally occurs because the rudder is used improperly or its effects and use are not fully understood.

        2) Maintain coordinated flight (i.e., keep the ball centered) throughout the maneuver.

            a) If the nose of the airplane starts to move before the bank starts, rudder is being applied too soon or too much.

            b) If the bank starts before the nose starts to move or if the nose moves in the opposite direction, the rudder is being used too late or too little.

    d. **Faulty altitude and bank control**

        1) Faulty altitude control is caused by incorrect interpretation of outside and instrument references.

            a) If the nose moves up or down in relation to the horizon, excessive or insufficient elevator pressure is being applied.

            b) Cross-check the outside references to the altimeter and vertical speed indicator.

        2) Poor bank control can be caused by incorrect interpretation of outside and instrument references.

            a) Cross-check the outside references to the attitude indicator.

3. **Demonstrate and simultaneously explain level turns from an instructional standpoint.**

    a. Your demonstration must be to commercial pilot skill level.

    b. You will be evaluated on your ability to explain and demonstrate simultaneously the key elements of level turns.

    c. Performing a level turn is not a separate task in either the private or commercial PTSs.

        1) Your performance of level turns should be "precise."

            a) This precise performance must be determined by your FAA inspector/examiner through the exercise of subjective judgment.

    d. If you do not perform the maneuver well, point out the deficiency to your FAA inspector/examiner as an example of an error, and explain how it should be corrected (including how and why it occurred).

4. **Analyze and correct simulated common errors related to level turns.**

    a. As your FAA inspector/examiner is performing this task, you must be able to recognize, analyze, and correct any error.

    b. Explain what the error is, and then explain what (s)he should do to correct the error.

END OF TASK

STRAIGHT CLIMBS AND CLIMBING TURNS

VIII.C. TASK: STRAIGHT CLIMBS AND CLIMBING TURNS
REFERENCES: FAA-H-8083-3; FAA-S-8081-14.

Objective. To determine that the applicant:

1. Exhibits instructional knowledge of the elements of straight climbs and climbing turns by describing --

   a. Effect and use of flight controls.

   b. The Integrated Flight Instruction method.

   c. Outside and instrument references used for pitch, bank, and power control; the cross-check and interpretation of those references; and the control procedure used.

   d. Trim procedure.

   e. Methods that can be used to overcome tenseness and overcontrolling.

2. Exhibits instructional knowledge of common errors related to straight climbs and climbing turns by describing --

   a. Failure to cross-check and correctly interpret outside and instrument references.

   b. Application of control movements rather than pressures.

   c. Improper correction for torque effect.

   d. Faulty trim procedure.

3. Demonstrates and simultaneously explains straight climbs and climbing turns from an instructional standpoint.

4. Analyzes and corrects simulated common errors related to straight climbs and climbing turns.

A. General Information

1. The objective of this task is to determine your instructional knowledge of the elements and common errors related to straight climbs and climbing turns. You must be able to demonstrate and simultaneously explain this task to your FAA inspector/examiner from an instructional standpoint.

2. Climbs and climbing turns are basic flight maneuvers in which the pitch attitude and power result in a gain in altitude. In a straight climb, the airplane gains altitude while traveling straight ahead. In climbing turns, the airplane gains altitude while turning.

3. There are various climb airspeeds that you will introduce to your student early in his/her flight training.

   a. **Best rate of climb (V_Y)** provides the greatest gain in altitude in the least amount of time.

   b. **Best angle of climb (V_X)** provides the greatest gain in altitude in a given distance.

   c. **Cruise climb** is used to climb to your desired altitude. This speed provides better engine cooling and forward visibility.

   d. These airspeeds are listed in the Pilot's Operating Handbook (POH).

B. Task Objectives

1. **Exhibit your instructional knowledge by explaining the following elements of straight climbs and climbing turns.**

   a. **Effect and use of flight controls**

      1) See Task II.D., Airplane Flight Controls, beginning on page 111, for a six-page discussion on the effect and use of the flight controls.

b. **The integrated flight instruction method**

1) The FAA recommends integrated flight instruction, which means that each flight maneuver (except those requiring ground references) should be learned first by outside visual references and then by instrument references only (i.e., flight instruments).

    a) Thus, instruction in the control of the airplane by outside visual references is **integrated** with instruction in the use of flight instrument indications for the same operations.

    b) This method will assist your student in developing a habit of monitoring the flight and engine instruments.

        i) Your student should be able to hold desired altitudes, control airspeed during various phases of flight, and maintain headings.

c. **Outside and instrument references used for pitch, bank, and power control; the cross-check and interpretation of those references; and the control procedure used**

1) To enter the climb, simultaneously advance the throttle and apply back elevator pressure.

2) As the power is increased to the climb setting, the airplane's nose will tend to rise to the climb attitude.

    a) In most trainer-type airplanes, the climb setting will be full power. Check the POH for information.

3) While the pitch attitude increases and airspeed decreases, progressively more right-rudder pressure must be used to compensate for torque effects (left-turning tendencies) and to maintain direction.

    a) Since the angle of attack is relatively high, the airspeed is relatively slow, and the power setting is high, the airplane will have a tendency to roll and yaw to the left.

        i) While right-rudder pressure will correct for the yaw, some aileron pressure may be required to keep the wings level.

    b) See Module 1.9, Torque (Left-Turning Tendency), in *Pilot Handbook* for a five-page discussion on torque (left-turning tendency) effect and correction.

4) When the climb is established, back elevator pressure must be maintained to keep the pitch attitude constant.

    a) As the airspeed decreases, the elevators may try to return to their streamline or neutral position, which will cause the nose to lower.

        i) Nose-up trim will need to be used.

    b) You will need to cross-check the airspeed indicator (ASI) since you want to climb at a specific airspeed, and the ASI will provide you an indirect indication of pitch attitude.

        i) If the airspeed is higher than desired, you need to use the outside references and attitude indicator to raise the nose.

        ii) If the airspeed is lower than desired, you need to use the outside references and attitude indicator to lower the nose.

    c) After the climbing attitude, power setting, and airspeed have been established, trim the airplane to relieve all pressures from the controls.

        i) If further adjustments are made in pitch, power, and/or airspeed, you must retrim the airplane.

d) If a straight climb is being performed, you need to maintain a constant heading with the wings level.

  i) If a climbing turn is being performed, maintain a constant angle of bank.

5) To return to straight-and-level flight from a climbing attitude, it is necessary to start the level-off a few feet below the desired altitude.

a) Start to level off a distance below the desired altitude equal to about 10% of the airplane's rate of climb as indicated on the vertical speed indicator.

  i) EXAMPLE: If you are climbing at 500 fpm, start to level off 50 ft. below your desired altitude.

b) To level off, the wings should be leveled and the nose lowered.

c) The nose must be lowered gradually, however, because a loss of altitude will result if the pitch attitude is decreased too abruptly before allowing the airspeed to increase adequately.

  i) As the nose is lowered and the wings are leveled, retrim the airplane.

  ii) When the airspeed reaches the desired cruise speed, reduce the throttle setting to appropriate cruise power setting, adjust the mixture control to the manufacturer's recommended setting, and trim the airplane.

6) Climbing turns. The following factors should be considered:

a) With a constant power setting, the same pitch attitude and airspeed cannot be maintained in a bank as in a straight climb due to the decrease in the vertical lift and airspeed during a turn.

  i) The loss of vertical lift becomes greater as the angle of bank is increased, so shallow turns may be used to maintain an efficient rate of climb. If a medium- or steep-banked turn is used, the airplane's rate of climb will be reduced.

  ii) The airplane will have a greater tendency towards nose-heaviness than in a straight climb, due to the decrease in the vertical lift.

b) As in all maneuvers, attention should be diverted from the airplane's nose and divided among all references equally.

7) There are two ways to establish a climbing turn: either establish a straight climb and then turn, or establish the pitch and bank attitudes simultaneously from straight-and-level flight.

a) The second method is usually preferred because you can more effectively check the area for other aircraft while the climb is being established.

d. **Trim procedure**

1) After the climbing attitude and power setting have been established, the airplane should be trimmed to relieve all pressures from the controls as the airspeed changes.

2) If further adjustments are made in the pitch attitude, power, or airspeed, the airplane must be retrimmed.

e. **Methods that can be used to overcome tenseness and overcontrolling**

   1) As an instructor, you can help your student to overcome tenseness and overcontrolling.

     a) Suggest that your student use only fingertip control to avoid gripping the yoke too tightly.

   2) Make the flight a pleasurable experience for your student, and make him/her feel capable.

   3) Demonstrate the maneuver to your student, and have your student put his/her hand on the control yoke and feet on the rudder pedals so (s)he can feel the control pressures you are using to make a straight climb or a climbing turn.

     a) During this time, instruct your student on the interpretation of outside references.

     b) Gradually release your pressure on the controls, and inform your student that (s)he is doing a great job of flying the airplane.

   4) This method is one way of demonstrating to your student that comfortable pressure on the controls and little control movement are needed to make a straight climb or a climbing turn.

   5) Help build confidence in your student to overcome his/her natural tenseness about these new experiences.

2. **Exhibit your instructional knowledge by explaining the following common errors related to straight climbs and turning climbs.**

a. **Failure to cross-check and correctly interpret outside and instrument references**

   1) This error is indicated by your student's having trouble maintaining the correct pitch attitude, bank attitude, or both.

   2) During your demonstration, you must explain what to look for as outside references and how to use the flight instruments during the maneuver.

     a) Then put these together, and have your student verbally interpret both outside and instrument references.

b. **Application of control movements rather than pressures**

   1) Applying control movements tends to jerk the airplane, while applying pressure will cause the airplane to react smoothly.

     a) Teach your student the importance of applying pressure to the controls.

c. **Improper correction for torque effect**

   1) This error normally occurs because your student is not cross-checking the ball of the inclinometer to ensure that the ball is centered.

   2) The flight controls must be used to remain in coordinated flight, in either a straight climb or a turning climb.

d. **Faulty trim procedure**

   1) Trim should be used, not to substitute for control with the control yoke and rudder, but to relieve pressures already held to stabilize attitude.

   2) Use trim frequently and in small amounts.

     a) Trim should be expected during any pitch, power, or airspeed change.

   3) Improper adjustment of seat or rudder pedals for comfortable positioning of legs and feet may contribute to trim errors.

     a) Tension in the ankles makes it difficult to relax rudder pressures.

3. **Demonstrate and simultaneously explain straight climbs and climbing turns from an instructional standpoint.**

    a. The demonstration must be to commercial pilot skill level.

    b. You will be evaluated on your ability to explain and demonstrate simultaneously the key elements of straight climbs and climbing turns.

    c. Performing climbs and climbing turns is not a separate task in either the private or commercial PTSs.

        1) Your performance of climbs and climbing turns should be "precise."

            a) This precise performance must be determined by your FAA inspector/examiner through the exercise of subjective judgment.

    d. If you do not perform the maneuver well, point out the deficiency to your FAA inspector/examiner as an example of an error, and explain how it should be corrected (including how and why it occurred).

4. **Analyze and correct simulated common errors related to straight climbs and climbing turns.**

    a. As your FAA inspector/examiner is performing this task, you must be able to recognize, analyze, and correct any error.

    b. Explain what the error is, and then explain what (s)he should do to correct the error.

END OF TASK

STRAIGHT DESCENTS AND DESCENDING TURNS

VIII.D. TASK: STRAIGHT DESCENTS AND DESCENDING TURNS

REFERENCES: FAA-H-8083-3; FAA-S-8081-14.

Objective. To determine that the applicant:

1. Exhibits instructional knowledge of the elements of straight descents and descending turns by describing --

   a. Effect and use of flight controls.

   b. The Integrated Flight Instruction method.

   c. Outside and instrument references used for pitch, bank, and power control; the cross-check and interpretation of those references; and the control procedure used.

   d. Trim procedure.

   e. Methods that can be used to overcome tenseness and overcontrolling.

2. Exhibits instructional knowledge of common errors related to straight descents and descending turns by describing --

   a. Failure to cross-check and correctly interpret outside and instrument references.

   b. Application of control movements rather than pressures.

   c. Uncoordinated use of flight controls.

   d. Faulty trim procedure.

   e. Failure to clear engine and use carburetor heat, as appropriate.

3. Demonstrates and simultaneously explains straight descents and descending turns from an instructional standpoint.

4. Analyzes and corrects simulated common errors related to straight descents and descending turns.

A. General Information

   1. The objective of this task is to determine your instructional knowledge of the elements and common errors related to straight descents and descending turns. You must be able to demonstrate and simultaneously explain this task to your FAA inspector/examiner from an instructional standpoint.

   2. A descent is a basic maneuver in which the airplane loses altitude in a controlled manner. Descents can be made

      a. With partial power, as used during an approach to a landing
      b. Without power, i.e., a glide
      c. At cruise airspeeds, during en route descents

B. Task Objectives

   1. **Exhibit your instructional knowledge by explaining the following elements of straight descents and descending turns.**

      a. **Effect and use of flight controls**

         1) See Task II.D., Airplane Flight Controls, beginning on page 111, for a six-page discussion on the effect and use of the flight controls.

      b. **The integrated flight instruction method**

         1) The FAA recommends integrated flight instruction, which means that each flight maneuver (except those requiring ground references) should be learned first by outside visual references and then by instrument references only (i.e., flight instruments).

            a) Thus, instruction in the control of the airplane by outside visual references is **integrated** with instruction in the use of flight instrument indications for the same operations.

    b)   This method will assist your student in developing a habit of monitoring the flight and engine instruments.

        i)   Your student should be able to hold desired altitudes, control airspeed during various phases of flight, and maintain headings.

c.   **Outside and instrument references used for pitch, bank, and power control; the cross-check and interpretation of those references; and the control procedure used**

    1)   To enter a descent, you should first apply carburetor heat (if recommended in the POH) and then reduce power to the desired setting or to idle.

        a)   Maintain a constant altitude by applying back elevator pressure as required until the airspeed decreases to the desired descent airspeed.

        b)   Once the descent airspeed has been reached, lower the nose attitude to maintain that airspeed and adjust the trim.

    2)   When the descent is established, cross-check the airspeed indicator (ASI) to ensure that you are descending at the desired airspeed.

        a)   If the airspeed is higher than desired, then slightly raise the nose and allow the airspeed to stabilize to confirm the adjustment.

        b)   If the airspeed is lower than desired, then slightly lower the nose and allow the airspeed to stabilize.

        c)   Once you are descending at the desired airspeed, note the position of the airplane's nose to the horizon and the position on the attitude indicator (AI).

            i)   Trim the airplane to relieve all control pressures.

        d)   Maintain either straight or turning flight, as desired.

    3)   The level-off from a descent must be started before reaching the desired altitude.

        a)   Begin the level-off at a distance equal to about 10% of the airplane's rate of descent as indicated on the vertical speed indicator (VSI).

            i)   EXAMPLE: If you are descending at 500 fpm, start the level-off 50 ft. above your desired altitude.

        b)   At the lead point, you should simultaneously raise the nose to a level attitude and increase power to the desired cruise setting.

            i)   The addition of power and the increase in airspeed will tend to raise the nose. You will need to apply appropriate elevator control pressure and make a trim adjustment to relieve some of the control pressures.

    4)   Turning descents

        a)   As with climbing turns, you can either enter the turn after the descent has been established or simultaneously adjust the bank and pitch attitudes.

        b)   At a desired power setting during a descending turn, maintain airspeed with pitch as you would in a straight descent.

d.   **Trim procedure**

    1)   After the descent airspeed has stabilized, it is important to retrim the airplane so that it will maintain the descent attitude without the need to hold pressure on the elevator control.

e.   **Methods that can be used to overcome tenseness and overcontrolling**

   1)   As an instructor, you can help your student to overcome tenseness and overcontrolling.

      a)   Suggest that your student use only fingertip control to avoid gripping the yoke too tightly.

   2)   Make the flight a pleasurable experience for your student, and make him/her feel capable.

   3)   Demonstrate the maneuver to your student, and have your student put his/her hand on the control yoke and feet on the rudder pedals so (s)he can feel the control pressures you are using to make a straight descent or a descending turn.

      a)   During this time, instruct your student on the interpretation of outside references.

      b)   Gradually release your pressure on the controls, and inform your student that (s)he is doing a great job of flying the airplane.

   4)   That method is one way of demonstrating to your student that comfortable pressure on the controls and little control movement are needed to make a straight descent or a descending turn.

   5)   Help build confidence in your student to overcome his/her natural tenseness about these new experiences.

2.   **Exhibit your instructional knowledge by explaining the following common errors related to straight descents and descending turns.**

   a.   **Failure to cross-check and correctly interpret outside and instrument references**

      1)   This error is indicated by your student's having trouble maintaining the correct pitch attitude, bank angle, or both.

      2)   During your demonstration, you must explain what to look for as outside references and how to use the flight instruments in a descent.

         a)   Then put these together, and have your student verbally interpret both outside and instrument references.

   b.   **Application of control movements rather than pressures**

      1)   Applying control movements tends to jerk the airplane, while applying pressure will cause the airplane to react smoothly.

         a)   Teach your student the importance of applying pressure to the controls.

   c.   **Uncoordinated use of flight controls**

      1)   Maintain a coordinated flight (i.e., keep the ball centered) throughout the maneuver.

         a)   Check the inclinometer.

   d.   **Faulty trim procedure**

      1)   Trim should be used, not to substitute for control with the control yoke and rudder, but to relieve pressures already held to stabilize attitude.

      2)   Use trim frequently and in small amounts.

         a)   Trim should be expected during any pitch, power, or airspeed change.

      3)   Improper adjustment of seat or rudder pedals for comfortable positioning of legs and feet may contribute to trim errors.

         a)   Tension in the ankles makes it difficult to relax rudder pressures.

e.   **Failure to clear engine and use carburetor heat, as appropriate**

1)   During prolonged power-off glides or low-power descents, the engine should be cleared and kept warm by applying cruise power momentarily.

2)   Carburetor heat may need to be on due to high humidity conditions and the possibility that air entering the carburetor may be near freezing.

a)   Consult the POH for operation of carburetor heat during descents and/or low power settings.

3.   **Demonstrate and simultaneously explain straight descents and descending turns from an instructional standpoint.**

a.   Your demonstration must be to commercial pilot skill level.

b.   You will be evaluated on your ability to explain and demonstrate simultaneously the key elements of straight descents and descending turns.

c.   If you do not perform the maneuver well, point out the deficiency to your FAA inspector/examiner as an example of an error, and explain how it should be corrected (including how and why it occurred).

4.   **Analyze and correct simulated common errors related to straight descents and descending turns.**

a.   As your FAA inspector/examiner is performing this task, you must be able to recognize, analyze, and correct any error.

b.   Explain what the error is, and then explain what (s)he should do to correct the error.

END OF TASK -- END OF STUDY UNIT

STUDY UNIT IX
PERFORMANCE MANEUVERS

This study unit explains the four tasks (A-D) of Performance Maneuvers. These tasks include both knowledge and skill. Your FAA inspector/examiner is required to test you on at least tasks A or B and C or D; therefore, you must be prepared for all four.

The performance maneuvers explained here are not necessarily operational in nature, but they are used effectively in pilot training to develop skills and safe habits in preparation for obtaining the best maneuvering performance from any airplane.

STEEP TURNS

IX.A. TASK: STEEP TURNS

> REFERENCES: FAA-H-8083-3; FAA-S-8081-12; FAA-S-8081-14; POH/AFM.

Objective. To determine that the applicant:

1. Exhibits instructional knowledge of the elements of steep turns by describing --

   a. Relationship of bank angle, load factor, and stalling speed.

   b. Overbanking tendency.

   c. Torque effect in right and left turns.

   d. Selection of a suitable altitude.

   e. Orientation, division of attention, and planning.

   f. Entry and rollout procedure.

   g. Coordination of flight and power controls.

   h. Altitude, bank, and power control during the turn.

   i. Proper recovery to straight-and-level flight.

2. Exhibits instructional knowledge of common errors related to steep turns by describing --

   a. Improper pitch, bank, and power coordination during entry and rollout.

   b. Uncoordinated use of flight controls.

   c. Improper procedure in correcting altitude deviations.

   d. Loss of orientation.

3. Demonstrates and simultaneously explains steep turns from an instructional standpoint.

4. Analyzes and corrects simulated common errors related to steep turns.

A. General Information

    1. The objective of this task is to determine your instructional knowledge of the elements and common errors related to steep turns. You will be required to simultaneously demonstrate and explain this task from an instructional standpoint.

    2. The maneuver is intended to develop smoothness, coordination, orientation, efficient division of attention, and polished control techniques.

        a. Theoretical and practical understanding of airplane stability and load factors is also required.

    3. Your airplane's turning performance is limited by the amount of power the engine is developing, its limit load factor (structural strength), and its aerodynamic characteristics.

    4. See Module 1.11, Loads and Load Factors, in *Pilot Handbook* for a five-page discussion on the relationship of bank angle, load factor, and stalling speed.

B. Task Objectives

    1. **Exhibit your instructional knowledge by explaining the following elements of steep turns.**

        a. **Relationship of bank angle, load factor, and stalling speed**

            1) A steep turn requires that a relatively large amount of vertical lift be transferred horizontally to turn the airplane.

                a) Since general aviation airplanes are normally designed to be flown nose-heavy, the loss of vertical lift will cause the nose to drop.

                b) You must respond by adding sufficient upward elevator pressure (stick or yoke back) to counteract the nose-down tendency.

                c) The additional upward elevator pressure increases the angle of attack and causes drag to increase.

                d) To counteract the additional drag, a power increase may be used to maintain a constant altitude throughout the turn.

2)   Theoretically, the steepest bank possible is that in which altitude can be maintained without approaching a stall with all available power.

   a)   Load factors begin to increase rapidly after passing a bank of approximately 45°. In a level turn, the airplane will develop 2 Gs with 60° of bank.

   b)   However, only 10° more of bank, up to 70°, will produce approximately 3 Gs. Loads build up progressively more rapidly after this.

3)   Stall speed increases with the square root of the load factor. If your airplane stalls at 60 kt. in level flight, it will stall at nearly 85 kt. in a 2-G turn (60° of bank).

b.   **Over-banking tendency**

1)   The so-called over-banking tendency is the result of the airplane being banked steeply enough to reach a condition of negative static stability about the longitudinal axis.

2)   Static stability can be positive, neutral, or negative. It is the tendency of the airplane, once displaced, to try to return to a stable condition as it was before being disturbed.

   a)   In a shallow turn, the airplane displays positive static stability and tries to return to a wings-level attitude.

   b)   In a medium bank turn, the airplane shows neutral static stability and will tend to remain in the medium bank, assuming calm air.

   c)   In a steep turn, the airplane demonstrates negative static stability and tries to steepen the bank rather than remain stable. This is the over-banking tendency.

3)   General aviation airplanes are designed with a limited amount of positive static stability around the longitudinal axis so that they will be easy to turn but will return to straight-and-level flight from shallow banks.

   a)   Lateral stability about the longitudinal axis is affected by the dihedral and sweepback angles and by keel effect.

   b)   **Dihedral** is the angle at which the wings are slanted upward from the root to the tip.

      i)   In a shallow turn, the increased angle of attack produces increased lift on the lower wing with a tendency to return the airplane to wings-level flight.

   c)   **Sweepback** is the angle at which the wings are slanted rearward from the root to the tip.

      i)   Sweepback produces the same effect on stability as dihedral, but the effect is not as pronounced.

      ii) It augments dihedral to achieve stability.

   d)   **Keel effect** depends upon the action of the relative wind on the side area of the fuselage.

      i)   The fuselage is forced by keel effect to parallel the relative wind.
      ii) This effect is more noticeable in turbulent air than in turns.

4) Why over-banking occurs: As the radius of the turn becomes smaller, a significant difference develops between the speed of the inside wing and the speed of the outside wing.

    a) The wing on the outside of the turn travels a longer circuit than the inside wing, yet both complete their respective circuits in the same length of time.

    b) Therefore, the outside wing must travel faster than the inside wing; as a result, it develops more lift. A slight differential between the lift of the inside and outside wings tends to further increase the bank.

    c) As a shallow bank changes to a medium bank and the radius of turn decreases, the airspeed of the wing on the outside of the turn increases in relation to the inside wing, but the force created exactly balances the force of the inherent lateral stability of the airplane so that, at a given speed, no aileron pressure is required to maintain that bank.

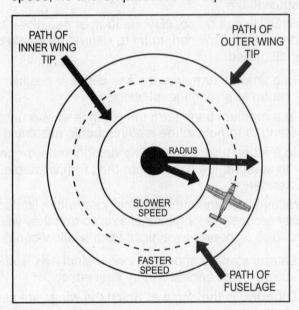

    d) As the radius decreases further when the bank progresses from a medium bank to a steep bank, the lift differential overbalances the lateral stability, and counteractive pressure on the ailerons is necessary to keep the bank from steepening.

c. **Torque effect in right and left turns**

1) The effects of torque (i.e., left-turning tendencies) should be anticipated.

    a) In a left turn, there is a tendency to develop a slight skid. You may need to add more right rudder (or less left rudder) to maintain coordinated flight (i.e., ball centered).

        i) In a left turn, adverse yaw (due to the left aileron being deflected downward as aileron pressure is applied to counteract over-banking tendency) acts in concert with other yawing moments to cause a skid.

    b) In a right turn, there may be a tendency to develop a slight slip. You may need to add right rudder to maintain coordinated flight.

        i) In a right turn, adverse yaw (due to the right aileron being deflected downward as aileron pressure is applied to counteract over-banking tendency) acts in opposition to other yawing moments to minimize the slipping tendency.

d.   **Selection of a suitable altitude**

1)   You should select an altitude that will allow this maneuver to be performed no lower than 1,500 ft. AGL or the airplane manufacturer's recommended altitude, whichever is higher.

2)   Select an altitude that is easy to read from your altimeter.

a)   If the terrain elevation is 300 ft. above sea level, the maneuver should be performed no lower than 1,800 ft. MSL (1,500 ft. AGL). Round this to the nearest 500-ft. increment (2,000 ft. MSL) to make it easier to identify on your altimeter.

b)   The FAA allows you to maintain altitude within 100 ft. Ensure that you are at least 1,600 ft. AGL.

e.   **Orientation, division of attention, and planning**

1)   Do not stare at any one object during this maneuver; i.e., divide your attention.

2)   To maintain orientation as well as altitude requires an awareness of the relative position of the nose, the horizon, the wings, and the amount of turn.

a)   If you watch only the nose of your airplane, you will have trouble holding altitude constant and remaining oriented in the turn.

b)   By watching all available visual and instrument references, you will be able to hold a constant altitude and remain oriented throughout the maneuver.

i)   Keep attitude indicator at 50° (commercial) or 45° (private).
ii)   Keep VSI at or near 0 fpm.
iii) Check heading and altitude.
iv)   Scan outside for traffic.

3)   You should select a prominent landmark (e.g., a road) on the horizon as a reference point.

a)   If there are no prominent landmarks, start the maneuver on a heading of 90°, 180°, 270°, or 360°.

4)   Plan ahead by determining the heading at which you will begin the rollout.

f.   **Entry and rollout procedure**

1)   Note your heading toward your reference point and smoothly roll into a coordinated turn with a 50° (commercial) or 45° (private) angle of bank, ±5°.

a)   As the turn is being established, back pressure on the elevator control should be smoothly increased to increase the angle of attack.

2)   As the bank steepens beyond 30°, you may find it necessary to hold a considerable amount of back elevator control pressure to maintain a constant altitude.

a)   Additional back elevator pressure increases the angle of attack, which increases drag.

i)   Additional power will be required to maintain entry altitude and airspeed.

b)   Retrim your airplane of excess control pressures, as appropriate.

i)   This will help you maintain a constant altitude.

3)   The rollout from the turn should be timed so that the wings reach level flight when your airplane is on the entry heading (i.e., toward your reference point).

    a)   Normally, you lead your desired heading by one-half of the number of degrees of bank, e.g., a 25° lead in a 50° bank.

    b)   While the rollout is being performed, back elevator pressure must be gradually released and power reduced as necessary to maintain the altitude and airspeed.

4)   As in all maneuvers, you will primarily use your outside (visual) references and cross-check bank control and altitude control with your instrument references.

g.   **Coordination of flight and power controls**

1)   During entry and rollout, keep the flight controls coordinated.

    a)   Keep the ball centered in the inclinometer.

2)   During the steep turn, check the inclinometer to ensure that coordinated flight is being maintained.

3)   Coordinate the use of pitch with the increase or decrease of bank angle, as appropriate.

    a)   While you are instructing, a student may attempt to raise the nose by applying top rudder pressure, thus causing a slip.

        i)   This error is corrected by proper use of the rudder and by checking the inclinometer.

        ii)   The adverse yaw produced by aileron drag has been a major consideration in design of the controls and, in most modern airplanes, has been significantly reduced. This reduction is accomplished by providing greater up travel than down travel in the ailerons.

        iii)   Also, during roll, lift is vectored aft on the rising wing and forward on the lowering wing, which also produces adverse yaw.

    b)   To counteract the yawing tendency, rudder pressure must be applied simultaneously in the desired direction of turn, thus producing a coordinated turn.

4)   The tendency to yaw and/or change bank must be controlled in very steep banks.

h.   **Altitude, bank, and power control during the turn**

1)   Maintain control of your airplane throughout the turn.

2)   To recover from an excessive nose-low attitude, you should first slightly reduce the angle of bank with coordinated aileron and rudder pressure.

    a)   Then, back elevator pressure should be used to raise your airplane's nose to the desired pitch attitude.

    b)   After completing this, reestablish the desired angle of bank.

    c)   Attempting to raise the nose first will usually cause a tight descending spiral and could lead to overstressing the airplane or loss of control.

3)   If your altitude increases, the bank should be increased to 50° by coordinated use of aileron and rudder.

4) Completion standards for this maneuver

    a) Commercial PTS

        i) ***Establish the manufacturer's recommended airspeed, or if one is not stated, a safe airspeed not to exceed V_A; roll into a coordinated 360° steep turn with at least a 50° bank, ±5°, followed by a 360° turn in the opposite direction; roll out on the entry heading, ±10°; and maintain entry altitude throughout the maneuver, ±100 ft., and airspeed, ±10 kt.***

    b) Private PTS

        i) During the steep turn, maintain a 45° bank, ±5°; roll out on the entry heading, ±10°; perform the task in the opposite direction when instructed; and maintain the entry altitude, ±100 ft., and airspeed, ±10 kt.

i. **Proper recovery to straight-and-level flight**

1) Begin the rollout to level flight with a lead equal to one-half the number of degrees of bank (e.g., use a 25° lead for a 50° bank).

2) Remember that as you roll the wings level, you will increase your vertical component of lift.

    a) This means that you must compensate with forward elevator pressure on the yoke to keep from climbing.

    b) If you added power or nose-up trim to remain level in the turn, reduce them to the appropriate cruise setting.

3) Do all three acts in one fluid motion, if possible.

2. **Exhibit your instructional knowledge by explaining the following common errors related to steep turns.**

a. **Improper pitch, bank, and power coordination during entry and rollout**

1) Do not overanticipate the amount of pitch change needed during entry and rollout.

    a) During entry, if the pitch is increased (nose up) before the bank is established, altitude will be gained.

    b) During recovery, if back pressure is not released, altitude will be gained.

2) Power should be added as required during entry and then reduced during rollout.

    a) Do not adjust power during transition to turn in opposite direction.

b. **Uncoordinated use of flight controls**

1) This error is normally indicated by a slip, especially in right-hand turns.

    a) Check the inclinometer.

2) If the airplane's nose starts to move before the bank starts, the rudder is being applied too soon.

3) If the bank starts before the nose starts turning, or if the nose moves in the opposite direction, the rudder is being used too late.

4) If the nose moves up or down when entering a bank, excessive or insufficient back elevator pressure is being applied.

c. **Improper procedure in correcting altitude deviations**

1) When altitude is lost, you may attempt to raise the nose first by increasing back elevator pressure without shallowing the bank. This error usually causes a tight descending spiral.

d.   **Loss of orientation**

1)   This error can be caused by forgetting the heading or reference point from which this maneuver was started.

2)   Select a prominent checkpoint to be used in this maneuver.

3)   During initial training on this maneuver, loss of orientation may be caused by spatial disorientation.

3.   **Demonstrate and simultaneously explain steep turns from an instructional standpoint.**

a.   Your demonstration must be to commercial pilot skill level.

b.   You will be evaluated on your ability to simultaneously explain and demonstrate the key elements of this task.

c.   If you do not perform the maneuver to FAA completion standards, point out the deficiency to your FAA inspector/examiner as an example of an error, and explain how it should be corrected (including how and why it occurred).

4.   **Analyze and correct simulated common errors related to steep turns.**

a.   As your FAA inspector/examiner is performing this task, you must be able to recognize, analyze, and correct any error.

b.   Explain what the error is, and then explain what (s)he should do to correct the error.

c.   If your FAA inspector/examiner does not perform the maneuver to FAA completion standards, point out the deficiency to him/her as an example of a student error, and explain how it should be corrected (including how and why it occurred).

END OF TASK

STEEP SPIRALS

IX.B. TASK: STEEP SPIRALS

REFERENCES: FAA-H-8083-3; FAA-S-8081-12.

Objective. To determine that the applicant:

1. Exhibits instructional knowledge of the elements of steep spirals by describing --

   a. Selection of entry altitude.

   b. Entry airspeed and power setting.

   c. Selection of a proper ground reference point.

   d. Division of attention and planning.

   e. Coordination of flight controls.

   f. Maintenance of constant radius around selected point.

   g. Maintenance of constant airspeed throughout maneuver.

2. Exhibits instructional knowledge of common errors related to steep spirals by describing --

   a. Improper pitch, bank, and power coordination during entry or completion.

   b. Uncoordinated use of flight controls.

   c. Improper planning and lack of maintenance of constant airspeed and radius.

   d. Failure to stay orientated to the number of turns and the rollout heading.

3. Demonstrates and simultaneously explains a steep spiral from an instructional standpoint.

4. Analyzes and corrects simulated common errors related to steep spirals.

A.   General Information

1.   The objective of this task is to determine your instructional knowledge of the elements and common errors related to steep spirals. You will be required to simultaneously demonstrate and explain this task to your FAA inspector/examiner.

B.   Task Objectives

1.   **Exhibit your instructional knowledge of the following elements of steep spirals**

   a.   **Selection of an entry altitude.**

      1)   Steep spirals should be begun at an altitude that will allow for at least three 360° turns to be completed. The maneuver should not be continued below 1,000 ft. above the surface unless you are performing the steep spiral in conjunction with an emergency approach and landing exercise.

         a)   As you practice the maneuver, note how much altitude is normally lost during a 360° turn in your airplane.

         b)   Multiply the altitude lost by 3 and add 1,000 ft. to determine the minimum AGL entry altitude.

            i)   Add the terrain elevation to this figure to determine the minimum indicated altitude at which the maneuver should be begun.

   b.   **Entry airspeed and power setting.**

      1)   Begin the maneuver by performing clearing turns to carefully scan for traffic, especially below you. Flaps and landing gear (if retractable) should be up unless otherwise specified by your examiner.

         a)   When you are established on the desired entry heading (downwind) and almost abeam the reference point, smoothly reduce engine power to idle (or the manufacturer's recommended minimum power setting) and maintain altitude using back pressure.

            i)   Allow your airspeed to bleed off to the selected glide airspeed (e.g., best glide speed).

    b)   When you are abeam the reference point, roll into approximately a 45° bank and relax the back pressure to maintain the desired airspeed.

        i)   Re-trim the airplane as needed.

c.   Selection of a proper ground reference point.

    1)   You should select a point that

        a)   Can be easily identified at a glance.

           i)   Do not select a vaguely-defined or non-unique point, such as one of several bales of hay scattered in a field, or you may lose track of exactly which point you are using.

        b)   Is located on or near a suitable emergency landing site.

           i)   Because the maneuver is normally completed at a low altitude, you must have emergency landing options available in case of an engine failure.

        c)   Is NOT in a congested area, near an open-air assembly of persons, on a federal airway, or near an area that might induce an undue hazard if a problem is experienced during the demonstration.

d.   Division of attention and planning.

    1)   As with other ground reference maneuvers, you will be required to divide your attention between following the proper ground track and maintaining control of your airplane.

        a)   Your attention must be divided among watching the ground reference point, maintaining the proper ground track, watching your flight instruments, and watching for other aircraft in your area.

e.   Coordination of flight controls.

    1)   Since you will be changing bank constantly throughout this maneuver, you must actively maintain coordinated flight (i.e., keep the ball centered) and airspeed (i.e., vary the pitch attitude as the bank angle changes).

        a)   Do not use rudder pressure alone to correct for wind drift.

           i)   Use coordinated aileron and rudder to increase or decrease the bank.

    2)   You must learn to divide your attention and not fixate on one item, such as your reference point.

f.   Maintenance of constant radius around selected point.

    1)   During the descent, you must maintain a constant radius from the selected point by making bank adjustments to account for the effects of the wind, just as you must do when performing a turn about a point.

        a)   On the downwind side of the maneuver, during which you will have a higher groundspeed, you must use a steeper bank angle in order to avoid being blown too far from the point (i.e., outside of the desired ground track).

           i)   You must use a lower pitch attitude to maintain the desired airspeed while in a steeper bank.

           ii)   Your bank angle should not exceed 60° at the steepest point in the turn.

        b)   On the upwind side of the maneuver, during which you will have a slower groundspeed, you must use a shallower bank angle in order to avoid being blown on top of the point.

            i)   You must use a higher pitch attitude to maintain the desired airspeed while in a shallower bank.

        c)   Note that the speed and direction of the wind may vary as you descend.

g.   **Maintenance of constant airspeed throughout the maneuver.**

    1)   The steep spiral should be performed at a safe selected airspeed, presumably your airplane's best glide speed.

        a)   The use of your airplane's best glide speed is consistent with the practical application of this maneuver, which is the dissipation of altitude over a selected landing site during the approach to a forced landing.

            i)   By spiraling down over your intended touchdown point, you can avoid accidentally gliding too far from the field and leaving insufficient altitude to successfully complete the approach.

        b)   A slight increase above best glide speed (e.g., 10 kt.) may be appropriate to avoid a stall during steep banks.

            i)   Note that the airplane's stall speed is not increased to the same degree in a steep spiral as it is in a steep turn because the airplane does not maintain a constant altitude.

    2)   Completion standards for this maneuver (commercial pilot only)

        a)   ***Maintain the specified airspeed, ±10 kt.; track a constant-radius circle around the reference point, with bank angle not to exceed 60° at the steepest point in the turn; roll out toward an object or on a specified heading, ±10°.***

2.   **Exhibit your instructional knowledge of common errors related to steep spirals by describing the following**

   a.   **Improper pitch, bank, and power coordination during entry or completion.**

    1)   This error is indicated by

        a)   Allowing the nose to drop excessively or maintaining an excessively nose-high attitude upon starting the descent.

            i)   If the nose is allowed to drop too low, your airspeed may exceed the allowable tolerances. If an excessively nose-high attitude is maintained, a stall may occur.

            ii)   You should determine the approximate correct pitch attitudes to establish for a straight glide and for beginning the descent during your training.

              •   Establish these attitudes during the entry and exit phases of the maneuver to avoid large airspeed changes; cross-check the airspeed indicator to fine tune your attitude.

        b)   Allowing the bank angle to exceed 60°. The bank angle may become excessive if

            i)   The turn radius is too small.

            ii)   You do not correct for the effects of overbanking tendency by applying aileron opposite the direction of the bank while in a steep turn.

    c) Forgetting to retard the throttle completely to idle (or to the manufacturer's recommended minimum power setting) before beginning the maneuver.

        i) At completion of the maneuver, remember to increase power to maintain level flight or begin a climb, as directed by the examiner

        ii) Do not allow a stall to occur.

b. **Uncoordinated use of flight controls.**

    1) This error is normally indicated by the development of a slip or a skid as the bank angle changes, or by variations in airspeed.

        a) Avoid this problem by frequently scanning the inclinometer (i.e., the ball) and cross-checking the airspeed indicator against the airplane's pitch attitude.

        b) Do not correct for wind drift using rudder pressure alone.

c. **Improper planning and lack of maintenance of constant airspeed and radius.**

    1) This error is indicated by

        a) The inability to complete three 360° turns before running out of altitude.

            i) Be sure that you have sufficient altitude above the terrain before beginning the maneuver.

        b) Losing track of the reference point during the maneuver.

            i) This may be caused by beginning the maneuver with the reference point on the right side of an airplane with side-by-side seating, or by choosing a vaguely-defined reference point.

                • Unless the examiner specifies otherwise, you should perform the maneuver to the left because you will have a better view of the point and its surroundings.

        c) The inability to maintain a constant distance from the reference point during the descent, or allowing the circle to drift downwind.

            i) Be sure to vary the bank angle using coordinated rudder and aileron inputs in order to maintain the desired ground track as the airplane's orientation to the wind changes.

                • While upwind, use a shallow bank; while downwind, use a steeper bank.

            ii) As with a turn about a point, while directly crosswind on the downwind side, the airplane's nose will be pointed inside the circle, and while directly crosswind on the upwind side, the airplane's nose will be pointed outside the circle.

                • Do not attempt to keep the wing on the reference point throughout the maneuver.

     d.   **Failure to stay orientated to the number of turns and the rollout heading.**

         1)   This error is indicated by completing fewer than three 360° turns or by rolling out on the incorrect heading or reference.

            a)   Avoid this error by starting the maneuver on a cardinal heading (i.e., north, south, east, or west) if possible.

               i)   If you cannot enter the maneuver downwind on a cardinal heading, try to use an entry heading that is shown on the heading indicator with digits, rather than tick marks (e.g., 150 instead of 165).

                  •   Use a heading bug, if available, or set the entry heading under the index of an unused VOR or ADF instrument to remind you.

               ii)   Alternatively, pick a prominent landmark to use as an entry/roll-out reference.

            b)   Count the number of turns out loud each time the airplane passes through the entry heading or the reference landmark.

            c)   At completion of the maneuver, remember to lead the roll-out by an amount equal to one-half the bank angle to avoid overshooting the desired heading.

3.   **Demonstrate and simultaneously explain a steep spiral from an instructional standpoint.**

     a.   The demonstration must be to commercial pilot skill level.

         1)   You will be evaluated on your ability to simultaneously explain and demonstrate the key elements of a steep spiral.

            a)   If you do not perform the maneuver to FAA completion standards, point out the deficiency to your FAA inspector/examiner as an example of an error, and explain how it should be corrected (including how and why it occurred).

4.   **Analyze and correct simulated common errors related to steep spirals.**

     a.   As your FAA inspector/examiner is performing a steep spiral, you must be able to recognize, analyze, and correct any error.

     b.   Explain what the error is, and then explain what (s)he should do to correct the error.

     c.   If your FAA inspector/examiner does not perform the maneuver to FAA completion standards, point out the deficiency to him/her as an example of a student error, and explain how it should be corrected (including how and why it occurred).

END OF TASK

CHANDELLES

IX.C. TASK: CHANDELLES

   REFERENCES: FAA-H-8083-3; FAA-S-8081-12.

Objective. To determine that the applicant:

1. Exhibits instructional knowledge of the elements of chandelles by describing --

   a. Selection of entry altitude.

   b. Entry airspeed and power setting.

   c. Division of attention and planning.

   d. Coordination of flight controls.

   e. Pitch and bank attitudes at various points during the maneuver.

   f. Proper correction for torque effect in right and left turns.

   g. Achievement of maximum performance.

   h. Completion procedure.

2. Exhibits instructional knowledge of common errors related to chandelles by describing --

   a. Improper pitch, bank, and power coordination during entry or completion.

   b. Uncoordinated use of flight controls.

   c. Improper planning and timing of pitch and bank attitude changes.

   d. Factors related to failure in achieving maximum performance.

   e. A stall during the maneuver.

3. Demonstrates and simultaneously explains chandelles from an instructional standpoint.

4. Analyzes and corrects simulated common errors related to chandelles.

A. General Information

   1. The objective of this task is to determine your instructional knowledge of the elements and common errors related to chandelles. You will be required to demonstrate and simultaneously explain this task from an instructional standpoint.

   2. Maneuver objectives

      a. Control coordination and smoothness.
      b. Develop positive control techniques at varying airspeeds and attitudes.
      c. Obtain maximum performance (altitude gain) from your airplane.

B. Task Objectives

   1. **Exhibit your instructional knowledge by explaining the following elements of chandelles.**

      a. **Selection of entry altitude**

         1) The commercial PTS requires that you select an altitude that will allow the chandelle to be performed no lower than 1,500 ft. AGL or the manufacturer's recommended altitude, whichever is higher.

         2) You may want to select an altitude that is easy to read from your altimeter.

            a) If the terrain elevation is 300 ft. MSL, the FAA requires the maneuver to be performed no lower than 1,800 ft. MSL (1,500 ft. AGL). Round this to the nearest 500-ft. increment (i.e., 2,000 ft. MSL) to make it easier to identify on your altimeter.

      b. **Entry airspeed and power setting**

         1) Prior to starting a chandelle, you should perform **clearing turns**, paying particular attention to possible traffic above and behind you.

            a) Even though a chandelle is a 180° turn, you must perform **clearing turns** and check for traffic carefully.

         2) The flaps and gear (if retractable) should be in the UP position.

         3) Power should be adjusted to enter the maneuver at V_A or the manufacturer's recommended entry speed, if specified in your POH.

4) Complete the following:

Entry	
Altitude	_____
Airspeed	_____
MP	_____
RPM	_____
Gear (if retractable)	UP
Flaps	UP

5) Some CFIs suggest selecting a reference line (e.g., road) that is parallel to the wind.

 a) As you fly over the road, you will enter the chandelle.

 b) Since the airplane will be in a nose-high pitch attitude and banked, it will be easier to determine your 90° and 180° point since you can look down to see your reference line.

 c) During your entry, make your turn toward the wind. This will keep you from traveling far from your starting point.

c. **Division of attention and planning**

 1) You must divide your attention and plan for each segment of the chandelle.

d. **Coordination of flight controls**

 1) Due to the constant changes in aileron, elevator, and rudder control pressures throughout the chandelle, it is important to use the flight controls in a smooth and coordinated manner.

e. **Pitch and bank attitudes at various points during the maneuver**

 1) A chandelle is illustrated below.

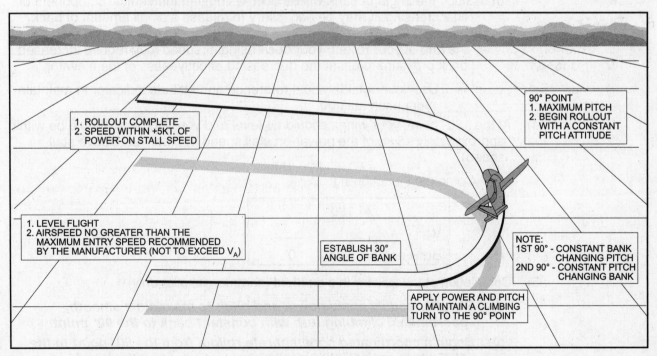

1. ROLLOUT COMPLETE
2. SPEED WITHIN +5KT. OF POWER-ON STALL SPEED

90° POINT
1. MAXIMUM PITCH
2. BEGIN ROLLOUT WITH A CONSTANT PITCH ATTITUDE

1. LEVEL FLIGHT
2. AIRSPEED NO GREATER THAN THE MAXIMUM ENTRY SPEED RECOMMENDED BY THE MANUFACTURER (NOT TO EXCEED V_A)

ESTABLISH 30° ANGLE OF BANK

NOTE:
1ST 90° - CONSTANT BANK CHANGING PITCH
2ND 90° - CONSTANT PITCH CHANGING BANK

APPLY POWER AND PITCH TO MAINTAIN A CLIMBING TURN TO THE 90° POINT

2) During the first 90° of turn, the pitch attitude increases at a constant rate to attain the highest pitch attitude at the 90° point, and the bank remains constant.

   a) Watch your bank angle because it may begin to increase without corrective control pressure. Remember, your rate-of-turn increases as your airspeed decreases.

   b) Some pilots use the airspeed indicator as a reference for the proper pitch attitude at the 90° point.

      i) The airspeed should be approximately at the midpoint between entry airspeed and the power-on stall speed of the airplane.

      ii) You should try this while practicing chandelles.

   c) Complete the following:

At 90°	
Airspeed	_____
Pitch attitude	_____
Bank angle	30°

3) When your airplane is headed toward your reference line (i.e., the 90° point), you should begin rolling out of the bank at a constant rate while maintaining a constant pitch attitude.

   a) The rate of rollout should be 10° of bank for each 30° of heading change.

   b) During the rollout, vertical lift increases, which tends to increase airspeed given a constant pitch attitude.

   c) This tendency to increase airspeed should only reduce the rate at which you decrease your bank. As you plan and execute chandelles, your 90°-point pitch attitude has to be sufficient to accommodate the effect of the increased vertical component of lift.

   d) Since the angle of bank will be decreasing and the vertical component of lift increasing, it **may** be necessary to release a small amount of back elevator pressure to prevent an increase in pitch attitude. Conversely, at the end of your rollout, your decreasing airspeed may require increased back pressure to offset the decreasing effectiveness of the elevator.

      i) Use your outside visual reference and attitude indicator to maintain your pitch attitude.

4) At the 180° point, your wings should be level and your airspeed should be within approximately +5 kt. of the power-on stall speed V$_{S1}$ (i.e., just above stall speed).

   a) Complete the following:

At 180°	
V$_{S1}$	_____
Bank angle	0°

5) Completion standards for this element (commercial pilot only)

   a) ***Simultaneously apply power and pitch to maintain a smooth, coordinated climbing turn with constant bank to the 90° point.***

   b) ***Begin a coordinated constant rate rollout from the 90° point to the 180° point, maintaining power and a constant pitch attitude.***

   c) ***Complete the rollout at the 180° point, ± 10°, and just above a stall airspeed; maintain that airspeed momentarily, avoiding a stall.***

f. **Proper correction for torque effect in right and left turns**

1) Torque has a significant effect on rudder use during this maneuver. Since the chandelle calls for a gradual decrease of airspeed during the climbing turn, the right rudder pressure must be adjusted for the amount of torque effect being experienced.

2) Depending on the direction of the turn, aileron drag from the down aileron may act either in concert with torque, thus adding to it, or against torque.

   a) To roll out of a left chandelle, the left wing must be raised. Thus, the left aileron is lowered and creates more drag than the aileron on the right wing, resulting in a tendency for the airplane to yaw to the left, in concert with the left torque effect. Therefore, significant right rudder correction is needed.

   b) To roll out of a right chandelle, the right wing must be raised. Thus, the right aileron is lowered and creates more drag on that wing and tends to make the airplane yaw to the right, acting in the opposite direction from torque. Therefore, less right rudder correction is needed and may result in a cross-control situation during the rollout from a right chandelle.

3) When the wings are leveled, the aileron drag is neutralized, and torque is acting alone again.

g. **Achievement of maximum performance**

1) A chandelle demands that the maximum flight performance of the airplane be obtained; that is, the airplane should gain the most altitude possible for a degree of bank and power setting without stalling.

   a) However, since numerous atmospheric variables beyond your control will affect the specific amount of altitude gained, the altitude gain is not a criterion of the quality of the chandelle.

2) Thus, the altitude gain is not a criterion of the quality, or an objective, of this maneuver.

h. **Completion procedure**

1) After the wings are level, you must continue to correct for torque, which is prominent due to the low airspeed and high power.

2) Gradually lower the nose to level flight while maintaining altitude.

   a) Right rudder pressure during the pitch decrease must be increased to counteract the additional torque caused by gyroscopic precession of the propeller.

3) As airspeed increases, adjust pitch, power, and trim for cruise flight.

4) Completion standard for this element (commercial pilot only)

   a) ***Resume straight-and-level flight with a minimum loss of altitude.***

2. **Exhibit your instructional knowledge by explaining the following common errors related to chandelles.**

a. **Improper pitch, bank, and power coordination during entry or completion**

1) If the pitch attitude is increased quickly, a stall will occur before the airplane reaches the 180° point.

   a) If the pitch attitude is increased too slowly or if pitch is allowed to decrease, you will complete the 180° turn well above the power-on stall speed.

      2) Rolling out too quickly will have the airplane at the 180° point at too high an airspeed.

        a) Rolling out too slowly will cause the airplane to stall before the 180° point.

      3) Power is used prior to entry to establish the proper airspeed and is increased only after the bank is established and pitch attitude has started to increase above the horizon.

        a) No other power adjustments are made during this maneuver.

  b. **Uncoordinated use of flight controls**

      1) Maintain coordinated flight and use of controls throughout the maneuver.

      2) Ensure that you compensate for the torque and aileron drag effect throughout the maneuver.

      3) Check the ball in the inclinometer during all phases of the maneuver.

  c. **Improper planning and timing of pitch and bank attitude changes**

      1) During the first 90° of the turn, the bank is constant (approximately 30°), and pitch gradually increases.

      2) At the 90° point, you should have maximum pitch attitude and begin rollout.

      3) During the second 90°, pitch attitude remains constant, and the bank is slowly rolled out.

      4) At the 180° point, the pitch attitude is constant, and the rollout to wings level flight is completed.

      5) You must plan and time the pitch and bank changes while dividing your attention to perform the maneuver properly.

        a) Subdividing pitch and bank changes will aid in this.

  d. **Factors related to failure in achieving maximum performance**

      1) Improper pitch throughout the maneuver
      2) Improper bank throughout the maneuver
      3) Lack of coordination, orientation, and planning

  e. **A stall during the maneuver**

      1) At no time during this maneuver should a stall develop.

      2) At the 180° point, your airspeed should be within approximately +5 kt. of the power-on stall speed.

3. **Demonstrate and simultaneously explain chandelles from an instructional standpoint.**

  a. Your demonstration must be to commercial pilot skill level.

  b. You will be evaluated on your ability to simultaneously explain and demonstrate the key elements for this task.

  c. If you do not perform the maneuver to FAA completion standards, point out the deficiency to your FAA inspector/examiner as an example of an error, and explain how it should be corrected (including how and why it occurred).

4. **Analyze and correct simulated common errors related to chandelles.**

  a. As your FAA inspector/examiner is performing this task, you must be able to recognize, analyze, and correct any error.

  b. Explain what the error is, and then explain what (s)he should do to correct the error.

  c. If your FAA inspector/examiner does not perform the maneuver to FAA completion standards, point out the deficiency to him/her as an example of a student error, and explain how it should be corrected (including how and why it occurred).

END OF TASK

LAZY EIGHTS

IX.D. TASK: LAZY EIGHTS

REFERENCES: FAA-H-8083-3; FAA-S-8081-12.

Objective. To determine that the applicant:

1. Exhibits instructional knowledge of the elements of lazy eights by describing --

   a. Selection of entry altitude.

   b. Selection of suitable reference points.

   c. Entry airspeed and power setting.

   d. Entry procedure.

   e. Orientation, division of attention, and planning.

   f. Coordination of flight controls.

   g. Pitch and bank attitudes at key points during the maneuver.

   h. Importance of consistent airspeed and altitude control at key points during the maneuver.

   i. Proper correction for torque effect in right and left turns.

   j. Loop symmetry.

2. Exhibits instructional knowledge of common errors related to lazy eights by describing --

   a. Poor selection of reference points.

   b. Uncoordinated use of flight controls.

   c. Unsymmetrical loops resulting from poorly planned pitch and bank attitude changes.

   d. Inconsistent airspeed and altitude at key points.

   e. Loss of orientation.

   f. Excessive deviation from reference points.

3. Demonstrates and simultaneously explains lazy eights from an instructional standpoint.

4. Analyzes and corrects simulated common errors related to lazy eights.

A. General Information

1. The objective of this task is to determine your instructional knowledge of the elements and common errors related to lazy eights. You will be required to simultaneously demonstrate and explain this task from an instructional standpoint.

2. The lazy eight consists of two symmetrical 180° turns flown in opposite directions while climbing and descending during the turns.

   a. The lazy eight gets its name from the flight pattern traced by the airplane. The maneuver is to be flown smoothly and appears very slow, i.e., lazy.

      1) An extension of the longitudinal axis from the nose of the airplane to the horizon looks like a figure "8" lying on its side, one-half above the horizon and one-half below the horizon.

3. The objective is to plan a complex maneuver accurately and then remain oriented while skillfully flying the airplane with positive, accurate control.

   a. You should use this maneuver to develop a fine feel for constant, gradually changing control forces.

4. You may wish to experiment with the following tutorial maneuver with your student.

   a. After clearing turns to locate all nearby traffic, fly straight-and-level at a low-cruise power setting.

b. Visualize a square in front of the airplane equally divided above and below by the horizon.

1) Initially, you are at "A." Raise the nose to point "B" with back pressure, and hold the nose at "B" momentarily.

2) Move the nose to "C" with right aileron while maintaining back pressure. When at "C," neutralize ailerons and hold the nose at "C" momentarily.

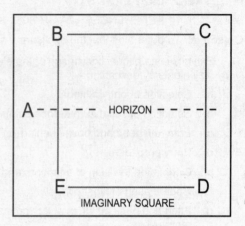

3) Move the nose to "D" with forward elevator pressure. Hold the nose at "D" momentarily.

4) Move the nose to "E" with left aileron pressure, and hold the nose at "E" momentarily.

c. Move the nose around the square once or twice more, stopping at each corner.

1) Pay attention to the feel, movement, and need for rudder coordination during the speed changes and momentary stop points.

d. Next, move the nose around the square without stopping at each corner.

e. Next, convert the square to a circle equidistant above and below the horizon.

1) Reverse the direction around the circle. Pay attention to the feel and movement and need for rudder coordination during airspeed changes and stop points.

2) Slow the movement of the track of the forward extension of the nose so the movement about the circle is "lazy."

f. Finally, convert the circles to a horizontally lying "8" equidistant above and below the horizon.

1) Again, this is where the lazy eight derives its name -- from the track of a forward extension of the longitudinal axis.

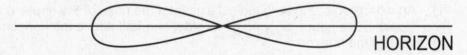

B. Task Objectives

1. **Exhibit your instructional knowledge by explaining the following elements of lazy eights.**

   a. **Selection of entry altitude**

   1) The commercial PTS requires that you select an altitude that will allow the maneuver to be performed no lower than 1,500 ft. AGL or the manufacturer's recommended altitude, whichever is higher.

   2) Select an altitude that is easy to read from your altimeter.

      a) EXAMPLE: If the terrain elevation is 300 ft., the FAA requires the maneuver to be performed no lower than 1,800 ft. MSL (1,500 ft. AGL). Round this up to the nearest 500-ft. increment (2,000 ft. MSL) to make it easier to identify on your altimeter.

    b.   **Selection of suitable reference points**

        1)   Select a prominent 90° reference point (or line) far away from the airplane (i.e., on or near the horizon) in the direction of your initial turn.

            a)   If possible, select additional reference points at the 45° and 135° points to assist you in remaining oriented during the maneuver.

        2)   You should enter perpendicular to the wind so that your turns are into the wind. This will allow you to perform the maneuver in a smaller area.

        3)   Perform **clearing turns** before starting this maneuver, especially in the airspace behind and above you.

    c.   **Entry airspeed and power setting**

        1)   You will normally use cruise power at the airspeed recommended in the POH or at V_A if no airspeed is specified.

        2)   Complete the following:

Entry	
Altitude	____
Airspeed	____
MP	____
RPM	____
Gear (if retractable)	UP
Flaps	UP

    d.   **Entry procedure**

        1)   The maneuver is started from straight-and-level flight with a gradual climbing turn to the 45° point. Pitch must be increased at a faster rate than bank.

        2)   The rate of rolling into the bank must be slow enough to prevent the rate of turn from becoming too rapid.

            a)   As you raise the nose of your airplane, the airspeed decreases, causing the rate of turn to increase. Since the bank is also increasing, it too causes the rate of turn to increase.

        3)   Unless the maneuver is begun with a slow rate of roll, the combination of increasing pitch and increasing bank will cause the rate of turn to be so rapid that the 45° point is reached before the highest pitch attitude is attained.

    e.   **Orientation, division of attention, and planning**

        1)   You must learn to divide your attention and plan for each segment of the lazy eight.

        2)   Maintain orientation by use of properly selected reference point(s).

    f.   **Coordination of flight controls**

        1)   Lazy eights require smooth, coordinated use of the flight controls.

        2)   At no time is the maneuver flown straight and level. The one point in the maneuver where a straight-and-level attitude is obtained is in the brief instant at the conclusion of each 180° turn just before the next turn is begun.

        3)   This maneuver requires constantly changing control pressures.

            a)   The control pressures need to change in accordance with changing combinations of climbing and descending turns at varying airspeeds.

g. **Pitch and bank attitudes at key points during the maneuver**

1) A lazy eight is illustrated below.

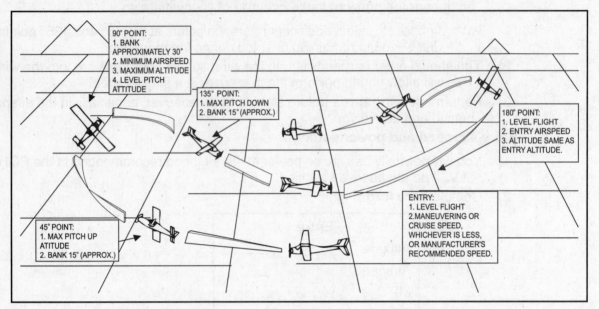

90° POINT:
1. BANK APPROXIMATELY 30°
2. MINIMUM AIRSPEED
3. MAXIMUM ALTITUDE
4. LEVEL PITCH ATTITUDE

135° POINT:
1. MAX PITCH DOWN
2. BANK 15° (APPROX.)

180° POINT:
1. LEVEL FLIGHT
2. ENTRY AIRSPEED
3. ALTITUDE SAME AS ENTRY ALTITUDE.

ENTRY:
1. LEVEL FLIGHT
2. MANEUVERING OR CRUISE SPEED, WHICHEVER IS LESS, OR MANUFACTURER'S RECOMMENDED SPEED.

45° POINT:
1. MAX PITCH UP ATTITUDE
2. BANK 15° (APPROX.)

2) At the 45° point, pitch is at its maximum, and the angle of bank should be at approximately 15°.

a) Complete the following:

At 45°	
Altitude	Increasing
Airspeed	Decreasing
Pitch attitude (max)	_____
Bank angle	approx. 15°

3) During the second 45° portion of the maneuver, the pitch attitude should decrease slowly, the airspeed should continue to decrease, and the bank should continue to increase.

a) Due to the decreasing airspeed, considerable right rudder pressure must be gradually applied to counteract torque.

i) More right rudder pressure will be needed during the climbing turn to the right because torque will be acting in the opposite direction of the turn.

ii) In the left turn, torque correction is reduced because torque is acting in the same direction as the turn.

b) The controls may be slightly crossed in the right climbing turn because you may use left aileron to prevent overbanking and right rudder to correct for torque.

4) At the 90° point, your airplane should be at 30° angle of bank, maximum altitude, and minimum airspeed (5 to 10 kt. above stall speed).

    a) It is at this time that an imaginary line, extending from your eyes and parallel to the longitudinal axis of your airplane, passes through your reference point at the 90° point.

    b) Complete the following:

At 90°	
Altitude (max.)	____
Airspeed (min.)	____
Pitch attitude	Level
Bank angle	30°

5) During the third 45° segment of the maneuver, the bank should be decreased gradually while the nose of the airplane is allowed to continue to lower.

    a) Remember to guide, not dive, the airplane during this segment.

6) At the 135° point, your airplane should be at the lowest pitch attitude and the bank angle should be approximately 15°.

    a) Complete the following:

At 135°	
Altitude	Decreasing
Airspeed	Increasing
Pitch attitude (min.)	_____
Bank angle	approx. 15°

7) During the last 45° of the turn, you need to plan your rate of rollout and pitch change so that the wings become level and the original airspeed is attained in level flight just as you reach the 180° point.

    a) During the descending turn, the airspeed will be increasing. Thus, it will be necessary to gradually relax rudder pressure (which was needed to correct for torque) as your airplane approaches your original airspeed.

    b) Complete the following:

At 180°	
Altitude (entry)	____
Airspeed (entry)	____
Pitch attitude	Level
Bank angle	0°

8) As you complete the 180° turn, you should immediately enter the maneuver to the opposite direction.

    a) That is, you are only straight-and-level for a moment at the beginning and end of each 180° turn.

h. **Importance of consistent airspeed and altitude control at key points during the maneuver**

1) The correct power setting for the lazy eight is that which will maintain the altitude for the maximum and minimum airspeeds used during the climbs and descents of the lazy eight.

   a) If excess power is used, either your airplane will have gained altitude when the maneuver is completed, or the airspeed will be greater than the entry airspeed if an attempt is made to complete the maneuver at entry altitude.

   b) If insufficient power is used, altitude will have been lost.

2) Completion standards for this element (commercial pilot only)

   a) *Altitude and airspeed consistent at the 180° points, ±100 ft. and ±10 kt., respectively*

   b) *Heading tolerance ±10° at each 180° point*

i. **Proper correction for torque effect in right and left turns**

1) The effect of torque (left-turning tendencies) was discussed in element g.

2) Throughout the maneuver, use proper rudder control to maintain coordinated flight.

   a) Check the inclinometer.

j. **Loop symmetry**

1) The nose of the airplane should travel the same distance below the horizon as above it.

2. **Exhibit your instructional knowledge by explaining the following common errors related to lazy eights.**

a. **Poor selection of reference points**

1) Select reference points that are easily identified.

2) If using a point, ensure that it is toward (or on) the horizon and not too close to your position.

b. **Uncoordinated use of flight controls**

1) Maintain coordinated flight and use of controls throughout this maneuver.

2) Ensure that you properly compensate for the torque effect.

3) Check the inclinometer.

c. **Unsymmetrical loops resulting from poorly planned pitch and bank attitude changes**

1) Excessive pitch-up attitude in the climbing turn may cause the airplane to stall before reaching the 90° point.

2) Excessive pitch-down attitude during the second part of the turn results in an excessive dive, which causes the airplane to exceed the entry airspeed at the 180° point.

3) Improper bank will cause you to hurry through the maneuver.

   a) A lazy eight should be a slow, lazy maneuver.

    d.   **Inconsistent airspeed and altitude at key points**

        1)   If you tend to climb and/or airspeed is becoming higher, reduce power slightly.

        2)   If you tend to lose altitude and/or airspeed is becoming slower, increase power slightly.

    e.   **Loss of orientation**

        1)   This error may be due to not properly dividing your attention among the elements of the lazy eight. You may be concentrating on pitch and bank attitude without watching outside for your reference point.

        2)   You may also forget what your reference point was, especially if you have selected poor reference points.

    f.   **Excessive deviation from reference points**

        1)   You must plan for the proper bank and pitch attitude, and aircraft heading.

        2)   Preplan the events in each 45° segment to help understand and anticipate the maneuver.

        3)   Get in the habit of talking through the maneuver; what are the targets for altitude, airspeed, pitch, and bank at the next 45° checkpoint?

3.   **Demonstrate and simultaneously explain lazy eights from an instructional standpoint.**

    a.   Your demonstration must be to commercial pilot skill level.

    b.   You will be evaluated on your ability to simultaneously explain and demonstrate the key elements of this task.

    c.   If you do not perform the maneuver to FAA completion standards, point out the deficiency to your FAA inspector/examiner as an example of an error, and explain how it should be corrected (including how and why it occurred).

4.   **Analyze and correct simulated common errors related to lazy eights.**

    a.   As your FAA inspector/examiner is performing this task, you must be able to recognize, analyze, and correct any error.

    b.   Explain what the error is, and then explain what (s)he should do to correct the error.

    c.   If your FAA inspector/examiner does not perform the maneuver to FAA completion standards, point out the deficiency to him/her as an example of a student error, and explain how it should be corrected (including how and why it occurred).

END OF TASK -- END OF STUDY UNIT

STUDY UNIT X
GROUND REFERENCE MANEUVERS

This study unit explains the four tasks (A-D) of Ground Reference Maneuvers. These tasks include both knowledge and skill. Your FAA inspector/examiner is required to test you on at least task D, Eights on Pylons, and one other task; therefore, you must be prepared for all four.

These maneuvers are designed to develop a student's ability to control the airplane and recognize and correct for the effect of wind while dividing attention among other matters. Developing these skills requires planning ahead of the airplane, maintaining orientation in relation to ground objects, flying appropriate headings to follow a desired ground track, and being cognizant of other air traffic in the immediate vicinity.

RECTANGULAR COURSE

> **X.A. TASK: RECTANGULAR COURSE**
>
> REFERENCES: FAA-H-8083-3; FAA-S-8081-14.
>
> **Objective.** To determine that the applicant:
>
> 1. Exhibits instructional knowledge of the elements of a rectangular course by describing--
>
>   a. How to select a suitable altitude.
>
>   b. How to select a suitable ground reference with consideration given to emergency landing areas.
>
>   c. Orientation, division of attention, and planning.
>
>   d. Configuration and airspeed prior to entry.
>
>   e. Relationship of a rectangular course to an airport traffic pattern.
>
>   f. Wind drift correction.
>
>   g. How to maintain desired altitude, airspeed, and distance from ground reference boundaries.
>
>   h. Timing of turn entries and rollouts.
>
>   i. Coordination of flight controls.
>
> 2. Exhibits instructional knowledge of common errors related to a rectangular course by describing--
>
>   a. Poor planning, orientation, or division of attention.
>
>   b. Uncoordinated flight control application.
>
>   c. Improper correction for wind drift.
>
>   d. Failure to maintain selected altitude or airspeed.
>
>   e. Selection of a ground reference where there is no suitable emergency landing area within gliding distance.
>
> 3. Demonstrates and simultaneously explains a rectangular course from an instructional standpoint.
>
> 4. Analyzes and corrects simulated common errors related to a rectangular course.

A. General Information

   1. The objective of this task is to determine your instructional knowledge of the elements and common errors related to a rectangular course. You will be required to demonstrate and simultaneously explain this task from an instructional standpoint.

   2. The rectangular course is a practice maneuver in which the ground track of the airplane is equidistant from all sides of a selected rectangular area on the ground.

      a. An objective is to develop your student's ability to recognize drift toward or away from a line parallel to the intended ground track.

         1) Development of this skill will assist your student in recognizing drift toward or away from an airport runway during the various legs of the traffic pattern.

B. Task Objectives

   1. **Exhibit your instructional knowledge by explaining the following elements of a rectangular course.**

      a. **How to select a suitable altitude**

         1) The Private Pilot PTS requires that this maneuver be performed at traffic pattern altitude.

         2) The FAA-recommended traffic pattern altitude is 1,000 ft. AGL, unless otherwise established.

         3) An altitude of 1,000 ft. AGL is a safe altitude for this maneuver.

            a) A smart pilot is always prepared for an emergency and minimizes low altitude activity that reflects poorly on aviation.

      b. **How to select a suitable ground reference with consideration given to emergency landing areas**

         1) You must determine the wind direction and estimate its speed before you begin any ground reference maneuver.

            a) You can determine the wind direction by observing the movement of smoke or dust, or the wave patterns on water or grain fields.

      b)   Another method is to make a 360° turn using a constant 30° angle of bank. By noting your ground track during the turn, you can determine wind direction and velocity.

          i)   Using a road intersection will provide you with a better starting point to begin the 360° turn.

2)   You need to select a rectangular field or an area bounded on four sides by section lines or roads.

      a)   The sides of the selected area should be approximately 1 mi. in length and well away from other air traffic.

      b)   The field should, however, be in an area away from communities, livestock, or groups of people on the ground to prevent possible annoyance or hazards to others.

3)   When selecting a suitable reference area for this maneuver, you must consider possible emergency landing areas.

      a)   There is little time available to search for a suitable field for landing in the event the need arises, e.g., during engine failure.

      b)   Select an area within gliding distance that meets the needs of the rectangular course and safe emergency landing areas.

4)   Check the area to ensure that no obstructions or other aircraft are in the immediate vicinity.

5)   When demonstrating this maneuver, identify your rectangular course and emergency landing area to your student (or FAA inspector/examiner).

c.   **Orientation, division of attention, and planning**

1)   As with other flight maneuvers by reference to ground objects, you are required to divide your attention between controlling your airplane and maintaining the desired ground track.

      a)   You will also need to plan for the next leg of the course.
      b)   Do not become focused on one item, e.g., watching the ground.

2)   While performing this maneuver, you must further divide your attention to watch for other aircraft in your area, i.e., collision avoidance.

d.   **Configuration and airspeed prior to entry**

1)   The airplane's configuration and airspeed should be that for normal cruise prior to entry.

e.   **Relationship of a rectangular course to an airport traffic pattern**

1)   The rectangular course is designed to be similar to the standard airport traffic pattern.

      a)   In fact, you will use the rectangular course to develop your student's skills needed to fly a proper traffic pattern.

2)   You should plan to enter this maneuver at traffic pattern altitude, 45° to the downwind leg, with the turns to the left (left-hand pattern).

      a)   This maneuver should also be done with turns to the right.

f.   **Wind drift correction**

1)   As soon as your airplane becomes airborne, it is free of ground friction. Its path is then affected by the air mass (wind) and will not always track along the ground in the exact direction that it is headed.

   a)   When flying with the longitudinal axis of your airplane aligned with a road, you may notice that you move closer to or farther from the road without any turn having been made.

      i)   This movement is an indication that the wind is moving sideward in relation to your airplane.

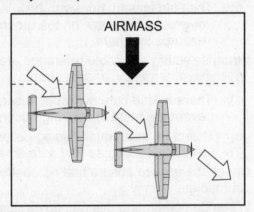

2)   In straight flight and following a selected ground track (e.g., a road), the preferred method of correcting for wind drift is to head (crab) your airplane into the wind to cause the airplane to move forward into the wind at the same rate that the wind is moving it sideways.

   a)   Depending on the wind velocity, correcting for drift may require a large crab angle or one of only a few degrees.

   b)   When the drift has been neutralized, the airplane will follow the desired ground track.

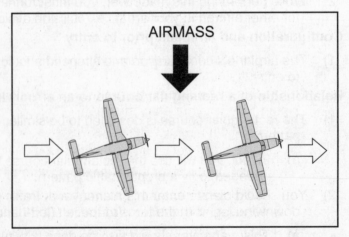

3)   In turning flight, the wind will be acting on your airplane from constantly changing angles.

   a)   The time it takes for the airplane to progress through any part of a turn is governed by the relative wing angle and speed.

   b)   When your airplane is headed into the wind, the groundspeed is decreased; when headed downwind, the groundspeed is increased.

   c) For you to fly a specific ground track, your rate of turn must be proportional to the groundspeed.

    i) When groundspeed is higher (tailwind), the rate of turn must be greater. To get a faster rate of turn, use a steeper bank.

    ii) Headwind results in a slower groundspeed, so use a lower rate of turn, i.e., less bank.

g. **How to maintain desired altitude, airspeed, and distance from ground reference boundaries**

  1) Enter the rectangular course 45° to, and at the midpoint of, the left downwind leg (i.e., the course will be to your left).

  2) Your entry on the downwind leg should place your airplane parallel to, and at a uniform distance (one-fourth to one-half mile) away from, the field boundary.

   a) You should be able to see the edges of the selected field while seated in a normal position and looking out the side of the airplane during either a left-hand or right-hand course.

    i) The distance of the ground track from the edges of the field should be the same regardless of the direction in which the course is flown.

   b) If you attempt to fly directly above the edges of the field, you will have no usable reference points to start and complete the turns.

   c) The closer the track of your airplane is to the field boundaries, the steeper the bank is required to be at the turning points.

    i) The maximum angle of bank is 45°.

  3) Throughout this maneuver, a constant altitude should be maintained.

   a) As the bank increases, you may need to increase back elevator pressure to pitch the airplane's nose up to maintain altitude.

    i) As the bank decreases, you may need to release some of the back elevator pressure to maintain altitude.

   b) Maintain pitch awareness by visual references, and use your altimeter to ensure that you are maintaining altitude.

  4) Normally, this maneuver is done at cruise airspeed in trainer-type airplanes.

   a) Check your airspeed indicator to ensure that you are maintaining your entry airspeed.

  5) During this maneuver, if you maintain your altitude, your airspeed should remain within 10 kt. of your entry airspeed.

   a) Use pitch to make altitude corrections.
   b) Make small power adjustments to make airspeed corrections, if necessary.

  6) Completion standard for this element (private pilot only)

   a) ***Maintain a constant ground track around the rectangular reference area; and maintain altitude, ±100 ft., and airspeed, ±10 kt.***

   b) Your performance is expected to be "more precise" than that required for a private pilot.

    i) This more precise performance must be determined by your FAA inspector/examiner through the exercise of subjective judgment.

h.   **Timing of turn entries and rollouts**

1)   All turns should be started when your airplane is abeam the corners of the field boundaries.

2)   This discussion begins with a downwind entry.

3)   While the airplane is on the downwind leg (similar to the downwind leg in a traffic pattern), observe the next field boundary as it approaches to plan the turn onto the crosswind leg.

a)   Maintain your desired distance from the edge of the course, and maintain entry altitude, i.e., 1,000 ft. AGL.

b)   Since you have a tailwind on this leg, your airplane has an increased groundspeed. During your turn to the next leg, the wind will tend to move your airplane away from the field.

i)   Thus, the turn must be entered with a fairly fast rate of roll-in with a relatively steep bank.

ii)   To compensate for the drift on the next leg, the amount of turn must be more than 90°.

c)   As the turn progresses, the tailwind component decreases, resulting in a decreasing groundspeed.

i)   Thus, the bank angle and rate of turn must be decreased gradually to assure that, upon completion of the turn, you will continue the crosswind ground track at the same distance from the field.

4)   The rollout onto this next leg (similar to the base leg in a traffic pattern) is such that, as the wings become level, your airplane is crabbed slightly toward the field and into the wind to correct for drift.

a)   The base leg should be continued at the same distance from the field boundary and at the entry altitude.

b)   While you are on the base leg, adjust the crab angle as necessary to maintain a uniform distance from the field.

c)   Since drift correction is being held on this leg, it is necessary to plan for a turn of less than 90° to align your airplane parallel to the upwind leg boundary.

d)   This turn should be started with a medium bank angle with a gradual reduction to a shallow bank as the turn progresses.

i)   This change is necessary due to the crosswind becoming a headwind, causing the groundspeed to decrease throughout the turn.

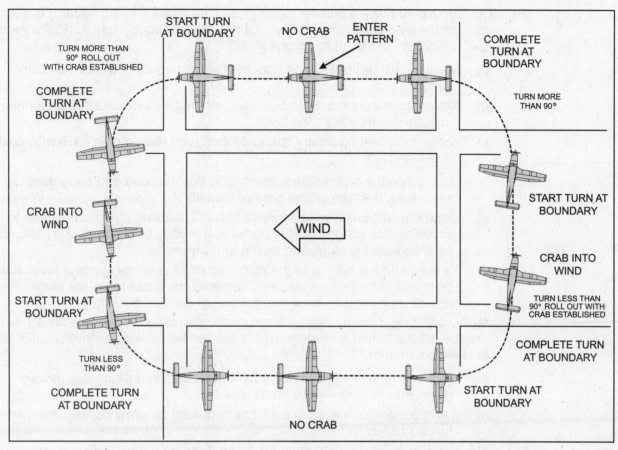

START TURN
AT BOUNDARY

NO CRAB

ENTER
PATTERN

COMPLETE
TURN AT
BOUNDARY

TURN MORE THAN
90° ROLL OUT
WITH CRAB ESTABLISHED

COMPLETE
TURN AT
BOUNDARY

TURN MORE
THAN 90°

CRAB INTO
WIND

WIND

START TURN AT
BOUNDARY

CRAB INTO
WIND

TURN LESS THAN
90° ROLL OUT WITH
CRAB ESTABLISHED

START TURN AT
BOUNDARY

COMPLETE TURN
AT BOUNDARY

TURN LESS
THAN 90°

COMPLETE TURN
AT BOUNDARY

START TURN AT
BOUNDARY

NO CRAB

5) The rollout onto this leg (similar to the final approach and upwind leg in a traffic pattern) should be timed to assure paralleling of the field as the wings become level.

    a) Maintain the same distance from the field boundary and maintain entry altitude.

    b) The next field boundary should be observed as it is being approached, to plan the turn onto the crosswind leg.

    c) Since the wind is a headwind on this leg, it is reducing your airplane's groundspeed and, during the turn onto the crosswind leg, will try to drift your airplane toward the field.

        i) Thus, the roll-in to the turn must be slow and the bank relatively shallow to counteract this effect.

        ii) As the turn progresses, the headwind component decreases, allowing the groundspeed to increase.

            • Consequently, the bank angle and rate of turn must be increased gradually to assure that, upon completion of the turn, you will continue the crosswind ground track at the same distance from the edge of the field.

        iii) To compensate for drift on the next leg, the amount of turn will be less than 90°.

6)   The rollout onto this leg (similar to the crosswind leg of a traffic pattern) is such that, as the wings become level, your airplane is crabbed slightly into the wind (i.e., away from the field) to correct for drift.

a)   Maintain the same distance from the field boundary and maintain entry altitude.

b)   While you are on this leg, adjust the crab angle as necessary to maintain a uniform distance from the field.

c)   As the next field boundary is approached, you should plan the turn onto the downwind leg.

i)   Since the crab angle is being held into the wind and away from the field, this turn will be greater than 90°.

d)   Since the crosswind will become a tailwind, causing the groundspeed to increase during this turn, the bank must initially be medium and then must be progressively increased as the turn proceeds.

e)   To complete the turn, time the rollout so that the wings become level at a point aligned with the crosswind corner of the field just as the longitudinal axis of your airplane becomes parallel to the field boundary.

7)   Ideally, drift should not be encountered on the downwind or the upwind leg, but it may be difficult to find a situation where the wind is blowing exactly parallel to the field boundaries.

a)   Since a wind blowing parallel to the boundaries is unlikely, it is usually necessary to crab slightly on all legs.

b)   It is important to anticipate the turns to correct for groundspeed, drift, and turning radius.

c)   You use these same techniques when flying an airport traffic pattern.

i.   **Coordination of flight controls**

1)   The airplane must remain in coordinated flight at all times.

a)   Do not use the rudder to correct for wind drift, but turn the airplane to establish the proper ground track by coordinated use of aileron and rudder.

2)   Hold altitude by maintaining level pitch attitude.

a)   The airplane should be trimmed for level flight before beginning this maneuver.

2.   **Exhibit your instructional knowledge by explaining the following common errors related to a rectangular course.**

a.   **Poor planning, orientation, or division of attention**

1)   Poor planning results in not beginning or ending the turns properly at the corners of the rectangular course. You must plan ahead and anticipate the effects of the wind.

2)   Poor orientation normally results in not being able to identify the wind direction, thus causing problems in your planning.

3)   Poor division of attention contributes to an inability to maintain a proper ground track, altitude, and/or airspeed.

a)   Also, you may not notice other aircraft that have entered the area near you.

b. **Uncoordinated flight control application**

1) This error normally occurs when you begin to fixate on the field boundaries and attempt to use only rudder pressure to correct for drift.

2) Use coordinated aileron and rudder in all turns and during necessary adjustments to the crab angle.

c. **Improper correction for wind drift**

1) This error occurs either from not fully understanding the effect the wind has on the ground track or from not dividing your attention to recognize the need for wind drift correction.

2) Once you recognize the need for a correction, take immediate steps to correct for wind drift with coordinated use of the flight controls.

d. **Failure to maintain selected altitude or airspeed**

1) Most student pilots will gain altitude during the initial training in this maneuver due to poor division of attention and/or a lack of proper pitch awareness.

    a) You must learn the visual references to maintain altitude.

2) Maintaining a constant altitude and not exceeding a 45° angle of bank will allow you to maintain your airspeed within ±10 kt.

e. **Selection of a ground reference where there is no suitable emergency landing area within gliding distance**

1) This is a part of poor planning, so stress to your student that (s)he should always be prepared for any type of emergency.

2) Demonstrate how to select a ground reference point that is within gliding distance to make a safe emergency landing.

3) When your student identifies a ground reference point, ask him/her to also identify his/her selection of an emergency landing area.

3. **Demonstrate and simultaneously explain a rectangular course from an instructional standpoint.**

a. Your demonstration must be to the commercial pilot skill level.

b. You will be evaluated on your ability to simultaneously explain and demonstrate the key elements of this task.

c. If you do not perform the maneuver to FAA completion standards, point out the deficiency to your FAA inspector/examiner as an example of an error, and explain how it should be corrected (including how and why it occurred).

4. **Analyze and correct simulated common errors related to a rectangular course.**

a. As your FAA inspector/examiner is performing this task, you must be able to recognize, analyze, and correct any error.

b. Explain what the error is and then explain what (s)he should do to correct the error.

c. If your FAA inspector/examiner does not perform the maneuver to FAA completion standards, point out the deficiency to him/her as an example of a student error, and explain how it should be corrected (including how and why it occurred).

END OF TASK

S-TURNS ACROSS A ROAD

X.B. TASK: S-TURNS ACROSS A ROAD

REFERENCES: FAA-H-8083-3; FAA-S-8081-14.

Objective. To determine that the applicant:

1. Exhibits instructional knowledge of the elements of S-turns across a road by describing--

   a. How to select a suitable altitude.

   b. How to select a suitable ground reference with consideration given to emergency landing areas.

   c. Orientation, division of attention, and planning.

   d. Configuration and airspeed prior to entry.

   e. Entry procedure.

   f. Wind drift correction.

   g. Tracking of semicircles of equal radii on either side of the selected ground reference line.

   h. How to maintain desired altitude and airspeed.

   i. Turn reversal over the ground reference line.

   j. Coordination of flight controls.

2. Exhibits instructional knowledge of common errors related to S-turns across a road by describing--

   a. Faulty entry procedure.

   b. Poor planning, orientation, or division of attention.

   c. Uncoordinated flight control application.

   d. Improper correction for wind drift.

   e. An unsymmetrical ground track.

   f. Failure to maintain selected altitude or airspeed.

   g. Selection of a ground reference line where there is no suitable emergency landing area within gliding distance.

3. Demonstrates and simultaneously explains S-turns across a road from an instructional standpoint.

4. Analyzes and corrects simulated common errors related to S-turns across a road.

A. General Information

   1. The objective of this task is to determine your instructional knowledge of the elements and common errors related to S-turns across a road. You will be required to demonstrate and simultaneously explain this task from an instructional standpoint.

   2. An S-turn across a road is a practice maneuver in which the airplane's ground track describes semicircles, each of an equal radius, on either side of a selected straight line on the ground.

   3. The objectives are

      a. To develop your ability to compensate for drift during turns
      b. To orient the flight path with ground references
      c. To divide your attention

B. Task Objectives

   1. **Exhibit your instructional knowledge by explaining the following elements of S-turns.**

      a. **How to select a suitable altitude**

         1) The Private Pilot PTS requires that this maneuver be performed at an altitude between 600 and 1,000 ft. AGL.

         2) When given an altitude window like this, you should always use the highest altitude, i.e., 1,000 ft. AGL.

            a) A smart pilot is always prepared for an emergency and minimizes low altitude activity that reflects poorly on aviation.

b. **How to select a suitable ground reference with consideration given to emergency landing areas**

1) You must determine the wind direction and estimate its speed before you begin any ground reference maneuver.

   a) You can determine the wind direction by observing the movement of smoke or dust, or the wave patterns on water or grain fields.

   b) Another method is to make a 360° turn using a constant medium angle of bank. By noting your ground track during the turn, you can determine wind direction and speed.

      i) Using a road intersection will provide you a better starting point to begin the 360° turn.

2) Before starting the maneuver, you must select a straight ground reference line.

   a) This line may be a road, fence, railroad, or section line that is easily identifiable to you.

   b) This line should be perpendicular (i.e., 90°) to the direction of the wind.

   c) The line should be a sufficient length for making a series of turns.

   d) The point should, however, be in an area away from communities, livestock, or groups of people on the ground to prevent possible annoyance or hazards to others.

3) When selecting a suitable ground reference line for this maneuver, you must also consider possible emergency landing areas.

   a) There is little time available to search for a suitable field for landing in the event the need arises, e.g., during engine failure.

   b) Select an area that meets the requirements of both S-turns and safe emergency landing areas.

4) Check the area to ensure that no obstructions or other aircraft are in the immediate vicinity.

5) When demonstrating this maneuver, identify your reference line and emergency landing area to your student (or FAA inspector/examiner).

c. **Orientation, division of attention, and planning**

1) As with other ground reference maneuvers, you will be required to divide your attention between following the proper ground track and maintaining control of your airplane.

   a) Your attention must be divided among watching the ground reference line, maintaining the proper ground track, watching your flight instruments, and watching for other aircraft in your area.

2) You must learn to divide your attention and not fixate on one item, such as the ground reference line.

d. **Configuration and airspeed prior to entry**

1) Prior to entry, the airplane should be in straight-and-level flight at 1,000 ft. AGL and in normal cruise configuration.

e. **Entry procedure**

1) Your airplane should be perpendicular to your ground reference line.

   a) Approach the reference line from the upwind side (i.e., so the airplane is heading downwind).

   b) Your airplane should be in the normal cruise configuration.

2)   When you are directly over the road, start the first turn immediately to the left.

a)   This normally means that when your airplane's lateral axis (i.e., wingtip-to-wingtip) is over the reference line, the first turn is started.

f.   **Wind drift correction**

1)   The maneuver consists of crossing a reference line on the ground at a 90° angle and immediately beginning a series of 180° turns of uniform radius in opposite directions, recrossing the road at a 90° angle just as each 180° turn is completed.

2)   Since turns to effect a constant radius on the ground track require a changing roll rate and angle of bank to establish the crab needed to compensate for the wind, both will increase or decrease as groundspeed increases or decreases.

a)   The bank must be steepest as the turn begins on the downwind side of the ground reference line and must be shallowed gradually as the turn progresses from a downwind heading to an upwind heading.

b)   On the upwind side, the turn should be started with a relatively shallow bank, which is gradually steepened as the airplane turns from an upwind heading to a downwind heading.

g.   **Tracking of semicircles of equal radii on either side of the selected ground reference line**

1)   With your airplane headed downwind, the groundspeed is the greatest, and the rate of departure from the road will be rapid.

a)   The roll into the steep bank (approximately 40°- 45°) must be fairly rapid to attain the proper crab angle.

i)   The proper crab angle prevents your airplane from flying too far from your selected reference line and from establishing a ground track of excessive radius.

b)   During the latter portion of the first 90° of turn when your airplane's heading is changing from a downwind heading to a crosswind heading, the groundspeed and the rate of departure from the reference line decrease.

i)   The crab angle will be at the maximum when the airplane is headed directly crosswind (i.e., parallel to the reference line).

2)   After you turn 90°, your airplane's heading becomes more of an upwind heading.

a)   The groundspeed will decrease, and the rate of closure with the reference will become slower.

i)   Thus, it will be necessary to gradually shallow the bank during the remaining 90° of the semicircle so that the crab angle is removed completely and the wings become level as the 180° turn is completed at the moment the reference line is reached.

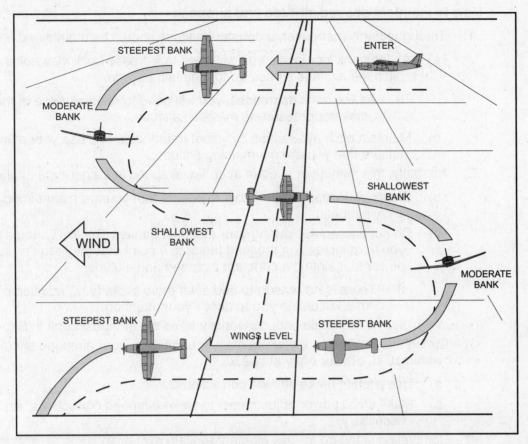

3)   Once you are over the reference line, you will start a turn in the opposite
     direction. Since your airplane is still flying into the headwind, the groundspeed
     is relatively slow.

   a)   The turn will have to be started with a shallow bank so as to avoid an
        excessive rate of turn, which would establish the maximum crab angle too
        soon.

   b)   The degree of bank should be that which is necessary to attain the proper
        crab so the ground track describes an arc the same size as the one
        established on the downwind side.

4)   Since your airplane is turning from an upwind to a downwind heading, the
     groundspeed will increase and, after you turn 90°, the rate of closure with the
     reference line will increase rapidly.

   a)   The angle of bank and rate of turn must be progressively increased so that
        your airplane will have turned 180° at the time it reaches the reference
        line.

5)   Throughout this maneuver, the bank angle should be changing constantly to
     track a constant radius turn on each side of the selected reference line.

   a)   There should not be any period of straight-and-level flight.

h. **How to maintain desired altitude and airspeed**

   1) Throughout this maneuver, a constant altitude should be maintained.

      a) As the bank increases, you will need to increase back elevator pressure to pitch the airplane's nose up to maintain altitude.

         i) As the bank decreases, you will need to release some of the back elevator pressure to maintain altitude.

      b) Maintain pitch awareness by visual references, and use your altimeter to ensure that you are maintaining altitude.

   2) Normally, this maneuver is done at cruise airspeed in trainer-type airplanes.

      a) Check your airspeed indicator to ensure that you are maintaining your entry airspeed.

      b) Do not exceed 45° during your steepest banks. This limit should prevent you from increasing the load factor to a point that may require additional power to maintain a constant airspeed and altitude.

         i) There is no reason to add even more tasks (e.g., addition of power) that will cause you to divide your attention.

         ii) A 45° angle of bank works well as the steepest bank in S-turns.

   3) During this maneuver, if you maintain your altitude, your airspeed should remain within 10 kt. of your entry airspeed.

      a) Use pitch to make altitude corrections.

      b) Make small power adjustments to make airspeed corrections, if necessary.

   4) Completion standard for this element (private pilot only)

      a) ***Maintain altitude, ±100 ft., and airspeed, ±10 kt.***

      b) Your performance is expected to be "more precise" than that required for a private pilot.

         i) This more precise performance must be determined by your FAA inspector/examiner through the exercise of subjective judgment.

i. **Turn reversal over the ground reference line**

   1) In this maneuver, the airplane should be rolled from one bank directly into the opposite bank just as the reference line on the ground is crossed.

j. **Coordination of flight controls**

   1) Since you will be changing bank constantly throughout this maneuver, you must maintain coordinated flight (i.e., keep the ball centered).

      a) Avoid using only the rudder to turn the airplane in order to arrive perpendicular over your reference line.

2. **While demonstrating and instructing your FAA inspector/examiner on the elements of S-turns across a road, you must be able to explain the common errors related to this task.**

   a. **Faulty entry procedure**

      1) You should enter this maneuver heading downwind, perpendicular to your selected reference line.

      2) As soon as your airplane's lateral axis is over the reference line, you must roll into your steepest bank at a fairly rapid rate.

         a) If the initial bank is too shallow, your airplane will be pushed too far from the reference line, thus establishing a ground track of excessive radius.

b. **Poor planning, orientation, or division of attention**

1) Poor planning results in a failure to maintain the constant change of bank required to effect a true semicircular ground track.

   a) If you do not change to the appropriate degree of bank, your airplane may be in straight-and-level flight before the reference line or still in a bank while crossing the reference line.

2) Poor orientation usually is the result of not selecting a good ground reference line and/or not identifying the wind direction.

3) Poor division of attention contributes to an inability to maintain a proper ground track, altitude, and/or airspeed.

   a) Also, you may not notice other aircraft that have entered the area near you.

c. **Uncoordinated flight control application**

1) This error normally occurs when you begin to fixate on the ground reference line and then forget to use the flight controls in a coordinated manner.

2) Do not use the rudder to yaw the nose of the airplane in an attempt to be directly over and perpendicular to the reference line.

3) Maintain a coordinated flight condition (i.e., keep the ball centered) throughout this maneuver.

d. **Improper correction for wind drift**

1) If a constant steep turn is maintained during the downwind side, the airplane will turn too quickly during the last 90° for the slower rate of closure and will be headed perpendicular to the reference line prematurely (i.e., wings level before you arrive over the reference line).

   a) To avoid this error, you must gradually shallow the bank during the last 90° of the semicircle so that the crab angle is removed completely as the wings become level directly over the reference line.

2) Often there is a tendency to increase the bank too rapidly during the initial part of the turn on the upwind side, which will prevent the completion of the 180° turn before recrossing the road.

   a) To avoid this error, you must visualize the desired half-circle ground track and increase the bank slowly during the early part of this turn.

      i) During the latter part of the turn, when approaching the road, you must judge the closure rate properly and increase the bank accordingly so as to cross the road perpendicular to it just as the rollout is completed.

e. **An unsymmetrical ground track**

1) Your first semicircle will establish the radii of the semicircles.

   a) You must be able to visualize your ground track and plan for the effect the wind will have on the ground track.

2) The bank of your airplane must be constantly changing (except in the case of no wind) in order to effect a true semicircular ground track.

f. **Failure to maintain selected altitude or airspeed**

1) Most student pilots will initially have trouble maintaining altitude or airspeed due to their inexperience in dividing their attention.

   a) Learn to divide your attention between the ground reference line and airplane control (e.g., pitch awareness).

2) By maintaining altitude, you should be able to maintain your selected airspeed when entering at normal power setting.

    g.   **Selection of a ground reference line where there is no suitable emergency landing area within gliding distance**

        1)   This error is a part of poor planning, so stress to your student that (s)he should always be prepared for any type of emergency.

        2)   Demonstrate how to select a ground reference point that is within gliding distance to make a safe emergency landing.

        3)   When your student identifies a ground reference point, ask him/her to also identify his/her selection of an emergency landing area.

3.   **Demonstrate and simultaneously explain S-turns across a road from an instructional standpoint.**

    a.   Your demonstration must be to the commercial pilot skill level.

    b.   You will be evaluated on your ability to simultaneously explain and demonstrate the key elements of this task.

    c.   If you do not perform the maneuver to FAA completion standards, point out the deficiency to your FAA inspector/examiner as an example of an error, and explain how it should be corrected (including how and why it occurred).

4.   **Analyze and correct simulated common errors related to S-turns across a road.**

    a.   As your FAA inspector/examiner is performing this task, you must be able to recognize, analyze, and correct any error.

    b.   Explain what the error is, and then explain what (s)he should do to correct the error.

    c.   If your FAA inspector/examiner does not perform the maneuver to FAA completion standards, point out the deficiency to him/her as an example of a student error, and explain how it should be corrected (including how and why it occurred).

END OF TASK

TURNS AROUND A POINT

X.C. TASK: TURNS AROUND A POINT

    REFERENCES: FAA-H-8083-3; FAA-S-8081-14.

Objective. To determine that the applicant:

1. Exhibits instructional knowledge of the elements of turns around a point by describing--

    a. How to select a suitable altitude.

    b. How to select a suitable ground reference point with consideration given to emergency landing areas.

    c. Orientation, division of attention, and planning.

    d. Configuration and airspeed prior to entry.

    e. Entry procedure.

    f. Wind drift correction.

    g. How to maintain desired altitude, airspeed, and distance from reference point.

    h. Coordination of flight controls.

2. Exhibits instructional knowledge of common errors related to turns around a point by describing--

    a. Faulty entry procedure.

    b. Poor planning, orientation, or division of attention.

    c. Uncoordinated flight control application.

    d. Improper correction for wind drift.

    e. Failure to maintain selected altitude or airspeed.

    f. Selection of a ground reference point where there is no suitable emergency landing area within gliding distance.

3. Demonstrates and simultaneously explains turns around a point from an instructional standpoint.

4. Analyzes and corrects simulated common errors related to turns around a point.

A. General Information

    1. The objective of this task is to determine your instructional knowledge of the elements and common errors related to turns around a point. You will be required to simultaneously demonstrate and explain this task from an instructional standpoint.

    2. In this training maneuver, the airplane is flown in two or more complete circles, each of a uniform radius or distance from a prominent ground reference point, using a maximum bank of approximately 45° while maintaining a constant altitude.

    3. The objectives are

        a. To develop your ability to compensate for drift during turns

        b. To orient the flight path with the ground reference point

        c. To divide your attention between the flight path and ground reference point and to watch for other air traffic in your area

B. Task Objectives

    1. **Exhibit your instructional knowledge by explaining the following elements of turns around a point.**

        a. **How to select a suitable altitude**

            1) The Private Pilot PTS requires that this maneuver be performed at an altitude between 600 and 1,000 ft. AGL.

            2) When given an altitude window like this, you should always use the highest altitude, i.e., 1,000 ft. AGL.

                a) A smart pilot is always prepared for an emergency and minimizes low altitude activity that reflects poorly on aviation.

b. **How to select a suitable ground reference point with consideration given to emergency landing areas**

1) You must determine the wind direction and estimate its speed before you begin any ground reference maneuver.

   a) You can determine the wind direction by observing the movement of smoke or dust, or the wave patterns on water or grain fields.

   b) Another method is to make a 360° turn using a constant medium angle of bank. By noting your ground track during the turn, you can determine wind direction and speed.

      i) Using a road intersection will provide you a better starting point to begin the 360° turn.

2) The point you select should be prominent, easily distinguished by you, and yet small enough to present a precise reference.

   a) Isolated trees, crossroads, or other similar landmarks are usually suitable.

   b) The point should, however, be in an area away from communities, livestock, or groups of people on the ground to prevent possible annoyance or hazards to others.

3) When selecting a suitable ground reference point, you must also consider possible emergency landing areas.

   a) There is little time available to search for a safe field for landing in the event the need arises, e.g., during an engine failure.

   b) Select an area that provides a usable ground reference point and the opportunity for a safe emergency landing.

4) Check the area to ensure that no obstructions or other aircraft are in the immediate vicinity.

5) When demonstrating this maneuver, identify your reference point and emergency landing area to your student (or FAA inspector/examiner).

c. **Orientation, division of attention, and planning**

1) As with other ground reference maneuvers, you will be required to divide your attention between following the proper ground track and maintaining control of your airplane.

   a) You must divide your attention among watching the ground reference point, maintaining the proper ground track, watching your flight instruments, and watching for other aircraft in your area.

2) You must learn to divide your attention and not fixate on any one item, such as the ground reference point.

d. **Configuration and airspeed prior to entry**

1) Prior to entry, the airplane should be in straight-and-level flight at 1,000 ft. AGL and in normal cruise configuration.

e. **Entry procedure**

1) To enter turns around a point, fly your airplane on a downwind heading to one side of the selected point at a distance equal to the desired radius of turn.

   a) To enter a left turn, keep the point to your left.

      i) To enter a right turn, keep the point to your right.

b) In a high-wing airplane (e.g., Cessna-152), the distance from the point must permit you to see the point throughout the maneuver even with the wing lowered in a bank.

i) If the radius is too large, the lowered wing will block your view of the point.

f. **Wind drift correction**

1) A constant radius around a point will, if any wind exists, require constantly changing the angle of bank and the angles of crab.

a) The closer your airplane is to a direct downwind heading where the groundspeed is greatest, the steeper the bank and the faster the rate of turn required to establish the proper crab.

b) The closer your airplane is to a direct upwind heading where the groundspeed is least, the shallower the bank and the slower the rate of turn required to establish the proper crab.

2) It thus follows that, throughout the maneuver, the bank and rate of turn must be gradually varied in proportion to the groundspeed.

g. **How to maintain desired altitude, airspeed, and distance from reference point**

1) With your airplane headed downwind, the groundspeed is the greatest. The steepest bank (about 45°) is used to attain the proper crab angle and to prevent your airplane from flying too far away from your reference point.

a) During the next 180° of turn (the downwind side), your airplane's heading is changing from a downwind to an upwind heading, and the groundspeed decreases.

b) During the downwind half of the circle, the nose of your airplane must be progressively crabbed toward the inside of the circle.

i) The crab angle will be at its maximum when the airplane is headed directly crosswind (i.e., at the 90° point).

ii) The crab is slowly taken out as your airplane progresses to a direct upwind heading.

c) Throughout the downwind side of the circle, your airplane goes from its steepest (directly downwind) to its shallowest (directly upwind) bank.

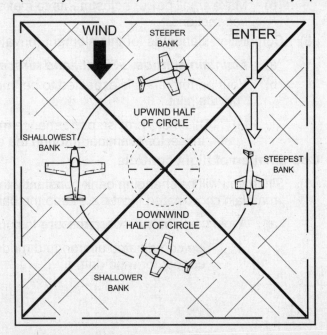

2) With your airplane headed upwind, the groundspeed is the least, requiring the shallowest bank.

   a) During the next 180° of turn (the upwind side), your airplane's heading is changing from an upwind to a downwind heading, and the groundspeed increases.

   b) During the upwind half of the circle, the nose of your airplane must be progressively crabbed toward the outside of the circle.

      i) The crab angle will be at its maximum when the airplane is headed directly crosswind.

      ii) The crab is slowly taken out as your airplane progresses to a direct downwind heading.

   c) Throughout the upwind side of the circle, your airplane goes from its shallowest (directly upwind) to its steepest (directly downwind) bank.

3) Throughout this maneuver, you should maintain a constant altitude. Since your bank will constantly change if a wind exists, you will need to adjust pitch attitude to maintain altitude.

   a) As bank increases, pitch attitude may need to be raised by back elevator pressure.

   b) As the bank decreases, release an appropriate amount of back elevator pressure to maintain altitude.

   c) Maintain pitch awareness by visual references, and use your altimeter to ensure that you are maintaining altitude.

4) Normally, this maneuver is done at cruise airspeed in trainer-type airplanes.

   a) Check your airspeed indicator to ensure that you are maintaining your entry airspeed.

   b) Using 45° as your steepest angle of bank, the load factors should not be great enough to require an addition of power to maintain a constant airspeed and altitude.

5) During this maneuver, if you maintain your altitude, your airspeed should remain within 10 kt. of your entry airspeed.

   a) Use pitch to make altitude corrections.

   b) Make small power adjustments to make any necessary airspeed corrections.

6) Completion standards for this element (private pilot only)

   a) ***Maintain altitude, ±100 ft., and airspeed, ±10 kt.***

   b) Your performance is expected to be "more precise" than that required for a private pilot.

      i) This more precise performance must be determined by your FAA inspector/examiner through the exercise of subjective judgment.

h. **Coordination of flight controls**

1) Since you will be changing bank constantly throughout this maneuver, you must maintain coordinated flight (i.e., keep the ball centered).

   a) Avoid using only rudder pressure to correct for wind drift.

      i) Use coordinated aileron and rudder to increase or decrease bank to correct for wind drift.

2. **Exhibit your instructional knowledge by explaining the following common errors related to turns around a point.**

    a. **Faulty entry procedure**

       1) Entry should be done on a downwind heading. By doing this, you can establish your steepest angle of bank at the start of the maneuver.

         a) If you attempt to enter this maneuver at any other point, the radius of the turn must be carefully selected, taking into account the wind velocity and groundspeed so that an excessive angle of bank is not required later on to maintain the proper ground track.

       2) When entering downwind, if the steepest bank is not used, the wind will blow your airplane too far from your reference point to maintain a constant radius.

    b. **Poor planning, orientation, or division of attention**

       1) Poor planning results in not changing the bank required to counteract drift to effect a circle of equal radius about a reference point.

       2) Poor orientation is usually the result of not selecting a prominent reference point, thus losing sight of the point.

         a) It may also be not knowing the wind direction, thus becoming disoriented as to an upwind and a downwind heading.

       3) Poor division of attention contributes to an inability to maintain a proper ground track, altitude, and/or airspeed.

         a) Also, you may not notice other aircraft that have entered the area near you.

    c. **Uncoordinated flight control application**

       1) This error normally occurs when you begin to fixate on your reference point and then forget to use the flight controls in a coordinated manner.

       2) Do not attempt to crab your airplane by using only rudder pressure.

       3) Maintain a coordinated flight condition (i.e., keep the ball centered) throughout this maneuver.

    d. **Improper correction for wind drift**

       1) You should use the steepest bank (i.e., 45°) when heading directly downwind.

         a) During the downwind side, the bank will gradually decrease as you approach an upwind heading.

           i) The nose of the airplane will be crabbed toward the inside of the circle.

       2) The bank should be the shallowest when you are heading directly upwind.

         a) During the upwind side, the bank will gradually increase as you approach a downwind heading.

           i) The nose of the airplane will be crabbed toward the outside of the circle.

       3) Do not attempt to keep the wing on the reference point throughout the maneuver.

    e. **Failure to maintain selected altitude or airspeed**

       1) Most student pilots will initially have trouble maintaining altitude or airspeed due to their inexperience in dividing their attention.

         a) Learn to divide your attention between the reference point and airplane control (e.g., pitch awareness).

       2) By maintaining altitude, you should be able to maintain your selected airspeed when entering at normal cruise power setting.

f.   **Selection of a ground reference point where there is no suitable emergency landing area within gliding distance**

1)   This is a part of poor planning, so stress to your student that (s)he should always be prepared for any type of emergency.

2)   Demonstrate how to select a ground reference point that is within gliding distance to make a safe emergency landing.

3)   When your student identifies a ground reference point, ask him/her to also identify his/her selection of an emergency landing area.

3.   **Demonstrate and simultaneously explain turns around a point from an instructional standpoint.**

a.   Your demonstration must be to the commercial pilot skill level.

b.   You will be evaluated on your ability to simultaneously explain and demonstrate the key elements of this task.

c.   If you do not perform the maneuver to FAA completion standards, point out the deficiency to your FAA inspector/examiner as an example of an error, and explain how it should be corrected (including how and why it occurred).

4.   **Analyze and correct simulated common errors related to turns around a point.**

a.   As your FAA inspector/examiner is performing this task, you must be able to recognize, analyze, and correct any error.

b.   Explain what the error is, and then explain what (s)he should do to correct the error.

c.   If your FAA inspector/examiner does not perform the maneuver to FAA completion standards, point out the deficiency to him/her as an example of a student error, and explain how it should be corrected (including how and why it occurred).

END OF TASK

EIGHTS ON PYLONS

X.D. TASK: EIGHTS ON PYLONS

REFERENCES: FAA-H-8083-3; FAA-S-8081-12.

Objective. To determine that the applicant:

1. Exhibits instructional knowledge of the elements of eights on pylons by describing--

   a. How to determine the approximate pivotal altitude.

   b. How to select suitable pylons with consideration given to emergency landing areas.

   c. Orientation, division of attention, and planning.

   d. Configuration and airspeed prior to entry.

   e. Relationship of groundspeed change to the performance of the maneuver.

   f. Pilot's "line-of-sight" reference to the pylon.

   g. Entry procedure.

   h. Procedure for maintaining "line-of-sight" on the pylon.

   i. Proper planning for turn entries and rollouts.

   j. How to correct for wind drift between pylons.

   k. Coordination of flight controls.

2. Exhibits instructional knowledge of common errors related to eights on pylons by describing--

   a. Faulty entry procedure.

   b. Poor planning, orientation, and division of attention.

   c. Uncoordinated flight control application.

   d. Use of an improper "line-of-sight" reference.

   e. Application of rudder alone to maintain "line-of-sight" on the pylon.

   f. Improper planning for turn entries and rollouts.

   g. Improper correction for wind drift between pylons.

   h. Selection of pylons where there is no suitable emergency landing area within gliding distance.

3. Demonstrates and simultaneously explains eights on pylons from an instructional standpoint.

4. Analyzes and corrects simulated common errors related to eights on pylons.

A. General Information

   1. The objective of this task is to determine your instructional knowledge of the elements and common errors related to eights on pylons. You will be required to simultaneously demonstrate and explain this task from an instructional standpoint.

   2. This is an advanced training maneuver that provides practice in developing coordination skills while the student's attention is directed at maintaining a pivotal position on a selected pylon.

   3. The objective of eights on pylons is to develop your ability to maneuver your airplane accurately while dividing your attention between the flight path and the selected points on the ground.

   4. This maneuver involves flying the airplane in circular paths, alternating left and right, in the form of a figure "8" around two selected points or pylons on the ground.

      a. Your student may benefit from practicing on one pylon in a continuous circle, a while on each side, before attempting to fly both pylons in the normal sequence.

B.   Task Objectives

1.   **Exhibit your instructional knowledge by explaining the following elements of eights on pylons.**

   a.   **How to determine the approximate pivotal altitude**

   1)   At any altitude above the pivotal altitude, your reference line will appear to move rearward in relation to the pylon.

   a)   When your airplane is below the pivotal altitude, your reference line will appear to move forward in relation to the pylon.

   2)   You can determine the approximate pivotal altitude by flying at an altitude well above the pivotal altitude.

   a)   Reduce power and begin a descent at cruise speed in a continuous medium bank turn around the pylon.

   i)   The apparent backward travel of your reference line with respect to the pylon will slow down as altitude is lost, stop for an instant, and then move forward if the descent is continued.

   b)   The altitude at which your reference line apparently ceases to move but actually pivots on the pylon is the pivotal altitude.

   c)   If your airplane descends below pivotal altitude, power should be added to maintain airspeed while altitude is regained to the pivotal altitude.

   3)   Some flight instructors recommend the following formula to determine the approximate pivotal altitude.

$$\frac{Groundspeed\ (kt.)^2}{11.3} = Pivotal\ Altitude$$

   a)   EXAMPLE: If your ground speed is 90 kt., then your pivotal altitude is

$$\frac{(90)^2}{11.3} = \frac{8100}{11.3} = 717\ ft.$$

   b)   If you are using miles per hour, use 15 instead of 11.3 as the denominator.

   b.   **How to select suitable pylons with consideration given to emergency landing areas**

   1)   Select two pylons along a line perpendicular to the wind direction. The pylons should be easily recognized during the maneuver and located away from populated areas. They should also be spaced far enough apart so your student has time to plan the turns.

   a)   The pylons should be about 1/2 SM apart, depending on your airplane's speed.

   b)   You should have enough space between the pylons to allow a few moments of straight-and-level flight as you transition from one pylon to the other. This length of time allows you to look for traffic and plan your next turn.

   2)   When selecting suitable pylons, consider possible emergency landing areas within gliding distance, such as open fields or other nonwooded areas.

   a)   There is little time available to search for a suitable field for landing in the event the need arises, e.g., an engine failure.

3) Perform **clearing turns** and carefully scan for other traffic.

   a) Look for obstacles, i.e., towers, since your altitude will be below 1,000 feet AGL.

c. **Orientation, division of attention, and planning**

1) You must maintain orientation of the location of your pylons and the direction of the wind relative to your heading.

2) As in any maneuver, you should always be planning ahead.

3) You must divide your attention between coordinated airplane control and outside visual references.

   a) Constantly be aware of the location of your line-of-sight reference in relation to the pylon.

4) While performing this maneuver, you are still responsible for collision avoidance.

d. **Configuration and airspeed prior to entry**

1) You should be in the specified configuration and at the airspeed listed in your POH.

   a) If none is listed, use your normal cruise configuration and airspeed.

2) Complete the following table with your CFI for eights on pylons planning:

Power MP: _____ RPM: _____	Airspeed: _____	Initial Altitude: _____

e. **Relationship of groundspeed change to the performance of the maneuver**

1) There is a specific altitude at which, when your airplane turns at a given groundspeed, a projection of your line-of-sight reference line to the selected point on the ground will appear to pivot on that point, i.e., pivotal altitude.

   a) The higher the groundspeed, the higher the pivotal altitude.
   b) The lower the groundspeed, the lower the pivotal altitude.

2) Note carefully that the pivotal altitude is NOT a function of the angle of bank.

   a) In calm air, with the groundspeed constant around the pylon, the pivotal altitude would be constant throughout the maneuver.

   b) You could start the maneuver closer to the pylon with a resulting higher bank angle, and the pivotal altitude would NOT be different.

3) The strength (or speed) of the wind will affect the altitude changes to maintain the pivotal altitude.

   a) In a strong wind, the altitude changes will be greater (e.g., 100 to 200 ft.).
   b) In a light wind, the altitude changes will be smaller (e.g., 50 to 100 ft.).

f.   **Pilot's "line-of-sight" reference to the pylon**

1)   In the correct performance of eights on pylons, you should use a sighting reference line which, from eye level, parallels the lateral axis of your airplane.

2)   High-wing, low-wing, swept-wing, and taper-wing airplanes, as well as those with tandem (fore and aft) or side-by-side seating, will all present different angles from your eyes to the wingtip.

   a)   Your line-of-sight reference, while not necessarily on the wingtip itself, may be positioned in relation to the wingtip (ahead, behind, above, or below), as shown below.

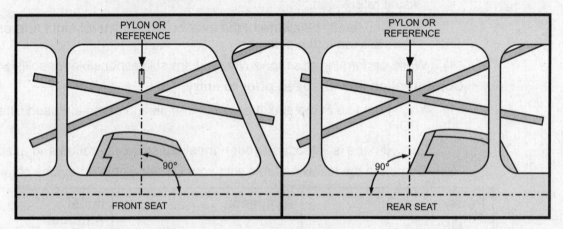

3)   The effects of using a correct and an incorrect line-of-sight reference to the pylon are shown below.

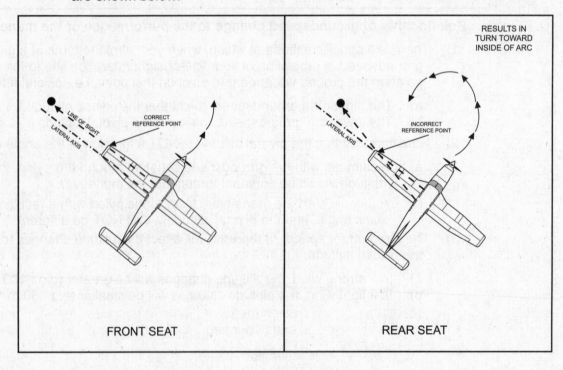

g. **Entry procedure**

1) Enter at the appropriate altitude, as calculated in your planning.

2) Enter the maneuver by flying diagonally crosswind between the pylons to a point downwind from the first pylon and then make the first turn into the wind.

   a) As your airplane approaches a position where the pylon appears to be just ahead of your line-of-sight reference, the turn should be started to keep the line-of-sight reference on the pylon.

h. **Procedure for maintaining "line-of-sight" on the pylon**

1) The airplane is flown at an altitude and airspeed such that a line parallel to the airplane's lateral axis and extending from your eye appears to pivot on each of the pylons, as shown below.

   a) No attempt is made to maintain a uniform distance from the pylon.

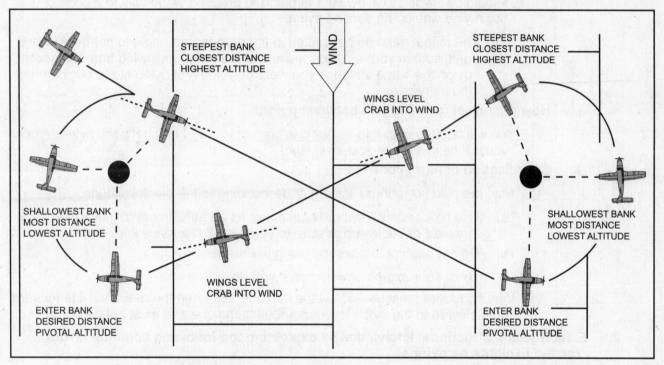

2) Since your headings throughout the turns continually vary from directly downwind to directly upwind, your groundspeed will constantly change.

   a) This will result in the proper pivotal altitude varying throughout the eight.

3) As your airplane heads into the wind, the groundspeed decreases; consequently, the pivotal altitude decreases.

   a) Thus, you must descend to hold your reference line on the pylon.

4) As the turn progresses on the upwind side of the pylon, the wind becomes more of a crosswind and drifts your airplane closer to the pylon.

   a) Since a constant distance from the pylon is not required, no correction to counteract drifting should be applied.

   b) Thus, with the airplane drifting closer to the pylon, the angle of bank must be increased to hold your reference line on the pylon.

5) If the reference line appears to move ahead of the pylon, you should increase altitude.

    a) If the reference line appears to move behind the pylon, you should decrease altitude.

6) Completion standard for this element (commercial pilot only)

    a) ***Enter the maneuver at the appropriate altitude and airspeed, such that the bank angle will be approximately 30° to 40° at the steepest point.***

    b) ***Apply the necessary corrections so that the line-of-sight reference line remains on the pylon.***

i. **Proper planning for turn entries and rollouts**

1) As your airplane turns toward a downwind heading, the rollout from the turn should be started to allow your airplane to proceed diagonally to a point on the downwind side of the second pylon.

    a) The rollout must be completed in the proper crab angle to correct for wind drift, so that your airplane will arrive at a point downwind from the second pylon the same distance you were from the first pylon at the beginning of the maneuver.

j. **How to correct for wind drift between pylons**

1) You will need to crab into the wind to correct for wind drift between pylons since you will be in straight-and-level flight.

k. **Coordination of flight controls**

1) Hold the pylon by either climbing or descending to the pivotal altitude.

    a) Use rudder only for coordination, not as an adjustment to move the pylon forward or backward relative to your visual reference line.

    b) Do not attempt to cheat by using the rudder.

        i) Your examiner will know if you do.

2) Varying rudder pressure to yaw the airplane and force the reference line forward or backward to the pylon is a dangerous technique and must not be done.

2. **Exhibit your instructional knowledge by explaining the following common errors related to eights on pylons.**

a. **Faulty entry procedure**

1) This error is normally due to poor planning, such as

    a) Not being at the approximate pivotal altitude;

    b) Rolling into the bank too early, which requires you to shallow the bank and then roll back in; and

    c) Being too close to the pylon and exceeding approximately 40° of bank.

b. **Poor planning, orientation, and division of attention**

1) Poor planning is recognized by faulty entry technique and/or lack of anticipation of the changes in groundspeed.

2) Poor orientation is recognized when you forget the wind direction or lose a pylon.

    a) Poor pylon selection will normally result in losing a pylon.

3) Poor division of attention is recognized by uncoordinated flight control applications and failure to watch for traffic.

   c. **Uncoordinated flight control application**

     1) This error is normally due to poor division of attention.

   d. **Use of improper "line-of-sight" reference**

     1) The proper line-of-sight reference is parallel to the lateral axis of the airplane to the pylon -- your eye level.

     2) This visual reference line will be different for pilots seated in different positions.

   e. **Application of rudder alone to maintain "line-of-sight" on the pylon**

     1) This is the most common error in attempting to hold a pylon.

     2) You should not press the inside rudder to yaw the wing backward if the pylon is behind the reference line.

        a) When the pylon is ahead of the reference line, you should not press the outside rudder to yaw the wing forward.

     3) Use rudder for coordination only.

   f. **Improper planning for turn entries and rollouts**

     1) This error is due primarily to poor planning.

     2) When the pylon is just ahead of the reference line, the entry is started.

     3) Rollout should be timed to allow the airplane to proceed diagonally to a point on the downwind side of the second pylon.

   g. **Improper correction for wind drift between pylons**

     1) Your student must track a line to the next pylon by imagining a reference line.

     2) An improper correction has occurred when your student arrives at the next pylon at a distance different from the distance at the first pylon.

   h. **Selection of pylons where there is no suitable emergency landing area within gliding distance**

     1) This is a part of poor planning, so stress to your student that (s)he should always be prepared for any type of emergency.

     2) Demonstrate how to select appropriate pylons that are within gliding distance to make a safe emergency landing.

     3) When your student identifies the pylons, ask him/her to also identify his/her selection of an emergency landing area.

3. **Demonstrate and simultaneously explain eights on pylons from an instructional standpoint.**

   a. Your demonstration must be to the commercial pilot skill level.

   b. You will be evaluated on your ability to simultaneously explain and demonstrate the key elements of this task.

   c. If you do not perform the maneuver to FAA completion standards, point out the deficiency to your FAA inspector/examiner as an example of an error, and explain how it should be corrected (including how and why it occurred).

4. **Analyze and correct simulated common errors related to eights on pylons.**

   a. As your FAA inspector/examiner is performing this task, you must be able to recognize, analyze, and correct any error.

   b. Explain what the error is, and then explain what (s)he should do to correct the error.

   c. If your FAA inspector/examiner does not perform the maneuver to FAA completion standards, point out the deficiency to him/her as an example of a student error, and explain how it should be corrected (including how and why it occurred).

END OF TASK -- END OF STUDY UNIT

STUDY UNIT XI
SLOW FLIGHT, STALLS, AND SPINS

This study unit explains the eight tasks (A-H) of Slow Flight, Stalls, and Spins. These tasks include both knowledge and skill. Your FAA inspector/examiner is required to examine you on at least one proficiency stall (Task B or C), at least one demonstration stall (Task D, E, F, or H), and Task G, Spins. At the discretion of your FAA inspector/examiner, your logbook endorsement attesting to your instructional competency in spins may be accepted in place of Task G. Task A is not required but is normally evaluated during your performance of the six other tasks in this area of operation.

The intent of Tasks B and C (proficiency) is to ensure that you are tested on proficiency for the purpose of teaching these tasks to students. These are the stalls (i.e., power-on and power-off) that your students will be required to perform on their practical test. The intent of Tasks D, E, F, and H (demonstration) is to ensure that you are knowledgeable about the maneuvers and can demonstrate them to students for both familiarization and stall/spin awareness purposes. Your students will not be required to perform these stalls on their practical test but should understand them for stall/spin awareness.

MANEUVERING DURING SLOW FLIGHT

XI.A. TASK: MANEUVERING DURING SLOW FLIGHT

    REFERENCES: FAA-H-8083-3; FAA-S-8081-12; FAA-S-8081-14, POH/AFM.

Objective. To determine that the applicant:

1. Exhibits instructional knowledge of the elements of maneuvering during slow flight by describing --

    a. Relationship of configuration, weight, center of gravity, maneuvering loads, angle of bank, and power to flight characteristics and controllability.

    b. Relationship of the maneuver to critical flight situations, such as go-around.

    c. Performance of the maneuver with selected landing gear and flap configurations during straight-and-level flight and level turns.

    d. Specified airspeed for the maneuver.

    e. Coordination of flight controls.

    f. Trim technique.

    g. Re-establishment of cruising flight.

2. Exhibits instructional knowledge of common errors related to maneuvering during slow flight by describing --

    a. Failure to establish specified gear and flap configuration.

    b. Improper entry technique.

    c. Failure to establish and maintain the specified airspeed.

    d. Excessive variations of altitude and heading when a constant altitude and heading are specified.

    e. Rough or uncoordinated control technique.

    f. Improper correction for torque effect.

    g. Improper trim technique.

    h. Unintentional stalls.

    i. Inappropriate removal of hand from throttles.

3. Demonstrates and simultaneously explains maneuvering during slow flight from an instructional standpoint.

4. Analyzes and corrects simulated common errors related to maneuvering during slow flight.

A. General Information

    1. The objective of this task is to determine your instructional knowledge of the elements and common errors related to maneuvering during slow flight. You will be required to demonstrate and simultaneously explain this task from an instructional standpoint.

B.   Task Objectives

1.   **Exhibit your instructional knowledge by explaining the following elements of maneuvering during slow flight.**

a.   **Relationship of configuration, weight, center of gravity, maneuvering loads, angle of bank, and power to flight characteristics and controllability**

1)   When performing slow flight at any speed, it is important to know the relationship between parasite drag, induced drag, and the power needed to maintain a given altitude (or climb angle or glide slope) at a selected airspeed.

a)   The following graph illustrates the variations in parasite, induced, and total drag with airspeed for a typical airplane in steady flight.

i)   The amount of drag present at a given airspeed is equal to the amount of thrust required to maintain level flight at that airspeed and angle of attack.

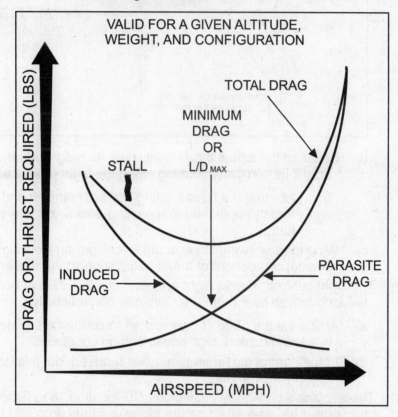

VALID FOR A GIVEN ALTITUDE, WEIGHT, AND CONFIGURATION

TOTAL DRAG

MINIMUM DRAG OR L/D$_{MAX}$

STALL

INDUCED DRAG

PARASITE DRAG

DRAG OR THRUST REQUIRED (LBS)

AIRSPEED (MPH)

b)   As the airspeed decreases from cruise to L/D$_{MAX}$, total drag and thrust required decrease to maintain a constant altitude, as shown in the figure above.

c)   As the airspeed decreases below L/D$_{MAX}$, additional power (thrust) is required to maintain a constant altitude.

i)   Here, total drag increases because induced drag increases faster (due to higher angle of attack) than parasite drag decreases.

ii)   This is known as the **backside of the power curve** or the **region of reverse command**.

•   The region of reverse command means that more power (not less) is required to fly at slower airspeeds while maintaining a constant altitude.

2) While straight-and-level flight is maintained at a constant airspeed, thrust is equal in magnitude to drag, and lift is equal in magnitude to weight, but some of these forces are separated into components.

   a) In slow flight, thrust no longer acts parallel to and opposite to the flight path and drag, as shown below. Note that thrust has two components:

      i) One acting perpendicular to the flight path in the direction of lift

      ii) One acting along the flight path

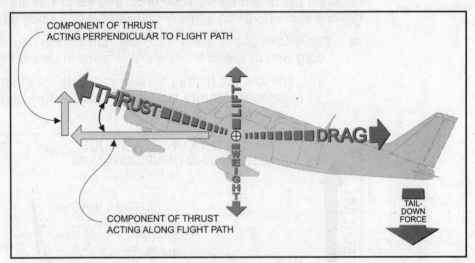

b) Because the actual thrust is inclined, its magnitude must be greater than drag if its component acting along the flight path is equal to drag.

      i) Note that the forces acting upward (wing lift and the component of thrust) equal the forces acting downward (weight and tail down force).

c) Wing loading (wing lift) is actually less during slow flight because the vertical component of thrust helps support the airplane.

3) The flight controls in slow flight are less effective than at normal cruise due to the reduced airflow over them (i.e., airplane controllability).

   a) Anticipate the need of right rudder to counteract the left-turning tendencies in a low airspeed, high power setting condition.

   b) Large control movements may be required, but this does not mean rough or jerky movements.

4) The airplane's configuration, weight, CG location, load factors, angle of bank, and power have an effect on the airplane's total drag and thus the power or lift required to maintain level flight.

   a) Additionally, these factors have the same effect on flight characteristics and controllability in slow flight as they do in effecting the stall speed of the airplane.

      i) For additional information on these effects, see Task XI.B. Power-On Stalls, beginning on page 416.

b. **Relationship of the maneuver to critical flight situations, such as go-around**

   1) This maneuver demonstrates the flight characteristics and degree of controllability of the airplane in slow flight.

      a) It is of great importance that you know the characteristic control responses of your airplane during slow flight.

      b) You must develop this awareness in order to avoid stalls in any airplane that you may fly at the slower airspeeds which are characteristic of takeoffs, climbs, landings, and go-arounds.

c. **Performance of the maneuver with selected landing gear and flap configurations during straight-and-level flight and level turns**

   1) Select an entry altitude that will allow this maneuver to be performed no lower than 1,500 ft. AGL or the manufacturer's recommended altitude, whichever is higher.

      a) Select an altitude that is easy to read from your altimeter.

         i) If the terrain elevation is 300 ft. above sea level, the FAA requires maneuver to be performed no lower than 1,800 ft. MSL (1,500 ft. AGL). Round this to the nearest 500-ft. increment (2,000 ft. MSL) to make it easier to identify on your altimeter.

      b) Before you begin this task, your examiner should specify the airplane configuration to use (i.e., full flaps, partial flaps, gear up, gear down). If not, then ask your examiner for the desired configuration

         i) During this task, your examiner may have you change the airplane configuration to evaluate your knowledge of the elements related to slow flight.

   2) Maintain your scan for other air traffic in your area, and perform **clearing turns**.

   3) Before you begin demonstrating this maneuver to your student (FAA inspector/ examiner), explain what airplane configuration you will use (i.e., full flaps, partial flaps, no flaps, gear up, gear down).

      a) Both the private and commercial PTSs state that the examiner will specify the configuration to use.

      b) When having your student perform this maneuver, ensure that (s)he asks what configuration to use.

   4) Begin slowing the airplane by gradually reducing power from the cruise power setting.

      a) While the airspeed is decreasing, the position of the nose in relation to the horizon should be noted and should be raised as necessary to maintain altitude.

   5) When the airspeed reaches the maximum allowable for landing gear operation (V_{LO}), the landing gear (if retractable) should be extended as appropriate.

      a) Perform all gear-down checks, e.g., three in green.
      b) In your airplane, V_{LO} _____.

   6) As the airspeed reaches the maximum allowable speed for flap operation (V_{FE}), full flaps should be incrementally lowered to a predetermined setting.

      a) This will allow you to maintain pitch control of your airplane as flaps are extended.

      b) In your airplane, V_{FE} _____.

   7) Additional power will be required as airspeed decreases below L/D_{MAX} to maintain altitude.

8) When the desired airspeed and pitch attitude have been established, it is important to continually cross-check the attitude indicator, altimeter, and airspeed indicator, as well as outside references, to ensure that accurate control is being maintained.

9) You should maintain straight-and-level flight and level turns at a constant altitude.

    a) During the turns, the pitch attitude and power may need to be increased to maintain airspeed and altitude.

10) The private PTS also requires the pilot to perform straight and turning climbs and descents. Thus, if you are demonstrating this maneuver to a student pilot, you must be able to meet these requirements also.

    a) Adjust the power to begin the desired climb or descent, and simultaneously adjust the pitch attitude as necessary to maintain the desired airspeed.

        i) Throughout the maneuver, remain in coordinated flight by using the necessary control pressures.

    b) Remember to avoid the natural tendency to pull back on the control during slow flight when more altitude is needed because the increase in the angle of attack may cause the airplane to stall.

        i) You will gain altitude by increasing power and adjusting pitch to maintain airspeed.

        • In some situations, you may actually pitch down to maintain airspeed in a climb.

11) Completion standards for this element

    a) Commercial PTS

        i) ***Accomplish coordinated straight-and-level flight, turns, climbs, and descents with landing gear and flap configurations specified by the examiner.***

        ii) ***Maintain the specified altitude, ±50 ft.***

        iii) ***Maintain the specified heading during straight flight, ±10°.***

        iv) ***Maintain the specified bank angle, ±5°, during turning flight.***

        v) ***Roll out on the specified headings, ±10°.***

    b) Private PTS

        i) Accomplish coordinated straight-and-level flight, turns, climbs, and descents with landing gear and flap configurations specified by the examiner.

        ii) Maintain the specified altitude, ±100 ft., the specified heading, ±10°, and the specified angle of bank, ±10°.

  d. Specified airspeed for the maneuver

1) Both the private and commercial PTSs state that the slow flight airspeed is an airspeed at which any further increase in angle of attack or load factor or any further reduction in power (while maintaining a constant altitude) will result in a stall.

    a) In your airplane, slow flight airspeed _____.

2) Completion standards for this element

    a) Commercial PTS

        i) ***Establish and maintain an airspeed at which any further increase in angle of attack, increase in load factor, or reduction in power would result in an immediate stall, ±5/–0 kt.***

    b) Private PTS

        i) Establish and maintain an airspeed at which any further increase in angle of attack, increase in load factor, or reduction in power would result in an immediate stall, +10/–0 kt.

e. **Coordination of flight controls**

1) Throughout the maneuver, remain in coordinated flight by using the necessary control pressures.

    a) Use the proper amount of rudder pressure to keep the ball of the inclinometer centered.

        i) Right rudder pressure will be necessary to counteract the left-turning tendencies.

2) When you are performing this maneuver, it is important to continually cross-check the attitude indicator, the altimeter, the airspeed indicator, and the ball of the turn coordinator, as well as outside references, to ensure that accurate control is being maintained.

    a) Do not become focused on one item, e.g., the altimeter.

3) You must also divide your attention to watch for other aircraft in your area, i.e., collision avoidance.

f. **Trim technique**

1) As the flight conditions change, it is important to retrim your airplane as often as necessary to compensate for changes in control pressures.

g. **Re-establishment of cruising flight**

1) To return to cruise flight, you should

    a) Increase power to the maximum allowable.

    b) Adjust pitch to maintain altitude.

    c) If at full flaps, partially raise the flaps. Use the setting recommended by the manufacturer for a go-around.

    d) Raise the landing gear, if retractable.

    e) Raise the flaps to 0°.

    f) Maintain coordinated flight throughout the airspeed transition.

    g) Adjust power to the cruise power setting at the appropriate airspeed.

    h) Use trim throughout the recovery to cruising flight.

2) Note that neither the private nor the commercial PTS requires the pilot to return to cruising flight from slow flight.

    a) This may be because most examiners will have the pilot go into slow flight and then execute a power-off stall.

2. **Exhibit your instructional knowledge by explaining the following common errors related to maneuvering during slow flight.**

    a. **Failure to establish specified gear and flap configuration**

        1) This maneuver can be performed in various configurations of landing gear (if retractable) and flaps.

        2) Your student should form a habit of repeating your instructions for all maneuvers to ensure that (s)he understands your instructions.

    b. **Improper entry technique**

        1) To begin this maneuver, reduce power and gradually raise the nose. Use carburetor heat, if applicable.

        2) When the desired airspeed is attained, increase power and adjust both power and pitch to maintain airspeed and altitude.

            a) Anticipate the need of right rudder to counteract the effect of torque as power is applied.

        3) Retrim the airplane as often as necessary.

    c. **Failure to establish and maintain the specified airspeed**

        1) This error is caused by the improper use of power and pitch adjustments.

    d. **Excessive variations of altitude and heading when a constant altitude and heading are specified**

        1) It is important to continually cross-check the attitude indicator, altimeter, and airspeed indicator, as well as outside references, to ensure that accurate control is being maintained.

    e. **Rough or uncoordinated control technique**

        1) A stall may occur as a result of abrupt or rough control movements.

        2) Uncoordinated control technique could risk the possibility of a crossed-control stall.

    f. **Improper correction for torque effect**

        1) Remember that torque effects are magnified at low airspeeds, high angles of attack, and high power settings; all of these conditions are present during slow flight.

        2) Accordingly, right rudder pressure will be required to keep the ball centered during straight flight, and may even be needed in a left turn.

    g. **Improper trim technique**

        1) Trim should be used to relieve control pressures.

        2) Faulty trim technique may be evidenced by poor altitude control and the student's tiring quickly.

    h. **Unintentional stalls**

        1) A stall may be caused by uneven or sudden control inputs.

        2) You must maintain your smooth control technique.

        3) Check airspeed frequently.

    i. **Inappropriate removal of hand from throttles**

        1) You should keep your hand on the throttle control at all times unless making an adjustment, such as trim.

3. **Demonstrate and simultaneously explain maneuvering during slow flight from an instructional standpoint.**

    a. Your demonstration must be to commercial pilot skill level.

    b. You will be evaluated on your ability to explain and demonstrate simultaneously the key elements of this task.

    c. If you do not perform the maneuver to FAA completion standards, point out the deficiency to your FAA inspector/examiner as an example of an error, and explain how it should be corrected (including how and why it occurred).

4. **Analyze and correct simulated common errors related to maneuvering during slow flight.**

    a. As your FAA inspector/examiner is performing this task, you must be able to recognize, analyze, and correct any error.

    b. Explain what the error is, and then explain what (s)he should do to correct the error.

    c. If your FAA inspector/examiner does not perform the maneuver to FAA completion standards, point out the deficiency to him/her as an example of a student error, and explain how it should be corrected (including how and why it occurred).

END OF TASK

POWER-ON STALLS (PROFICIENCY)

XI.B. TASK: POWER-ON STALLS (PROFICIENCY)

REFERENCES: AC 61-67; FAA-H-8083-3; FAA-S-8081-12; FAA-S-8081-14, POH/AFM.

Objective. To determine that the applicant:

1. Exhibits instructional knowledge of the elements of power-on stalls, in climbing flight (straight or turning) with selected landing gear and flap configurations by describing --

    a. Aerodynamics of power-on stalls.

    b. Relationship of various factors such as landing gear and flap configuration, weight, center of gravity, load factor, and bank angle to stall speed.

    c. Flight situations where unintentional power-on stalls may occur.

    d. Entry technique and minimum entry attitude.

    e. Performance of power-on stalls in climbing flight (straight or turning).

    f. Coordination of flight controls.

    g. Recognition of the first indications of power-on stalls.

    h. Recovery technique and minimum recovery altitude.

2. Exhibits instructional knowledge of common errors related to power-on stalls, in climbing flight (straight or turning), with selected landing gear and flap configurations by describing --

    a. Failure to establish specified landing gear and flap configuration prior to entry.

    b. Improper pitch, heading, and bank control during straight ahead and turning stalls.

    c. Improper pitch and bank control during turning stalls.

    d. Rough or uncoordinated control procedure.

    e. Failure to recognize the first indications of a stall.

    f. Failure to achieve a stall.

    g. Improper torque correction.

    h. Poor stall recognition and delayed recovery.

    i. Excessive altitude loss or excessive airspeed during recovery.

    j. Secondary stall during recovery.

3. Demonstrates and simultaneously explains power-on stalls, in climbing flight (straight or turning), with selected landing gear and flap configurations, from an instructional standpoint.

4. Analyzes and corrects simulated common errors related to power-on stalls, in climbing flight (straight or turning), with selected landing gear and flap configurations.

A. General Information

1. The objective of this task is to determine your instructional knowledge of the elements and common errors related to power-on stalls in climbing flight (straight or turning), with selected landing gear and flap configurations. You will be required to demonstrate and simultaneously explain this task from an instructional standpoint.

2. The objectives in performing stalls are to

    a. Familiarize your student with the conditions that produce stalls
    b. Assist in stall/spin awareness
    c. Develop the habit of taking prompt preventive or corrective action

3. The practice of stall recovery and the development of awareness of stalls with a student is of primary importance to you as an instructor.

B.   Task Objectives

1.   **Exhibit your instructional knowledge by explaining the following elements of power-on stalls in climbing flight (straight or turning), with selected landing gear and flap configurations.**

a.   **Aerodynamics of power-on stalls**

1)   A stall is a loss of lift and an increase in drag occurring when an aircraft is flown at an angle of attack greater than the angle for maximum lift. The angle of attack for maximum lift is also called the **critical angle of attack**.

a)   Thus, a stall occurs whenever the critical angle of attack is exceeded.

2)   When the angle of attack is increased to approximately 18° to 20° on most airfoils, the airstream can no longer follow the upper curvature of the wing because of the excessive change in direction. This is the critical angle of attack.

a)   As the critical angle of attack is approached, the airstream begins separating from the rear of the upper wing surface. As the angle of attack is further increased, the airstream is forced to flow straight back, away from the top surface of the wing and from the area of highest camber. See the figure below.

b)   This causes a swirling or burbling of the air as it attempts to follow the upper surface of the wing. When the critical angle of attack is reached, the turbulent airflow, which appeared near the trailing edge of the wing at lower angles of attack, quickly spreads forward over the entire upper wing surface.

c)   This results in a sudden increase in pressure on the upper wing surface and a considerable loss of lift. Due to both this loss of lift and the increase in form drag (a larger area of the wing and fuselage is exposed to the airstream), the remaining lift is insufficient to support the airplane, and the wing stalls.

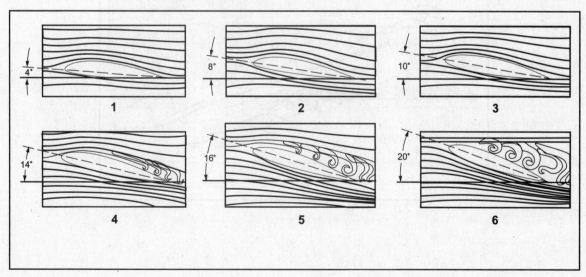

3)   Most airplanes are designed so that the wings will stall progressively outward from the wing roots to the wingtips.

a)   The wings are designed with washout; i.e., the wingtips have less angle of incidence than the wing roots.

i)   The **angle of incidence** is the angle between the chord line of the wing and the longitudinal axis of the airplane.

b)   Thus, the tips of such wings have a smaller angle of attack than the wing roots during flight.

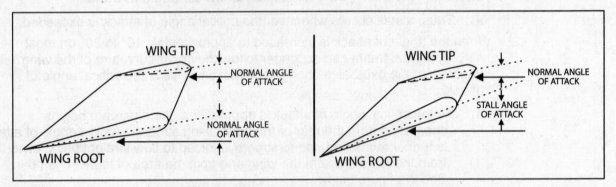

c)   A stall is caused by exceeding the critical angle of attack. Since the wing roots of an airplane will exceed the critical angle before the wingtips, the roots will stall first. The wings are designed in this manner so that control of the ailerons (which are located toward the tips of the wings) will be available at high angles of attack and give the airplane more stable stalling characteristics.

b.   **Relationship of various factors such as landing gear and flap configuration, weight, center of gravity, load factor, and bank angle to stall speed**

1)   Airspeed is controlled primarily by the elevator or longitudinal control position for a given configuration and power.

a)   If the speed is too slow, the angle of attack required for level flight will be so large that the air can no longer follow the upper curvature of the wing. The result is a separation of airflow from the wing, loss of lift, a large increase in drag, and eventually a stall if the angle of attack is not reduced.

b)   A stall can occur at ANY AIRSPEED, in ANY ATTITUDE, and at ANY POWER SETTING.

i)   Remember that a stall is the result of excessive angle of attack, not airspeed.

c)   The stall speed is the speed at which the critical angle of attack is exceeded.

2)   Configuration. Flaps, landing gear (if retractable), and other configuring devices can affect your airplane's stall speed. Flap extension will generally increase the lifting ability of the wings, thus reducing the stall speed.

a)   The effect can be seen by markings on the airspeed indicator, where the lower airspeed limit of the white arc (V_{S0}, power-off stall speed with gear and flaps in the landing configuration) is less than the lower airspeed limit of the green arc (V_{S1}, power-off stall speed with the flaps and gear up).

3)   Weight. Although the distribution of weight has the most direct effect on stability, increased gross weight can also have an effect on an airplane's flight characteristics, regardless of the CG location.

a)   The increased weight requires a higher angle of attack at a given airspeed to produce additional lift to support the weight.

b)   Thus, the critical angle of attack will be exceeded (causing the airplane to stall) at a higher airspeed.

4)   Center of gravity. Because the CG location affects both angle of attack and stability, it has a significant effect on stall speed and ease of recovery.

a)   As the CG is moved aft, the airplane flies at a lower angle of attack at a given airspeed.

i)   Thus, the critical angle of attack will be exceeded (causing the airplane to stall) at a lower airspeed.

ii)   However, the airplane is less stable because less elevator force is needed to disturb its equilibrium.

b)   With an extremely aft CG, the airplane loses its natural tendency to pitch nose down, making stall recovery more difficult.

i)   If a spin is entered, the balance of forces on the airplane may result in a flat spin.

ii)   Recovery from a flat spin may be impossible.

c)   A forward CG location will often cause the critical angle of attack to be reached (and the airplane to stall) at a higher airspeed.

i)   However, stall recovery is easier because the airplane has a greater tendency to pitch nose down.

5) Load factor. The airplane's stall speed increases in proportion to the square root of the load factor.

a) Load factor is the ratio of the lifting force produced by the wings to the actual weight of the airplane and its contents, usually expressed in Gs.

i) EXAMPLE: An airplane with a normal unaccelerated stall speed of 45 kt. can be stalled at 90 kt. when subjected to a load factor of 4 Gs.

b) A stall entered from straight-and-level flight or from an unaccelerated straight climb will not produce additional load factors.

c) In a constant altitude turn, increased load factors will cause your airplane's stall speed to increase as the angle of bank increases.

i) See Module 1.11, Loads and Load Factors, in *Pilot Handbook* for a six-page discussion on the effect of turns (bank angles) on load factor.

6) Snow, ice, or frost on the wings. Even a small accumulation of snow, ice, or frost on the airplane can cause an increase in the stall speed.

a) Such accumulation changes the shape of the wing, disrupting the smooth airflow over the surface and thus increasing drag and decreasing lift.

7) Turbulence. Turbulence can cause the airplane to stall at a significantly higher airspeed than in stable conditions.

a) A vertical gust or wind shear can cause a sudden change in the relative wind and result in an abrupt increase in angle of attack.

i) Even though a gust may not be maintained long enough for a stall to develop, the airplane may stall while you attempt to control the flight path, especially during an approach in gusty conditions.

b) When flying in moderate to severe turbulence or strong crosswinds, a higher-than-normal approach speed should be maintained.

i) In cruise flight, maintain an airspeed well above the indicated stall speed and below V_A (maneuvering speed).

c. **Flight situations where unintentional power-on stalls may occur**

1) Many stall/spin accidents have occurred during the takeoff and climbout phases of flight, particularly during go-arounds.

a) A causal factor in go-arounds has been the pilot's failure to maintain positive control due to a nose-high trim setting or premature flap retraction.

2) Failure to maintain positive control during short-field takeoffs has also been an accident factor.

d. **Entry technique and minimum entry altitude**

1) Select an entry altitude that will allow this maneuver to be completed no lower than 1,500 ft. AGL or the manufacturer's recommended altitude, whichever is higher.

a) Select an altitude that is easy to read from your altimeter.

i) If the terrain elevation is 300 ft. above sea level, the FAA requires a recovery no lower than 1,800 ft. MSL (1,500 ft. AGL). Round this to the nearest 500-ft. increment (2,000 ft. MSL) to make it easier to identify on your altimeter.

      b) Do not let yourself be rushed into performing this maneuver. If you do not feel that you can recover before 1,500 AGL, explain this to your examiner and proceed to climb to a higher altitude.

        i) During your training, you will learn how much altitude you need to perform this maneuver.

2) Perform **clearing turns** to ensure the area is clear of other traffic.

3) Power-on stalls should be performed in a takeoff configuration (e.g., for a short-field takeoff) or in a normal climb configuration (flaps and/or gear retracted).

      a) Use the recommended takeoff or normal climb configuration in your POH, unless specified otherwise by your FAA inspector/examiner.

4) Ensure that you understand which configuration your FAA inspector/examiner wants you to use. If you have any doubt about what (s)he wants, ask for clarification.

      a) At this time, you should also confirm with your FAA inspector/examiner the altitude, heading, and airspeed that you should return to after recovering from the stall.

5) Reduce power to achieve the desired airspeed while establishing the desired configuration. Maintain a constant altitude while you are slowing your airplane.

      a) The commercial PTS requires a pilot to establish the takeoff or departure configuration and slow the airplane to normal lift-off speed.

        i) In your airplane, V_R _____.

      b) The private PTS requires a pilot to establish the takeoff or departure configuration, airspeed, and power, as specified by the examiner.

      NOTE: The purpose of reducing the speed before the throttle is advanced to the recommended setting is to avoid an excessively steep nose-up attitude for a long period before your airplane stalls.

6) At this point, your airplane will be in a nose-high or climb attitude, and you should set the power at takeoff or climb power while maintaining a climb attitude.

      a) Both the private and commercial PTS require a pilot to set power to no less than 65% available power.

      b) Instructors should be particularly alert when performing power-on stalls. In some aircraft, this can result in an excessively nose high attitude which could result in a tail-slide as the stall occurs.

        i) If you are unfamiliar with the stall characteristics of a particular aircraft, seek advice from an instructor or other qualified pilot who is familiar with the aircraft.

        ii) Alternatively, you may attempt the power-on stall by selecting a much lower power setting (e.g., 50% power), establishing the aircraft on the proper attitude, then gradually applying power, up to 65%, while delaying the stall.

      c) Always adhere to any and all limitations found in the aircraft's Pilot's Operating Handbook (POH).

7) After the climb has been established, the nose is then brought smoothly upward to an attitude obviously impossible for the airplane to maintain (i.e., greater than V_X pitch attitude) and is held at that attitude until the stall occurs.

      a) Increased back elevator pressure will be necessary to maintain this attitude as the airspeed decreases.

      b) Do not use an extreme pitch attitude, which could result in loss of control.

e.   **Performance of power-on stalls in climbing flight (straight or turning)**

1)   Power-on stalls are practiced to simulate takeoff and climbout conditions and configurations.

2)   Both the private and commercial PTSs no longer differentiate between imminent and full stalls. You are required to announce the onset of a stall (i.e., imminent) and promptly recover as the stall occurs (i.e., full).

a)   When the stall occurs, you must promptly recover so your airplane does not remain in a stalled condition.

3)   Power-on stalls should be performed in straight flight or turning flight using a 20° angle of bank.

f.   **Coordination of flight controls**

1)   Throughout this maneuver, you should maintain coordinated flight.

2)   In straight flight, maintain directional control with the rudder, the wings held level with the ailerons, and a constant pitch attitude with the elevator.

a)   In turning flight, maintain coordinated flight while using a bank angle of 20°.

b)   Increasing right rudder pressure will be required during this maneuver as the airspeed decreases to counteract torque.

3)   The likelihood of stalling in uncoordinated flight is increased during a power-on stall due to the greater torque from high pitch attitude, high power setting, and low airspeed.

a)   A power-on stall will often result in one wing dropping.

b)   Maintaining directional control with the rudder is vital to avoiding a spin.

4)   Completion standards for this element

a)   Commercial PTS

i)   ***Maintain a specified heading, ±5°, in straight flight; maintain a specified angle of bank, not to exceed a 20° angle of bank, ±10°, in turning flight, while inducing the stall.***

b)   Private PTS

i)   Maintain a specified heading, ±10°, in straight flight; maintain a specified angle of bank not to exceed 20°, ±10°, in turning flight, while inducing the stall.

g.   **Recognition of the first indications of power-on stalls**

1)   Both the private and commercial PTSs require the pilot to recognize and announce the first aerodynamic indications of the oncoming stall, i.e., buffeting or decay of control effectiveness.

a)   Buffeting is caused by the turbulent air from the wings flowing back over the fuselage and horizontal stabilizer.

b)   The feeling of control pressures is also very important. As speed is reduced, the "live" resistance to pressures on the controls becomes progressively less.

i)   The airplane controls become less and less effective as one approaches the critical angle of attack.

ii)   In a complete stall, all controls can be moved with almost no resistance and with little immediate effect on the airplane.

2) Other signs used in stall recognition include vision, hearing, kinesthesia, and stall warning indicators in the airplane.

    a) Vision is useful in detecting a stall condition by noting the attitude of the airplane and the airspeed approaching stall speed. This sense can be fully relied on only when the stall is the result of an intentional unusual attitude of the airplane.

    b) Hearing is also helpful in sensing a stall condition since the tone level and intensity of sounds incidental to flight decrease as the airspeed decreases.

    c) Kinesthesia, or the mind's sensing of changes in direction or speed of motion, is probably the most important and the best indicator to the trained and experienced pilot. If this sensitivity is properly developed, it will warn of a decrease in speed or the beginning of a settling or "mushing" of the airplane.

    d) Many airplanes are equipped with stall warning devices (e.g., a horn) to alert the pilot when the airflow over the wing(s) approaches a point that will not allow lift to be sustained.

h. Recovery technique and minimum recovery altitude

1) Though the recovery actions must be taken in a coordinated manner, they are broken down into three steps here for explanation purposes.

2) First, the key factor in recovering from a stall is regaining positive control of your airplane by reducing the angle of attack.

    a) Since the basic cause of a stall is always an excessive angle of attack, the cause must be eliminated by releasing the back elevator pressure that was necessary to attain that angle of attack or by moving the elevator control forward.

        i) Each airplane may require a different amount of forward pressure.

        ii) Too much forward pressure can hinder the recovery by imposing a negative load on the wing.

    b) The object is to reduce the angle of attack but only enough to allow the wing to regain lift. Remember that you want to minimize your altitude loss.

3) Second, promptly and smoothly apply maximum allowable power to increase airspeed and to minimize the loss of altitude. In most airplanes, this will be full power, but do not exceed the RPM red line speed.

    a) Since the throttle is already at the takeoff or climb power setting, the addition of power will be relatively slight, if any.

        i) Use this step to confirm that you have maximum allowable power.

4) Third, straight-and-level flight should be established with coordinated use of the controls.

    a) At this time, the wings should be leveled, if they were previously banked.

    b) Do not attempt to deflect the ailerons until the angle of attack has been reduced.

        i) The adverse yaw caused by the downward aileron may place the airplane in uncoordinated flight, and if the airplane is still in a stalled condition, a spin could be induced.

      c)   To maintain level flight normally requires the nose of your airplane to be in a nose-high attitude.

         i)   Thus, after the initial reduction in the angle of attack, the pitch should be adjusted to that required for a climb at V_X or V_Y, as appropriate.

      d)   Maintain coordinated flight throughout the recovery.

   5)   Normally the wing flaps will be set to simulate a stall during a short-field takeoff or retracted to simulate a stall during a normal takeoff and/or climb.

      a)   If flaps are extended, retract them to the setting recommended by your POH.

         i)   Do not extend the flaps if they are retracted.

      b)   Make the final flap retraction only after your airplane has accelerated to V_Y.

   6)   Normally a power-on stall is performed with the landing gear retracted (if retractable).

      a)   If you have the gear down, it should be retracted only after you have established a positive rate of climb on the vertical speed indicator.

   7)   Completion standards for this element

      a)   Commercial PTS

         i)   *Recognize and recover promptly as the stall occurs by simultaneously reducing the angle of attack, increasing power to maximum allowable, and leveling the wings to return to a straight-and-level flight attitude, with a minimum loss of altitude appropriate for the airplane.*

         ii)   *Retract the flaps to the recommended setting and retract the landing gear, if retractable, after a positive rate of climb is established.*

         iii)   *Accelerate to V_X or V_Y speed before the final flap retraction; return to the altitude, heading, and airspeed specified by the examiner.*

      b)   Private PTS

         i)   Recognize and recover promptly after the stall occurs by simultaneously reducing the angle of attack, increasing power as appropriate, and leveling the wings to return to a straight-and-level flight attitude with a minimum loss of altitude appropriate for the airplane.

         ii)   Retract the flaps to the recommended setting; retract the landing gear, if retractable, after a positive rate of climb is established; accelerate to V_X or V_Y speed before the final flap retraction; and return to the altitude, heading, and airspeed as specified by the examiner.

2.   **Exhibit your instructional knowledge by explaining the following common errors related to power-on stalls.**

   a.   **Failure to establish the specified landing gear and flap configuration prior to entry**

      1)   Repeat the instructions that your FAA inspector/examiner gave to you regarding the airplane configuration for the stall.

         a)   Your airplane will be configured for a takeoff or a normal departure climb.

      2)   Remember to perform the required clearing turns.

b. **Improper pitch, heading, and bank control during straight ahead and turning stalls**

   1) Use your visual and instrument references as in straight and turning climbs but with a pitch attitude that will induce a stall.

   2) Maintain heading and wings level during the straight ahead stall.

   3) Maintain the specified bank angle and keep the ball centered during the turning stall.

   4) Use rudder pressure to counteract the torque effects.

c. **Improper pitch and bank control during turning stalls**

   1) Use your visual and instrument references as in a turning climb but with a pitch attitude that will induce a stall.

   2) Use whatever control pressure is necessary to maintain a specified bank angle of not more than 20°, in coordinated flight.

d. **Rough or uncoordinated control procedure**

   1) As the airplane approaches the stall, the controls will become increasingly sluggish, and you may assume that the controls need to be moved in a rough or jerky manner.

      a) Maintain smooth control applications at all times.

      b) Do not try to muscle your way through this maneuver.

   2) Keep your airplane in coordinated flight (i.e., the ball centered), even if the controls feel crossed.

      a) If a power-on stall is not properly coordinated, one wing will often drop before the other wing, and the nose will yaw in the direction of the low wing during the stall.

e. **Failure to recognize the first indications of a stall**

   1) The first indication of a stall is signaled by the first buffeting or decay of control effectiveness.

f. **Failure to achieve a stall**

   1) You must maintain sufficient elevator back pressure to induce a stall.

   2) A full stall is evident by such clues as

      a) Full back elevator pressure

      b) High sink rate

      c) Nose-down pitching

      d) Possible buffeting

g. **Improper torque correction**

   1) Since the airspeed is decreasing with a high power setting and a high angle of attack, the effect of torque becomes more prominent. Right rudder pressure must be used to counteract torque.

h. **Poor stall recognition and delayed recovery**

   1) Some pilots may attempt to hold a stall attitude because they are waiting for a particular event to occur, e.g., an abrupt pitch-down attitude.

      a) While waiting for this to occur, the airplane is losing altitude from the high sink rate of a stalled condition.

   2) Delayed recovery aggravates the stall situation and, if you do not remain in coordinated flight, the airplane is likely to enter a spin.

   3) Recognition and recovery must be immediate and prompt.

    i.   **Excessive altitude loss or excessive airspeed during recovery**

        1)   Do not maintain a pitch-down attitude during recovery.

           a)   Move the control yoke forward to reduce the angle of attack; then smoothly adjust the pitch to the desired climb attitude.

    j.   **Secondary stall during recovery**

        1)   This stall happens when you rush your stall recovery to straight-and-level flight or climb before the airplane has realigned itself with the flight path (relative wind).

3.   **Demonstrate and simultaneously explain power-on stalls in climbing flight (straight or turning), with selected landing gear and flap configurations, from an instructional standpoint.**

    a.   Your demonstration must be to commercial pilot skill level.

    b.   You will be evaluated on your ability to explain and demonstrate simultaneously the key elements of this task.

    c.   If you do not perform the maneuver to FAA completion standards, point out the deficiency to your FAA inspector/examiner as an example of an error, and explain how it should be corrected (including how and why it occurred).

4.   **Analyze and correct simulated common errors related to power-on stalls in climbing flight (straight or turning), with selected landing gear and flap configurations.**

    a.   As your FAA inspector/examiner is performing this task, you must be able to recognize, analyze, and correct any error.

    b.   Explain what the error is, and then explain what (s)he should do to correct the error.

    c.   If your FAA inspector/examiner does not perform the maneuver to FAA completion standards, point out the deficiency to him/her as an example of a student error, and explain how it should be corrected (including how and why it occurred).

END OF TASK

POWER-OFF STALLS (PROFICIENCY)

XI.C. TASK: POWER-OFF STALLS (PROFICIENCY)

REFERENCES: FAA-H-8083-3; FAA-S-8081-12; FAA-S-8081-14, POH/AFM.

Objective. To determine that the applicant:

1. Exhibits instructional knowledge of the elements of power-off stalls, in descending flight (straight or turning), with selected landing gear and flap configurations by describing --

   a. Aerodynamics of power-off stalls.

   b. Relationship of various factors, such as landing gear and flap configuration, weight, center of gravity, load factor, and bank angle to stall speed.

   c. Flight situations where unintentional power-off stalls may occur.

   d. Entry technique and minimum entry altitude.

   e. Performance of power-off stalls in descending flight (straight or turning).

   f. Coordination of flight controls.

   g. Recognition of the first indications of power-off stalls.

   h. Recovery technique and minimum recovery altitude.

2. Exhibits instructional knowledge of common errors related to power-off stalls, in descending flight (straight or turning), with selected landing gear and flap configurations by describing --

   a. Failure to establish the specified landing gear and flap configuration prior to entry.

   b. Improper pitch, heading, and bank control during straight-ahead stalls.

   c. Improper pitch and bank control during turning stalls.

   d. Rough or uncoordinated control technique.

   e. Failure to recognize the first indications of a stall.

   f. Failure to achieve a stall.

   g. Improper torque correction.

   h. Poor stall recognition and delayed recovery.

   i. Excessive altitude loss or excessive airspeed during recovery.

   j. Secondary stall during recovery.

3. Demonstrates and simultaneously explains power-off stalls, in descending flight (straight or turning), with selected landing gear and flap configurations, from an instructional standpoint.

4. Analyzes and corrects simulated common errors related to power-off stalls, in descending flight (straight or turning), with selected landing gear and flap configurations.

A. General Information

1. The objective of this task is to determine your instructional knowledge of the elements and common errors related to power-off stalls in descending flight (straight or turning), with selected landing gear and flap configurations. You will be required to demonstrate and simultaneously explain this task from an instructional standpoint.

2. The objectives in performing stalls are to

   a. Familiarize your student with the conditions that produce stalls
   b. Assist in stall/spin awareness
   c. Develop the habit of taking prompt preventive or corrective action

3. The practice of stall recovery and the development of stall/spin awareness with a student are of primary importance to you as an instructor.

B. Task Objectives

1. **Exhibit your instructional knowledge by explaining the following elements of power-off stalls in descending flight (straight or turning), with selected landing gear and flap configurations.**

    a. **Aerodynamics of power-off stalls**

        1) A stall occurs because the critical angle of attack was exceeded. For a detailed discussion on the aerodynamics of a stall, see Task XI.B., Power-On Stalls, beginning on page 416.

    b. **Relationship of various factors such as landing gear and flap configuration, weight, center of gravity, load factor, and bank angle to stall speed**

        1) See the discussion on the relationship of various factors on stall speed in Task XI.B., Power-On Stalls, beginning on page 416.

    c. **Flight situations in which unintentional power-off stalls may occur**

        1) Power-off stalls occur when the airplane is in the normal approach-to-landing conditions and configuration and also in an emergency approach and landing situation.

        2) Many stall/spin accidents have occurred in these power-off situations, including

            a) Crossed-control turns (aileron pressure in one direction, rudder pressure in the opposite direction) from base leg to final approach which result in a skidding or slipping (uncoordinated) turn

            b) Attempting to recover from a high sink rate on final approach by only increasing pitch attitude

            c) Improper airspeed control on final approach or in other segments of the traffic pattern

            d) Attempting to "stretch" a glide in a power-off approach

    d. **Entry technique and minimum entry altitude**

        1) Select an entry altitude that will allow this maneuver to be completed no lower than 1,500 ft. AGL or the manufacturer's recommended altitude, whichever is higher.

            a) Select an altitude that is easy to read from your altimeter.

                i) If the terrain elevation is 300 ft. above sea level, the FAA requires a recovery no lower than 1,800 ft. MSL (1,500 ft. AGL). Round this to the nearest 500-ft. increment (2,000 ft. MSL) to make it easier to identify on your altimeter.

            b) Do not let yourself be rushed into performing this maneuver. If you do not feel that you can recover before 1,500 ft. AGL (or a higher manufacturer's recommended altitude), explain this to your examiner and proceed to climb to a higher altitude.

                i) During your training, you will learn how much altitude you need to perform this maneuver.

        2) Perform **clearing turns** to ensure that the area is clear of other traffic.

        3) Use the same procedure that you use to go into slow flight in the landing configuration (i.e., landing gear and flaps extended).

        4) Maintain a constant altitude and heading while you are slowing your airplane to the normal approach speed.

            a) In your airplane, normal approach speed _____.

5) As your airplane approaches the normal approach speed, adjust pitch and power to establish a stabilized approach (i.e., descent).

6) The commercial PTS requires a pilot to slow the airplane to normal approach speed and establish the approach and landing configuration and then to set the power to approach power while establishing the approach attitude.

   a) The private PTS requires a pilot to establish a stabilized approach in the approach or landing configuration as specified by the examiner.

7) Once established in a stabilized approach, you should smoothly raise the airplane's nose to an attitude that will induce a stall.

   a) Neither the private nor the commercial PTS requires that the throttle be moved to the idle position.

   b) You should discuss with your FAA inspector/examiner that this maneuver can be done at approach power or with the power at idle.

e. **Performance of power-off stalls in descending flight (straight or turning)**

   1) Power-off stalls are practiced to simulate approach and landing conditions and are usually performed with landing gear and flaps fully extended, i.e., in landing configuration.

   2) Both the private and commercial PTSs no longer differentiate between imminent and full stalls. You are required to announce the onset of a stall (i.e., imminent) and promptly recover as the stall occurs (i.e., full).

      a) When the stall occurs, you must promptly recover so your airplane does not remain in a stalled condition.

   3) Power-off stalls should be performed in straight flight and turning flight.

f. **Coordination of flight controls**

   1) Throughout this maneuver, you should maintain coordinated flight.

      a) In straight flight, maintain directional control with the rudder, the wings held level with the ailerons, and a constant pitch attitude with the elevator.

      b) In turning flight, maintain coordinated flight with the rudder, bank angle with the ailerons, and a constant pitch attitude with the elevator.

         i) No attempt should be made to stall your airplane on a predetermined heading.

   2) In most training airplanes, the elevator should be smoothly brought fully back.

   3) The hazard of stalling during uncoordinated flight is that you may enter a spin.

      a) Often a wing will drop at the beginning of a stall, and the nose of your airplane will attempt to move (yaw) in the direction of the low wing.

         i) The correct amount of opposite rudder must be applied to keep the nose from yawing toward the low wing.

      b) If you maintain directional control (coordinated flight), the wing will not drop further before the stall is broken, thus preventing a spin.

   4) Completion standards for this element (commercial and private PTSs)

      a) Commercial PTS

         i) *Maintain a specified heading, ±10°, in straight flight; maintain a specified angle of bank, not to exceed 20°, ± 5°, in turning flight, while inducing the stall.*

      b) Private PTS

         i) The Private PTS is the same except that the specified angle of bank must be maintained ± 10°.

g.   **Recognition of the first indications of power-off stalls**

1)   Both the private and commercial PTSs require the pilot to recognize and announce the first aerodynamic indications of the oncoming stall, i.e., buffeting or decay of control effectiveness.

a)   Buffeting is caused by the turbulent air from the wings flowing back over the fuselage and horizontal stabilizer.

b)   The feeling of control pressures is also very important. As speed is reduced, the "live" resistance to pressures on the controls becomes progressively less.

i)   The airplane controls become less and less effective as one approaches the critical angle of attack.

ii)   In a complete stall, all controls can be moved with almost no resistance and with little immediate effect on the airplane.

2)   Other signs used in stall recognition include vision, hearing, kinesthesia, and stall warning indicators in the airplane.

a)   Vision is useful in detecting a stall condition by noting the attitude of the airplane and the airspeed approaching stall speed. This sense can be fully relied on only when the stall is the result of an intentional unusual attitude of the airplane.

b)   Hearing is also helpful in sensing a stall condition since the tone level and intensity of sounds incidental to flight decrease as the airspeed decreases.

c)   Kinesthesia, or the mind's sensing of changes in direction or speed of motion, is probably the most important and the best indicator to the trained and experienced pilot. If this sensitivity is properly developed, it will warn of a decrease in speed or the beginning of a settling or "mushing" of the airplane.

d)   Many airplanes are equipped with stall warning devices (e.g., a horn) to alert the pilot when the airflow over the wing(s) approaches a point that will not allow lift to be sustained.

h.   **Recovery technique and minimum recovery altitude**

1)   Though the recovery actions must be taken in a coordinated manner, they are broken down into three steps here for explanation purposes.

2)   First, the key factor in recovering from a stall is regaining positive control of your airplane by reducing the angle of attack.

a)   Since the basic cause of a stall is always an excessive angle of attack, the cause must be eliminated by releasing the back elevator pressure that was necessary to attain that angle of attack or by moving the elevator control forward.

i)   Each airplane may require a different amount of forward pressure.

ii)   Too much forward pressure can hinder the recovery by imposing a negative load on the wing.

b)   The object is to reduce the angle of attack but only enough to allow the wing to regain lift. Remember that you want to minimize your altitude loss.

3) Second, promptly and smoothly apply the manufacturer's maximum allowable power to increase airspeed and to minimize the loss of altitude. In most airplanes, the maximum allowable power will be full power, but do not exceed the RPM red line speed.

   a) If carburetor heat is on, you need to turn it off.

   b) Right rudder pressure will be necessary to overcome the torque effect as power is advanced and the nose is being lowered.

   c) Flaps should be partially retracted to reduce drag during recovery from the stall.

      i) Follow the procedures in your POH.

4) Third, straight-and-level flight should be established with coordinated use of the controls.

   a) At this time, the wings should be leveled, if they were previously banked.

   b) Do not attempt to deflect the ailerons until the angle of attack has been reduced.

      i) The adverse yaw caused by the downward aileron may place the airplane in uncoordinated flight, and if the airplane is still in a stalled condition, a spin could be induced.

   c) To maintain level flight normally requires the nose of your airplane to be in a nose-high attitude.

      i) Thus, after the initial reduction in the angle of attack, the pitch should be adjusted to that required for a climb at V_Y.

   d) Maintain coordinated flight throughout the recovery.

5) Completion standards for this element

   a) Commercial PTS

      i) *Recognize and recover promptly as the stall occurs by simultaneously reducing the angle of attack, increasing power to maximum allowable, and leveling the wings to return to a straight-and-level flight attitude with a minimum loss of altitude appropriate for the airplane.*

      ii) *Retract the flaps to the recommended setting, and retract the landing gear, if retractable, after a positive rate of climb is established.*

      iii) *Accelerate to V_X or V_Y speed before the final flap retraction.*

      iv) *Return to the altitude, heading, and airspeed specified by the examiner.*

   b) Private PTS

      i) Recognize and recover promptly after the stall occurs by simultaneously reducing the angle of attack, increasing power as appropriate, and leveling the wings to return to a straight-and-level flight attitude with a minimum loss of altitude appropriate for the airplane.

      ii) Retract the flaps to the recommended setting; retract the landing gear, if retractable, after a positive rate of climb is established; accelerate to V_X or V_Y speed before the final flap retraction; and return to the altitude, heading, and airspeed specified by the examiner.

        • The landing gear (if retractable) should be retracted after a positive rate of climb has been established on the vertical speed indicator.

2.   **Exhibit your instructional knowledge by explaining the following common errors related to power-off stalls in descending flight (straight or turning).**

a.   **Failure to establish the specified landing gear and flap configuration prior to entry**

1)   While maintaining altitude, reduce airspeed to slow flight with wing flaps and landing gear (if retractable) extended to the landing configuration.

a)   Use the normal landing configuration unless otherwise specified by you or your FAA inspector/examiner.

2)   Remember to perform the required **clearing turns**.

b.   **Improper pitch, heading, and bank control during straight-ahead stalls**

1)   Use your visual and instrument references as in straight descents but with an increasing pitch attitude to induce a stall.

2)   Maintain directional control with the rudder and wings level with the ailerons.

c.   **Improper pitch and bank control during turning stalls**

1)   Use your visual and instrument references as in turning descents but with an increasing pitch attitude to induce a stall.

2)   Use whatever control pressure is necessary to maintain the specified angle of bank and coordinated flight.

d.   **Rough or uncoordinated control technique**

1)   As the airplane approaches a stall, the controls become increasingly sluggish, and you may assume that the flight controls need to be moved in a rough or jerky manner.

a)   Maintain smooth control application at all times.

2)   Keep your airplane in coordinated flight, even if the controls feel crossed.

a)   If a power-off stall is not properly coordinated, one wing will often drop before the other, and the nose will yaw in the direction of the low wing during the stall.

e.   **Failure to recognize the first indications of a stall**

1)   The first indication of a stall is signaled by the first buffeting or decay of control effectiveness.

f.   **Failure to achieve a stall**

1)   You must maintain sufficient elevator back pressure to induce a stall.

2)   A full stall is evidenced by such clues as

a)   Full back elevator pressure
b)   High sink rate
c)   Nose-down pitching
d)   Possible buffeting

g.   **Improper torque correction**

1)   During recovery, right rudder pressure is necessary to overcome the torque effects (left-turning tendencies) as power is advanced and the nose is being lowered.

2)   You must cross-check outside references with the turn coordinator to ensure that the ball remains centered.

h.   **Poor stall recognition and delayed recovery**

   1)   Some pilots may attempt to hold a stall attitude because they are waiting for a particular event to occur, e.g., an abrupt pitch-down attitude.

      a)   While waiting for this to occur, the airplane is losing altitude from the high sink rate of a stalled condition.

   2)   Delayed recovery aggravates the stall situation and, if you do not remain in coordinated flight, the airplane is likely to enter a spin.

   3)   Recognition and recovery must be immediate and prompt.

i.   **Excessive altitude loss or excessive airspeed during recovery**

   1)   Do not maintain a pitch-down attitude during recovery.

      a)   Move the control yoke forward to reduce the angle of attack; then smoothly adjust the pitch to the desired attitude.

j.   **Secondary stall during recovery**

   1)   This stall happens because you have hastened to complete your stall recovery (return to straight-and-level flight or climb) before the airplane has realigned itself with the flight path (relative wind).

   2)   Your student may experience a secondary stall if his/her recovery technique is incomplete or deficient in a primary stall.

3.   **Demonstrate and simultaneously explain power-off stalls in descending flight (straight or turning), with selected landing gear and flap configurations, from an instructional standpoint.**

   a.   Your demonstration must be to commercial pilot skill level.

   b.   You will be evaluated on your ability to explain and demonstrate simultaneously the key elements of this task.

   c.   If you do not perform the maneuver to FAA completion standards, point out the deficiency to your FAA inspector/examiner as an example of an error, and explain how it should be corrected (including how and why it occurred).

4.   **Analyze and correct simulated common errors related to power-off stalls in descending flight (straight or turning), with selected landing gear and flap configurations.**

   a.   As your FAA inspector/examiner is performing this task, you must be able to recognize, analyze, and correct any error.

   b.   Explain what the error is, and then explain what (s)he should do to correct the error.

   c.   If your FAA inspector/examiner does not perform the maneuver to FAA completion standards, point out the deficiency to him/her as an example of a student error, and explain how it should be corrected (including how and why it occurred).

END OF TASK

CROSSED-CONTROL STALLS (DEMONSTRATION)

XI.D. TASK: CROSSED-CONTROL STALLS (DEMONSTRATION)

REFERENCES: FAA-H-8083-3; FAA-S-8081-12; FAA-S-8081-14, POH/AFM.

Objective. To determine that the applicant:

1. Exhibits instructional knowledge of the elements of crossed-control stalls, with the landing gear extended by describing --

   a. Aerodynamics of crossed-control stalls.

   b. Effects of crossed controls in gliding or reduced airspeed descending turns.

   c. Flight situations where unintentional crossed-control stalls may occur.

   d. Entry procedure and minimum entry altitude.

   e. Recognition of crossed-control stalls.

   f. Recovery procedure and minimum recovery altitude.

2. Exhibits instructional knowledge of common errors related to crossed-control stalls, with the landing gear extended by describing --

   a. Failure to establish selected configuration prior to entry.

   b. Failure to establish a crossed-control turn and stall condition that will adequately demonstrate the hazards of a crossed-control stall.

   c. Improper or inadequate demonstration of the recognition and recovery from a crossed-control stall.

   d. Failure to present simulated student instruction that emphasizes the hazards of a crossed-control condition in a gliding or reduced airspeed condition.

3. Demonstrates and simultaneously explains crossed-control stall, with the landing gear extended, from an instructional standpoint.

4. Analyzes and corrects simulated common errors related to a crossed-control stall with the landing gear extended.

A. General Information

   1. The objective of this task is to determine your instructional knowledge of the elements and common errors related to crossed-control stalls with the landing gear extended. You will be required to demonstrate and simultaneously explain this task from an instructional standpoint.

   2. The objective of this demonstration maneuver is to show the effect of improper control technique and to emphasize the importance of using coordinated control pressures whenever making turns.

      a. Your students must be able to recognize when this stall is imminent and must take immediate action to prevent a completely stalled condition.

B. Task Objectives

   1. **Exhibit your instructional knowledge by explaining the following elements of crossed-control stalls with the landing gear extended.**

      a. **Aerodynamics of crossed-control stalls**

         1) See the two-page discussion on the aerodynamics of a stall in Task XI.B., Power-On Stalls, beginning on page 416.

         2) Crossed-control means that aileron pressure is applied in one direction, and rudder pressure is applied in the opposite direction.

         3) Normally, a crossed-control stall occurs during a skidding turn because the forces the pilot feels (causing him/her to be pushed to the outside of the turn) are not uncomfortable since these forces are what (s)he feels when making a turn in a car.

b. **Effects of crossed controls in gliding or reduced airspeed descending turns**

1) During a descending turn when a pilot attempts to maintain a constant bank while using rudder to increase the rate of the turn, the airplane will be in a skid.

2) During the skidding turn, the outside wing will speed up and produce more lift than the inside wing.

   a) This difference in lift causes the airplane to increase the bank, and the pilot will use the opposite aileron to prevent the increase in the bank angle.

3) The down aileron on the lowered (inside) wing produces more drag and reduces the lift more; thus even more opposite aileron is required.

   a) The additional drag and increased difference in each wing's lift component may cause the rate of roll to increase so fast that it will be past 90° angle of bank before it can be stopped.

4) Additionally, the down aileron increases the lift coefficient, and if the pitch is increased, the inside wing will reach the maximum lift coefficient (i.e., critical angle of attack) before the outside wing.

c. **Flight situations where unintentional crossed-control stalls may occur**

1) A crossed-control stall normally occurs when a pilot turns late onto final, which results in overshooting the extended centerline of the runway. Factors which contribute include a tailwind on the base leg and distractions.

   a) A tailwind on the base leg will increase the airplane's groundspeed and provide the pilot with less time to react.

      i) The pilot may not properly plan for the effect of the wind on the turn to final.

   b) Distractions (e.g., conflicting traffic, problems in the cockpit, etc.) may cause the pilot to concentrate on the distractions; i.e., there is a lack of division of attention.

d. **Entry procedure and minimum entry altitude**

1) Before demonstrating this maneuver to your student, you must be at a safe altitude that will allow a recovery no lower than 1,500 ft. AGL. Remember the possible extreme nose-down attitude and loss of altitude that may result.

2) Select a reference line (e.g., road) to use as the runway, and approach the reference line from a left-base leg.

3) Perform **clearing turns** to ensure that the area (especially below you) is clear of other traffic.

4) Reduce power to the approach power setting, and perform the before-landing checklist while maintaining altitude.

   a) Landing gear (if retractable) should be extended.

   b) Since there is a possibility of exceeding the airplane's limitations, flaps should not be extended.

   c) Trim the airplane as necessary.

5) As the airplane approaches the normal approach airspeed, close the throttle and adjust pitch to establish a stabilized glide.

   a) Adjust trim as necessary.

6) Begin the turn toward the reference line which will cause you to overshoot the reference line.

   a) Use a bank angle of approximately 20°.

7) During the turn, apply excessive rudder pressure in an attempt to increase the rate of turn (i.e., in a left turn, use left rudder).

   a) Use aileron pressure to maintain a constant bank angle. This will require aileron pressure opposite the rudder pressure.

   b) Use back elevator pressure to keep the nose of the airplane from lowering.

8) Increase all these control pressures until the airplane stalls.

e. **Recognition of crossed-control stalls**

   1) In a crossed-control stall, the airplane often stalls with little warning.

      a) The nose may pitch down, the inside wing may suddenly drop, and the airplane may continue to roll to an inverted position.

      b) These events usually indicate the beginning of a spin.

f. **Recovery procedure and minimum recovery altitude**

   1) Though the recovery actions must be taken in a coordinated manner, they are broken down into three steps here for explanation purposes.

   2) First, the key factor in all stalls is to reduce the angle of attack by relaxing enough back pressure to allow the wing to regain lift.

   3) Second, neutralize the ailerons and rudder.

      a) Stop any rotation with opposite rudder.

   4) Third, recover to straight-and-level flight by coordinated use of the controls.

      a) The airplane will be in a nose-down attitude, and speed will quickly increase.

      b) As the nose of the airplane is raised to level flight, add maximum available power.

         i) Applying power when the nose is pointed toward the ground will result in a greater loss of altitude, with the possibility of impacting the ground if the stall occurs during an actual landing approach.

   5) You should demonstrate using both skidding and slipping turns.

   6) You must recover from this maneuver no lower than 1,500 ft. AGL.

2. **Exhibit your instructional knowledge by explaining the following common errors related to crossed-control stalls with the landing gear extended.**

   a. **Failure to establish selected configuration prior to entry**

      1) Ensure that the landing gear is extended and the flaps are up.

   b. **Failure to establish a crossed-control turn and stall condition that will adequately demonstrate the hazards of a crossed-control stall**

      1) While in the turn, apply excessive rudder pressure and maintain the bank angle and pitch attitude to ensure a crossed-control situation.

      2) Steadily increase the control pressures, and remember that you are demonstrating a turn onto final approach and overshooting the centerline of the runway.

      3) Increase control pressures until the airplane stalls.

    c. **Improper or inadequate demonstration of the recognition or recovery from a crossed-control stall**

       1) Demonstrate that the airplane may stall with little warning.

       2) At the stall, return to straight-and-level flight by coordinated use of the controls and power.

          a) This action will be sufficient if recovery can be made before your airplane enters an abnormal attitude.

       3) If your airplane reaches an abnormal or inverted attitude, release the control pressures to break the stall, and allow the roll to continue until the airplane reaches straight-and-level flight.

          a) Do not add power with the nose pointed toward the ground.

    d. **Failure to present simulated student instruction that emphasizes the hazards of a crossed-control condition in a gliding or reduced airspeed condition**

       1) Explain the importance of proper control techniques and using coordinated control pressures when making a turn.

       2) Explain the flight conditions in which this stall will most likely occur, and why.

       3) Demonstrate a crossed-control stall.

3. **Demonstrate and simultaneously explain a crossed-control stall with the landing gear extended from an instructional standpoint.**

    a. Your demonstration must be to commercial pilot skill level.

    b. You will be evaluated on your ability to explain and demonstrate simultaneously the key elements of this task.

    c. If you do not perform the maneuver to FAA completion standards, point out the deficiency to your FAA inspector/examiner as an example of an error, and explain how it should be corrected (including how and why it occurred).

4. **Analyze and correct simulated common errors related to a crossed-control stall with the landing gear extended.**

    a. As your FAA inspector/examiner is performing this task, you must be able to recognize, analyze, and correct any error.

    b. Explain what the error is, and then explain what (s)he should do to correct the error.

    c. If your FAA inspector/examiner does not perform the maneuver to FAA completion standards, point out the deficiency to him/her as an example of a student error, and explain how it should be corrected (including how and why it occurred).

END OF TASK

ELEVATOR TRIM STALLS (DEMONSTRATION)

XI.E. TASK: ELEVATOR TRIM STALLS (DEMONSTRATION)

     REFERENCES: FAA-H-8083-3; FAA-S-8081-12; FAA-S-8081-14, POH/AFM.

Objective. To determine that the applicant:

1. Exhibits instructional knowledge of the elements of elevator trim stalls, in selected landing gear and flap configurations by describing --

    a. Aerodynamics of elevator trim stalls.

    b. Hazards of inadequate control pressures to compensate for thrust, torque, and up-elevator trim during go-arounds and other related maneuvers.

    c. Entry procedure and minimum entry altitude.

    d. Recognition of elevator trim stalls.

    e. Importance of recovering from an elevator trim stall immediately upon recognition.

2. Exhibits instructional knowledge of common errors related to elevator trim stalls, in selected landing gear and flap configurations by describing --

    a. Failure to present simulated student instruction that adequately emphasizes the hazards of poor correction for torque and up-elevator trim during go-arounds and other maneuvers.

    b. Failure to establish selected configuration prior to entry.

    c. Improper or inadequate demonstration of the recognition of and the recovery from an elevator trim stall.

3. Demonstrates and simultaneously explains elevator trim stalls, in selected landing gear and flap configurations, from an instructional standpoint.

4. Analyzes and corrects simulated common errors related to elevator trim stalls in selected landing gear and flap configurations.

A. General Information

    1. The objective of this task is to determine your instructional knowledge of the elements and common errors related to elevator trim stalls, in selected landing gear and flap configurations. You will be required to demonstrate and simultaneously explain this task from an instructional standpoint.

    2. The objective of this demonstration maneuver is to show the importance of making smooth power applications, overcoming strong trim forces and maintaining positive control of the airplane to hold safe flight attitudes, and using proper and timely trim techniques during go-arounds and other related maneuvers.

B. Task Objectives

    1. **Exhibit your instructional knowledge by explaining the following elements of elevator trim stalls, in selected landing gear and flap configurations.**

        a. **Aerodynamics of elevator trim stalls**

            1) See the two-page discussion on the aerodynamics of a stall in Task XI.B., Power-On Stalls, beginning on page 416.

            2) In the event of a go-around, as maximum allowable power is applied, the combined forces of thrust, torque, and back-elevator (nose-up) trim will tend to make the nose rise sharply and turn to the left.

        b. **Hazards of inadequate control pressures to compensate for thrust, torque, and up-elevator trim during go-arounds and other related maneuvers**

            1) Failure to maintain adequate control pressures to compensate for thrust, torque, and up-elevator trim can cause the airplane to exceed its critical angle of attack, and to stall.

            2) This stall normally occurs close to the ground, as in a go-around procedure.

            3) Your student must learn to recognize when the stall is imminent and must take prompt corrective action.

c. **Entry procedure and minimum entry altitude**

1) Before demonstrating this maneuver, you must be at a safe altitude that will allow a recovery no lower than 1,500 ft. AGL.

2) While this stall can be demonstrated at various landing gear and flap configurations, we will cover one method.

3) Perform **clearing turns** to ensure that the area is clear of other traffic.

4) Reduce power to the approach power setting, and complete the before-landing checklist while maintaining altitude.

   a) Your airplane should be in the landing configuration, i.e., landing gear (if retractable) and flaps extended.

   b) Once approach airspeed is attained, establish an approach descent, and trim the airplane (nose-up trim).

5) During this simulated final-approach glide, the throttle is then advanced smoothly to maximum allowable power as would be done in a go-around procedure.

6) The combined forces of thrust, torque, and up-elevator trim will tend to make the nose rise sharply and turn to the left.

7) To demonstrate what could occur if positive control of the airplane were not maintained, no immediate attempt should be made to correct these forces.

d. **Recognition of elevator trim stalls**

1) When the throttle is fully advanced and the pitch attitude increases above the normal climbing attitude, you should recognize when the stall is imminent.

   a) This is evident by the buffeting.

e. **Importance of recovery from an elevator trim stall immediately upon recognition**

1) It is imperative that a full stall not occur during an actual go-around from a landing approach.

   a) The amount of pitch change necessary for recovery may be such that the airplane may strike the ground, if the stall occurred at a low altitude.

2. **Exhibit your instructional knowledge by explaining the following common errors related to elevator trim stalls.**

a. **Failure to present simulated student instruction that adequately emphasizes the hazards of poor correction for torque and up-elevator trim during go-arounds and other maneuvers**

1) Provide realistic demonstrations of elevator trim stalls.

2) Emphasize the importance of correct attitude control, application of control pressures, and proper trim during go-arounds and other maneuvers.

b. **Failure to establish selected configuration prior to entry**

1) This maneuver can be demonstrated in various landing gear and flap configurations.

2) Ensure that the airplane is in the appropriate configuration before you demonstrate this maneuver to your student (FAA inspector/examiner).

c. **Improper or inadequate demonstration of the recognition of and recovery from an elevator trim stall**

1) Allow your airplane to increase above the normal climbing attitude, and do not attempt to correct the forces acting on the airplane.

2) Talk through the demonstration of the effects of thrust, torque, and up-elevator trim.

3) Point out the excessive pitch attitude, the yaw to the left, and the imminent stall.

4) Apply the correct recovery techniques to prevent a full stall.

3. **Demonstrate and simultaneously explain elevator trim stalls in selected landing gear and flap configurations from an instructional standpoint.**

    a. Your demonstration must be to commercial pilot skill level.

    b. You will be evaluated on your ability to explain and demonstrate simultaneously the key elements of this task.

    c. If you do not perform the maneuver to FAA completion standards, point out the deficiency to your FAA inspector/examiner as an example of an error, and explain how it should be corrected (including how and why it occurred).

4. **Analyze and correct simulated common errors related to elevator trim stalls in selected landing gear and flap configurations.**

    a. As your FAA inspector/examiner is performing this task, you must be able to recognize, analyze, and correct any error.

    b. Explain what the error is, and then explain what (s)he should do to correct the error.

    c. If your FAA inspector/examiner does not perform the maneuver to FAA completion standards, point out the deficiency to him/her as an example of a student error, and explain how it should be corrected (including how and why it occurred).

END OF TASK

SECONDARY STALLS (DEMONSTRATION)

XI.F. TASK: SECONDARY STALLS (DEMONSTRATION)

REFERENCES: FAA-H-8083-3; FAA-S-8081-12; FAA-S-8081-14, POH/AFM.

Objective. To determine that the applicant:

1. Exhibits instructional knowledge of the elements of secondary stalls, in selected landing gear and flap configurations by describing --

   a. Aerodynamics of secondary stalls.

   b. Flight situations where secondary stalls may occur.

   c. Hazards of secondary stalls during normal stall or spin recovery.

   d. Entry procedure and minimum entry altitude.

   e. Recognition of a secondary stall.

   f. Recovery procedure and minimum recovery altitude.

2. Exhibits instructional knowledge of common errors related to secondary stalls, in selected landing gear and flap configurations by describing --

   a. Failure to establish selected configuration prior to entry.

   b. Improper or inadequate demonstration of the recognition of and recovery from a secondary stall.

   c. Failure to present simulated student instruction that adequately emphasizes the hazards of poor procedure in recovering from a primary stall.

3. Demonstrates and simultaneously explains secondary stalls, in selected landing gear and flap configurations, from an instructional standpoint.

4. Analyzes and corrects simulated common errors related to secondary stalls in selected landing gear and flap configurations.

A. General Information

   1. The objective of this task is to determine your instructional knowledge of the elements and common errors related to secondary stalls, in selected landing gear and flap configurations. You will be required to demonstrate and simultaneously explain this task from an instructional standpoint.

B. Task Objectives

   1. **Exhibit your instructional knowledge by explaining the following elements of secondary stalls, in selected landing gear and flap configurations.**

      a. **Aerodynamics of secondary stalls**

         1) A secondary stall is caused by attempting to hasten the completion of a stall recovery before the airplane has regained sufficient flying speed.

         2) Recoveries from stalls and spins involve a tradeoff -- a loss of altitude (and an increase in airspeed) and an increase in load factors in the pullup.

            a) These increased load factors cause an increase in the stall speed.

               i) If the airspeed is less than the stall speed, the airplane will enter a secondary stall.

      b. **Flight situations where secondary stalls may occur**

         1) A secondary stall usually occurs when a pilot becomes too anxious to return to straight-and-level flight after a stall (of any type) or spin recovery.

         2) In an actual flight situation, a stall may surprise the pilot, causing him/her to hurry through an initial stall recovery.

c. **Hazards of secondary stalls during normal stall or spin recovery**

   1) Significant load factor increases are sometimes induced during pullup from a stall or spin recovery.

      a) Structural damage to the airplane can result from the high load factors imposed by intentional stalls practiced above the airplane's maneuvering speed (V_A).

d. **Entry procedure and minimum entry altitude**

   1) Before demonstrating this maneuver, you must be at a safe altitude that will allow a recovery from the secondary stall to be completed no lower than 1,500 ft. AGL.

   2) Perform **clearing turns** to ensure that the area is clear of other traffic.

   3) A secondary stall may be demonstrated during recovery from either a power-off or a power-on stall.

      a) Ensure that your student (FAA inspector/examiner) understands which stall you are doing, and also explain why a secondary stall may develop from that situation.

      b) Configure your airplane as appropriate before entering the maneuver.

   4) At the stall, reduce the angle of attack; then abruptly pull back on the yoke.

      a) Ensure that this action is taken before the airplane exceeds V_A.

e. **The recognition of a secondary stall**

   1) Indications of a secondary stall include buffeting, loss of control effectiveness, full-up elevator, high sink rate, nose-down pitching.

f. **Recovery procedure and minimum recovery altitude**

   1) When a secondary stall occurs, the back elevator pressure should again be released just as in a normal stall recovery. When the angle of attack has been reduced and sufficient airspeed has been regained, the airplane can be returned to straight-and-level flight attitude.

   2) Recovery should be made no lower than 1,500 ft. AGL.

2. **Exhibit your instructional knowledge by explaining the following common errors of secondary stalls.**

   a. **Failure to establish selected configuration prior to entry**

      1) This maneuver can be demonstrated in various landing gear and flap configurations.

      2) Ensure that the airplane is in the appropriate configuration before you demonstrate this maneuver to your student (FAA inspector/examiner).

   b. **Improper or inadequate demonstration of the recognition of and recovery from a secondary stall**

      1) Explain while you demonstrate what you are doing that will produce a secondary stall and why.

      2) Explain the recognition and recovery from a secondary stall.

   c. **Failure to present simulated student instruction that adequately emphasizes the hazards of poor procedure in recovering from a primary stall**

      1) Provide realistic demonstrations of a secondary stall.

      2) Demonstrate the problems that can occur when attempting to hasten the recovery from a primary stall.

      3) Develop your student's awareness of proper stall recoveries.

3. **Demonstrate and simultaneously explain secondary stalls in selected landing gear and flap configurations from an instructional standpoint.**

   a. Your demonstration must be to commercial pilot skill level.

   b. You will be evaluated on your ability to explain and demonstrate simultaneously the key elements of this task.

   c. If you do not perform the maneuver to FAA completion standards, point out the deficiency to your FAA inspector/examiner as an example of an error, and explain how it should be corrected (including how and why it occurred).

4. **Analyze and correct simulated common errors related to secondary stalls in selected landing gear and flap configurations.**

   a. As your FAA inspector/examiner is performing this task, you must be able to recognize, analyze, and correct any error.

   b. Explain what the error is, and then explain what (s)he should do to correct the error.

   c. If your FAA inspector/examiner does not perform the maneuver to FAA completion standards, point out the deficiency to him/her as an example of a student error, and explain how it should be corrected (including how and why it occurred).

END OF TASK

SPINS

XI.G. TASK: SPINS

NOTE: At the discretion of the examiner, a logbook record attesting applicant instructional competency in spin entries, spins, and spin recoveries may be accepted in lieu of this TASK. The flight instructor who conducted the spin instruction shall certify the logbook record.

REFERENCES: 14 CFR Part 23; Type Certificate Data Sheet; AC 61-67, FAA-H-8083-3; POH/AFM.

Objective. To determine that the applicant:

1.   Exhibits instructional knowledge of the elements of spins by describing --

    a.   Anxiety factors associated with spin instruction.

    b.   Aerodynamics of spins.

    c.   Airplanes approved for the spin maneuver based on airworthiness category and type certificate.

    d.   Relationship of various factors such as configuration, weight, center of gravity, and control coordination to spins.

    e.   Flight situations where unintentional spins may occur.

    f.   How to recognize and recover from imminent, unintentional spins.

    g.   Entry procedure and minimum entry altitude for intentional spins.

    h.   Control procedure to maintain a stabilized spin.

    i.   Orientation during a spin.

    j.   Recovery procedure and minimum recovery altitude for intentional spins.

2.   Exhibits instructional knowledge of common errors related to spins by describing --

    a.   Failure to establish proper configuration prior to spin entry.

    b.   Failure to achieve and maintain a full stall during spin entry.

    c.   Failure to close throttle when a spin entry is achieved.

    d.   Failure to recognize the indications of an imminent, unintentional spin.

    e.   Improper use of flight controls during spin entry, rotation, or recovery.

    f.   Disorientation during a spin.

    g.   Failure to distinguish between a high-speed spiral and a spin.

    h.   Excessive speed or accelerated stall during recovery.

    i.   Failure to recover with minimum loss of altitude.

    j.   Hazards of attempting to spin an airplane not approved for spins.

3.   Demonstrates and simultaneously explains spin (one turn) from an instructional standpoint.

4.   Analyzes and corrects simulated common errors related to spins.

A.   General Information

    1.   The objective of this task is to determine your instructional knowledge of the elements and common errors related to spins. You will be required to demonstrate and simultaneously explain a spin (one turn) from an instructional standpoint.

    2.   At the discretion of your FAA inspector/examiner, a logbook endorsement attesting to your instructional competency in spin entries, spins, and spin recoveries may be accepted as successful completion of this task.

        a.   Your logbook must be endorsed by the flight instructor who conducted your spin instruction.

B.   Task Objectives

   1.   **Exhibit your instructional knowledge by explaining the following elements of spins.**

     a.   **Anxiety factors associated with spin instruction**

       1)   Fear of spins is deeply rooted in the public's mind, and many pilots have a subconscious (or even conscious) aversion to them.

         a)   Learning the cause of a spin and the proper techniques to prevent and/or recover from the spin removes mental anxiety and many causes of unintentional spins.

         b)   The use of spin instruction will not only increase your stall/spin awareness but will add confidence in your ability to be a safe pilot.

     b.   **Aerodynamics of spins**

       1)   A spin is an aggravated stall that results in what is termed autorotation, in which the airplane follows a corkscrew path in a downward direction.

       2)   For a spin to occur, two conditions must exist.

         a)   The airplane must be in a stall.
         b)   The airplane must be in uncoordinated flight; i.e., the ball is not centered.

       3)   If the nose of the airplane is allowed to yaw at the beginning of a stall, the wing will drop in the direction of the yaw.

         a)   Unless rudder is applied to keep the nose from yawing, the airplane begins to slip toward the lowered wing.

         b)   This slip causes the airplane to weathervane into the relative wind, i.e., toward the lowered wing, thus continuing the yaw.

       4)   At the same time, the airplane continues to roll toward the lowered wing.

         a)   The lowered wing has an increasingly greater angle of attack, due to the upward motion of the relative wind against its surfaces.

           i)   It is then well beyond the critical angle of attack and suffers an extreme loss of lift and an increase in drag.

         b)   The rising wing, since the relative wind is striking it at a smaller angle, has a smaller angle of attack than the opposite wing.

           i)   The rising wing, in effect, becomes less stalled and thus develops some lift so that the airplane continues to roll.

         c)   This autorotation, combined with the effects of centrifugal force and the different amounts of drag on the two wings, becomes a spin, and the airplane descends, rolling and yawing, until recovery is effected.

       5)   Remember that, in order to spin, both of the airplane's wings must first be stalled; then one wing becomes less stalled than the other.

       6)   A spin may be broken down into three phases.

         a)   The incipient phase is the transient period between a stall and a fully developed spin, when a final balancing of aerodynamic and inertial forces has not yet occurred.

         b)   The steady state phase is that portion of the spin in which it is fully developed and the aerodynamic forces are in balance.

         c)   The recovery phase begins when controls are applied to stop the spin and ends when level flight is attained.

       7)   Use of the rudder is important during a stall. The correct amount of rudder must be applied to keep the nose from yawing. By maintaining directional control and not allowing the nose to yaw, the wing will not drop any more before the stall is broken. Thus, a spin will be averted.

c. **Airplanes approved for the spin maneuver based on airworthiness category and type certificate**

   1) The INTENTIONAL SPINNING of an airplane for which the spin maneuver is not specifically approved is NOT encouraged and is NOT authorized either by this study manual or by FARs.

   2) Official sources for determining if the spin maneuver is approved for a specific airplane are

      a) The airplane's type certificate and data sheets

      b) A statement on a placard located in clear view of the pilot in the airplane, e.g., "NO ACROBATIC MANEUVERS INCLUDING SPINS APPROVED"

      c) The airplane's POH

d. **Relationship of various factors such as configuration, weight, center of gravity, and control coordination to spins**

   1) Since a stall is a prerequisite for a spin, the factors that affect a stall will affect a spin.

      a) See the discussion on the factors affecting stall speed in Task XI.B., Power-On Stalls, beginning on page 416.

e. **Flight situations where unintentional spins may occur**

   1) The primary cause of an inadvertent spin is stalling the airplane while executing a turn with excessive or insufficient rudder.

   2) The critical phases of flight for stall/spin accidents are

      a) Takeoff and departure
      b) Approach and landing and go-around
      c) Engine failure

   3) Spins can occur when practicing stalls with

      a) Uncoordinated flight control input
      b) Aileron deflection at critical angles of attack

f. **How to recognize and recover from imminent, unintentional spins**

   1) Continued practice in stalls will help you to develop a more instinctive and prompt reaction in recognizing an approaching spin.

   2) It is essential to learn to apply immediate corrective action anytime it is apparent that your airplane is near a spin condition.

   3) If an unintentional spin can be prevented, it should be.

      a) Avoiding a spin shows sound pilot judgment and is a positive indication of alertness.

   4) To recover from an imminent spin, release the spin inducing controls by applying opposite rudder and forward elevator pressure.

g. **Entry procedure and minimum entry altitude for intentional spins**

   1) Spin avoidance, imminent spins, and actual spin entry, spin, and spin recovery techniques should be practiced from an altitude above 3,500 ft. AGL.

   2) As a rough estimate, an altitude loss of approximately 500 ft. per each 3-sec. turn can be expected in most small airplanes in which spins are authorized.

      a) Greater losses can be expected at higher density altitudes.

3) How to enter a spin

    a) Apply the entry procedure for power-off stall as described in Task XI.C., Power-Off Stalls, beginning on page 427.

    b) Just before the airplane stalls, smoothly apply full rudder in the direction of desired spin rotation, and as the nose drops continue to apply back elevator smoothly to the limit of travel.

        i) Applying rudder pressure will cause the inside wing to stall and the nose to drop as the airplane yaws.

    c) The ailerons should be neutral.

h. **Control procedure to maintain a stabilized spin**

    1) To allow a spin to be developed and maintained, the following control techniques may be used.

    a) Maintain full-back elevator pressure to keep the wings stalled.
    b) Maintain full rudder, in the direction of the spin, to keep yawing.
    c) Maintain neutral ailerons.

i. **Orientation during a spin**

    1) Maintain orientation by selecting an outside reference point and using your turn coordinator.

    2) Your gyroscopic instruments (attitude indicator, heading indicator) may tumble during a spin and may be misleading.

    a) In an inadvertent spin, when the direction of spin may not be obvious, the turn coordinator can provide you with the best reference of the spin direction.

j. **Recovery procedure and minimum recovery altitude for intentional spins**

    1) In the absence of specific recovery techniques in your airplane's POH, the following technique is suggested for spin recovery.

    2) The first corrective action taken during any power-on spin is to close the throttle completely to eliminate power and minimize loss of altitude.

    a) Power aggravates the spin characteristics and causes an abnormal loss of altitude in the recovery.

    3) To recover from the spin, you should neutralize the ailerons, determine the direction of the turn, and apply full opposite rudder.

    a) Opposite rudder should be maintained until the rotation stops. Then the rudder should be neutralized. Continue to use the rudder for directional control.

    b) If the rudder is not neutralized at the proper time, the ensuing increased airspeed acting upon the fully deflected rudder will cause an excessive and unfavorable yawing effect. This effect places great strain on the airplane and may cause a secondary spin in the opposite direction.

    4) When the rotation slows, apply brisk, positive straight-forward movement of the elevator control (forward of the neutral position). The control should be held firmly in this position.

    a) The forceful movement of the elevator will decrease the excessive angle of attack and will break the stall.

5)   Once the stall is broken, the spinning will stop. When the rudder is neutralized, gradually apply enough back elevator pressure to return to level flight.

    a)   Too much or abrupt back elevator pressure and/or application of rudder and ailerons during the recovery can result in a secondary stall and possibly another spin.

6)   The airplane's engine may stop developing power due to centrifugal force acting on the fuel in the fuel tanks, causing fuel interruption.

    a)   Thus, the assumption that power is not available is recommended when practicing spin recovery.

7)   See the figure below.

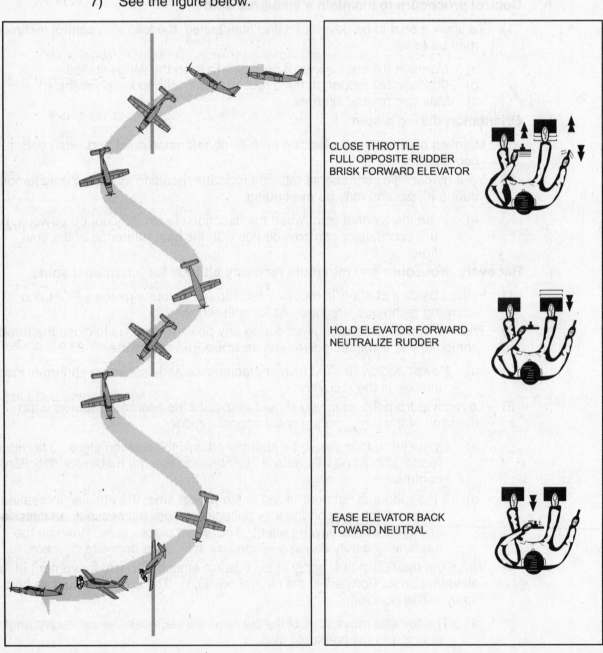

CLOSE THROTTLE
FULL OPPOSITE RUDDER
BRISK FORWARD ELEVATOR

HOLD ELEVATOR FORWARD
NEUTRALIZE RUDDER

EASE ELEVATOR BACK
TOWARD NEUTRAL

2. **Exhibit your instructional knowledge by explaining the following common errors related to spins.**

    a. **Failure to establish proper configuration prior to spin entry**

        1) The proper configuration for your make and model of airplane (that is approved for spins) will be found in the airplane's POH.

    b. **Failure to achieve and maintain a full stall during spin entry**

        1) The wings must be stalled to enter a spin; otherwise, a spiral will develop.

    c. **Failure to close the throttle when a spin entry is achieved**

        1) The throttle must be closed when a spin entry is achieved to prevent an abnormal loss of altitude.

        2) Power may aggravate the spin characteristics.

    d. **Failure to recognize the indications of an imminent, unintentional spin**

        1) Remember that a spin is from a stalled condition and improper directional control (rudder), allowing the nose of the airplane to yaw.

    e. **Improper use of flight controls during spin entry, rotation, or recovery**

        1) During entry at the stall, apply full rudder pressure in the desired direction of the spin.

        2) During rotation, keep the control yoke full back, ailerons neutral, and full rudder (as in the entry).

        3) During recovery, apply full opposite rudder to stop rotation, apply forward pressure on the control yoke to break the stall, and use coordinated controls to return to straight-and-level flight.

    f. **Disorientation during a spin**

        1) During demonstration of a spin, select an outside reference point to help in orientation.

    g. **Failure to distinguish between a high-speed spiral and a spin**

        1) A high-speed spiral is evidenced by a nose-low attitude, the wings not stalled, airspeed increasing rapidly, and a high rate of descent (which may be increasing).

        2) A spin will have a nose-down attitude, continuous rotation, possible buffeting, an almost constant low airspeed, wings stalled, and a steady rate of descent.

    h. **Excessive speed or accelerated stall during recovery**

        1) Excessive speed will occur if you are too cautious in applying back elevator pressure.

        2) Accelerated stall occurs when you are too anxious to stop the descent.

            a) This condition is actually a secondary stall and may result in another spin.

    i. **Failure to recover with minimum loss of altitude**

        1) When recovery from a spin is initiated, brisk and positive control applications provide for a more positive recovery with minimum loss of altitude.

j.   **Hazards of attempting to spin an airplane not approved for spins**

1)   Some pilots reason that the airplane was spin tested during its certification process and therefore no problem should result from demonstrating or practicing spins.

a)   Actually, certification in the normal category requires only that the airplane recover from a one-turn spin in no more than one additional turn or three seconds, whichever takes longer.

b)   This same test of controllability can also be used in certificating an airplane in the utility category [FAR 23.221(b)].

2)   THE PILOT OF AN AIRPLANE PLACARDED AGAINST INTENTIONAL SPINS SHOULD ASSUME THAT THE AIRPLANE MAY BECOME UNCONTROLLABLE IN A SPIN.

3.   **Demonstrate and simultaneously explain a spin (one turn) from an instructional standpoint.**

a.   Your demonstration must be to commercial pilot skill level.

b.   You will be evaluated on your ability to explain and demonstrate simultaneously the key elements of this task.

c.   If you do not perform the maneuver to FAA completion standards, point out the deficiency to your FAA inspector/examiner as an example of an error, and explain how it should be corrected (including how and why it occurred).

4.   **Analyze and correct simulated common errors related to spins.**

a.   As your FAA inspector/examiner is performing this task, you must be able to recognize, analyze, and correct any error.

b.   Explain what the error is, and then explain what (s)he should do to correct the error.

c.   If your FAA inspector/examiner does not perform the maneuver to FAA completion standards, point out the deficiency to him/her as an example of a student error, and explain how it should be corrected (including how and why it occurred).

END OF TASK

ACCELERATED MANEUVER STALLS (DEMONSTRATION)

XI.H. TASK: ACCELERATED MANEUVER STALLS (DEMONSTRATION)

NOTE: This TASK shall be completed by oral examination or demonstration at discretion of examiner.

REFERENCES: FAA-H-8083-3; POH/AFM.

Objective. To determine that the applicant:

1. Exhibits instructional knowledge of the elements of accelerated maneuver stalls by describing --

   a. Aerodynamics of accelerated maneuver stalls.

   b. Flight situations where accelerated maneuver stalls may occur.

   c. Hazards of accelerated stalls during stall or spin recovery.

   d. Entry procedure and minimum entry altitude.

   e. Recognition of the accelerated stall.

   f. Recovery procedure and minimum recovery altitude.

2. Demonstrates and simultaneously explains accelerated maneuver stall, from an instructional standpoint.

3. Exhibits instructional knowledge of common errors related to accelerated maneuver stalls by describing --

   a. Failure to establish proper configuration prior to entry.

   b. Improper or inadequate demonstration of the recognition and recovery from an accelerated maneuver stall.

   c. Failure to present simulated student instruction that adequately emphasizes the hazards of poor procedures in recovering from an accelerated stall.

4. Analyzes and corrects simulated common errors related to accelerated stalls.

A. General Information

   1. The objective of this task is to determine your instructional knowledge of the elements and common errors related to accelerated maneuver stalls. You will be required to demonstrate and simultaneously explain this task from an instructional standpoint.

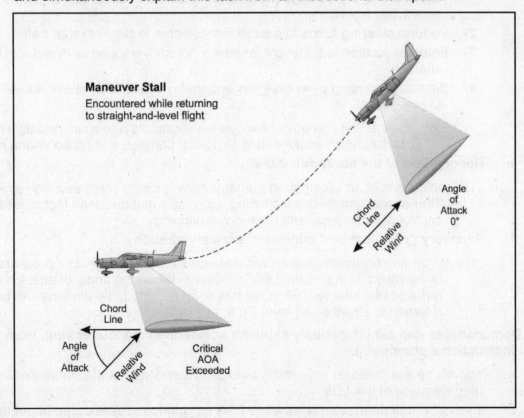

Maneuver Stall
Encountered while returning to straight-and-level flight

Angle of Attack 0°

Chord Line

Relative Wind

Chord Line

Angle of Attack

Relative Wind

Critical AOA Exceeded

B.   Task Objectives

1.   **Exhibit your instructional knowledge of the elements of accelerated maneuver stalls.**

a.   **Aerodynamics of accelerated maneuver stalls**

1)   An accelerated maneuver stall occurs when the airplane's angle of attack exceeds its critical angle attack while recovering from a steep descent too sharply.

a)   The relative wind may be aligned with the airplane's descent angle while recovering from the descent.

i)   This can cause some airplanes to stall at level, or even slightly nose-low attitudes.

b.   **Flight situations where accelerated maneuver stalls may occur**

1)   An accelerated maneuver stall occurs when a pilot pulls out of a steep descent too rapidly.

2)   This can cause excessive load factors on the airplane.

c.   **Hazards of accelerated stalls during stall or spin recovery**

1)   Significant load factor increases are sometimes induced during pullup from a stall or spin recovery.

a)   Structural damage to the airplane can result from the high load factors imposed by intentional stalls practiced above the airplane's maneuvering speed (V_A).

d.   **Entry procedure and minimum entry altitude**

1)   Before demonstrating this maneuver, you must be at a safe altitude that will allow a recovery from the accelerated maneuver stall to be completed no lower than 1,500 ft. AGL.

2)   Perform **clearing turns** to ensure that the area is clear of other traffic.

3)   Enter an accelerated descent, ensuring not to overspeed or overstress the airplane.

4)   Smoothly but rapidly increase the airplane's attitude to or above a level attitude to enter a stall.

a)   Ensure that you do not exceed the airplane's maximum negative load factor, which could lead to structural damage and cause undue hazard.

e.   **Recognition of the accelerated stall**

1)   Indications of an accelerated maneuver stall include continued high sink rate after descent, nose-down pitching, extremely negative load factor, loss of control effectiveness, and nose-down pitching.

f.   **Recovery procedure and minimum recovery altitude**

1)   When an accelerated maneuver stall occurs, the back elevator pressure should be released as in a normal stall recovery. When the angle of attack has been reduced and sufficient airspeed has been regained, the airplane can be returned to straight-and-level flight.

2.   **Demonstrates and simultaneously explains accelerated maneuver stall, from an instructional standpoint.**

a.   You will be evaluated on your ability to explain and demonstrate simultaneously the key elements of this task.

b.   If you do not perform the maneuver to FAA completion standards, point out the deficiency to your FAA inspector/examiner as an example of an error, and explain how it should be corrected (including how and why it occurred).

3. **Exhibit your instructional knowledge by explaining the following common errors related to accelerated maneuver stalls.**

    a. **Failure to establish proper configuration prior to entry**

        1) This maneuver is normally demonstrated in the clean configuration (flaps and gear up) unless otherwise instructed by the aircraft manufacturer.

        2) Ensure that the airplane is in the appropriate configuration before you demonstrate this maneuver to your student (FAA inspector/examiner).

    b. **Improper or inadequate demonstration of the recognition of and recovery from an accelerated maneuver stall**

        1) Explain while you demonstrate what you are doing that will produce an accelerated maneuver stall and why.

        2) Explain the recognition and recovery from an accelerated maneuver stall.

    c. **Failure to present simulated student instruction that adequately emphasizes the hazards of poor procedures in recovering from an accelerated stall**

        1) Provide realistic demonstrations of an accelerated maneuver stall.

        2) Demonstrate problems that can occur when attempting to descend too steeply or recover from a descent too rapidly.

        3) Develop your student's awareness of proper stall recoveries.

4. **Analyze and correct simulated common errors related to accelerated stalls.**

    a. As your FAA inspector/examiner is performing this task, you must be able to recognize, analyze, and correct any error.

    b. Explain what the error is, and then explain what (s)he should do to correct that error.

    c. If your FAA inspector/examiner does not perform the maneuver to FAA completion standards, point out the deficiency to him/her as an example of a student error, and explain how it should be corrected (including how and why it occurred).

END OF TASK -- END OF STUDY UNIT

STUDY UNIT XII
BASIC INSTRUMENT MANEUVERS

This study unit explains the five tasks (A-E) of Basic Instrument Maneuvers. These tasks include both knowledge and skill. Your FAA inspector/examiner is required to test you on at least one of the five tasks; thus, you must be prepared for all five.

Accident investigations reveal that weather continues to be cited as a factor in general aviation accidents more frequently than any other cause. The data also show that weather-involved accidents are more likely to result in fatalities than are other accidents. Low ceilings, rain, and fog continue to head the list in the fatal, weather-involved, general aviation accidents. This type of accident is usually the result of inadequate preflight preparation and/or planning, continued VFR flight into adverse weather conditions, and attempted operation beyond the pilot's experience/ability level.

Pilots cannot cope with flight when external visual references are obscured unless visual reference is transferred to the flight instruments. The motion sensing by the inner ear in particular tends to confuse us. False sensations often are generated, leading you to believe the attitude of your airplane has changed when, in fact, it has not. These sensations result in spatial disorientation.

This training in the use of flight instruments does not prepare a private pilot for unrestricted operations in instrument weather conditions. It is intended as an emergency measure only (although it is also excellent training in the smooth control of an airplane). Intentional flight in such conditions should be attempted only by those who have been thoroughly trained and hold their instrument rating.

The objective of learning basic instrument maneuvers as part of a private pilot (VFR) training syllabus is to allow a pilot to return to VFR conditions should (s)he inadvertently/accidentally fly into instrument conditions. Having some experience flying by instruments and entering instrument conditions briefly will prepare a student for this eventuality as a private pilot.

VIEW LIMITING DEVICES

A.   In order to instruct your student to fly by instrument reference only, you will use an easily
      removable device (e.g., a hood, an extended visor cap, or foggles) that will limit your student's
      vision to the instrument panel.

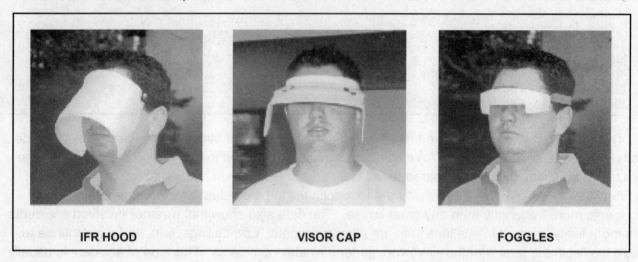

IFR HOOD   VISOR CAP   FOGGLES

B.   These view-limiting devices obviously require acclimation. You should have your student spend a
      few minutes in the airplane with the device on before you meet for your first flight lesson that
      prescribes flying by instrument reference only. This added familiarity with (1) the view-limiting
      device and (2) the location of the instruments and their appearance will make it easier for your
      student to concentrate on flight maneuvers once in the air.

ATTITUDE INSTRUMENT FLYING

A.   Attitude instrument flying may be defined in general terms as the control of an airplane's spatial position by use of instruments rather than by outside visual reference. Thus, proper interpretation of the flight instruments provides the same information as visual references outside the airplane.

1.   Attitude control is stressed in this book (and by the FAA) in terms of pitch control, bank control, power control, and trim control. Instruments are divided into the following three categories:

a.   Pitch instruments

1)   Attitude indicator (AI)
2)   Altimeter (ALT)
3)   Airspeed indicator (ASI)
4)   Vertical speed indicator (VSI)

b.   Bank instruments

1)   Attitude indicator (AI)
2)   Heading indicator (HI)
3)   Turn coordinator (TC) or turn-and-slip indicator (T&SI)
4)   Magnetic compass

c.   Power instruments

1)   Manifold pressure gauge (MP), if equipped
2)   Tachometer (RPM)
3)   Airspeed indicator (ASI)

2.   Write the name of each instrument and the related abbreviation, while thinking about how the instrument looks and what information it provides.

3.   For a particular maneuver or condition of flight, the pitch, bank, and power control requirements are most clearly indicated by certain key instruments.

a.   Those instruments which provide the most pertinent and essential information are referred to as primary instruments.

b.   Supporting instruments back up and supplement the information shown on the primary instruments.

c.   For each maneuver, there will be one primary instrument from each of the above categories. There may be several supporting instruments from each category.

4.   This concept of primary and supporting instruments in no way lessens the value of any particular instrument.

a.   The AI is the basic attitude reference, just as the real horizon is used in visual conditions. It is the only instrument that portrays instantly and directly the actual flight attitude.

1)   It should always be used, when available, in establishing and maintaining pitch and bank attitudes.

5.   Remember, the primary instruments (for a given maneuver) are the ones that will show the greatest amount of change over time if the maneuver is being improperly controlled (pitch, bank, power).

B.   During your attitude instrument training, you should develop three fundamental skills involved in all instrument flight maneuvers: instrument cross-check, instrument interpretation, and airplane control. Trim technique is a skill that should be refined.

   1.   **Cross-checking** (also called scanning) is the continuous and logical observation of instruments for attitude and performance information.

      a.   You will maintain your airplane's attitude by reference to instruments that will produce the desired result in performance. Your author suggests always knowing

         1)   Your airplane's pitch and bank (AI)
         2)   Your present heading (HI)

            a)   And your desired heading

         3)   Your present altitude

            a)   And your desired altitude (ALT)

> The instruments below show straight-and-level flight.

      b.   Since your AI is your reference instrument for airplane control and provides you with a quick reference as to your pitch and bank attitude, it should be your start (or homebase) for your instrument scan. You should begin with the AI and scan one instrument (e.g., the HI) and then return to the AI before moving to a different instrument, as shown below.

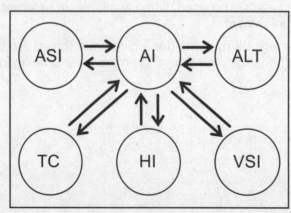

        1) Thus, you continuously visualize your present attitude, heading, and altitude in conjunction with your intended heading and altitude.

        2) Last and certainly not least, interrupt your flight instrument scan every few minutes to review all your other instruments, including

            a) Compass to HI for precession (resetting HI as necessary)
            b) Engine RPM and/or MP, as appropriate
            c) Engine temperatures (oil, cylinder head, and EGT)
            d) Oil pressure
            e) Fuel level
            f) Vacuum pressure
            g) Ammeter

        3) Your CFI will have his/her suggested approach to the instrument scan.

        4) You should write down (using pencil and paper) your scan -- what you do and why.

            a) This will force you to think "what and why" and avoid haphazard scanning of your instruments.

    c. Frequent cross-check faults are

        1) Fixation, or staring at a single instrument

        2) Omission of an instrument from cross-check

        3) Emphasis on a single instrument, instead of a combination of instruments necessary for attitude information

2. **Instrument interpretation** requires you to understand each instrument's construction, operating principle, and relationship to the performance of your airplane.

    a. This understanding enables you to interpret the indication of each instrument during the cross-check.

3. **Airplane control** requires you to maintain your airplane's attitude or change it by interpretation of the instruments. It is composed of three elements.

    a. **Pitch control** is controlling the rotation of your airplane about the lateral axis by movement of the elevators.

        1) After interpreting the pitch attitude from the proper flight instruments, you will exert control pressures to effect the desired pitch with reference to the AI.

    b. **Bank control** is controlling the angle made by the wing and the horizon.

        1) After interpreting the bank attitude from the appropriate instruments, you will exert the necessary pressures to move the ailerons and roll your airplane about the longitudinal axis with reference to the AI.

        2) The rudder should be used as necessary to maintain coordinated flight.

    c. **Power control** is used when interpretation of the flight instruments indicates a need for a change with reference to the RPM.

4. **Trim** is used to relieve all possible control pressures held after a desired attitude has been attained.

    a. The pressures you feel on the controls must be those that you apply while controlling a planned change in airplane attitude, not pressures held because you let the airplane control you.

    b. An improperly trimmed airplane requires constant control pressures, produces tension, distracts your attention from cross-checking, and contributes to abrupt and erratic attitude control.

STRAIGHT-AND-LEVEL FLIGHT

XII.A. TASK: STRAIGHT-AND-LEVEL FLIGHT

REFERENCES: FAA-H-8083-3; FAA-H-8083-15; FAA-S-8081-14.

Objective. To determine that the applicant:

1. Exhibits instructional knowledge of the elements of straight-and-level flight solely by reference to instruments by describing--

   a. Instrument cross-check, instrument interpretation, and aircraft control.

   b. Instruments used for pitch, bank, and power control, and how those instruments are used to maintain altitude, heading, and airspeed.

   c. Trim procedure.

2. Exhibits instructional knowledge of common errors related to straight-and-level flight solely by reference to instruments by describing--

   a. "Fixation," "omission," and "emphasis" errors during instrument cross-check.

   b. Improper instrument interpretation.

   c. Improper control applications.

   d. Failure to establish proper pitch, bank, or power adjustments during altitude, heading, or airspeed corrections.

   e. Faulty trim procedure.

3. Demonstrates and simultaneously explains straight-and-level flight, solely by reference to instruments, from an instructional standpoint.

4. Analyzes and corrects simulated common errors related to straight-and-level flight, solely by reference to instruments.

A.   General Information

1.   The objective of this task is to determine your instructional knowledge of the elements and common errors related to straight-and-level flight solely by reference to instruments. You will be required to demonstrate and simultaneously explain this task from an instructional standpoint.

B.   Task Objectives

1.   **Exhibit your instructional knowledge by explaining the following elements of straight-and-level flight solely by reference to instruments.**

   a.   **Instrument cross-check, instrument interpretation, and aircraft control**

   1)   See the two-page discussion on instrument cross-checking, instrument interpretation, and airplane control beginning on page 458.

   b.   **Instruments used for pitch, bank, and power control, and how those instruments are used to maintain altitude, heading, and airspeed**

   1)   The figure on the opposite page illustrates the instrument indications for straight-and-level flight.

   2)   Maintain straight flight by holding the wings level on the AI and maintaining your heading on the HI.

   a)   Since you want to maintain a specific heading, the HI is primary for bank.

   i)   The supporting instruments for bank are the AI and the TC.

   b)   If you deviate from your heading, use the AI to level your wings and ensure the ball of the TC is centered.

   i)   Determine the direction you must turn to return to your desired heading, and use the AI to establish a bank in the proper direction.

   •   Use an angle of bank no greater than the number of degrees to be turned, but limit the bank angle to that required for a standard-rate turn.

   ii)   Use coordinated aileron and rudder.

c) The ball of the TC should be centered. If not, you may be holding rudder pressure, or your airplane is improperly trimmed (if rudder trim is available).

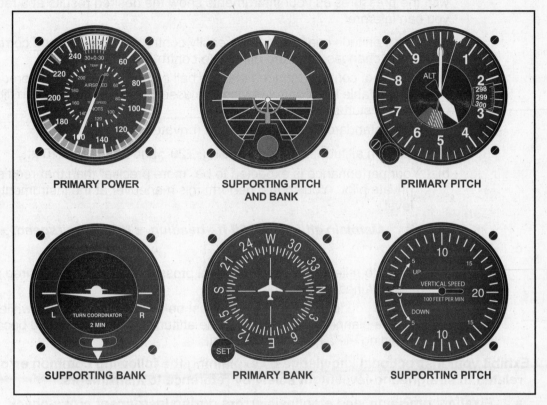

PRIMARY POWER   SUPPORTING PITCH   PRIMARY PITCH
                            AND BANK

SUPPORTING BANK   PRIMARY BANK   SUPPORTING PITCH

3) Maintain level flight by adjusting your pitch as necessary on the AI to maintain your altitude.

   a) Since you want to maintain a specific altitude, the ALT is primary for pitch.

      i) The supporting instruments for pitch are the AI and VSI.

         • As a trend instrument, the VSI will show immediately, even before your ALT, the initial vertical movement of your airplane.

   b) If you deviate from your altitude, use the AI to return to level flight, and determine if you need to climb or descend to return to your desired altitude.

      i) Use the AI to make a small pitch adjustment in the proper direction, and use the VSI to ensure that you are moving in the proper direction.

      ii) Small altitude deviations (i.e., 100 ft. or less) should be corrected with pitch only, using a rate of approximately 200 fpm on the VSI.

      iii) Large altitude deviations (i.e., greater than 100 ft.) may be more easily corrected by adjusting both pitch and power, using a greater rate of return to altitude (approximately double your error in altitude).

   c) The VSI becomes the primary pitch instrument while returning to altitude after a deviation is noticed during level flight.

4) During straight-and-level flight, you should maintain a constant airspeed; thus the ASI is the primary power instrument. Maintain airspeed with power.

5) You will need to learn to overcome a natural tendency to make a large control movement for a pitch change, and learn to apply small control pressures smoothly, cross-checking rapidly for the results of the change and continuing with the pressures as your instruments show the desired results at a rate that you can interpret.

    a) Small attitude changes can be easily controlled, stopped, and corrected.

    b) Large changes are more difficult to control.

6) Coordination of controls requires that the ball of the TC be kept centered and that the available trim control devices be used whenever a change in flight conditions disturbs the existing trim.

7) Completion standards for this maneuver (private pilot only)

    a) Maintain altitude, ±200 ft.; heading, ±20°; and airspeed, ±10 kt.

    b) Your performance is expected to be "more precise" than that required for a private pilot. You should perform this maneuver at the instrument pilot level.

        i) *Maintain altitude, ±100 ft.; heading, ±10°; and airspeed, ±10 kt.*

c. Trim procedure

1) Trim is used to relieve all possible control pressures held after a desired attitude has been attained.

2) The pressure you feel on the controls must be that which you apply while controlling a planned change in airplane attitude, not pressure held because you are letting the airplane control you.

2. Exhibit your instructional knowledge by explaining the following common errors related to straight-and-level flight solely by reference to instruments.

a. Fixation, omission, and emphasis errors during instrument cross-check

1) Fixation, or staring at a single instrument, usually occurs for a good reason, but with poor results.

    a) You may stare at (or fixate on) the ALT, which reads 200 ft. below assigned altitude, wondering how the needle got there. During that time, perhaps with increasing tension on the controls, a heading change occurs unnoticed, and more errors accumulate.

    b) It may not be entirely a cross-checking error. It may be related to difficulties with one or both of the other fundamental skills (i.e., interpretation and control).

2) Omission of an instrument from the cross-check may be caused by failure to anticipate significant instrument indications following attitude changes.

    a) All instruments should be included in the scan.

3) Emphasis on a single instrument, instead of on the combination of instruments necessary for attitude information, is normal during the initial stages of instrument training.

    a) You may tend to rely on the instrument that you understand the best, e.g., the ALT.

    b) The VSI can give more immediate pitch information than the ALT.

b. Improper instrument interpretation

1) This error may indicate that you do not fully understand each instrument's operating principle and relationship to the performance of your airplane.

2) You must be able to interpret small changes in your instrument indications from your cross-checking.

    c.   **Improper control applications**

        1)   This error normally occurs when you incorrectly interpret the instruments and then apply the improper controls to obtain a desired performance, e.g., using rudder pressure to correct for a heading error.

        2)   It may also occur when you apply control inputs (pitch and bank) without referring to the AI.

    d.   **Failure to establish proper pitch, bank, or power adjustments during altitude, heading, or airspeed corrections**

        1)   You must understand which instruments provide information for pitch, bank, and power.

           a)   The AI is the only instrument for pitch and bank control inputs.

        2)   This error may indicate that you do not fully understand instrument cross-check, interpretation, and/or control.

    e.   **Faulty trim procedure**

        1)   Trim should be used, not to substitute for control with the control yoke and rudder, but to relieve pressures already held to stabilize attitude.

        2)   Use trim frequently and in small amounts.

        3)   Improper adjustment of seat or rudder pedals for comfortable positioning of legs and feet may contribute to trim errors.

           a)   Tension in the ankles makes it difficult to relax rudder pressures.

3.   **Demonstrate and simultaneously explain straight-and-level flight solely by reference to instruments from an instructional standpoint.**

    a.   Your demonstration must be to instrument pilot skill level.

    b.   You will be evaluated on your ability to explain and demonstrate simultaneously the key elements of this task.

    c.   If you do not perform the maneuver to FAA completion standards, point out the deficiency to your FAA inspector/examiner as an example of an error, and explain how it should be corrected (including how and why it occurred).

4.   **Analyze and correct simulated common errors related to straight-and-level flight solely by reference to instruments.**

    a.   As your FAA inspector/examiner is performing this task, you must be able to recognize, analyze, and correct any error.

    b.   Explain what the error is, and then explain what (s)he should do to correct the error.

    c.   If your FAA inspector/examiner does not perform the maneuver to FAA completion standards, point out the deficiency to him/her as an example of a student error, and explain how it should be corrected (including how and why it occurred).

END OF TASK

CONSTANT AIRSPEED CLIMBS

XII.B. TASK: CONSTANT AIRSPEED CLIMBS

REFERENCES: FAA-H-8083-3; FAA-H-8083-15; FAA-S-8081-14.

Objective. To determine that the applicant:

1. Exhibits instructional knowledge of the elements of straight and turning, constant airspeed climbs, solely by reference to instruments by describing--

   a. Instrument cross-check, instrument interpretation, and aircraft control.

   b. Instruments used for pitch, bank, and power control during entry, during the climb, and during level-off, and how those instruments are used to maintain climb heading and airspeed.

   c. Trim procedure.

2. Exhibits instructional knowledge of common errors related to straight and turning, constant airspeed climbs, solely by reference to instruments by describing--

   a. "Fixation," "omission," and "emphasis" errors during instrument cross-check.

   b. Improper instrument interpretation.

   c. Improper control applications.

   d. Failure to establish proper pitch, bank, or power adjustments during heading and airspeed corrections.

   e. Improper entry or level-off procedure.

   f. Faulty trim procedure.

3. Demonstrates and simultaneously explains a straight and turning, constant airspeed climb, solely by reference to instruments, from an instructional standpoint.

4. Analyzes and corrects simulated common errors related to straight and turning, constant airspeed climbs, solely by reference to instruments.

A. General Information

   1. The objective of this task is to determine your instructional knowledge of the elements and common errors related to constant airspeed climbs, solely by reference to instruments. You will be required to demonstrate and simultaneously explain this task from an instructional standpoint.

   2. When adverse weather is encountered, a climb by reference to instruments may be required to assure clearance of obstructions or terrain, or to climb above a layer of fog, haze, or low clouds.

   3. For a constant airspeed climb with a given power setting, a single pitch attitude will maintain the desired airspeed.

      a. For some airspeeds, such as V_X or V_Y, the climb power setting and airspeed that will determine this climb attitude are given in the performance data found in your POH.

         1) Most trainer-type airplane manufacturers recommend using maximum power.

   4. Flying straight means to maintain a constant heading on the HI, which is done by keeping the wings level on the AI and the ball centered on the TC.

   5. To perform a turning constant airspeed climb, combine the information from this task with the discussion of Task XII.D., Turns to Headings, beginning on page 476.

B. Task Objectives

1. **Exhibit your instructional knowledge by explaining the following elements of straight and turning, constant airspeed climbs, solely by reference to instruments.**

   a. **Instrument cross-check, instrument interpretation, and aircraft control**

   1) See the two-page discussion on instrument cross-checking, instrument interpretation, and airplane control beginning on page 458.

   b. **Instruments used for pitch, bank, and power control during entry, during the climb, and during level-off, and how those instruments are used to maintain climb heading and airspeed**

   1) The private pilot PTS states that the examiner will specify the climb configuration.

      a) Normally this maneuver is done in the clean configuration using climb power and a cruise climb airspeed.

   2) To enter a constant airspeed climb, use the AI to raise the nose to the approximate pitch attitude for the desired climb speed. Thus, during entry, the AI is primary for pitch.

      a) As the airspeed approaches the desired climb speed, advance the power to the climb power setting (e.g., full power).

      b) In a straight climb, the primary instrument for bank is the HI.

         i) In a turning climb, the primary bank instrument is the TC.

      c) As you establish the climb, you must increase your rate of instrument cross-check and interpretation.

      d) You will need to learn to overcome a natural tendency to make a large control movement for a pitch change, and learn to apply small control pressures smoothly, cross-checking rapidly for the results of the change and continuing with the pressures as your instruments show the desired results at a rate that you can interpret.

         i) Small pitch changes can be easily controlled, stopped, and corrected.

         ii) Large changes are more difficult to control.

      e) Coordination of controls requires that the ball of the TC be kept centered and that the available trim control devices be used whenever a change in flight conditions disturbs the existing trim.

3)   The figure below illustrates the instrument indications for constant airspeed climbs.

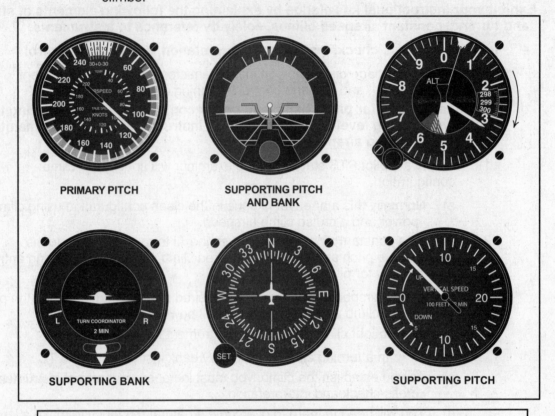

PRIMARY PITCH

SUPPORTING PITCH AND BANK

SUPPORTING BANK

SUPPORTING PITCH

Constant Airspeed Climbs

1.   AI indicates a nose-up pitch attitude.
2.   ALT indicates a climb (from 5,200 ft. MSL toward 5,500 ft. MSL).
3.   VSI indicates a 500-fpm rate of climb.

---

a)   During a constant airspeed climb, the ASI becomes the primary pitch instrument.

   i)   If the airspeed is higher than desired, the pitch must be increased. Use the AI to make a small increase in pitch, and then check the ASI to determine if additional corrections are necessary.

   ii)   If the airspeed is lower than desired, the pitch must be decreased. Use the AI to make a small decrease in pitch, and check the ASI.

b)   The RPM remains the primary power instrument, which is used to ensure that the proper climb power is maintained.

4)   To level off from a climb, it is necessary to start the level-off before reaching the desired altitude. An effective practice is to lead the altitude by 10% of the vertical speed (e.g., at 500 fpm, the lead would be 50 ft.).

a)   Apply smooth, steady forward elevator pressure toward level flight attitude for the speed desired. As the AI shows the pitch change, the VSI will move toward zero, the ALT will move more slowly, and the ASI will increase.

   b) Once the ALT, AI, and VSI show level flight, constant changes in pitch and application of nose-down trim will be required as the airspeed increases.

   c) Maintain straight flight by holding the wings level on the AI and maintaining your heading on the HI.

   i) Maintain turning flight by keeping the miniature airplane's wing tip on the TC on the standard rate mark.

   d) Once again, increase the rate of your cross-check and interpretation during level-off until straight-and-level flight is resumed at cruise airspeed and power.

5) Completion standards for this maneuver (private pilot only)

   a) Level off at the assigned altitude and maintain that altitude, ±200 ft; maintain heading, ±20°; maintain airspeed, ±10 kt.

   b) Your performance is expected to be "more precise" than that required for a private pilot. You should perform this maneuver at the instrument pilot level.

   i) ***Level off at the assigned altitude, ±100 ft.; maintain heading, ±10°; maintain airspeed, ±10 kt.; maintain bank angle, ±5°.***

c. **Trim procedure**

1) Trim is used to relieve all possible control pressures held after a desired attitude has been attained.

2) The pressure you feel on the controls must be that which you apply while controlling a planned change in airplane attitude, not pressure held because you are letting the airplane control you.

2. **Exhibit your instructional knowledge by explaining the following common errors related to straight and turning, constant airspeed climbs, solely by reference to instruments.**

   a. **Fixation, omission, and emphasis errors during instrument cross-check**

   1) Fixation, or staring at a single instrument, usually occurs for a good reason, but with poor results.

      a) You may stare at the ASI, which reads 20 kt. below assigned airspeed, wondering how the needle got there. During that time, perhaps with increasing tension on the controls, a heading change occurs unnoticed, and more errors accumulate.

      b) It may not be entirely a cross-checking error. It may be related to difficulties with one or both of the other fundamental skills (i.e., interpretation and control).

   2) Omission of an instrument from the cross-check may be caused by failure to anticipate significant instrument indications following attitude changes.

      a) All instruments should be included in the scan.

   3) Emphasis on a single instrument, instead of on the combination of instruments necessary for attitude information, is normal during the initial stages of instrument training.

      a) You may tend to rely on the instrument that you understand the best, e.g., the AI.

      b) The ALT will be changing; however, the ASI is primary for pitch during this maneuver.

b. **Improper instrument interpretation**

    1) This error may indicate that you do not fully understand each instrument's operating principle and relationship to the performance of the airplane.

    2) You must be able to interpret even the slightest changes in your instrument indications from your cross-checking.

c. **Improper control applications**

    1) This error occurs when you incorrectly interpret the instruments and/or apply the improper controls to obtain a desired performance, e.g., using power instead of pitch to correct a minor airspeed error.

d. **Failure to establish proper pitch, bank, or power adjustments during heading and airspeed corrections**

    1) You must understand which instruments provide information for pitch, bank, and power.

        a) The AI is the only instrument for pitch and bank control inputs.

    2) This error may indicate that you do not fully understand instrument cross check, interpretation, and/or control.

e. **Improper entry or level-off procedure**

    1) Until you learn and use the proper pitch attitudes in climbs, you may tend to make larger than necessary pitch adjustments.

        a) You must restrain the impulse to change a flight attitude until you know what the result will be.

            i) Do not chase the needles.

            ii) The rate of cross-check must be varied during speed, power, or attitude changes.

            iii) During leveling off, you must note the rate of climb to determine the proper lead.

                • Failure to do this will result in overshooting or undershooting the desired altitude.

    2) You must maintain an accelerated cross-check until straight-and-level flight is positively established.

f. **Faulty trim procedure**

    1) Trim should be used, not to substitute for control with the control yoke and rudder, but to relieve pressures already held to stabilize attitude.

    2) Use trim frequently and in small amounts.

    3) Improper adjustment of seat or rudder pedals for comfortable positioning of legs and feet may contribute to trim errors.

        a) Tension in the ankles makes it difficult to relax rudder pressures.

3. **Demonstrate and simultaneously explain a straight and turning, constant airspeed climb solely by reference to instruments from an instructional standpoint.**

    a. Your demonstration must be to the instrument pilot skill level.

    b. You will be evaluated on your ability to explain and demonstrate simultaneously the key elements of this task.

    c. If you do not perform the maneuver to FAA completion standards, point out the deficiency to your FAA inspector/examiner as an example of an error, and explain how it should be corrected (including how and why it occurred).

4. **Analyze and correct simulated common errors related to straight and turning, constant airspeed climbs solely by reference to instruments.**

    a. As your FAA inspector/examiner is performing this task, you must be able to recognize, analyze, and correct any error.

    b. Explain what the error is, and then explain what (s)he should do to correct the error.

    c. If your FAA inspector/examiner does not perform the maneuver to FAA completion standards, point out the deficiency to him/her as an example of a student error, and explain how it should be corrected (including how and why it occurred).

END OF TASK

CONSTANT AIRSPEED DESCENTS

XII.C. TASK: CONSTANT AIRSPEED DESCENTS

   REFERENCES: FAA-H-8083-3; FAA-H-8083-15; FAA-S-8081-14.

Objective. To determine that the applicant:

1. Exhibits instructional knowledge of the elements of straight and turning, constant airspeed descents, solely by reference to instruments by describing --

   a. Instrument cross-check, instrument interpretation, and aircraft control.

   b. Instruments used for pitch, bank, and power control during entry, during the descent, and during level-off, and how those instruments are used to maintain descent heading and airspeed.

   c. Trim procedure.

2. Exhibits instructional knowledge of common errors related to straight and turning, constant airspeed descents, solely by reference to instruments by describing --

   a. "Fixation," "omission," and "emphasis" errors during instrument cross-check.

   b. Improper instrument interpretation.

   c. Improper control applications.

   d. Failure to establish proper pitch, bank, or power adjustments during heading and airspeed corrections.

   e. Improper entry or level-off procedure.

   f. Faulty trim procedure.

3. Demonstrates and simultaneously explains a straight and turning, constant airspeed descent, solely by reference to instruments, from an instructional standpoint.

4. Analyzes and corrects simulated common errors related to straight and turning, constant airspeed descents, solely by reference to instruments.

A. General Information

   1. The objective of this task is to determine your instructional knowledge of the elements and common errors related to constant airspeed descents, solely by reference to instruments. You will be required to demonstrate and simultaneously explain this task from an instructional standpoint.

   2. When unexpected adverse weather is encountered, the most likely situation is that of being trapped in or above a broken or solid layer of clouds or haze, requiring that a descent be made to an altitude where you can reestablish visual reference to the ground.

   3. A descent can be made at a variety of airspeeds and attitudes by reducing power, adding drag, and lowering the nose to a predetermined attitude. Sooner or later, the airspeed will stabilize at a constant value (i.e., a single pitch attitude will maintain the desired airspeed).

   4. To perform a turning constant airspeed descent, combine the information from this task with the discussion of Task XII.D., Turns to Headings, beginning on page 476.

B.   Task Objectives

   1.   **Exhibit your instructional knowledge by explaining the following elements of straight and turning, constant airspeed descents, solely by reference to instruments.**

      a.   **Instrument cross-check, instrument interpretation, and airplane control**

         1)   See the two-page discussion on instrument cross-checking, instrument interpretation, and airplane control beginning on page 458.

      b.   **Instruments used for pitch, bank, and power control during entry, during the descent, and during level-off, and how those instruments are used to maintain climb heading and airspeed**

         1)   The private pilot PTS states that the examiner will specify the descent configuration.

            a)   Establishing the descent configuration before starting the descent will permit a more stabilized descent and require less division of your attention once the descent is started.

         2)   While a constant airspeed descent can be done at cruise speed, above cruise speed, or below cruise speed, we will limit our discussion to a descent airspeed that is below cruise speed.

            a)   Remember that this instrument training for the private pilot certificate is to prepare your student for emergency situations, and a descent at a speed below cruise speed is easier to control.

            b)   To enter a constant airspeed descent at an airspeed lower than cruise, use the following method:

               i)   Reduce power to a predetermined setting for the descent; thus MP or RPM is the primary power instrument.

               ii)   Maintain straight-and-level flight as the airspeed decreases.

               iii)   As the airspeed approaches the desired speed for the descent, lower the nose on the AI to maintain constant airspeed, and trim off control pressures. The ASI is now the primary pitch instrument.

            c)   In a straight descent, the HI is the primary bank instrument.

               i)   In a turning descent, the primary bank instrument is the TC.

            d)   As you establish the descent, you must increase your rate of instrument cross-check and interpretation.

            e)   You will need to learn to overcome a natural tendency to make a large control movement for a pitch change. You should also learn to apply small control pressures smoothly, cross-checking rapidly for the results of the change and continuing with the pressures as your instruments show the desired results at a rate that you can interpret.

               i)   Small pitch changes can be easily controlled, stopped, and corrected.

               ii)   Large changes are more difficult to control.

            f)   Coordination of controls requires that the ball of the TC be kept centered and that the available trim control devices be used whenever a change in flight conditions disturbs the existing trim.

3) The figure below illustrates the instrument indications for constant airspeed descent, which is similar to that of a climb except the ALT and VSI will indicate a descent.

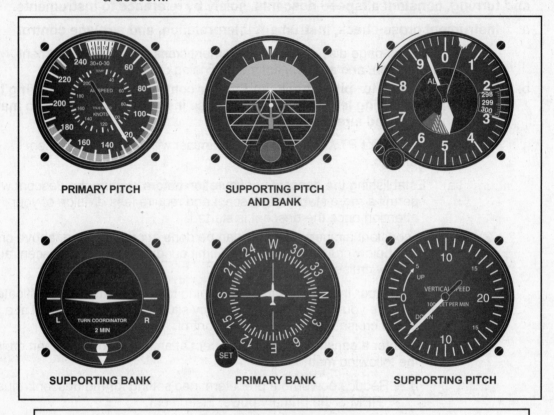

PRIMARY PITCH

SUPPORTING PITCH AND BANK

SUPPORTING BANK

PRIMARY BANK

SUPPORTING PITCH

Constant Airspeed Descent
1. AI indicates a slight nose-down pitch attitude.
2. ALT indicates a descent from 5,200 ft. MSL toward 5,500 ft. MSL.
3. VSI indicates a 500-fpm rate of descent.

---

a) During the constant airspeed descent, the ASI becomes the primary pitch instrument. Make small pitch changes to maintain the desired airspeed.

　　i) If the airspeed is higher than desired, the pitch must be increased. Use the AI to raise the nose, and then check the ASI to determine if additional corrections are necessary.

　　ii) If the airspeed is lower than desired, the pitch must be decreased. Use the AI to lower the nose, and then check the ASI.

b) The MP or RPM remains the primary power instrument, which is used to ensure that the proper descent power is maintained.

      4)   The level-off from a descent must be started before you reach the desired altitude. Assuming a 500-fpm rate of descent, lead the altitude by 100 to 150 ft. for a level-off at an airspeed higher than descending airspeed (i.e., to level off at cruise airspeed).

           a)   At the lead point, add power to the appropriate level flight cruise setting. Since the nose will tend to rise as the airspeed increases, hold forward elevator pressure to maintain the descent until approximately 50 ft. above the altitude; then smoothly adjust pitch to the level flight attitude.

                 i)   Application of trim will be required as you resume normal cruise airspeed.

           b)   Increase your rate of instrument cross-check and interpretation throughout the level-off.

                 i)   Maintain a constant heading by using the HI.

                 ii)   Maintain turning flight by keeping the miniature airplane's wing tip on the TC on the standard rate mark.

      5)   Completion standards for this maneuver (private pilot only)

           a)   Level off at the assigned altitude and maintain that altitude, ±200 ft; maintain heading, ±20°; maintain airspeed, ±10 kt.

           b)   Your performance is expected to be "more precise" than that required for a private pilot. You should perform this maneuver at the instrument pilot level.

                 i)   ***Level off at the assigned altitude, ±100 ft.; maintain heading, ±10°; maintain airspeed, ±10 kt.; maintain bank angle, ±5°.***

  c.   **Trim procedure**

      1)   Trim is used to relieve all possible control pressures held after a desired attitude has been attained.

      2)   The pressure you feel on the controls must be that which you apply while controlling a planned change in airplane attitude, not pressure held because you are letting the airplane control you.

2.   **Exhibit your instructional knowledge by explaining the following common errors related to straight and turning, constant airspeed descents, solely by reference to instruments.**

  a.   **Fixation, omission, and emphasis errors during instrument cross-check**

      1)   Fixation, or staring at a single instrument, usually occurs for a good reason, but with poor results.

           a)   You may stare at the ASI, which reads 20 kt. below assigned airspeed, wondering how the needle got there. During that time, perhaps with increasing tension on the controls, a heading change occurs unnoticed, and more errors accumulate.

           b)   It may not be entirely a cross-checking error. It may be related to difficulties with one or both of the other fundamental skills (i.e., interpretation and control).

      2)   Omission of an instrument from the cross-check may be caused by failure to anticipate significant instrument indications following attitude changes.

           a)   All instruments should be included in the scan.

3)   Emphasis on a single instrument, instead of on the combination of instruments necessary for attitude information, is normal during the initial stages of instrument training.

   a)   You may tend to rely on the instrument that you understand the best, e.g., the AI.

   b)   The ALT will be changing; however, the ASI is primary for pitch during this maneuver.

b.   **Improper instrument interpretation**

1)   This error may indicate that you do not fully understand each instrument's operating principle and relationship to the performance of the airplane.

2)   You must be able to interpret even the slightest changes in your instrument indications from your cross-checking.

c.   **Improper control applications**

1)   This error occurs when you incorrectly interpret the instruments and/or apply the improper controls to obtain a desired performance, e.g., using power instead of pitch to correct a minor airspeed error.

d.   **Failure to establish proper pitch, bank, or power adjustments during heading and airspeed corrections**

1)   You must understand which instruments provide information for pitch, bank, and power.

   a)   The AI is the only instrument for pitch and bank control inputs.

2)   This error may indicate that you do not fully understand instrument cross check, interpretation, and/or control.

e.   **Improper entry or level-off procedure**

1)   Until you learn and use the proper power setting and pitch attitudes in descents, you may tend to make larger than necessary pitch adjustments.

   a)   You must restrain the impulse to change a flight attitude until you know what the result will be.

      i)   Do not chase the needles.

      ii)   The rate of cross-check must be varied during speed, power, or attitude changes on descents.

      iii)   During leveling off, you must note the rate of descent to determine the proper lead.

         •   Failure to do this will result in overshooting or undershooting the desired altitude.

2)   "Ballooning" (allowing the nose to pitch up) on level-off results when descent attitude with forward elevator pressure is not maintained as power is increased.

3)   You must maintain an accelerated cross-check until straight-and-level flight is positively established.

    f.   **Faulty trim procedure**

       1) Trim should be used, not to substitute for control with the control yoke and rudder, but to relieve pressures already held to stabilize attitude.

       2) Use trim frequently and in small amounts.

          a) Trim should be expected during any pitch, power, or airspeed change.

       3) Improper adjustment of seat or rudder pedals for comfortable positioning of legs and feet may contribute to trim errors.

          a) Tension in the ankles makes it difficult to relax rudder pressures.

3. **Demonstrate and simultaneously explain a straight and turning, constant airspeed descent, solely by reference to instruments from an instructional standpoint.**

    a. Your demonstration must be to the instrument pilot skill level.

    b. You will be evaluated on your ability to explain and demonstrate simultaneously the key elements of this task.

    c. If you do not perform the maneuver to FAA completion standards, point out the deficiency to your FAA inspector/examiner as an example of an error, and explain how it should be corrected (including how and why it occurred).

4. **Analyze and correct simulated common errors related to straight and turning, constant airspeed descents, solely by reference to instruments.**

    a. As your FAA inspector/examiner is performing this task, you must be able to recognize, analyze, and correct any error.

    b. Explain what the error is, and then explain what (s)he should do to correct the error.

    c. If your FAA inspector/examiner does not perform the maneuver to FAA completion standards, point out the deficiency to him/her as an example of a student error, and explain how it should be corrected (including how and why it occurred).

END OF TASK

TURNS TO HEADINGS

XII.D. TASK: TURNS TO HEADINGS

REFERENCES: FAA-H-8083-3; FAA-H-8083-15; FAA-S-8081-14.

Objective. To determine that the applicant:

1. Exhibits instructional knowledge of the elements of turns to headings, solely by reference to instruments by describing--

   a. Instrument cross-check, instrument interpretation, and aircraft control.

   b. Instruments used for pitch, bank, and power control during turn entry, during the turn, and during the turn rollout, and how those instruments are used.

   c. Trim procedure.

2. Exhibits instructional knowledge of common errors related to turns to headings, solely by reference to instruments by describing--

   a. "Fixation," "omission," and "emphasis" errors during instrument cross-check.

   b. Improper instrument interpretation.

   c. Improper control applications.

   d. Failure to establish proper pitch, bank, and power adjustments during altitude, bank, and airspeed corrections.

   e. Improper entry or rollout procedure.

   f. Faulty trim procedure.

3. Demonstrates and simultaneously explains a turn to a heading, solely by reference to instruments, from an instructional standpoint.

4. Analyzes and corrects simulated common errors related to turns to a heading, solely by reference to instruments.

A. General Information

   1. The objective of this task is to determine your instructional knowledge of the elements and common errors related to turns to headings, solely by reference to instruments. You will be required to demonstrate and simultaneously explain this task from an instructional standpoint.

   2. Sometimes upon encountering adverse weather conditions, it is advisable for you to use radio navigation aids or to obtain directional guidance from ATC facilities.

      a. Such guidance usually requires that you make turns and/or maintain specific headings.

   3. When making turns in adverse weather conditions, you gain nothing by maneuvering your airplane faster than your ability to keep up with the changes that occur in the flight instrument indications.

      a. You should limit all turns to a standard rate, which is a turn during which the heading changes 3° per sec.

         1) A standard-rate turn is shown on a TC when the wing tip of the representative airplane is opposite the standard rate marker.

         2) On T&SIs, a standard-rate turn is shown when the needle is deflected to the doghouse marker.

         3) Most training airplanes require no more than 15° to 20° of bank for a standard-rate turn.

      b. For small heading changes (less than 15° to 20°), use a bank angle no greater than the number of degrees of turn desired.

         1) The rate at which a turn should be made is dictated generally by the amount of turn desired.

B. Task Objectives

    1. **Exhibit your instructional knowledge by explaining the following elements of turns to headings, solely by reference to instruments.**

        a. **Instrument cross-check, instrument interpretation, and aircraft control**

            1) See the two-page discussion on instrument cross-checking, instrument interpretation, and airplane control beginning on page 458.

        b. **Instruments used for pitch, bank, and power control during turn entry, during the turn, and during the turn rollout, and how those instruments are used**

            1) Before starting the turn to a new heading, you should hold the airplane straight and level and determine in which direction the turn is to be made. Then decide the rate or angle of bank required to reach the new heading.

            2) To enter a turn, use coordinated aileron and rudder pressure to establish the desired bank angle on the AI. If using a standard-rate turn, check the miniature airplane of the TC for the standard rate indication.

                a) Control pitch attitude and altitude throughout the turn as previously described in Task XII.A., Straight-and-Level Flight, beginning on page 460.

            3) To roll out on a desired heading, apply coordinated aileron and rudder pressure to level the wings on the AI and stop the turn.

                a) Begin the rollout about 10° before the desired heading (less for small heading changes).

                b) Adjust elevator pressure referencing the AI to maintain altitude on the ALT.

4)   The figure below illustrates the instrument indications while in a left turn.

PRIMARY POWER   PRIMARY BANK INITIALLY   PRIMARY PITCH
                               SUPPORTING PITCH

PRIMARY BANK AS   SUPPORTING PITCH
TURN IS ESTABLISHED

Turns to a Heading

1.   AI indicates a 20° bank turn to the left.
2.   HI indicates a left turn (from 360° toward 270°).
3.   TC indicates a standard rate turn to the left.

5)   Coordination of controls requires that the ball of the TC be kept centered and
      that the available trim control devices be used whenever a change in flight
      conditions disturbs the existing trim.

6)   Completion standards for this maneuver (private pilot only)

      a)   Demonstrate turns to headings solely by reference to instruments; maintain
            altitude, ±200 ft; maintain a standard-rate turn and roll out on the
            assigned heading, ±20°; maintain airspeed, ±10 kt.

      b)   Your performance is expected to be "more precise" than that required for a
            private pilot. You should perform this maneuver at the instrument pilot
            level.

            i)   *Maintain altitude, ±100 ft.; roll out on the assigned heading,*
                  ±10°; maintain airspeed, ±10 kt.

c. **Trim procedure**

    1)   Trim is used to relieve all possible control pressures held after a desired attitude
          has been attained.

    2)   The pressure you feel on the controls must be that which you apply while
          controlling a planned change in airplane attitude, not pressure held because
          you are letting the airplane control you.

2. **Exhibit your instructional knowledge by explaining the following common errors related to turns to headings, solely by reference to instruments.**

    a. **Fixation, omission, and emphasis errors during instrument cross-check**

        1) Fixation, or staring at a single instrument, usually occurs for a good reason, but with poor results.

            a) You may stare at the TC to maintain a standard-rate turn. During this time, an altitude change occurs unnoticed, and more errors accumulate.

        2) Omission of an instrument from your cross-check may be caused by failure to anticipate significant instrument indications following attitude changes.

            a) All instruments should be included in your scan.

        3) Emphasis on a single instrument, instead of on the combination of instruments necessary for attitude information, is normal in your initial stages of flight solely for reference to instruments.

            a) You will tend to rely on the instrument that you understand the best, e.g., the AI.

    b. **Improper instrument interpretation**

        1) You can avoid this error by understanding each instrument's operating principle and relationship to the performance of the airplane.

    c. **Improper control applications**

        1) Before you start your turn, look at the HI to determine your present heading and the desired heading.

        2) Decide in which direction to turn and how much bank to use; then apply control pressure to turn the airplane in that direction.

        3) Do not rush yourself.

    d. **Failure to establish proper pitch, bank, and power adjustments during altitude, bank, and airspeed corrections**

        1) You must understand which instruments provide information for pitch, bank, and power.

            a) The AI is the only instrument for pitch and bank control inputs.

        2) As control pressures change with bank changes, your instrument cross-check must be increased and pressure adjusted.

    e. **Improper entry or rollout procedure**

        1) This error is caused by overcontrolling, resulting in overbanking on turn entry, and overshooting and undershooting headings on rollout.

            a) Enter and roll out at the rate of your ability to cross check and interpret the instruments.

        2) Maintain coordinated flight by keeping the ball centered.

        3) Remember the heading you are turning to.

    f. **Faulty trim procedure**

        1) Trim should not be used as a substitute for control with the control yoke and rudder pedals, but to relieve pressures already held to stabilize attitude.

        2) Use trim frequently and in small amounts.

        3) You cannot feel control pressures with a tight grip on the control yoke.

            a) Relax and learn to control with the eyes and the brain instead of only the muscles.

3. **Demonstrate and simultaneously explain a turn to a heading solely by reference to instruments from an instructional standpoint.**

   a. Your demonstration must be to the instrument pilot skill level.

   b. You will be evaluated on your ability to explain and demonstrate simultaneously the key elements of this task.

   c. If you do not perform the maneuver to FAA completion standards, point out the deficiency to your FAA inspector/examiner as an example of an error, and explain how it should be corrected (including how and why it occurred).

4. **Analyze and correct simulated common errors related to turns to headings solely by reference to instruments.**

   a. As your FAA inspector/examiner is performing this task, you must be able to recognize, analyze, and correct any error.

   b. Explain what the error is, and then explain what (s)he should do to correct the error.

   c. If your FAA inspector/examiner does not perform the maneuver to FAA completion standards, point out the deficiency to him/her as an example of a student error, and explain how it should be corrected (including how and why it occurred).

END OF TASK

RECOVERY FROM UNUSUAL FLIGHT ATTITUDES

XII.E. TASK: RECOVERY FROM UNUSUAL FLIGHT ALTITUDES

REFERENCES: FAA-H-8083-3; FAA-H-8083-15; FAA-S-8081-14.

Objective. To determine that the applicant:

1. Exhibits instructional knowledge of the elements of recovery from unusual flight attitudes by describing--

   a. Conditions and situations that may result in unusual flight attitudes.

   b. The two basic unusual flight attitudes -- nose-high (climbing turn) and nose-low (diving spiral).

   c. How unusual flight attitudes are recognized.

   d. Control sequence for recovery from a nose-high attitude and the reasons for that sequence.

   e. Control sequence for recovery from a nose-low attitude and the reasons for that sequence.

   f. Reasons why the controls should be coordinated during unusual flight attitude recoveries.

2. Exhibits instructional knowledge of common errors related to recovery from unusual flight attitudes by describing--

   a. Failure to recognize an unusual flight attitude.

   b. Consequences of attempting to recover from an unusual flight attitude by "feel" rather than by instrument indications.

   c. Inappropriate control applications during recovery.

   d. Failure to recognize from instrument indications when the airplane is passing through a level flight attitude.

3. Demonstrates and simultaneously explains a recovery from nose-high and nose-low flight attitude from an instructional standpoint.

4. Analyzes and corrects simulated common errors related to recovery from unusual flight attitudes.

A. General Information

   1. The objective of this task is to determine your instructional knowledge of the elements and common errors related to recovery from unusual flight attitudes. You will be required to demonstrate and simultaneously explain this task from an instructional standpoint.

B. Task Objectives

   1. **Exhibit your instructional knowledge by explaining the following elements of recovery from unusual flight attitudes.**

      a. **Conditions and situations that may result in unusual flight attitudes**

         1) When visual references are inadequate or lost, a pilot may unintentionally let the airplane enter a critical (unusual) attitude. Since such attitudes are unintentional and unexpected, the inexperienced pilot may react incorrectly and stall or overstress the airplane.

         2) Unusual flight attitudes may result from a number of conditions, such as turbulence, disorientation, instrument failure, confusion, preoccupation with cockpit duties, errors in cross-checking or instrument interpretation, or lack of proficiency in airplane control.

         3) Since such flight attitudes are not intentional, they are often unexpected, and the reaction of an inexperienced or inadequately trained pilot is usually instinctive rather than intelligent and deliberate. With practice, the techniques for rapid and safe recovery from these unusual flight attitudes can be learned.

      b. **The two basic unusual flight attitudes -- nose-high (climbing turn) and nose-low (diving spiral)**

         1) If the airspeed is decreasing rapidly, the altimeter indication is increasing faster than desired, and the turn coordinator (or turn needle) indicates a bank, the airplane is **nose-high** (climbing turn).

         2) If the airspeed is increasing rapidly, the altimeter indication is decreasing faster than desired, and the turn coordinator (or turn needle) indicates a bank, the airplane is **nose-low** (diving spiral).

c.   **How unusual flight attitudes are recognized**

1)   As a general rule, any time there is an instrument rate of movement or indication other than those associated with basic instrument flight maneuvers, assume an unusual flight attitude, and increase the speed of cross-check to confirm the attitude, instrument error, or instrument malfunction.

2)   When a critical attitude is noted on the flight instruments, the immediate priority is to recognize what the airplane is doing and decide how to return it to straight-and-level flight as quickly as possible.

3)   In order to avoid aggravating the critical attitude with a control application in the wrong direction, the initial interpretation of the instruments must be accurate.

4)   Nose-high attitudes are shown by the rate and direction of movement of the ALT, VSI, and ASI needles, in addition to the obvious pitch and bank attitude on the AI (see the figure below).

Unusual Attitude -- Nose High

1.   ASI is decreasing from 140 kt. down to 75 kt.
2.   ALT is increasing from 4,500 ft. toward 5,000 ft.
3.   TC indicates a right turn.
4.   HI indicates a right turn from 270° toward 360°.
5.   VSI indicates a positive rate of climb.

5) Nose-low attitudes are shown by the same instruments, but in the opposite direction, as shown in the figure below.

Unusual Attitude -- Nose Low

1. ASI is increasing from 170 kt. to 225 kt.
2. ALT is decreasing from 6,500 ft. to 6,000 ft.
3. TC indicates a right turn.
4. HI indicates a right turn from 270° toward 330°.
5. VSI indicates a negative vertical speed (i.e., a descent).

d. **Control sequence for recovery from a nose-high attitude and the reasons for that sequence**

1) Nose-high unusual attitude is indicated by

   a) Nose high and wings banked on AI
   b) Decreasing airspeed
   c) Increasing altitude
   d) A turn on the TC

2) Take action in the following sequence:

    a) Add power. If the airspeed is decreasing or below the desired airspeed, increase power (as necessary in proportion to the observed deceleration).

    b) Reduce pitch. Apply forward elevator pressure to lower the nose on the AI and prevent a stall.

        i) If no AI is available, sufficient forward pressure should be applied to stop the movement of the pointer on the ALT.

            • The point of reversal is level attitude for that airspeed.

        ii) Deflecting ailerons to level the wings before the angle of attack is reduced could result in a spin.

    c) Level the wings. Correct the bank (if any) by applying coordinated aileron and rudder pressure to level the miniature airplane of the AI and center the ball of the TC.

3) The corrective control applications should be made almost simultaneously but in the sequence above.

4) After initial control has been applied, continue with a fast cross-check for possible overcontrolling, since the necessary initial control pressures may be large.

    a) As the rate of movement of the ALT and VSI needles decrease, the attitude is approaching level flight. When the needles stop and reverse direction, your airplane is passing through level flight for that airspeed.

5) When airspeed increases to normal speed, set cruise power.

e. Control sequence for recovery from a nose-low attitude and the reasons for that sequence

1) Nose-low unusual attitude is indicated by

    a) Nose low and wings banked on AI

    b) Increasing airspeed

    c) Decreasing altitude

    d) A turn on the TC

2) Take action in the following sequence:

    a) Reduce power to idle. If the airspeed is increasing or above the desired speed, reduce power to prevent excessive airspeed (e.g., approaching V_{NE} and loss of altitude).

    b) Level the wings. Correct the bank attitude with coordinated aileron and rudder pressure to straight flight by referring to the AI and TC.

        i) Increasing elevator back pressure before the wings are leveled will tend to increase the bank and make the situation worse.

        ii) Excessive G-loads may be imposed, resulting in structural failure.

    c) Raise the nose. Smoothly apply back elevator pressure to raise the nose on the AI to level flight.

        i) With the higher-than-normal airspeed, it is vital to raise the nose very smoothly to avoid overstressing the airplane.

3) The corrective control applications should be made almost simultaneously but in the previous sequence.

       4)   After initial control has been applied, continue with a fast cross-check for possible overcontrolling, since the necessary initial control pressures may be large.

          a)   As the rate of movement of the ALT and VSI needles decrease, the attitude is approaching level flight. When the needles stop and reverse direction, your airplane is passing through level flight.

       5)   When airspeed decreases to normal speed, set cruise power.

  f.   **Coordination of the controls during recovery**

       1)   As the indications of the ALT, TC, and ASI stabilize, the AI and TC should be checked to determine coordinated straight flight; i.e., the wings are level and the ball is centered.

          a)   Slipping or skidding sensations can easily aggravate disorientation and retard recovery.

          b)   You should return to your last assigned altitude after stabilizing in straight-and-level flight.

       2)   Unlike the control applications in normal maneuvers, larger control movements in recoveries from unusual attitudes may be necessary to bring the airplane under control.

          a)   Nevertheless, such control applications must be smooth, positive, prompt, and coordinated.

          b)   Once the airplane is returned to approximately straight-and-level flight, control movements should be limited to small adjustments.

       3)   Completion standards for this maneuver (private pilot only).

          a)   ***Recognize unusual flight attitudes solely by reference to instruments and recover promptly to a stabilized level flight attitude using proper instrument cross-check and interpretation and smooth, coordinated control application in the correct sequence.***

          b)   Your performance is expected to be "more precise" than that required for a private pilot.

             i)   Since the instrument rating PTS does not have a different (tougher) standard, this more precise performance must be determined by your FAA inspector/examiner through the exercise of subjective judgment.

2. **Exhibit your instructional knowledge by explaining the following common errors related to recovery from unusual flight attitudes.**

  a.   **Failure to recognize an unusual flight attitude**

       1)   This error is due to poor instrument cross-check and interpretation.

       2)   Once you are in an unusual attitude, determine how to return to straight-and-level flight, NOT how your airplane got there.

       3)   Unusually loud or soft engine and wind noise may provide and indication.

b. **Attempting to recover from an unusual flight attitude by "feel" rather than by instrument indications**

1) The most hazardous illusions that lead to spatial disorientation are created by information received by our motion sensing system, located in each inner ear.

2) The motion sensing system is not capable of detecting a constant velocity or small changes in velocity, nor can it distinguish between centrifugal force and gravity.

3) The motion sensing system, functioning normally in flight, can produce false sensations.

4) During unusual flight attitudes, you must believe and interpret the flight instruments because spatial disorientation is normal in unusual flight attitudes.

c. **Inappropriate control applications during recovery**

1) Accurately interpret the initial instrument indications before recovery is started.

2) Follow the recovery steps in sequence.

3) Control movements may be larger, but they must be smooth, positive, prompt, and coordinated.

d. **Failure to recognize from instrument indications when the airplane is passing through a level flight attitude**

1) With an operative attitude indicator, level flight attitude exists when the miniature airplane is level with the horizon.

2) Without an attitude indicator, a level flight is indicated by the reversal and stabilization of the airspeed indicator and altimeter needles.

3. **Demonstrate and simultaneously explain a recovery from a nose-high and a nose-low flight attitude from an instructional standpoint.**

a. Your demonstration must be to the instrument pilot skill level.

b. You will be evaluated on your ability to explain and demonstrate simultaneously the key elements of this task.

c. If you do not perform the maneuver to FAA completion standards, point out the deficiency to your FAA inspector/examiner as an example of an error, and explain how it should be corrected (including how and why it occurred).

4. **Analyze and correct simulated common errors related to recovery from unusual flight attitudes.**

a. As your FAA inspector/examiner is performing this task, you must be able to recognize, analyze, and correct any error.

b. Explain what the error is, and then explain what (s)he should do to correct the error.

c. If your FAA inspector/examiner does not perform the maneuver to FAA completion standards, point out the deficiency to him/her as an example of a student error, and explain how it should be corrected (including how and why it occurred).

END OF TASK -- END OF STUDY UNIT

STUDY UNIT XIII
EMERGENCY OPERATIONS

This study unit explains the four tasks (A-D) of Emergency Operations. These tasks include both knowledge and skill. Your FAA inspector/examiner is required to test you on at least Task A, Emergency Approach and Landing (Simulated), and Task B, Systems and Equipment Malfunctions. You will also need to be prepared to discuss Tasks C and D. You will be tested on these emergency operations in the complex airplane you are using for your practical test.

There are several factors that may interfere with your ability to act promptly and properly when faced with an emergency.

1. Reluctance to accept the emergency situation: Allowing your mind to become paralyzed by the emergency may lead to failure to maintain flying speed, delay in choosing a suitable landing area, and indecision in general.

2. Desire to save the airplane: If you have been conditioned to expect to find a suitable landing area whenever your instructor simulated a failed engine, you may be apt to ignore good procedures in order to avoid rough terrain where the airplane may be damaged. There may be times that the airplane will have to be sacrificed so that you and your passengers can walk away.

3. Undue concern about getting hurt: Fear is a vital part of self-preservation, but it must not lead to panic. You must maintain your composure and apply the proper concepts and procedures.

Emergency operations require that you maintain situational awareness of what is happening. You must help your student to develop an organized process for decision making that can be used in all situations. One method is to use **DECIDE**:

D etect a change -- Recognize immediately when indications, whether visual, aural, or intuitive, are different from those expected.

E stimate need to react -- Determine whether these different indications constitute an adverse situation, and, if so, what sort of action, if any, will be required to deal with it.

C hoose desired outcome -- Decide how, specifically, you would like the current situation altered.

I dentify actions to control change -- Formulate a definitive plan of action to remedy the situation.

D o something positive -- Even if no ideal plan of action presents itself, something can always be done to improve things at least.

E valuate the effects -- Have you solved the predicament, or is further action required?

The following are ideas about good judgment and sound operating practice as you prepare your student to meet emergencies:

1. All pilots hope to be able to act properly and efficiently when the unexpected occurs. As a safe pilot, you should cultivate coolness in an emergency.

2. Know the airplane well enough to interpret the indications correctly before taking corrective action. For this purpose, study the airplane's POH regularly.

3. While difficult, make a special effort to remain proficient in procedures seldom, if ever, used.

4. Do not be reluctant to accept the fact that you have an emergency. Take appropriate action immediately without overreacting. Explain your problem to ATC so they can help you plan alternatives and be in a position to grant you priority.

5. Assume that an emergency will occur every time you take off; i.e., expect the unexpected. If it does not happen, you have a pleasant surprise. If it does, you will be in the correct mind-set to recognize the problem and handle it in a safe and efficient manner.

6. Avoid putting yourself into a situation where you have no alternatives. Be continuously alert for suitable emergency landing spots.

EMERGENCY APPROACH AND LANDING (SIMULATED)

XIII.A. TASK: EMERGENCY APPROACH AND LANDING (SIMULATED)

NOTE: The examiner shall NOT simulate a power failure by placing the fuel selector to the "off" position or by placing the mixture control in the "idle-cutoff" position. No simulated emergency approach shall be continued below 500 feet AGL, unless over an area where a safe landing can be accomplished in compliance with 14 CFR section 91.119.

REFERENCES: FAA-H-8083-3; FAA-S-8081-12; FAA-S-8081-14; POH/AFM.

Objective. To determine that the applicant:

1. Exhibits instructional knowledge of the elements related to an emergency approach and landing by describing --

   a. Prompt establishment of the best glide airspeed and the recommended configuration.

   b. How to select a suitable emergency landing area.

   c. Planning and execution of approach to the selected landing area.

   d. Use of emergency checklist.

   e. Importance of attempting to determine reason for the malfunction.

   f. Importance of dividing attention between flying the approach and accomplishing emergency checklist.

   g. Procedures that can be used to compensate for undershooting or overshooting selected emergency landing area.

2. Exhibits instructional knowledge of common errors related to an emergency approach and landing by describing --

   a. Improper airspeed control.

   b. Poor judgment in the selection of an emergency landing area.

   c. Failure to estimate the approximate wind speed and direction.

   d. Failure to fly the most suitable pattern for existing situation.

   e. Failure to accomplish the emergency checklist.

   f. Undershooting or overshooting selected emergency landing area.

3. Demonstrates and simultaneously explains an emergency approach with a simulated engine failure from an instructional standpoint.

4. Analyzes and corrects simulated common errors related to an emergency approach with a simulated engine failure.

A. General Information

1. The objective of this task is to determine your instructional knowledge of the elements and common errors related to an emergency approach and landing. You will be required to demonstrate and simultaneously explain this task from an instructional standpoint.

   a. The objective of a simulated emergency approach and landing is to develop your student's accuracy, judgment, planning, technique, and confidence when little or no power is available.

2. Emergency approaches and landings can be the result of a complete engine failure, a partial power loss, or a system and/or equipment malfunction that requires an emergency landing during which you may have engine power.

3. During actual forced landings, it is recommended that you maneuver your airplane to conform to a normal traffic pattern as closely as possible.

4. If engine power is lost during the takeoff roll, pull the throttle to idle, apply the brakes, and slow the airplane to a stop.

   a. If you are just lifting off the runway and you lose your engine power, land the airplane straight ahead.

5. If an actual engine failure should occur immediately after takeoff and before a safe maneuvering altitude (at least 500 ft. AGL) is attained, it is usually inadvisable to attempt to turn back to the runway from which the takeoff was made.

    a. Instead, it is generally safer to establish the proper glide attitude immediately and select a field directly ahead or slightly to either side of the takeoff path.

    b. The decision to continue straight ahead is often a difficult one to make unless you consider the problems involved in turning back.

        1) First, the takeoff was in all probability made into the wind. To get back to the runway, you must make a downwind turn, which will increase your groundspeed and rush you even more in the performance of emergency procedures and in planning the approach.

        2) Next, your airplane will lose considerable altitude during the turn and might still be in a bank when the ground is contacted, thus resulting in the airplane cartwheeling.

        3) Last, but not least, after you turn downwind, the apparent increase in groundspeed could mislead you into attempting to slow down your airplane prematurely, thus causing it to stall.

    c. Continuing straight ahead or making only a slight turn allows you more time to establish a safe landing attitude, and the landing can be made as slowly as possible.

        1) Importantly, the airplane can be landed while under control.

6. The main objective when a forced landing is imminent is to complete a safe landing (for you and your passengers) in the largest and best field available.

    a. This objective involves getting the airplane on the ground in as near a normal landing attitude as possible without hitting obstructions.

    b. The airplane may suffer damage, but as long as you keep the airplane under control, you and your passengers should survive.

7. You must stress the need to know and understand the procedures discussed in Section 3, Emergency Procedures, of the POH.

8. You (as a CFI) will normally simulate a complete power loss with the airplane in any configuration above a safe altitude (at least 500 ft. AGL). This simulation is accomplished by reducing power to idle and informing your student that (s)he has just experienced a complete loss of power.

B. Task Objectives

    1. **Exhibit your instructional knowledge by explaining the following elements related to an emergency approach and landing.**

        a. **Prompt establishment of the best glide airspeed and the recommended configuration**

            1) Your first reaction to a complete loss of power should be to establish the best glide attitude immediately and ensure that the landing gear and flaps are retracted (if so equipped).

                a) The best glide airspeed is indicated in the POH.

                    i) In your airplane, best glide airspeed _____.

b)   If the airspeed is above the best glide airspeed, altitude should be maintained, and the airspeed allowed to dissipate to the best glide airspeed.

    i)   When the best glide airspeed is attained, the nose of the airplane should be lowered to maintain that airspeed and the airplane trimmed for the glide.

2)   A constant gliding speed and pitch attitude should be maintained because variations of gliding speed will disrupt any attempt at accuracy in judgment of gliding distance and the landing spot.

3)   Completion standards for this element

a)   Commercial PTS

    i)   ***Establish and maintain the recommended best-glide airspeed, ±10 kt.***

b)   Private PTS

    i)   Establish and maintain the recommended best-glide airspeed, ±10 kt.

b.   **How to select a suitable emergency landing area**

1)   Select a suitable field that is within gliding distance.

2)   Many pilots select from locations in front of them when there may be a perfect site just behind them or on the other side of the airplane. Remember that you may want to perform a 180° turn to look for a suitable field if altitude permits and if you do not have a suitable field in sight.

3)   Be aware of wind direction and velocity both for the desired landing direction and for their effect on glide distance.

4)   Check for traffic, and ask your FAA inspector/examiner to check for traffic.

a)   Inform your FAA inspector/examiner that you, as a CFI, would instruct your student to ask his/her passengers (if appropriate), especially one sitting in the right front seat, to assist in looking for other traffic and pointing it out to him/her.

5)   You should always be aware of suitable forced-landing fields. The perfect field would be an established airport or a hard-packed, long, smooth field with no high obstacles on the approach end. Select the best field available.

a)   A forced landing is a soft-field touchdown without power.

b)   Attempt to land into the wind, although other factors may dictate a crosswind or downwind landing.

    i)   Insufficient altitude may make it inadvisable or impossible to attempt to maneuver into the wind.

    ii)   Ground obstacles may make landing into the wind impractical or inadvisable because they shorten the effective length of the available field.

    iii)   The distance from a suitable field upwind from the present position may make it impossible to reach the field from the altitude at which the engine failure occurs.

    iv)   The best available field may be on a hill and at such an angle to the wind that a downwind landing uphill would be preferable and safer.

    v)   See the top figure on the next page.

c)   Choose a smooth, grassy field if possible. If landing in a cultivated field, land parallel to the furrows. See the bottom figure below.

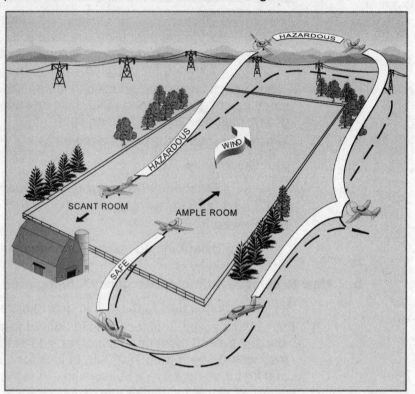

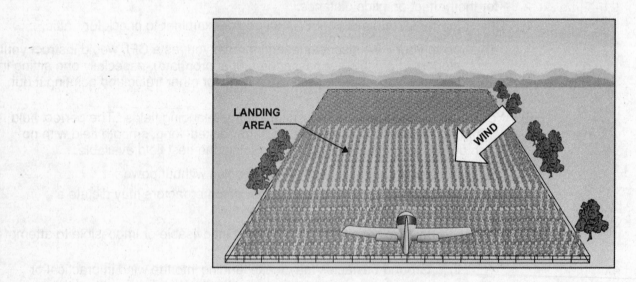

6) Roads should be used only as a last resort. Frequently, power lines cross them, which cannot be seen until you are committed to the road.

    a) Wires often are not seen at all, and the airplane just goes out of control, to the surprise of the pilot.

    b) The presence of wires can be assumed if you see telephone or power poles.

    c) Also, roads must be wide (e.g., four lanes) because of fences, adjacent trees, and road signs.

    d) Use roads only if clear of BOTH traffic and electric/telephone wires.

7) Your altitude at the time of engine failure will determine

    a) The number of alternative landing sites available
    b) The type of approach pattern
    c) The amount of time available to determine and correct the engine problem

8) Identify a suitable landing site, point it out, and explain why that site was chosen.

c. **Planning and execution of approach to the selected landing area**

1) During your selection of a suitable landing area, you should have taken into account your altitude, the wind speed and direction, the terrain, obstructions, and other factors.

    a) Now you must finalize your plan and follow your flight pattern to the landing area.

    b) This step involves performing and explaining what you have decided.

2) You can utilize any combination of normal gliding maneuvers, from wings level to spirals.

    a) You should eventually arrive at the normal "key" position at a normal traffic pattern altitude for the selected field, i.e., abeam the touchdown point on downwind.

        i) If you arrive at the key position significantly higher than pattern altitude, it is recommended that you circle the intended landing point until near pattern altitude.

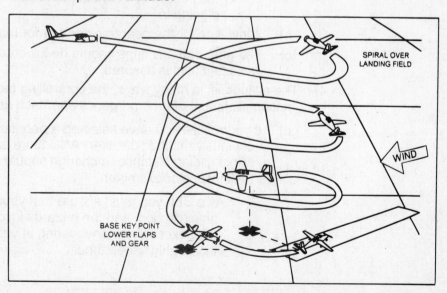

SPIRAL OVER LANDING FIELD

WIND

BASE KEY POINT
LOWER FLAPS
AND GEAR

• Avoid extending your downward leg too far from your landing site.

b)   From this point on, your approach should be similar to a soft-field power-off approach.

   i)   Plan your turn onto the final approach, as shown below.

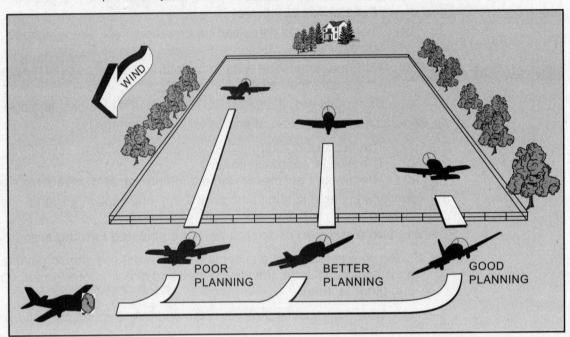

3)   You need to make a decision as to whether to land with the gear up or down (if retractable).

   a)   When the field is smooth and firm, and long enough to bring the airplane to a stop, a gear down landing is appropriate.

      i)   If the field has stumps, rocks, or other large obstacles, the gear down will better protect your student and his/her passengers.

      ii)   If you suspect the field to be excessively soft, wet, short, or snow-covered, a gear-up landing will normally be safer to eliminate the possibility of the airplane nosing over as a result of the wheels digging in.

   b)   Allow time for the gear to extend or for manual extension of the gear.

   c)   The gear and any flaps should be lowered only after a landing at the selected field is assured.

4)   The altitude is, in many ways, the controlling factor in the successful accomplishment of an emergency approach and landing.

   a)   If you realize you have selected a poor landing area (one that would obviously result in disaster) AND there is a more advantageous field within gliding distance, a change should be made and explained to your FAA inspector/examiner.

      i)   As a CFI, you must ensure that your students understand that this is an exception, and the hazards involved in last-minute decisions (i.e., excessive maneuvering at very low altitudes) must be thoroughly understood.

    d.   **Use of emergency checklist**

        1)   Follow the appropriate emergency checklist as shown in the POH, i.e., engine failure.

        2)   You should be in the habit of performing from memory the first few critical steps necessary to get the engine operating again.

            a)   If there is sufficient altitude, use a printed emergency checklist.

                i)   Select the correct checklist and read each item out loud. Comment on your action as you perform the task.

        3)   Simulate reporting to ATC, "Mayday, mayday, mayday."

            a)   The communication radio should be tuned to 121.5 if you are unable to contact an ATC or FSS facility that you were using or monitoring.

            b)   Once communication is established, provide ATC with the airplane call sign, type of airplane, position, problem, and intentions.

            c)   Squawk "7700" on the transponder.

        4)   Once committed to the forced landing, reduce the chance of fire by completing the appropriate checklist in the POH. Normally you should

            a)   Turn the fuel vale, the fuel pump (if electric), and the ignition switch to "OFF," and move the mixture to the idle cut-off position.

                i)   Turn off the master switch after electronically driven flaps and/or landing gear have been extended.

            b)   Wedge the door open to prevent it from being jammed shut upon impact.

                i)   Protect passengers from head injuries with pillows, coats, or other padded items.

        5)   For training or demonstrating emergencies, only simulate these procedures.

    e.   **Importance of attempting to determine reason for the malfunction**

        1)   As you are performing the prescribed emergency checklist, you will be attempting to determine the reason for the loss of engine power and to restart the engine.

            a)   EXAMPLE: If power is restored after switching fuel tanks, check the gauge to see if the tank is empty. If the fuel gauge indicates fuel, the problem may be poor fuel management, faulty gauge, or problems in the fuel line.

        2)   If power is regained, level off and continue circling the selected landing area until you are assured that the problem has been corrected.

    f.   **Importance of dividing attention between flying the approach and accomplishing the emergency checklist**

        1)   You must maintain positive control of your airplane from the time you lose power (simulated or actual) to the minimum safe altitude (simulated) or to the end of the landing roll (actual).

        2)   Maintain your best glide airspeed. Avoid a stall/spin situation at all costs, in both simulated and actual emergencies.

        3)   During the simulation, you (as a CFI) are responsible for maintaining the engine.

            a)   The engine must be kept warm and cleared by gentle bursts of power.

        4)   During an actual forced landing, you need to land as you would on a soft field.

            a)   When in ground effect, slow the airplane by increasing back elevator pressure while attempting to prevent the main gear from contacting the ground.

g. **Procedures that can be used to compensate for overshooting or undershooting the selected emergency landing area**

1) Attempt to fly a base to final approach (altitude permitting).

2) If it appears that you will overshoot the landing area

a) Lengthen the base leg, if the landing area permits.
b) Widen the base leg by heading slightly away from the field.
c) Extend flaps.
d) Use a forward slip on final.

3) If it appears that you will undershoot the landing area

a) Shorten the base leg.
b) Start the turn toward the field to shorten the final approach.
c) Delay in using flaps until the landing area is assured.
d) Do not attempt to raise the nose to extend a glide.
e) In an airplane that has a controllable propeller that uses oil pressure to change the prop from a low-pitch to a high-pitch, the prop control can be pulled back to a high-pitch, low RPM position to extend the glide.

i) This is used as a last resort.

2. **Exhibit your instructional knowledge by explaining the following common errors related to an emergency approach and landing.**

a. **Improper airspeed control**

1) Eagerness to get down is one of the most common errors in this maneuver.

a) In the rush to get down, a pilot may forget about maintaining airspeed and arrive at the edge of the landing area with too much speed to permit a safe landing.

2) While performing the emergency checklist, a pilot may tense and add back pressure on the control to bring the airplane close to a stall.

a) Stress establishing the best glide airspeed and trim the aircraft for hands-off flying.

b. **Poor judgment in the selection of an emergency landing area**

1) Always be aware of suitable fields.

2) At higher altitudes, a pilot may delay in making a decision, and errors in maneuvering and estimating glide distance may develop.

3) As a CFI, demonstrate to your student how to descend from a cruising altitude to a field below the airplane.

a) The use of a steep spiral is not recommended, but rather a gliding descent with shallow turns to remain oriented with the landing area.

b) This method will keep you over a suitable field, whereas gliding in one direction will require you to find another suitable field, if one is available.

c. **Failure to estimate the approximate wind speed and direction**

1) Use all available means to determine wind speed and direction.

a) Smoke, trees, windsocks, and/or wind lines on water are good indicators of surface winds.

b) Be aware of the crab angle you are maintaining for drift correction.

2) Failure to know the wind speed and direction will lead to problems during the approach to your selected field.

    d.   **Failure to fly the most suitable pattern for existing situation**

        1) Constantly evaluate the airplane's position relative to the intended spot for landing.

        2) Attempt to fly as much of a normal traffic pattern as possible since that is familiar and allows you to make needed decisions.

        3) Do not rush to the landing spot and do not attempt to extend a glide to get to a particular spot unless the situation dictates (i.e., the spot selected is without question the only available landing area).

    e.   **Failure to accomplish the emergency checklist**

        1) The checklist is important from the standpoint that it takes you through all the needed procedures to regain power.

        2) If power is not restored, the checklist will prepare you and your airplane for the landing.

    f.   **Undershooting or overshooting selected emergency landing area**

        1) This error is due to poor planning and not constantly evaluating and making the needed corrections during the approach.

        2) Familiarity with glide characteristics and the effects of forward slips, flaps, and gear (if retractable) is essential.

3.   **Demonstrate and simultaneously explain an emergency approach with a simulated engine failure from an instructional standpoint.**

    a.   Your demonstration must be to the instrument pilot skill level.

    b.   You will be evaluated on your ability to explain and demonstrate simultaneously the key elements of this task.

    c.   If you do not perform the maneuver to FAA completion standards, point out the deficiency to your FAA inspector/examiner as an example of an error, and explain how it should be corrected (including how and why it occurred).

4.   **Analyze and correct simulated common errors related to an emergency approach with a simulated engine failure.**

    a.   As your FAA inspector/examiner is performing this task, you must be able to recognize, analyze, and correct any error.

    b.   Explain what the error is, and then explain what (s)he should do to correct the error.

    c.   If your FAA inspector/examiner does not perform the maneuver to FAA completion standards, point out the deficiency to him/her as an example of a student error, and explain how it should be corrected (including how and why it occurred).

END OF TASK

SYSTEMS AND EQUIPMENT MALFUNCTIONS

XIII.B. TASK: SYSTEMS AND EQUIPMENT MALFUNCTIONS

  REFERENCES: FAA-H-8083-3; FAA-S-8081-12; FAA-S-8081-14; POH/AFM.

NOTE: The examiner shall not simulate a system or equipment malfunction in a manner that may jeopardize safe flight or result in possible damage to the airplane.

Objective. Exhibits instructional knowledge of at least five (5) of the systems and equipment malfunctions, appropriate to the airplane used for the practical test by describing recommended pilot action for:

1. Smoke, fire, or both, during ground or flight operations.

2. Rough running engine or partial power loss.

3. Loss of engine oil pressure.

4. Fuel starvation.

5. Engine overheat.

6. Hydraulic malfunction.

7. Electrical malfunction.

8. Carburetor or induction icing.

9. Door or window opening in flight.

10. Inoperative or "runaway" trim.

11. Landing gear or flap malfunction.

12. Pressurization malfunction.

A. General Information

  1. The objective of this task is to determine your instructional knowledge of the elements related to systems and equipment malfunctions, appropriate to the complex airplane you are using for this practical test.

  2. This task will require you to be able to explain causes of, indications of, and recommended pilot actions for at least five systems and equipment malfunctions. Because you do not know which systems and equipment malfunctions will be tested, you must be prepared to discuss all that apply to your airplane.

    a. You must be able to instruct your students in the proper use of Section 3, Emergency Procedures, of the POH.

      1) Your instruction should include both the checklist and amplified procedures.

    b. Ensure that you have these checklists within easy access in the cockpit while flying.

      1) Ensure that you know and use these checklists.

  3. To best prepare for this task, you must have a good working knowledge of all the systems and equipment in your airplane.

  4. The following items are airplane and model specific, so you will need to research each item in your POH.

    a. Our discussion is general and is not a substitute for the manufacturer's recommended procedures.

B. Task Elements

  1. **Smoke, fire, or both, during ground or flight operations**

    a. The presence of fire is noted through smoke, smell, and possible heat in the cabin. It is essential that you check for the source first.

      1) Engine fires are rare in flight.

        a) Perform the checklist in the POH.

        b) Since you will be turning off the fuel to starve the fire, you will be forced to execute an emergency landing.

        c) The use of a slip can help keep the smoke away from the cabin.

2) A fire in the cabin can become a hazardous situation. Know the procedure that is in your POH.

    a) With a fire, you may need to close all vents so as to prevent any drafts.

        i) If available, use an appropriately rated fire extinguisher.

            • During preflight, check the fire extinguisher for serviceability and the types of fires on which it can be used.

        ii) After discharging the extinguisher, ventilate the cabin.

3) If the smoke in the cabin indicates an electrical fire, immediately turn off the master switch.

    a) Then follow the checklist in the POH.

        i) Some airplane manufacturers recommend that the pilot turn on one electrical system item at a time to isolate the cause of the electrical fire.

        ii) Other manufacturers recommend that the pilot keep the electrical system off and land as soon as possible to determine the cause of the electrical fire.

        iii) Your choice of procedures will hinge on the conditions in which you are flying. IMC would lead you to act differently than would VMC.

    b) Land as soon as possible.

4) If smoke is coming through the airplane's vents, turn off any environmental system, i.e., heat, air conditioner.

    a) Close all vents, and then check by opening to determine the source of the smoke.

    b) Land at the nearest airport to investigate the cause.

b. Engine fires during start are normally caused by overpriming.

1) Some airplane manufacturers suggest that you shut off the fuel and attempt to draw the fire back into the engine by cranking.

    a) If the engine has already started, continue operating to attempt to pull the fire into the engine.

    b) Attempt this procedure for only a few seconds before extinguishing the fire by the best available external means.

2) Other manufacturers suggest an immediate shutdown and extinguishing the fire by external means, e.g., fire extinguisher.

3) If smoke is present on the ground, attempt to determine the cause, shut the airplane down, and have a qualified mechanic check it out.

2. **Rough running engine or partial power loss**

    a.   Engine roughness may be caused by magneto, induction system icing, spark plug fouling, or clogged fuel injectors in a fuel-injected engine.

        1) First adjust the mixture.

            a) The engine will run rough if the mixture is too rich or too lean.

        2) Sudden roughness or misfiring is usually caused by magneto problems.

            a) Switch from BOTH to either L or R ignition switch position to identify the bad magneto.

            b) Your student should try different settings and enrich the mixture to determine if (s)he can operate on BOTH magnetos.

            c) If not, operate on the good magneto and proceed to the nearest airport for repairs.

        3) Induction system icing could be a problem. Apply carburetor heat or use the alternate air source in fuel-injected engines. Icing could be the cause of an unexplained drop in RPM or MP and eventual roughness.

        4) Slight engine roughness may be caused by spark plug fouling.

            a) Switch the ignition from BOTH to L or R. An obvious power loss is a sign of fouling or a magneto problem.

            b) Normally, it is due to fouling.

            c) Adjust the mixture to attempt to produce a smoother operation.

        5) Consult your POH for more specific procedures.

    b.   The most common cause of partial power loss or engine roughness is carburetor icing, or induction icing in a fuel-injected engine (see item 8. beginning on page 502).

    c.   Another reason for a partial power loss is an over-lean mixture. In this case, adjust the mixture to a richer setting; then lean the mixture according to you POH.

        1) Also check your fuel system if the engine is running rough, i.e., switch tanks, turn the fuel pump on, and adjust mixture to a richer setting, to see if that solves the problem.

3. **Loss of engine oil pressure**

    a.   Partial loss of oil pressure is normally a result of a malfunction in the oil pressure regulating system.

        1) Land at the nearest airport to investigate the cause and prevent engine damage.

    b.   Complete loss of oil pressure can indicate oil exhaustion or a faulty gauge.

        1) If a complete loss of oil pressure is accompanied by a rise in oil temperature, you should prepare for an engine failure.

4. **Fuel starvation**

    a.   Fuel starvation is normally indicated by a rough engine. It can be caused by an obstruction in the fuel line or by running a fuel tank empty.

        1) Check the fuel-flow gauge (if equipped) for an indication of fuel flowing to the engine.

        2) Adjust the mixture to full rich.

        3) With the electric fuel pump on, check the fuel pressure gauge for normal pressure.

        4) Switch tanks.

    b.   In some airplanes, if you run a fuel tank completely dry so that the engine quits, it may take as long as 10 seconds before fuel is in the fuel lines to restart the engine.

    c.   Fuel starvation is normally due to poor fuel management/planning.

5.   **Engine overheat**

    a.   The oil temperature gauge is the primary engine instrument for determining if the engine is starting to overheat.

        1)   Remember that the gauge will provide a delayed indication.

        2)   High oil temperature can be caused by

            a)   Low oil level

            b)   Obstruction in the oil cooler

            c)   Damaged or improper baffle seals

            d)   Defective gauge

        3)   A rapid rise in temperature is an indication of trouble. Land at the nearest airport for repairs.

    b.   Monitor the airplane's cylinder head and/or exhaust temperature gauges, if equipped.

        1)   If these are showing a higher temperature than normal, increase the mixture to see if the temperature decreases.

        2)   Cowl flaps (if equipped) should be opened to provide cooling.

        3)   During a climb, lower the nose to provide a higher airspeed, thus cooling the engine with an increased flow of air.

        4)   Reduce power during cruise to reduce combustion heating.

6.   **Hydraulic malfunction**

    a.   The hydraulic system is used on some airplanes for the extension and retraction of the landing gear and/or flaps.

    b.   If the hydraulic pump were to fail, there are manual means to raise the gear and/or flaps.

        1)   On some airplanes, there is a system that allows the hydraulic fluid to by-pass the pump in order to lower the landing gear.

    c.   If the system malfunctions because of a leak, some airplanes' gear will free-fall because hydraulic pressure holds the gear up.

        1)   More complex aircraft may have a CO_2 system to lower the gear in the event of a hydraulics malfunction.

7.   **Electrical malfunction**

    a.   Electrical system malfunction can be detected by monitoring the airplane's ammeter or loadmeter gauge.

    b.   A broken alternator belt is normally the cause of an alternator failure.

        1)   Test the alternator by turning on the airplane's landing light. If there is no increase in the ammeter, assume an alternator failure.

        2)   If a failure has occurred, reduce the electrical load by turning off all unnecessary electrical systems.

    c.   With only the battery supplying power, plan about 30 min. of supply. Save as much power as possible, especially if the airplane's landing gear extension system uses an electrically driven pump.

    d.   An excess rate of charge can also have an adverse effect on the airplane's electrical equipment.

        1)   Turn off all unnecessary equipment, and land as soon as possible.

    e.   Follow the procedures in your POH.

8. **Carburetor or induction icing**

    a.   Carburetor icing is one cause of engine failure. The vaporization of fuel, combined with the decreasing air pressure as it flows through the carburetor, causes a sudden cooling of the mixture.

         1)   The temperature of the air passing through the carburetor may drop as much as 15°C (27°F) within a fraction of a second. Water vapor in the air is "squeezed out" by this cooling.

             a)   If the temperature in the carburetor reaches 0°C (32°F) or below, the moisture will be deposited as frost or ice inside the carburetor passages.

             b)   Even a slight accumulation of ice will reduce power and may lead to complete engine failure, particularly when the throttle is partially or fully closed.

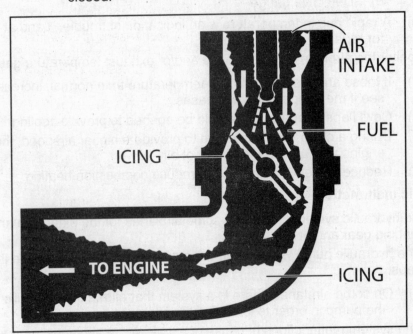

         2)   Conditions conducive to carburetor icing. On dry days, or when the temperature is well below freezing, there is generally not enough moisture in the air to cause trouble. But if the temperature is between −7°C (20°F) and 21°C (70°F), with visible moisture or high humidity, your student should be constantly on the alert for carburetor ice.

         3)   Indications of carburetor icing. For airplanes with fixed-pitch propellers, the first indication of carburetor icing is a loss of RPM. For airplanes with controllable pitch (constant-speed) propellers, the first indication is usually a drop in manifold pressure.

             a)   In both cases, a roughness in engine operation may develop later.

             b)   There will be no reduction in RPM in airplanes with constant-speed propellers, since propeller pitch is automatically adjusted to compensate for the loss of power, thus maintaining constant RPM.

4) Use of carburetor heat. The carburetor heater is an anti-icing/deicing device that preheats the air before it reaches the carburetor. It can be used to melt any ice or snow entering the intake, to melt ice that forms in the carburetor passages (provided the accumulation is not too great), and to keep the fuel mixture above the freezing temperature to prevent formation of carburetor ice.

a) When conditions are conducive to carburetor icing during flight, periodic checks should be made to detect its presence. If detected, FULL carburetor heat should be applied immediately, and the carburetor heater should remain in the "on" position until the pilot is certain that all the ice has been removed. If ice is present, applying partial heat or leaving heat on for an insufficient time might aggravate the condition.

b) When carburetor heat is first applied, there will be a drop in RPM in airplanes equipped with fixed-pitch propellers. There will be a drop in manifold pressure in airplanes equipped with controllable-pitch propellers.

   i) If there is no carburetor ice present, there will be no further change in RPM or manifold pressure until the carburetor heat is turned off. Then the RPM or manifold pressure will return to the original reading before heat was applied.

   ii) If carburetor ice is present, there will normally be a further drop in RPM or manifold pressure after the initial drop (often accompanied by intermittent engine roughness). This drop is due to the ingestion of the water from the melted ice. When the carburetor heat is turned off, the RPM or manifold pressure will rise to a setting greater than that before application of the heat. The engine should also run more smoothly after the ice melts.

c) Whenever the throttle is closed during flight, the engine cools rapidly and vaporization of the fuel is less complete than if the engine is warm. In this condition, the engine is more susceptible to carburetor icing.

   i) If your student suspects carburetor-icing conditions and anticipates closed-throttle operation, the carburetor heat should be turned to "full-on" before closing the throttle and left on during the closed-throttle operation. The heat will aid in vaporizing the fuel and preventing carburetor ice.

   ii) Periodically, the throttle should be opened smoothly for a few seconds to keep the engine warm; otherwise the carburetor heater may not provide enough heat to prevent icing.

d) Use of carburetor heat tends to reduce the output of the engine and also to increase the operating temperature.

   i) Therefore, carburetor heat should not be used when full power is required (as during takeoff) or during normal engine operation except to check for the presence or removal of carburetor ice.

   ii) In extreme cases of carburetor icing, after the ice has been removed, it may be necessary to apply just enough carburetor heat to prevent further ice formation. However, this must be done with caution.

   iii) If carburetor ice still forms, apply FULL heat to remove it.

   NOTE: Partial use of carburetor heat may raise the temperature of the induction air into the range that is likely for the formation of ice, thus increasing the risk of icing.

e) Check the engine manufacturer's recommendations for the correct use of carburetor heat.

      5)   Carburetor air temperature gauge. Some airplanes are equipped with this gauge useful in detecting potential icing conditions.

         a)   Usually, the face of the gauge is calibrated in degrees Celsius. A yellow arc indicates the carburetor air temperatures at which icing may occur. This yellow arc ranges between −15°C and +5°C.

         b)   If the air temperature and moisture content of the air are such that carburetor icing is improbable, the engine can be operated with the indicator in the yellow range with no adverse effects.

         c)   However, if the atmospheric conditions are conducive to carburetor icing, the indicator must be kept outside the yellow arc by application of carburetor heat.

      6)   Outside air temperature gauge (OAT). Most airplanes are equipped with this gauge calibrated in degrees both Celsius and Fahrenheit. It is useful for obtaining the outside or ambient air temperature for calculating true airspeed and also in detecting potential icing conditions.

   b.   Induction system icing. This kind of icing forms under a variety of conditions on both piston and turbine aircraft. As air is ingested through the engine intakes, the moisture can freeze inside the induction system, reducing or stopping the flow of combustible air into the engine. The ice can also form on the exterior of the airplane and clog the air intake openings. It is possible in engines equipped with either carburetors or fuel injectors.

      1)   The air intake for the fuel injection system is somewhat similar to that used in the carburetor system. The fuel injection system, however, is equipped with an alternate air source located within the engine cowling. It is used if the external air source is obstructed by ice or other matter. The alternate air source is usually operated automatically but has a backup manual system that can be used if the automatic feature malfunctions.

9.   **Door or window opening in flight**

   a.   Door opening during flight can be recognized by an increase in noise and/or a change in the cabin airflow.

      1)   Check your POH for the proper procedures for closing a door in flight. If there are none, land as soon as possible to correct the situation.

         a)   Maintain control of the airplane before attempting to close the door.

10.   **Inoperative or "runaway" trim**

   a.   An inoperative trim will not allow you to reduce the control pressures.

      1)   Plan for this during descent and especially during approach and landing.

      2)   You may need to plan to land at a higher airspeed and a longer landing distance if you are unable to apply enough control pressure to keep the nose up.

   b.   If using an electric-powered trim, disengage the electric connection, and determine if you can manually trim the airplane.

   c.   Follow the recommended procedures in the POH.

11. **Landing gear or flap malfunction**

    a.   Landing gear or flap malfunction can occur during retraction or extension or may be caused by another system malfunction, e.g., electrical failure.

    b.   Landing gear malfunction is normally shown by an indicator or a lack of indication on the instrument panel.

        1)   The gear lever may be down, but there is no green-light indication that the gear is down and locked.

        2)   Check the indicators to ensure that a bulb is not burned out.

        3)   You can manually lower the landing gear in some airplanes, whereas in others you can release hydraulic pressure to allow the gear to fall.

        4)   A thorough knowledge of the system will allow you to analyze the problem. Follow the procedures in the POH.

    c.   Wing flap malfunction is normally caused by an obstruction or a broken control.

        1)   Understanding the effect of wing flaps on the airflow will help you through this situation.

        2)   You can attempt to cycle the flaps by using the flap control to lower and raise the flaps.

        3)   Leave the flap selector in the up position to relieve any mechanical pressure holding the flaps down. This will possibly allow the airflow to reduce the angle of flap.

            a)   Prior to landing, set the flap lever to the setting at which the flaps are stuck. This will preclude any dramatic aerodynamic changes should the flaps retract during the landing sequence.

        4)   Operate below the maximum flap extension airspeed.

        5)   Follow the checklist in the POH and land as soon as possible.

12. **Pressurization malfunction**

    a.   A loss of pressurization (decompression) is the inability of the airplane's pressurization system to maintain its design pressure differential.

        1)   This problem is caused by a malfunction in the pressurization system or by structural damage to the airplane.

        2)   You must monitor the systems instruments for signs of problems.

    b.   During a rapid decompression, there may be a noise, and you may feel dazed. The cabin air may fill with fog, dust, or flying debris, depending on how rapidly the decompression occurs.

        1)   The primary danger of decompression is hypoxia. You must descend or use supplemental oxygen, especially if it is a rapid decompression.

    c.   Follow the recommended procedures in the POH.

13. This task does not require simultaneous demonstration and explanation of systems and equipment malfunctions.

    a.   Nonetheless, you should be able to explain and demonstrate the proper procedures to be used in this task.

    b.   Explain any deficiency in describing the recommended pilot actions of this task to your FAA inspector/examiner as an example of an error, and then provide the information as to where you can locate the correct procedure.

14. This task does not require you to analyze and correct common errors related to systems and equipment malfunctions, but you can and will (as a CFI) critique your student's emergency procedures.

END OF TASK

EMERGENCY EQUIPMENT AND SURVIVAL GEAR

XIII.C. TASK: EMERGENCY EQUIPMENT AND SURVIVAL GEAR

REFERENCES: FAA-H-8083-3; FAA-S-8081-12; FAA-S-8081-14; POH/AFM.

Objective. To determine that the applicant exhibits instructional knowledge of the elements related to emergency equipment and survival gear appropriate to the airplane used for the practical test by describing:

1. Equipment and gear appropriate for operation in various climates, over various types of terrain, and over water.

2. Purpose, method of operation or use, servicing and storage of appropriate equipment.

A. General Information

1. The objective of this task is to determine your instructional knowledge of the elements related to emergency equipment and survival gear appropriate to the complex airplane you are using for this practical test.

2. This is specific for your airplane. You should know what emergency equipment and survival gear are on your airplane and be able to instruct a student on their use.

B. Task Elements

1. **Equipment and gear appropriate for operation in various climates, over various types of terrain, and over water**

   a. Survival kits should have appropriate equipment and gear for the climate and terrain over which your flight will be conducted

   b. Different items are needed for cold versus hot weather and mountainous versus flat terrain.

      1) Survival manuals that are published commercially and by the government suggest items to be included.

   c. While no FAR requires any type of survival gear for over-water operations under Part 91 (other than large and turbine-powered multiengine airplanes), it is a good operating practice to provide a life preserver and a life raft to accommodate everyone on the airplane.

   d. It is best to be prepared for an emergency.

2. **Purpose, method of operation or use, servicing and storage of appropriate equipment**

   a. Purpose

      1) An ELT will transmit on the emergency frequencies 121.5 and 243.0 and is used as a homing beacon for search airplanes.

      2) Fire extinguishers are used for cabin and possibly engine fires.

         a) You must check the fire extinguisher for the types of fires on which it can be used.

      3) Survival gear is used to help in signaling search airplanes and to support short-term survival.

   b. Method of operation or use.

      1) ELTs are normally automatically activated upon an impact of sufficient force (approximately 5 Gs).

         a) There may be a switch in the cockpit with which you can manually activate the ELT.

            i) If not, you can access the ELT and manually activate it.

            ii) You should show your students how to do this.

      b)   If you must make an emergency landing, you will want to use the manual switch to activate the ELT since your landing may not cause the ELT to activate automatically.

    2)   Follow the instructions for operation and use of the fire extinguisher and any survival gear.

c.   Servicing

    1)   The batteries used in the ELT must be replaced (or recharged, if the batteries are rechargeable) when the ELT has been in use for more than 1 cumulative hour or when 50% of their useful life (as established by the manufacturer) has expired (FAR 91.207).

        a)   The new expiration date for replacing (or recharging) the battery must be marked on the outside of the transmitter and entered in the airplane's maintenance record.

        b)   The ELT must be inspected annually and this date entered in the airplane's maintenance record.

    2)   A fire extinguisher will normally have a gauge by the handle to indicate if it is properly charged and a card attached to tell when the next inspection is required.

        a)   The gauge and the card should be checked during your visual inspection.

        b)   Most fire extinguishers should be checked and serviced by an authorized person.

    3)   Periodically remove the items in your survival kit and check them for serviceability.

d.   Storage

    1)   Ensure that the ELT and fire extinguisher are stored and approximately secured in the airplane.

    2)   While in the airplane, your survival gear should be easily accessible and secured by tie-down or safety belts.

        a)   When you are not flying, your survival gear should be stored in a cool, dry place.

3.   This task does not require simultaneous demonstration and explanation of emergency equipment and survival gear.

    a.   Nonetheless, you should be able to explain and demonstrate (if appropriate) emergency equipment and survival gear.

    b.   Explain any deficiency in your discussion of emergency equipment and survival gear to your FAA inspector/examiner as an example of an error, and then provide the correct information or where to locate the correct information.

4.   This task does not require you to analyze and correct common errors relating to emergency equipment and survival gear, but you can and will (as a CFI) critique your student's knowledge of this task.

END OF TASK

EMERGENCY DESCENT

XIII.D. TASK: EMERGENCY DESCENT

REFERENCES: FAA-H-8083-3; FAA-S-8081-12; FAA-S-8081-14; POH/AFM.

Objective. To determine that the applicant exhibits instructional knowledge of the elements related to emergency descents appropriate to the airplane flown by describing:

1. Exhibits instructional knowledge of the elements related to an emergency descent by describing --

 a. Situations that require an emergency descent.

 b. Proper use of the prescribed emergency checklist to verify accomplishment of procedures for initiating the emergency descent.

 c. Proper use of clearing procedures before initiating and during the emergency descent.

 d. Procedures for recovering from emergency descent.

 e. Manufacturer's procedures.

2. Exhibits instructional knowledge of common errors related to an emergency descent by describing --

 a. The consequences of failing to identify reason for executing an emergency descent.

 b. Improper use of the prescribed emergency checklist to verify accomplishment of procedures for initiating the emergency descent.

 c. Improper use of clearing procedures before initiating and during the emergency descent.

 d. Improper procedures for recovering from an emergency descent.

3. Demonstrates and simultaneously explains an approach and landing with a simulated inoperative engine from an instructional standpoint.

4. Analyzes and corrects simulated common errors related to an approach and landing with an inoperative engine.

A. General Information

 1. The objective of this task is for you to demonstrate your ability to teach students to recognize the need for and perform an emergency descent.

B. Task Objectives

 1. **Exhibit your instructional knowledge related to an emergency descent by describing the following elements.**

 a. **Situations that require an emergency descent**

 1) There are some emergencies for which you may choose an emergency descent is your best course of action, such as:

 a) Decompression
 b) Cockpit smoke and/or fire

 b. **Proper use of the prescribed emergency checklist to verify accomplishment of procedures for initiating the emergency descent**

 1) Follow the procedures outlined in Section 3, Emergency Procedures, of your airplane's Pilot's Operating Handbook (POH).

 c. **Proper use of clearing procedures before initiating and during the emergency descent**

 1) Before beginning the maneuver, clear the area visually.

 a) This should include **clearing turns**.

 2) Before beginning this maneuver, if equipped, broadcast your intentions on a local practice area frequency or multicom.

 3) During the maneuver, continue to scan for traffic.

    d. **Procedures for recovering from an emergency descent**

       1) Procedures differ between aircraft.

       2) A general method might be:

          a) Slowly, smoothly return the airplane to a level attitude.

          b) Apply an appropriate throttle setting for level flight.

          c) As appropriate, retract any gear, flaps, and/or spoilers that were extended during the maneuver.

          d) As appropriate, trim the airplane and lean the mixture.

          e) Perform a post-maneuver checklist.

    e. **Manufacturer's procedures**

       1) The airplane manufacturer's recommended procedures will be located in Section 3, Emergency Procedures, of the airplane's Pilot's Operating Handbook (POH).

       2) Regardless of whether a procedure is given, ensure that you **do not exceed V_{NE}, V_{LE}, or V_{FE}**.

       3) Further, ensure that you maintain **positive control** throughout the maneuver.

          a) If you deem safety may be (or is nearly) compromised, abort the maneuver.

2. **Exhibit your instructional knowledge of emergency descent by describing the following common errors.**

    a. **The consequences of failing to identify reasons for executing an emergency descent**

       1) You should only begin an emergency descent when you have established it is necessary.

    b. **Improper use of the prescribed emergency checklist to verify accomplishment of procedures for initiating the emergency descent**

       1) You must use checklists before beginning this maneuver to ensure the area is clear and the airplane is configured for the maneuver.

    c. **Improper use of clearing procedures before initiating and during the emergency descent.**

       1) Before and during this maneuver, you should clear the area both visually and by multicom broadcast (where applicable).

    d. **Improper procedures for recovering from an emergency descent.**

       1) Recovering from the descent requires a transition between flight phases that can be dangerous.

       2) Make your actions smooth to ensure you do not exceed the airplane's critical load factor.

       3) If applicable, follow manufacturer's procedures.

3. **Demonstrate and simultaneously explain an approach and landing with a simulated inoperative engine from an instructional standpoint.**

    a. Your demonstration must be to commercial pilot skill level.

    b. You will be evaluated on your ability to explain and demonstrate simultaneously the key elements of this task.

    c. If you do not perform the maneuver to FAA completion standards, point out the deficiency to your FAA inspector/examiner as an example of an error, and explain how it should be corrected (include how and why it occurred).

4. **Analyze and correct simulated common errors related to an approach and landing with an inoperative engine.**

   a. As your FAA inspector/examiner is performing this task, you must be able to recognize, analyze, and correct any error.

   b. Explain what the error is, and then explain what (s)he should do to correct the error.

   c. If your FAA inspector/examiner does not perform the maneuver to FAA completion standards, point out the deficiency to him/her as an example of a student error, and explain how it should be corrected (including how and why it occurred).

END OF TASK -- END OF STUDY UNIT

STUDY UNIT XIV
POSTFLIGHT PROCEDURES

This study unit explains the only task of Postflight Procedures. This task includes both knowledge and skill. Your FAA inspector/examiner is required to test you on this task.

POSTFLIGHT PROCEDURES

XIV.A. TASK: POSTFLIGHT PROCEDURES

　　REFERENCES: FAA-H-8083-3; FAA-S-8081-12; FAA-S-8081-14; POH/AFM.

Objective. To determine that the applicant:

1. Exhibits instructional knowledge of the elements of postflight procedures by describing --

    a. Parking procedure.
    b. Engine shutdown and securing cockpit.
    c. Deplaning passengers.
    d. Securing airplane.
    e. Postflight inspection.
    f. Refueling.

2. Exhibits instructional knowledge of common errors related to postflight procedures by describing --

    a. Hazards resulting from failure to follow recommended procedures.

    b. Poor planning, improper procedure, or faulty judgment in performance of postflight procedures.

A. General Information

1. The objective of this task is to determine your instructional knowledge of the elements and common errors related to postflight procedures. You will be required to demonstrate and simultaneously explain this task from an instructional standpoint.

2. A flight is never complete until the airplane is parked, the engine shut down, and the airplane secured.

B.   Task Objectives

1.   **Exhibit your instructional knowledge of postflight procedures by describing the following elements.**

a.   **Parking procedure**

1)   You should select a parking spot based on airport custom and be considerate of other pilots and airport personnel.

2)   When you are taxiing on a ramp, a lineman may give you hand signals to tell you where to taxi and/or park your airplane.

a)   Hand signals are used by all ground personnel and are similar at all airports; i.e., this is an international language.

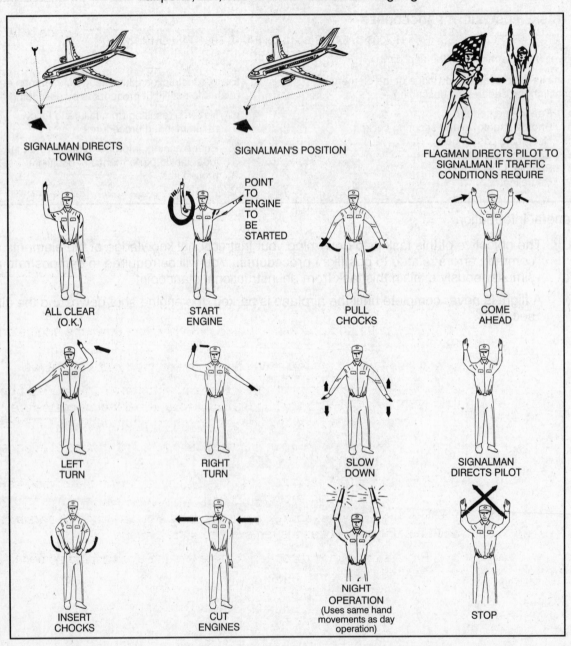

3) The airplane should be parked on the ramp in such a way as to facilitate taxiing and parking by other airplanes and to avoid being struck by other airplanes or their prop/jet wash.

    a) Frequently, airport ramps are marked with painted lines that indicate where and how to park. At other airports, airplane tiedown ropes (or chains) mark parking spots.

        i) However, these markings do not guarantee adequate spacing from other parked airplanes. You must ensure proper spacing.

    b) Almost always, there are three ropes provided for each airplane: one rope positioned for the middle of each wing and one rope to tie the tail. If the ramp is not paved, each of the tiedown ropes (chains) is usually marked by a tire.

4) You should chock and/or tie down the airplane so it cannot roll or be blown into another aircraft or other object.

    a) Chocks are usually blocks of wood placed both in front of and behind a tire to keep the airplane from rolling.

    b) On an unfamiliar ramp, place the front chock an inch or so ahead of the tire. In this way it will become evident whether the airplane will roll forward later as you prepare to depart.

5) Ensure that the nose or tailwheel is straight when your student parks the airplane.

6) At most transient ramps, do not use the parking brake because the FBO personnel frequently move aircraft.

    a) The normal procedure is to lock the airplane with parking brakes off -- wheel chocks or tiedowns secure the airplane.

    b) In many airplanes, leaving the brake on is not recommended because it may cause the hydraulic lines to burst.

b. **Engine shutdown and securing cockpit**

1) Follow the prescribed checklist in the POH for shutting down the engine and securing the cockpit.

2) Instruct your student to read each item aloud and then perform that task.

    a) While the engine is running, the ignition switch should be moved from BOTH to OFF and back to BOTH to ensure that the magnetos are properly grounded when the ignition switch is turned to the OFF position.

        i) This magneto grounding check should be performed at every engine shutdown.

    b) Ensure that the magnetos are turned off by removing the ignition key immediately upon shutdown.

3) Once the engine is shut down, secure the cockpit by gathering all personal items and ensuring that all trash is removed from the airplane.

    a) Professionalism and courtesy dictate that the airplane should be left as it was found.

c.   **Deplaning passengers**

    1)   Ensure that your passengers remain seated with their safety belts fastened until the engine is shut down. Then they should gather all personal belongings and deplane in a safe manner.

    2)   Inform passengers of the safe exit from the ramp area to the terminal or have them remain next to the airplane while you finish conducting the postflight procedures.

        a)   At that time, you can safely escort them off the ramp area.

d.   **Securing airplane**

    1)   Obviously, hangar storage is the best means of protecting aircraft from the elements, flying debris, vehicles, vandals, etc. Even in hangars, airplanes should be chocked to avoid scrapes and bumps from rolling.

    2)   Airplanes stored outside are normally tied down.

        a)   Chains or ropes are used to secure the airplane to the ground from three locations on the airplane: usually, the midpoint of each wing and the tail.

        b)   Tiedown hooks or eyelets are provided at these locations on most airplanes.

    3)   Tiedown knot for ropes

        a)   Pull the rope through the hook or eyelet, and then use two half-hitches as illustrated below.

            i)   The first set of half-hitches should be tied approximately 6 in. from the hook or eyelet, and the second set of half-hitches should be about 1 ft. away from the first set.

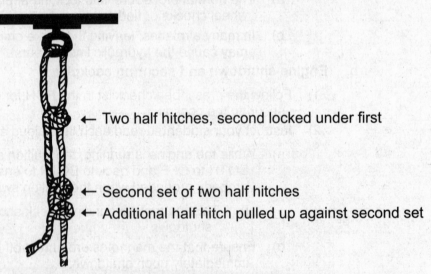

← Two half hitches, second locked under first

← Second set of two half hitches
← Additional half hitch pulled up against second set

        b)   Tie one half-hitch; then the second half-hitch should lock under the first half-hitch.

            i)   After you tie the second set of half-hitches, tie an additional half-hitch and pull it up against the second set as a safety.

        c)   This knot is relatively quick and easy to tie and untie.

        d)   Show your student how to tie the knot, and have him/her practice it five or six times.

        e)   Make sure the ropes are properly secured to the ground at appropriate intervals.

4) When leaving the airplane tied down for an extended period of time or when expecting windy weather, install the control or gust locks to hold the control yoke stationary so the control surfaces cannot bang back and forth in the wind.

    a) On older airplanes, this precaution is sometimes accomplished by clamping the aileron, elevator, and rudder to adjacent stationary surfaces so they cannot move.

    b) Alternatively, the control yoke (or stick in older airplanes) can be secured tightly with a seatbelt.

e. **Postflight inspection**

1) Inspect the outside of the airplane for any damage that may have occurred during the flight.

2) Also inspect the underside of the fuselage to note any excessive oil being blown out of the engine.

3) Note any malfunctions (discrepancies) in the proper logbooks, and signal to other pilots when an unairworthy condition exists. Always take the airplane out of service if there is an airworthiness problem.

4) Complete the appropriate checklists in the POH for postflight procedures.

f. **Refueling**

1) Instruct your student that the airplane should be refueled after each flight or at least after the last flight of the day.

    a) Refueling will prevent moisture condensation within the tank since no airspace will be left.

2) Avoid refueling the airplane from cans or drums because this increases the risk of fuel contamination.

3) Static electricity created by the friction of air passing over the surfaces of an airplane in flight and by the flow of fuel through the hose and nozzle creates a fire hazard during refueling.

    a) A ground wire should be attached to the airplane before the fuel tank cap is removed.

    b) The refueling nozzle should be grounded to the airplane before refueling is begun and throughout the process.

2. **Exhibit your instructional knowledge by explaining the following common errors related to postflight procedures.**

a. **Hazards resulting from failure to follow recommended procedures**

1) The checklist for postflight procedures is as important as those for any other situation. You must follow recommended procedures to prevent creating unsafe situations.

b. **Poor planning, improper procedure, or faulty judgment in performance of postflight procedures**

1) Just because this is the end of a flight, do not let your student get rushed or practice bad habits in conducting postflight procedures.

2) This task must be approached in the same professional manner as the preflight and flying procedures.

3.   This task does not require you to demonstrate and simultaneously explain postflight procedures from an instructional standpoint, but you can and will (as a CFI) demonstrate and explain postflight procedures to your students.

   a.   Your demonstration must be to commercial pilot skill level.

   b.   Point out any deficiency in your postflight procedures to your FAA inspector/examiner as an example of an error, and explain how it should be corrected (including how and why it occurred).

4.   This task does not require you to analyze and correct simulated common errors related to postflight procedures, but you can and will (as a CFI) analyze and correct your students' errors related to postflight procedures.

   a.   As your student is performing this task, you should be able to recognize, analyze, and correct any error.

   b.   Explain what the error is, and then explain what (s)he should do to correct the error.

   c.   If you do not perform these procedures to FAA completion standards, point out the deficiency to your FAA inspector/examiner as an example of a student error, and explain how it should be corrected (including how and why it occurred).

END OF TASK

The Orlando Flight Standards District Office (FSDO) compiled the following list. See page 55 for information on the Orlando FSDO CFI Special Emphasis Program.

COMMON WEAK AREAS OBSERVED ON INITIAL CFI PRACTICAL TESTS

AERODYNAMICS:

1. How lift, weight, thrust, and drag act on aircraft in a climb or descent (PH Ch. 1)
2. Why maneuvering speed varies with weight (PH Ch. 1)
3. Explanation of the relationship between center of pressure and center of gravity (PH Ch. 1)
4. Explanation of trim devices (p. 114)
5. Explanation of spin entries, spins, and spin recovery techniques (p. 444)

AIRSPACE:

1. Explanation of communication requirements at an airport with an operating control tower located within Class G airspace (p. 155)
2. Explanation of VFR weather minimums and operational requirements at an airport having Class E airspace beginning at the surface (p. 154)
3. Explanation of Class E airspace starting at other than the surface, 700', or 1,200' (PH Ch. 3)
4. How to obtain a special VFR clearance (p. 153)
5. Explanation of operations within a Terminal Radar Service Area (p. 156)

ENDORSEMENTS:

1. Knowledge of logbook endorsements necessary to be eligible for a practical test (p. 168)
2. Instructor endorsements required from first solo through private pilot certification (p. 165)
3. Instructor logbook entries for flight and ground instruction given (p. 162)

FLIGHT OPERATIONS:

1. Aircraft preflight, including detailed knowledge of systems and components (pp. 185, 206, Section 7 of Pilot's Operating Handbook)
2. Landing gear operation, including malfunction troubleshooting (p. 187, Sections 3 and 7 of Pilot's Operating Handbook)
3. Steep turns: maintaining angle of bank within 50° ±5° throughout turn (p. 357)
4. Power on stalls: recover as the stall occurs (p. 416)
5. Power off stalls: enter from a descent, and recover as the stall occurs (p. 427)
6. Effective demonstration of crossed control stalls (p. 434)
7. Effective demonstration of spin recovery, if required (p. 447)
8. Demonstration of successful simulated emergency approach and landing (p. 489)
9. Short field, soft field, and cross-wind takeoffs and landings (pp. 256, 264, 270, 275, 309, 316)
10. Slips (p. 296)

MISCELLANEOUS:

1. Instructional ability and presentation skills (all of Part 1, Study Unit 2, beginning on p. 7, and all of Part 2, Study Unit I, beginning on p. 65)
2. Aircraft maintenance records review (FAR Part 43)
3. Medical certificate requirements for flight instructors (p. 177)
4. Limitations on performing flight instruction without a medical certificate (p. 177)
5. Explanation of the FAA Pilot Proficiency Awards Program (Wings Program) (PH Appendix A)

END OF STUDY UNIT

Appendix A
FAA FLIGHT INSTRUCTOR
PRACTICAL TEST STANDARDS REPRINTED
(FAA-S-8081-6C)
AIRPLANE SINGLE ENGINE LAND ONLY

The purpose of this appendix is to reproduce verbatim what you would get in PTS reprint books that are normally sold for $5.00 at FBOs.

All of these PTSs are reproduced (and explained, discussed, and illustrated!!) elsewhere throughout this book.

NOTE: The INTRODUCTION of this practical test book includes information that is pertinent to **all** flight instructor practical test standards in addition to that which applies specifically to FAA-S-8081-6C, Flight Instructor - Airplane (Single-engine and Multiengine) Practical Test Standards.

FOREWARD

The Flight Instructor—Airplane Practical Test Standards book has been published by the Federal Aviation Administration (FAA) to establish the standards for the flight instructor certification practical tests for the airplane category and the single-engine and multiengine classes. FAA inspectors and designated pilot examiners shall conduct practical tests in compliance with these standards. Flight instructors and applicants should find these standards helpful in practical test preparation.

/s/ 9/22/2006

Joseph K. Tintera, Manager
Regulatory Support Division
Flight Standards Service

INTRODUCTION

General Information

The Flight Standards Service of the Federal Aviation Administration (FAA) has developed this practical test as the standard to be used by examiners[1] when conducting flight instructor airplane practical tests. Instructors are expected to address all of the elements contained in this practical test standard (PTS) when preparing applicants for practical tests. Applicants should be familiar with this PTS and refer to these standards during their training.

The FAA gratefully acknowledges the valuable assistance provided by many individuals, companies, and organizations throughout the aviation community who have contributed their time and talent in assisting with the revision of this practical test standard.

This book may be purchased from the Superintendent of Documents, U.S. Government Printing Office (GPO), Washington, DC 20402-9325, or from GPO's web site.

http://bookstore.gpo.gov

[1] The word "examiner" denotes either the FAA inspector, FAA designated pilot examiner, or other authorized person who conducts the practical test.

This PTS is also available for download, in pdf format, from the Flight Standards Service web site.

www.faa.gov

This PTS is published by the U.S. Department of Transportation, Federal Aviation Administration, Airman Testing Standards Branch, AFS-630, P.O. Box 25082, Oklahoma City, OK 73125.

Comments regarding this publication should be sent, in e-mail form, to the following address.

AFS630comments@faa.gov

Practical Test Standards Concept

Title 14 of the Code of Federal Regulations (14 CFR) part 61 specifies the AREAS OF OPERATION in which knowledge and skill shall be demonstrated by the applicant before the issuance of a flight instructor certificate with the associated category and class ratings. The CFRs provide the flexibility that permits the FAA to publish practical test standards containing the AREAS OF OPERATION and specific TASKs in which competency must be demonstrated. The FAA shall revise this book whenever it is determined that changes are needed in the interest of safety. **Adherence to the provisions of regulations and the practical test standards is mandatory for the evaluation of flight instructor applicants.**

Flight Instructor Practical Test Book Description

This book contains the practical test standards for Flight Instructor—Airplane (single-engine and multiengine). Other flight instructor practical test books include:

- FAA-S-8081-7, Flight Instructor—Rotorcraft (Helicopter and Gyroplane)
- FAA-S-8081-8, Flight Instructor—Glider
- FAA-S-8081-9, Flight Instructor—Instrument (Airplane and Helicopter)

The Flight Instructor Practical Test Standards include the AREAS OF OPERATION and TASKs for the issuance of an initial flight instructor certificate, for the addition of category and/or class ratings to that certificate, and for renewal or reinstatement of a certificate or rating by a practical test.

Flight Instructor Practical Test Standards Description

AREAS OF OPERATION are phases of the practical test. In this practical test book, the first AREA OF OPERATION is Fundamentals of Instructing; the last is Postflight Procedures. The examiner may conduct the practical test in any sequence that will result in a complete and efficient test; **however, the ground portion of the practical test shall be completed prior to the flight portion.**

TASKs are titles of knowledge areas, flight procedures, or maneuvers appropriate to an AREA OF OPERATION. The abbreviation(s) within parentheses immediately following a TASK refer to the category and/or class aircraft appropriate to that TASK. The meaning of each abbreviation is as follows:

ASEL	Airplane—Single-Engine Land
AMEL	Airplane—Multiengine Land
ASES	Airplane—Single-Engine Sea
AMES	Airplane—Multiengine Sea

NOTE: When administering a test based on sections 1 and 2 of this PTS, the TASKs appropriate to the class airplane (ASEL, ASES, AMEL, or AMES) used for the test shall be included in the plan of action. The absence of a class indicates the TASK is for all classes.

NOTE is used to emphasize special considerations required in the AREA OF OPERATION or TASK.

REFERENCE identifies the publication(s) that describe(s) the TASK. Descriptions of TASKs and maneuver tolerances are not included in these standards because this information can be found in the current issue of the listed reference. Publications other than those listed may be used for references if their content conveys substantially the same meaning as the referenced publications.

These practical test standards are based on the following references.

14 CFR part 1	Definitions and Abbreviations
14 CFR part 23	Airworthiness Standards: Normal, Utility, Acrobatic, and Commuter Category Airplanes
14 CFR part 39	Airworthiness Directives
14 CFR part 43	Maintenance, Preventative Maintenance, Rebuilding, and Alteration
14 CFR part 61	Certification: Pilots and Flight Instructors
14 CFR part 67	Medical Standards and Certification
14 CFR part 91	General Operating and Flight Rules
NTSB part 830	Notification and Reporting of Aircraft Accidents and Incidents
AC 00-6	Aviation Weather
AC 00-45	Aviation Weather Services
AC 60-22	Aeronautical Decision Making
AC 60-28	English Language Skill Standards as Required by 14 CFR parts 61, 63, and 65
AC 61-65	Certification: Pilots and Flight Instructors
AC 61-67	Stall and Spin Awareness Training
AC 61-84	Role of Preflight Preparation
AC 61-94	Pilot Transition Course for Self-Launching or Powered Sailplanes (Motorgliders)
AC 61-107	Operations of Aircraft at Altitude Above 25,000 feet MSL and/or Mach Numbers (Mmo) Greater than .75
AC 90-42	Traffic Advisory Practices at Airport Without Operating Control Towers
AC 90-48	Pilots' Role in Collision Avoidance
AC 90-66	Recommended Standard Traffic Patterns for Aeronautical Operations at Airports Without Operating Control Towers
AC 91-13	Cold Weather Operation of Aircraft
AC 91-55	Reduction of Electrical System Failures Following Aircraft Engine Starting
FAA-H-8083-1	Aircraft Weight and Balance Handbook
FAA-H-8083-3	Airplane Flying Handbook

FAA-H-8083-9	Aviation Instructor's Handbook
FAA-S-8081-4	Instrument Rating Practical Test Standards
FAA-S-8081-12	Commercial Pilot Practical Test Standards
FAA-S-8081-14	Private Pilot Practical Test Standards
FAA-H-8083-15	Instrument Flying Handbook
FAA-H-8083-23	Seaplane, Skiplane, and Float/Ski Equipped Helicopter Operations Handbook
FAA-H-8083-25	Pilot's Handbook of Aeronautical Knowledge
Order 8080.6	Conduct of Airman Knowledge Tests
AC 150/5340-1	Standards for Airport Markings
AC 150/5340-18	Standards for Airport Sign Systems
AC 150/5340-30	Design and Installation Details for Airport Visual Aids
AIM	Aeronautical Information Manual
A/FD	Airport/Facility Directory
NOTAMs	Notices to Airmen
POH/AFM	Pilot Operating Handbooks and FAA-Approved Airplane Flight Manuals
USCG	Navigation Rule: International - Inland

The Objective lists the elements that must be satisfactorily performed to demonstrate competency in a TASK. The Objective includes:

1. specifically what the applicant should be able to do;
2. conditions under which the TASK is to be performed; and
3. acceptable performance standards.

The examiner determines that the applicant meets the TASK Objective through the demonstration of competency in all elements of knowledge and/or skill unless otherwise noted. The Objectives of TASKs in certain AREAS OF OPERATION, such as Fundamentals of Instructing and Technical Subjects, include only knowledge elements. Objectives of TASKs in AREAS OF OPERATION that include elements of skill, as well as knowledge, also include common errors, which the applicant shall be able to describe, recognize, analyze, and correct.

The Objective of a TASK that involves pilot skill consists of four parts. The four parts include determination that the applicant exhibits:

1. instructional knowledge of the elements of a TASK. This is accomplished through descriptions, explanations, and simulated instruction;
2. instructional knowledge of common errors related to a TASK, including their recognition, analysis, and correction;
3. the ability to demonstrate and simultaneously explain the key elements of a TASK. The TASK demonstration must be to the COMMERCIAL PILOT skill level[2]; and
4. the ability to analyze and correct common errors related to a TASK.

Abbreviations

14 CFR	Title 14 of the Code of Federal Regulations
AC	Advisory Circular
ADM	Aeronautical Decision Making
AGL	Above Ground Level
AIRMETS	Airman's Meteorological Information
AME	Airplane Multiengine
AMEL	Airplane Multiengine Land
AMES	Airplane Multiengine Sea
ASEL	Airplane Single-Engine Land
ASES	Airplane Single-Engine Sea
ATC	Air Traffic Control
ATIS	Automatic Terminal Information Service
CFIT	Controlled Flight into Terrain
CRM	Crew Resource Management

[2] The teaching techniques and procedures should conform to those set forth in FAA-H-8083-25, Pilot's Handbook of Aeronautical Knowledge; FAA-H-8083-9, Aviation Instructor's Handbook; FAA-H-8083-3, Airplane Flying Handbook; and FAA-H-8083-15, Instrument Flying Handbook

DME	Distance Measuring Equipment
FAA	Federal Aviation Administration
FDC	Flight Data Center
FSDO	Flight Standards District Office
G	Glider
GPO	Government Printing Office
GPS	Global Positioning System
IA	Instrument Airplane
IH	Instrument Helicopter
LAHSO	Land and Hold Short Operations
LORAN	Long Range Navigation
MEL	Minimum Equipment List
Mmo	Means Maximum Operating Limit Speed
NAVAID	Navigation Aid
NDB	Non Directional Beacon
NOTAM	Notice to Airmen
NTSB	National Transportation Board
NWS	National Weather Service
PC	Proficiency Check
PTS	Practical Test Standard
RG	Rotorcraft Gyroplane
RH	Rotorcraft Helicopter
SIGMETS	Significant Meteorological Advisory
SUA	Special Use Airspace
TFR	Temporary Flight Restriction
VFR	Visual Flight Rules
VHF	Very High Frequency
VOR	Very High Frequency Omnirange
V_{MC}	Minimum Control with the Critical Engine Inoperative
V_X	Best Angle of Climb
V_Y	Best Rate of Climb
V_{SSE}	Safe Single Engine Speed
V_{YSE}	Single-Engine Best Rate of Climb

Use of the Practical Test Standards Book

The Flight Instructor Practical Test Standards are designed to evaluate competency in both knowledge and skill.

The FAA requires that all Flight Instructor practical tests be conducted in accordance with the appropriate Flight Instructor Practical Test Standards and the policies set forth in the INTRODUCTION. The flight instructor applicant must be prepared to demonstrate the ability to instruct effectively in **ALL** TASKs included in the AREAS OF OPERATION of the appropriate practical test standards, unless otherwise noted.

All of the procedures and maneuvers in the Private Pilot and Commercial Pilot Practical Test Standards have been included in the Flight Instructor Practical Test Standards. **However, the flight instructor PTS allows the examiner to select one or more TASKs in each AREA OF OPERATION therefore allowing the practical test for initial certification to be completed within a reasonable time frame.** In certain AREAS OF OPERATION, there are **required** TASKs, which the examiner must select. These **required** TASKs are identified by **NOTES** immediately following the AREA OF OPERATION titles.

The term "instructional knowledge" means the instructor applicant is capable of using the appropriate reference to provide the "application or correlative level of knowledge" of a subject matter topic, procedure, or maneuver. It also means that the flight instructor applicant's discussions, explanations, and descriptions should follow the recommended teaching procedures and techniques explained in FAA-H-8083-9, Aviation Instructor's Handbook.

In preparation for the practical test, the examiner shall develop a written "plan of action" for each practical test. The plan of action is a tool, for the sole use of the examiner, to be used in evaluating the applicant. The plan of action need not be grammatically correct or in any formal format. The plan of action for an initial certification test shall include one or more TASKs in each AREA OF OPERATION and shall **always** include the required TASKs. The plan of action must incorporate one or more scenarios that will be used during the practical test. The examiner should try to include as many of the TASKs into the scenario portion of the test as possible, but maintain the flexibility to change due to unexpected situations as they arise and still result in an efficient and valid test. **Any TASK selected for evaluation during a practical test must be evaluated in its entirety.** If the applicant is unable to perform a TASK listed in the "plan of action" due to circumstances beyond his/her control, the examiner may substitute another TASK from the applicable AREA OF OPERATION.

The examiner is not required to follow the precise order in which the AREAS OF OPERATION and TASKs appear in this book. The examiner may change the sequence or combine TASKs with similar objectives to have an orderly and efficient flow of the practical test.

The "plan of action" for a test administered *for the addition of an aircraft category and/or class rating* to a flight instructor certificate shall include the required AREAS OF OPERATION as indicated in the table at the beginning of each section. The required TASKs appropriate to the rating(s) sought must also be included. In some instances, notes identify additional required TASKs. **Any TASK selected shall be evaluated in its entirety.**

NOTE: AREA OF OPERATION XI, Slow Flight, Stalls, and Spins, contains TASKs referred to as "proficiency" and "demonstration." The intent of TASKs A and B (proficiency) is to ensure that the flight instructor applicant is tested on proficiency for the purpose of teaching to students these TASKs that are required for pilot certification. The intent of TASKs C, D, E, G, and H (demonstration) is to ensure that the flight instructor applicant is knowledgeable and proficient in these maneuvers and can teach them to students for both familiarization and stall/spin awareness purposes.

With the exception of the **required** TASKs, the examiner shall not tell the applicant in advance, which TASKs will be included in the "plan of action." The applicant should be well prepared in **all** knowledge and skill areas included in the standards. Throughout the flight portion of the practical test, the examiner will evaluate the applicant's ability to simultaneously demonstrate and explain procedures and maneuvers, and to give flight instruction to students at various stages of flight training and levels of experience.

The purpose for including common errors in certain TASKs is to assist the examiner in determining that the flight instructor applicant has the ability to recognize, analyze, and correct such errors. The common errors listed in the TASK Objectives may or may not be found in the TASK References. However, the FAA considers their frequency of occurrence justification for their inclusion in the TASK Objectives.

The examiner is expected to use good judgment in the performance of simulated emergency procedures. The use of the safest means for simulation is expected. Consideration must be given to local conditions, both meteorological and topographical, at the time of the test, as well as the applicant's workload, and the condition of the aircraft used. If the procedure being evaluated would jeopardize safety, it is expected that the applicant will simulate that portion of the maneuver.

Special Emphasis Areas

Examiners shall place special emphasis upon areas of aircraft operation considered critical to flight safety. Among these are:

1. positive aircraft control;
2. positive exchange of the flight controls procedure;
3. stall/spin awareness;
4. collision avoidance;
5. wake turbulence avoidance;
6. LAHSO;
7. runway incursion avoidance;
8. CFIT;
9. ADM and risk management;
10. wire strike avoidance;
11. checklist usage;
12. temporary flight restrictions (TFRs);
13. special use airspace (SUA);
14. aviation security; and
15. other areas deemed appropriate to any phase of the practical test.

Although these areas may not be specifically addressed under each TASK, they are essential to flight safety and will be evaluated during the practical test. In all instances, the applicant's actions will relate to the complete situation.

Practical Test Prerequisites

An applicant for a flight instructor—initial certification practical test is required by 14 CFR part 61 to:

1. be at least 18 years of age;
2. be able to read, speak, write, and understand the English language. If there is a doubt, use AC 60-28, English Language Skill Standards;
3. hold either a commercial/instrument pilot or airline transport pilot certificate with an aircraft category rating appropriate to the flight instructor rating sought;
4. have an endorsement from an authorized instructor on the fundamentals of instructing appropriate to the required knowledge test;
5. have passed the appropriate flight instructor knowledge test(s) since the beginning of the 24th month before the month in which he or she takes the practical test. Knowledge test validity can be verified in FAA Order 8080.6, Conduct of Airman Knowledge Tests, Chapter 7, Eligibility Requirements; and
6. have an endorsement from an authorized instructor certifying that the applicant has been given flight training in the AREAS OF OPERATION listed in 14 CFR part 61, section 61.187 and a written statement from an authorized flight instructor within the preceding 60 days, in accordance with 14 CFR part 61, section 61.39, that instruction was given in preparation for the practical test. The endorsement shall also state that the instructor finds the applicant prepared for the required practical test, and that the applicant has demonstrated satisfactory knowledge of the subject area(s) in which the applicant was deficient on the airman knowledge test.

An applicant holding a flight instructor certificate who applies for an **additional** rating on that certificate is required by 14 CFR to:

1. hold a valid pilot certificate with ratings appropriate to the flight instructor rating sought;
2. have at least 15 hours as pilot-in-command in the category and class aircraft appropriate to the rating sought;
3. have passed the appropriate knowledge test prescribed for the issuance of a flight instructor certificate with the rating sought since the beginning of the 24th month before the month in which he/she takes the practical test; and

4. have an endorsement from an authorized instructor certifying that the applicant has been given flight training in the AREAS OF OPERATION listed in 14 CFR part 61, section 61.187 and a written statement from an authorized flight instructor within the preceding 60 days, in accordance with 14 CFR part 61, section 61.39, that instruction was given in preparation for the practical test. The endorsement shall also state that the instructor finds the applicant prepared for the required practical test, and that the applicant has demonstrated satisfactory knowledge of the subject area(s) in which the applicant was deficient on the airman knowledge test.

If there are questions concerning English language requirements, refer to your local FSDO or to AC 60-28, English Language Skill Standards Required by 14 CFR parts 61, 63, and 65. English language requirements should be determined to be met prior to beginning the practical test.

Aircraft and Equipment Required for the Practical Test

The flight instructor applicant is required by 14 CFR part 61, section 61.45 to provide an airworthy, certificated aircraft for use during the practical test. This section further requires that the aircraft must:

1. be of U.S., foreign or military registry of the same category, class, and type for the certificate and/or rating for which the applicant is applying;
2. have fully functioning dual controls except as provided in 14 CFR section 61.45(c) and (e); and
3. be capable of performing all appropriate TASKs for the flight instructor rating sought and have no operating limitations, which prohibit the performance of those TASKs. A complex airplane must be furnished for the performance of takeoff and landing maneuvers, and appropriate emergency procedures. A complex landplane is one having retractable gear, flaps, and controllable propeller. A complex seaplane is one having flaps and controllable propeller.

Flight Instructor Responsibility

An appropriately rated flight instructor is responsible for training the flight instructor applicant to acceptable standards in **all** subject matter areas, procedures, and maneuvers included in the TASKs within each AREA OF OPERATION in the appropriate flight instructor practical test standard.

Because of the impact of their teaching activities in developing safe, proficient pilots, flight instructors should exhibit a high level of knowledge, skill, and the ability to impart that knowledge and skill to students. The flight instructor must certify that the applicant is:

1. able to make a practical application of the fundamentals of instructing;
2. competent to teach the subject matter, procedures, and maneuvers included in the standards to students with varying backgrounds and levels of experience and ability;
3. able to perform the procedures and maneuvers included in the standards to at least the COMMERCIAL PILOT skill level while giving effective flight instruction; and
4. competent to pass the required practical test for the issuance of the flight instructor certificate with the associated category and class ratings or the addition of a category and/or class rating to a flight instructor certificate.

Throughout the applicant's training, the flight instructor is responsible for emphasizing the performance of, and the ability to teach, **effective visual scanning, runway incursion avoidance, collision avoidance procedures, and Land and Hold Short Operations (LAHSO)**. The flight instructor applicant should develop and use scenario based teaching methods particularly on special emphasis areas. These areas are covered in AC 90-48, Pilot's Role in Collision Avoidance; FAA-H-8083-3, Airplane Flying Handbook; FAA-H-8083-25, Pilot's Handbook of Aeronautical Knowledge; and the current Aeronautical Information Manual.

Examiner Responsibility

The examiner conducting the practical test is responsible for determining that the applicant meets acceptable standards of teaching ability, knowledge, and skill in the selected TASKs. The examiner makes this determination by accomplishing an Objective that is appropriate to each selected TASK, and includes an evaluation of the applicant's:

1. ability to apply the fundamentals of instructing;
2. knowledge of, and the ability to teach, the subject matter, procedures, and maneuvers covered in the TASKs;
3. ability to perform the procedures and maneuvers included in the standards to the COMMERCIAL PILOT skill level while giving effective flight instruction; and
4. ability to analyze and correct common errors related to the procedures and maneuvers covered in the TASKs.

It is intended that oral questioning be used at any time during the ground or flight portion of the practical test to determine that the applicant can instruct effectively and has a comprehensive knowledge of the TASKs and their related safety factors.

During the flight portion of the practical test, the examiner shall act as a student during selected maneuvers. This will give the examiner an opportunity to evaluate the flight instructor applicant's ability to analyze and correct simulated common errors related to these maneuvers. The examiner will place special emphasis on the applicant's use of visual scanning and collision avoidance procedures, and the applicant's ability to teach those procedures.

Examiners should to the greatest extent possible test the applicant's application and correlation skills. When possible scenario based questions should be used during the practical test. The examiner will evaluate the applicant's ability to teach visual scanning, runway incursion avoidance, collision avoidance procedures, and Land and Hold Short Operations (LAHSO).

If the examiner determines that a TASK is incomplete, or the outcome uncertain, the examiner may require the applicant to repeat that TASK, or portions of that TASK. This provision has been made in the interest of fairness and does not mean that instruction, practice, or the repeating of an unsatisfactory task is permitted during the certification process. When practical, the remaining TASKs of the practical test phase should be completed before repeating the questionable TASK.

On multiengine practical tests where the failure of the most critical engine after lift off is required, the instructor applicant and examiner must give consideration to local atmospheric conditions, terrain and type of aircraft used. However the failure of an engine shall not be simulated until attaining at least Vsse/Vyse and at an altitude not lower than 200 feet AGL.

During simulated engine failures on multiengine practical tests, after simulated feathering of the propeller the engine shall be set to zero thrust. The examiner shall require the instructor applicant to simultaneously demonstrate and explain procedures for landing with a simulated feathered propeller with the engine set to zero thrust. **The examiner must not simulate any conditions that may jeopardize safe flight or result in possible damage to the aircraft.**

Satisfactory Performance

The practical test is passed if, in the judgment of the examiner, the applicant demonstrates satisfactory performance with regard to:

1. knowledge of the fundamentals of instructing;
2. knowledge of the technical subject areas;
3. knowledge of the flight instructor's responsibilities concerning the pilot certification process;
4. knowledge of the flight instructor's responsibilities concerning logbook entries and pilot certificate endorsements;
5. ability to demonstrate the procedures and maneuvers selected by the examiner to at least the COMMERCIAL PILOT skill level while giving effective instruction;
6. competence in teaching the procedures and maneuvers selected by the examiner;
7. competence in describing, recognizing, analyzing, and correcting common errors simulated by the examiner; and
8. knowledge of the development and effective use of a course of training, a syllabus, and a lesson plan.

Unsatisfactory Performance

If, in the judgment of the examiner, the applicant does not meet the standards of performance of any TASK performed, the applicable AREA OF OPERATION is considered unsatisfactory and therefore, the practical test is failed. The examiner or applicant may discontinue the test at any time when the failure of an AREA OF OPERATION makes the applicant ineligible for the certificate or rating sought. **The test will be continued only with the consent of the applicant.** If the test is discontinued, the applicant is entitled credit for only those AREAS OF OPERATION and their associated TASKs satisfactorily performed; however, during the retest and at the discretion of the examiner, any TASK may be re-evaluated, including those previously considered satisfactory. Specific reasons for disqualification are:

1. failure to perform a procedure or maneuver to the COMMERCIAL PILOT skill level while giving effective flight instruction;
2. failure to provide an effective instructional explanation while demonstrating a procedure or maneuver (explanation during the demonstration must be clear, concise, technically accurate, and complete with no prompting from the examiner);
3. any action or lack of action by the applicant which requires corrective intervention by the examiner to maintain safe flight;
4. failure to use proper and effective visual scanning techniques to clear the area before and while performing maneuvers.

When a Disapproval Notice is issued, the examiner must record the applicant's unsatisfactory performance in terms of AREA OF OPERATIONS and specific TASKs not meeting the standard appropriate to the practical test conducted. If the applicant fails the practical test because of a special emphasis area, the Notice of Disapproval shall indicate the associated TASK. An example would be: AREA OF OPERATION IX, Maneuvering During Slow Flight, failure to teach proper collision avoidance procedures.

Letter of Discontinuance

When a practical test is discontinued for reasons other than unsatisfactory performance (i.e., equipment failure, weather, or illness) FAA Form 8700-1, Airman Certificate and/or Rating Application, and, if applicable, the Airman Knowledge Test Report, is to be returned to the applicant. The examiner at that time prepares, signs, and issues a Letter of Discontinuance to the applicant. The Letter of Discontinuance should identify the AREAS OF OPERATION and their associated TASKs of the practical test that were successfully completed. The applicant should be advised that the Letter of Discontinuance must be presented to the examiner when the practical test is resumed, and made part of the certification file.

Aeronautical Decision Making And Risk Management

Throughout the practical test, the examiner evaluates the applicant's ability to use good aeronautical decision-making procedures in order to identify risks. The examiner accomplishes this requirement by developing scenarios that incorporate as many TASKs as possible to evaluate the applicants risk management in making safe aeronautical decisions. For example, the examiner may develop a scenario that incorporates weather decisions and performance planning.

The applicant's ability to utilize all the assets available in making a risk analysis to determine the safest course of action is essential for satisfactory performance. The scenarios should be realistic and within the capabilities of the aircraft used for the practical test.

Single-Pilot Resource Management

Single-Pilot Resource Management refers to the effective use of ALL available resources: human resources, hardware, and information. It is similar to Crew Resource Management (CRM) procedures that are being emphasized in multi-crewmember operations except that only one crewmember (the pilot) is involved. Human resources "...includes all other groups routinely working with the pilot who are involved in decisions that are required to operate a flight safely. These groups include, but are not limited to: dispatchers, weather briefers, maintenance personnel, and air traffic controllers." Pilot Resource Management is not a single TASK; it is a set of skill competencies that must be evident in all TASKs in this practical test standard as applied to single-pilot operation.

Applicant's Use of Checklists

Throughout the practical test, the instructor applicant is evaluated on the use and teaching of an appropriate checklist. Proper use is dependent on the specific TASK being evaluated. The situation may be such that the use of the checklist, while accomplishing elements of an Objective, would be either unsafe or impractical, especially in a single-pilot operation. In this case, a review of the checklist after the elements have been accomplished would be appropriate. Division of attention and proper visual scanning should be considered when using a checklist.

Use of Distractions During Practical Tests

Numerous studies indicate that many accidents have occurred when the pilot has been distracted during critical phases of flight. To evaluate the applicant's ability to utilize proper control technique while dividing attention both inside and outside the cockpit, the examiner shall cause realistic distractions during the flight portion of the practical test to evaluate the applicant's ability to divide attention while maintaining safe flight.

Positive Exchange of Flight Controls

During flight training, there must always be a clear understanding between students and flight instructors of who has control of the aircraft. Prior to flight, a briefing should be conducted that includes the procedure for the exchange of flight controls. A positive three-step process in the exchange of flight controls between pilots is a proven procedure and one that is strongly recommended.

When the instructor wishes the student to take control of the aircraft, he or she will say, "You have the flight controls." The student acknowledges immediately by saying, "I have the flight controls." The flight instructor again says, "You have the flight controls." When control is returned to the instructor, follow the same procedure. A visual check is recommended to verify that the exchange has occurred. There should never be any doubt as to who is flying the aircraft. The instructor applicant is expected to teach proper positive exchange of flight controls during the practical test.

Initial Flight Instructor Certification

An applicant who seeks initial flight instructor certification will be evaluated in all AREAS OF OPERATION of the standards appropriate to the rating(s) sought. The examiner shall refer to the **NOTE** in the front of the AREA OF OPERATION to determine which and how many TASKs shall be tested.

Addition of Aircraft Category and/or Class Ratings to a Flight Instructor Certificate

An applicant who holds a flight instructor certificate and seeks an additional aircraft category and/or class rating will be evaluated in at least the AREAS OF OPERATION and TASKs that are unique and appropriate to the rating(s) sought (see table at the beginning of each section). At the discretion of the examiner, the applicant's competence in **ALL** AREAS OF OPERATION may be evaluated.

Renewal or Reinstatement of a Flight Instructor Certificate

14 CFR part 61, sections 61.197(a)(1) and 61.199(a) allows an individual that holds a flight instructor certificate to renew or reinstate that certificate by passing a practical test. The examiner shall develop a plan of action that includes the AREAS OF OPERATION and at least the minimum number of TASKs prescribed in the table at the beginning of each section. The Renewal or Reinstatement of one rating on a Flight Instructor Certificate renews or reinstates all privileges existing on the certificate.

SECTION 1
FLIGHT INSTRUCTOR
AIRPLANE—SINGLE-ENGINE
Practical Test Standards

CONTENTS

Airplane Single-Engine Land

CHECKLISTS

Applicant's Practical Test Checklist
Examiner's Checklist

ADDITIONAL RATING TASK TABLE

RENEWAL OR REINSTATEMENT OF A FLIGHT INSTRUCTOR TABLE

AREAS OF OPERATION:

I. FUNDAMENTALS OF INSTRUCTING

- A. The Learning Process
- B. Human Behavior and Effective Communication
- C. The Teaching Process
- D. Teaching Methods
- E. Critique and Evaluation
- F. Flight Instructor Characteristics and Responsibilities
- G. Planning Instructional Activity

II. TECHNICAL SUBJECT AREAS

- A. Aeromedical Factors
- B. Visual Scanning and Collision Avoidance
- C. Principles of Flight
- D. Airplane Flight Controls
- E. Airplane Weight and Balance
- F. Navigation and Flight Planning
- G. Night Operations
- H. High Altitude Operations
- I. Federal Aviation Regulations and Publications
- J. National Airspace System
- K. Navigation Systems and Radar Services
- L. Logbook Entries and Certificate Endorsements

III. PREFLIGHT PREPARATION

- A. Certificates and Documents
- B. Weather Information
- C. Operation of Systems
- D. Performance and Limitations
- E. Airworthiness Requirements

IV. PREFLIGHT LESSON ON A MANEUVER TO BE PERFORMED IN FLIGHT

- A. Maneuver Lesson

V. PREFLIGHT PROCEDURES

- A. Preflight Inspection
- B. Single-Pilot Crew Resource Management
- C. Engine Starting
- D. Taxiing—Landplane
- G. Before Takeoff Check

VI. AIRPORT OPERATIONS

- A. Radio Communications and ATC Light Signals
- B. Traffic Patterns
- C. Airport, Runway and Taxiway Signs, Markings, and Lighting

VII. TAKEOFFS, LANDINGS, AND GO-AROUNDS

- A. Normal and Crosswind Takeoff and Climb
- B. Short-Field Takeoff and Maximum Performance Climb
- C. Soft-Field Takeoff and Climb
- F. Normal and Crosswind Approach and Landing
- G. Slip to a Landing
- H. Go-Around/Rejected Landing
- I. Short-Field Approach and Landing
- J. Soft-Field Approach and Landing
- K. Power-Off 180° Accuracy Approach and Landing

VIII. FUNDAMENTALS OF FLIGHT

- A. Straight-and-Level Flight
- B. Level Turns
- C. Straight Climbs and Climbing Turns
- D. Straight Descents and Descending Turns

IX. PERFORMANCE MANEUVERS

- A. Steep Turns
- B. Steep Spirals
- C. Chandelles
- D. Lazy Eights

X. GROUND REFERENCE MANEUVERS

- A. Rectangular Course
- B. S-Turns Across a Road
- C. Turns Around a Point
- D. Eights on Pylons

XI. SLOW FLIGHT, STALLS, AND SPINS

- A. Maneuvering During Slow Flight
- B. Power-On Stalls (Proficiency)
- C. Power-Off Stalls (Proficiency)
- D. Crossed-Control Stalls (Demonstration)
- E. Elevator Trim Stalls (Demonstration)
- F. Secondary Stalls (Demonstration)
- G. Spins
- H. Accelerated Maneuver Stalls (Demonstration)

XII. BASIC INSTRUMENT MANEUVERS

- A. Straight-and-Level Flight
- B. Constant Airspeed Climbs
- C. Constant Airspeed Descents
- D. Turns to Headings
- E. Recovery from Unusual Flight Attitudes

XIII. EMERGENCY OPERATIONS

- A. Emergency Approach and Landing (Simulated)
- B. Systems and Equipment Malfunctions
- C. Emergency Equipment and Survival Gear
- D. Emergency Descent

XIV. POSTFLIGHT PROCEDURES

- A. Postflight Procedures

APPLICANT'S PRACTICAL TEST CHECKLIST

APPOINTMENT WITH INSPECTOR OR EXAMINER:

NAME: _____

TIME/DATE: _____

ACCEPTABLE AIRCRAFT

- ☐ View-Limiting Device (if applicable)
- ☐ Aircraft Documents:
  Airworthiness Certificate
  Registration Certificate
  Operating Limitations
- ☐ Aircraft Maintenance Records:
  Airworthiness Inspections
- ☐ Pilot's Operating Handbook and FAA-Approved Airplane Flight Manual

PERSONAL EQUIPMENT

- ☐ Current Aeronautical Charts
- ☐ Computer and Plotter
- ☐ Flight Plan Form
- ☐ Flight Logs
- ☐ Current AIM
- ☐ Current Airport Facility Directory

PERSONAL RECORDS

- ☐ Pilot Certificate
- ☐ Medical Certificate
- ☐ Completed FAA Form 8710-1, Airman Certificate and/or Rating Application
- ☐ Airman Knowledge Test Report
- ☐ Logbook with Instructor's Endorsement
- ☐ Letter of Discontinuance (if applicable)
- ☐ Notice of Disapproval (if applicable)
- ☐ Approved School Graduation Certificate (if applicable)
- ☐ Examiner's Fee (if applicable)

EXAMINER'S CHECKLIST
FLIGHT INSTRUCTOR—AIRPLANE
(SINGLE-ENGINE)

APPLICANT'S NAME: _____

EXAMINER'S NAME: _____

DATE: _____   TYPE CHECK: _____

TYPE AIRPLANE: _____

AREA OF OPERATION:

I.   FUNDAMENTALS OF INSTRUCTING

- ☐ **A.**   The Learning Process
- ☐ **B.**   Human Behavior and Effective Communication
- ☐ **C.**   The Teaching Process
- ☐ **D.**   Teaching Methods
- ☐ **E.**   Critique and Evaluation
- ☐ **F.**   Flight Instructor Characteristics and Responsibilities
- ☐ **G.**   Planning Instructional Activity

II.   TECHNICAL SUBJECT AREAS

- ☐ **A.**   Aeromedical Factors
- ☐ **B.**   Visual Scanning and Collision Avoidance
- ☐ **C.**   Principles of Flight
- ☐ **D.**   Airplane Flight Controls
- ☐ **E.**   Airplane Weight and Balance
- ☐ **F.**   Navigation and Flight Planning
- ☐ **G.**   Night Operations
- ☐ **H.**   High Altitude Operations
- ☐ **I.**   Federal Aviation Regulations and Publications
- ☐ **J.**   National Airspace System
- ☐ **K.**   Navigation Systems and Radar Services
- ☐ **L.**   Logbook Entries and Certificate Endorsements

III.   PREFLIGHT PREPARATION

- ☐ **A.**   Certificates and Documents
- ☐ **B.**   Weather Information
- ☐ **C.**   Operation of Systems
- ☐ **D.**   Performance and Limitations
- ☐ **E.**   Airworthiness Requirements

IV.   PREFLIGHT LESSON ON A MANEUVER TO BE PERFORMED IN FLIGHT

- ☐ **A.**   Maneuver Lesson

V.   PREFLIGHT PROCEDURES

- ☐ **A.**   Preflight Inspection
- ☐ **B.**   Single-Pilot Resource Management
- ☐ **C.**   Engine Starting
- ☐ **D.**   Taxiing—Landplane
- ☐ **G.**   Before Takeoff Check

VI.   AIRPORT OPERATIONS

- ☐ **A.**   Radio Communications and ATC Light Signals
- ☐ **B.**   Traffic Patterns
- ☐ **C.**   Airport, Runway and Taxiway Signs, Markings, and Lighting

VII. TAKEOFFS, LANDINGS, AND GO-AROUNDS

☐ A. Normal and Crosswind Takeoff and Climb
☐ B. Short-Field Takeoff and Maximum Performance Climb
☐ C. Soft-Field Takeoff and Climb
☐ F. Normal and Crosswind Approach and Landing
☐ G. Slip to a Landing
☐ H. Go-Around/Rejected Landing
☐ I. Short-Field Approach and Landing
☐ J. Soft-Field Approach and Landing
☐ K. Power-Off 180° Accuracy Approach and Landing

VIII. FUNDAMENTALS OF FLIGHT

☐ A. Straight-and-Level Flight
☐ B. Level Turns
☐ C. Straight Climbs and Climbing Turns
☐ D. Straight Descents and Descending Turns

IX. PERFORMANCE MANEUVERS

☐ A. Steep Turns
☐ B. Steep Spirals
☐ C. Chandelles
☐ D. Lazy Eights

X. GROUND REFERENCE MANEUVERS

☐ A. Rectangular Course
☐ B. S-Turns Across a Road
☐ C. Turns Around a Point
☐ D. Eights on Pylons

XI. SLOW FLIGHT, STALLS, AND SPINS

☐ A. Maneuvering During Slow Flight
☐ B. Power-On Stalls (Proficiency)
☐ C. Power-Off Stalls (Proficiency)
☐ D. Crossed-Control Stalls (Demonstration)
☐ E. Elevator Trim Stalls (Demonstration)
☐ F. Secondary Stalls (Demonstration)
☐ G. Spins
☐ H. Accelerated Maneuver Stalls (Demonstration)

XII. BASIC INSTRUMENT MANEUVERS

☐ A. Straight-and-Level Flight
☐ B. Constant Airspeed Climbs
☐ C. Constant Airspeed Descents
☐ D. Turns to Headings
☐ E. Recovery from Unusual Flight Attitudes

XIII. EMERGENCY OPERATIONS

☐ A. Emergency Approach and Landing (Simulated)
☐ B. Systems and Equipment Malfunctions
☐ C. Emergency Equipment and Survival Gear
☐ D. Emergency Descent

XIV. POSTFLIGHT PROCEDURES

☐ A. Postflight Procedures

ADDITIONAL RATING TASK TABLE

ADDITION OF A SINGLE-ENGINE CLASS RATING (AND AN AIRPLANE CATEGORY RATING, IF APPROPRIATE) TO A FLIGHT INSTRUCTOR CERTIFICATE						
REQUIRED AREAS OF OPERATION	FLIGHT INSTRUCTOR CERTIFICATE AND RATING HELD					
	AME	RH	RG	G	IA	IH
I	NONE	NONE	NONE	NONE	NONE	NONE
II	NONE	C, D	C, D	C, D	C, D	C, D
III	NONE	C, D	C, D	C, D	C, D	C, D
IV	NONE	NONE	NONE	NONE	NONE	NONE
V	NONE	*	*	*	*	*
VI	NONE	*	NONE	*	*	*
VII	*	*	*	*	*	*
VIII	NONE	*	*	*	*	*
IX	*	*	*	*	*	*
X	D	*	*	*	*	*
XI	*	*	*	*	*	*
XII	NONE	*	*	*	NONE	*
XIII	*	*	*	*	*	*
XIV	NONE	*	*	*	*	*

NOTE: If an applicant holds more than one rating on a flight instructor certificate and the table indicates both a "NONE" and a "SELECT ONE" for a particular AREA OF OPERATION, the "NONE" entry applies. This is logical since the applicant has satisfactorily accomplished the AREA OF OPERATION on a previous flight instructor practical test. At the discretion of the examiner, the applicant's competence in **any** AREAS OF OPERATION may be evaluated.

* Refer to NOTE under AREA OF OPERATION for TASK requirements.

RENEWAL OR REINSTATEMENT OF A FLIGHT INSTRUCTOR TABLE

Airplane Single-Engine Category

REQUIRED AREAS OF OPERATION	NUMBER OF TASKS
II	TASK L and 1 other TASK
III	1
IV	1
V	1
VI	2 Takeoffs and 2 Landings
IX	1
X	1
XI	2
XIII	1
XIV	1

The Renewal or Reinstatement of one rating on a Flight Instructor Certificate renews or reinstates all privileges existing on the certificate. (14 CFR part 61, sections 61.197 and 61.199)

I.   AREA OF OPERATION: FUNDAMENTALS OF INSTRUCTING

NOTE: The examiner shall select TASK F and one other TASK.

A.   TASK: THE LEARNING PROCESS

REFERENCE: FAA-H-8083-9.

Objective. To determine that the applicant exhibits instructional knowledge of the elements of the learning process by describing:

1.   Learning theory.
2.   Characteristics of learning.
3.   Principles of learning.
4.   Levels of learning.
5.   Learning physical skills.
6.   Memory.
7.   Transfer of learning.

B.   TASK: HUMAN BEHAVIOR AND EFFECTIVE COMMUNICATION

REFERENCE: FAA-H-8083-9.

Objective. To determine that the applicant exhibits instructional knowledge of the elements of the teaching process by describing:

1.   Human behavior--

    a.   control of human behavior.
    b.   human needs.
    c.   defense mechanisms.
    d.   the flight instructor as a practical psychologist.

2.   Effective communication--

    a.   basic elements of communication.
    b.   barriers of effective communication.
    c.   developing communication skills.

C.   TASK: THE TEACHING PROCESS

REFERENCE: FAA-H-8083-9.

Objective. To determine that the applicant exhibits instructional knowledge of the elements of the teaching process by describing:

1.   Preparation of a lesson for a ground or flight instructional period.
2.   Presentation methods.
3.   Application, by the student, of the material or procedure presented.
4.   Review and evaluation of student performance.

D.   TASK: TEACHING METHODS

REFERENCE: FAA-H-8083-9.

Objective. To determine that the applicant exhibits instructional knowledge of the elements of teaching methods by describing:

1.   Material organization.
2.   The lecture method.
3.   The cooperative or group learning method.
4.   The guided discussion method.
5.   The demonstration-performance method.
6.   Computer-based training method.

E.   TASK: CRITIQUE AND EVALUATION

REFERENCE: FAA-H-8083-9.

Objective. To determine that the applicant exhibits instructional knowledge of the elements of critique and evaluation by explaining:

1.   Critique--

    a.   purpose and characteristics of an effective critique.
    b.   methods and ground rules for a critique.

2.   Evaluation--

    a.   characteristics of effective oral questions and what types to avoid.
    b.   responses to student questions.
    c.   characteristics and development of effective written questions.
    d.   characteristics and uses of performance test, specifically, the FAA practical test standards.

F.   TASK: FLIGHT INSTRUCTOR CHARACTERISTICS AND RESPONSIBILITIES

REFERENCE: FAA-H-8083-9.

Objective. To determine that the applicant exhibits instructional knowledge of the elements of flight instructor characteristics and responsibilities by describing:

1.   Aviation instructor responsibilities in--

    a.   providing adequate instruction.
    b.   establishing standards of performance.
    c.   emphasizing the positive.

2.   Flight instructor responsibilities in--

    a.   providing student pilot evaluation and supervision.
    b.   preparing practical test recommendations and endorsements.
    c.   determining requirements for conducting additional training and endorsement requirements.

3.   Professionalism as an instructor by--

    a.   explaining important personal characteristics.
    b.   describing methods to minimize student frustration.

G.   TASK: PLANNING INSTRUCTIONAL ACTIVITY

REFERENCE: FAA-H-8083-9.

Objective. To determine that the applicant exhibits instructional knowledge of the elements of planning instructional activity by describing:

1.   Developing objectives and standards for a course of training.
2.   Theory of building blocks of learning.
3.   Requirements for developing a training syllabus.
4.   Purpose and characteristics of a lesson plan.

II.   AREA OF OPERATION: TECHNICAL SUBJECT AREAS

NOTE: The examiner shall select TASK L and at least one other TASK.

A.   TASK: AEROMEDICAL FACTORS

REFERENCES: FAA-H-8083-3; FAA-S-8081-12, FAA-S-8081-14; AIM.

Objective. To determine that the applicant exhibits instructional knowledge of the elements related to aeromedical factors by describing:

1.   How to obtain an appropriate medical certificate.
2.   How to obtain a medical certificate in the event of a possible medical deficiency.
3.   The causes, symptoms, effects, and corrective action of the following medical factors--

    a.   hypoxia.
    b.   hyperventilation.
    c.   middle ear and sinus problems.
    d.   spatial disorientation.
    e.   motion sickness.
    f.   carbon monoxide poisoning.
    g.   fatigue and stress.
    h.   dehydration.

4. The effects of alcohol and drugs, and their relationship to flight safety.
5. The effect of nitrogen excesses incurred during scuba dives and how this affects pilots and passengers during flight.

B. TASK: VISUAL SCANNING AND COLLISION AVOIDANCE

REFERENCES: FAA-H-8083-3, FAA-H-8083-25; AC 90-48; AIM.

Objective. To determine that the applicant exhibits instructional knowledge of the elements of visual scanning and collision avoidance by describing:

1. Relationship between a pilot's physical condition and vision.
2. Environmental conditions that degrade vision.
3. Vestibular and visual illusions.
4. "See and avoid" concept.
5. Proper visual scanning procedure.
6. Relationship between poor visual scanning habits and increased collision risk.
7. Proper clearing procedures.
8. Importance of knowing aircraft blind spots.
9. Relationship between aircraft speed differential and collision risk.
10. Situations that involve the greatest collision risk.

C. TASK: PRINCIPLES OF FLIGHT

REFERENCES: FAA-H-8083-3, FAA-H-8083-25.

Objective. To determine that the applicant exhibits instructional knowledge of the elements of principles of flight by describing:

1. Airfoil design characteristics.
2. Airplane stability and controllability.
3. Turning tendency (torque effect).
4. Load factors in airplane design.
5. Wingtip vortices and precautions to be taken.

D. TASK: AIRPLANE FLIGHT CONTROLS

REFERENCES: FAA-H-8083-3, FAA-H-8083-25.

Objective. To determine that the applicant exhibits instructional knowledge of the elements related to the airplane flight controls by describing the purpose, location, direction of movement, effect, and proper procedure for use of the:

1. Primary flight controls.
2. Trim control(s).
3. Wing flaps.

E. TASK: AIRPLANE WEIGHT AND BALANCE

REFERENCES: FAA-H-8083-1, FAA-H-8083-3, FAA-H-8083-25.

Objective. To determine that the applicant exhibits instructional knowledge of the elements of airplane weight and balance by describing:

1. Weight and balance terms.
2. Effect of weight and balance on performance.
3. Methods of weight and balance control.
4. Determination of total weight and center of gravity and the changes that occur when adding, removing, or shifting weight.

F. TASK: NAVIGATION AND FLIGHT PLANNING

REFERENCES: FAA-H-8083-3, FAA-H-8083-25.

Objective. To determine that the applicant exhibits instructional knowledge of the elements of navigation and flight planning by describing:

1. Terms used in navigation.
2. Features of aeronautical charts.
3. Importance of using the proper and current aeronautical charts.
4. Method of plotting a course, selection of fuel stops and alternates, and appropriate actions in the event of unforeseen situations.
5. Fundamentals of pilotage and dead reckoning.
6. Fundamentals of radio navigation.
7. Division to an alternate.
8. Lost procedures.
9. Computation of fuel consumption.
10. Importance of preparing and properly using a flight log.
11. Importance of a weather check and the use of good judgment in making a "go/no-go" decision.
12. Purpose of and procedure used in, filing a flight plan.

G. TASK: NIGHT OPERATIONS

REFERENCES: FAA-H-8083-3, FAA-H-8083-25; FAA-S-8081-12, FAA-S-8081-14; AIM.

Objective. To determine that the applicant exhibits instructional knowledge of the elements of night operations by describing:

1. Factors related to night vision.
2. Disorientation and night optical illusions.
3. Proper adjustment of interior lights.
4. Importance of having a flashlight with a red lens.
5. Night preflight inspection.
6. Engine starting procedures, including use of position and anticollision lights prior to start.
7. Taxiing and orientation on an airport.
8. Takeoff and climb-out.
9. In-flight orientation.
10. Importance of verifying the airplane's attitude by reference to flight instruments.
11. Night emergencies procedures.
12. Traffic patterns.
13. Approaches and landings with and without landing lights.
14. Go-around.

H. TASK: HIGH ALTITUDE OPERATIONS

REFERENCES: 14 CFR part 91; AC 61-107; FAA-H-8083-3; FAA-S-8081-12; POH/AFM, AIM.

Objective. To determine that the applicant exhibits instructional knowledge of the elements of high altitude operations by describing:

1. Regulatory requirements for use of oxygen.
2. Physiological hazards associated with high altitude operations.
3. Characteristics of a pressurized airplane and various types of supplemental oxygen systems.
4. Importance of "aviators breathing oxygen."
5. Care and storage of high-pressure oxygen bottles.
6. Problems associated with rapid decompression and corresponding solutions.
7. Fundamental concept of cabin pressurization.
8. Operation of cabin pressurization system.

I. TASK: FEDERAL AVIATION REGULATIONS AND PUBLICATIONS

REFERENCES: 14 CFR parts 1, 61, 91; NTSB part 830; AC 00-2; FAA-H-8083-25; POH/AFM, AIM.

Objective. To determine that the applicant exhibits instructional knowledge of the elements related to Federal Aviation Regulations and publications:

1. Availability and method of revision of 14 CFR parts 1, 61, 91, and NTSB part 830 by describing--

   a. purpose.
   b. general content.

2. Availability of flight information publications, advisory circulars, practical test standards, pilot operating handbooks, and FAA-approved airplane flight manuals by describing--

    a. availability.
    b. purpose.
    c. general content.

J. TASK: NATIONAL AIRSPACE SYSTEM

REFERENCES: 14 CFR part 91; FAA-S-8081-12, FAA-S-8081-14; AIM.

Objective. To determine that the applicant exhibits instructional knowledge of the elements of the national airspace system by describing:

1. Basic VFR Weather Minimums—for all classes of airspace.
2. Airspace classes—the operating rules, pilot certification, and airplane equipment requirements for the following--

    a. Class A.
    b. Class B.
    c. Class C.
    d. Class D.
    e. Class E.
    f. Class G.

3. Special use airspace (SUA).
4. Temporary flight restrictions (TFR).

K. TASK: NAVIGATION SYSTEMS AND RADAR SERVICES

REFERENCES: FAA-H-8083-3, FAA-H-8083-15; FAA-S-8081-12, FAA-S-8081-14; AIM.

Objective. To determine that the applicant exhibits instructional knowledge of the elements related to navigation systems and radar service by describing:

1. One ground-based navigational system (VOR/VORTAC, NDB, DME, and LORAN).
2. Satellite-based navigation system.
3. Radar service and procedures.
4. Global positioning system (GPS).

L. TASK: LOGBOOK ENTRIES AND CERTIFICATE ENDORSEMENTS

REFERENCES: 14 CFR part 61; AC 61-65.

Objective. To determine that the applicant exhibits instructional knowledge of the elements related to logbook entries and certificate endorsements by describing:

1. Required logbook entries for instruction given.
2. Required student pilot certificate endorsements, including appropriate logbook entries.
3. Preparation of a recommendation for a pilot practical test, including appropriate logbook entry for--

    a. initial pilot certification.
    b. additional pilot certification.
    c. additional aircraft qualification.

4. Required endorsement of a pilot logbook for the satisfactory completion of the required FAA flight review.
5. Required flight instructor records.

III. AREA OF OPERATION: PREFLIGHT PREPARATION

NOTE: The examiner shall select at least one TASK.

A. TASK: CERTIFICATES AND DOCUMENTS

REFERENCES: 14 CFR parts 23, 43, 61, 67, 91; FAA-H-8083-3, FAA-H-8083-25; FAA-S-8081-12, FAA-S-8081-14; POH/AFM.

Objective. To determine that the applicant exhibits instructional knowledge of the elements related to certificates and documents by describing:

1. The training requirements for the issuance of a recreational, private, and commercial pilot certificate.
2. The privileges and limitations of pilot certificates and ratings at recreational, private, and commercial levels.
3. Class and duration of medical certificates.
4. Recent pilot flight experience requirements.
5. Required entries in pilot logbook or flight record.

B. TASK: WEATHER INFORMATION

REFERENCES: AC 00-6, AC 00-45; FAA-H-8083-25; FAA-S-8081-12, FAA-S-8081-14.

Objective. To determine that the applicant exhibits instructional knowledge of the elements related to weather information by describing:

1. Importance of a thorough preflight weather briefing.
2. Various means and sources of obtaining weather information.
3. Use of real time weather reports, forecasts, and charts for developing scenario based training.
4. In-flight weather advisories.
5. Recognition of aviation weather hazards to include wind shear.
6. Factors to be considered in making a "go/no-go" decision.

C. TASK: OPERATION OF SYSTEMS

REFERENCES: FAA-H-8083-23, FAA-H-8083-25, FAA-H-8083-3; FAA-S-8081-12, FAA-S-8081-14; POH/AFM.

Objective. To determine that the applicant exhibits instructional knowledge of the elements related to the operation of systems, as applicable to the airplane used for the practical test, by describing the following systems:

1. Primary flight controls and trim.
2. Flaps, leading edge devices, and spoilers.
3. Water rudders (ASES).
4. Powerplant and propeller.
5. Landing gear.
6. Fuel, oil, and hydraulic.
7. Electrical.
8. Avionics.
9. Pitot static, vacuum/pressure, and associated instruments.
10. Environmental.
11. Deicing and anti-icing.

D. TASK: PERFORMANCE AND LIMITATIONS

REFERENCES: FAA-H-8083-3, FAA-H-8083-23, FAA-H-8083-25; AC 61-84; FAA-S-8081-12, FAA-S-8081-14; POH/AFM.

Objective. To determine that the applicant exhibits instructional knowledge of the elements related to performance and limitations by describing:

1. Determination of weight and balance condition.
2. Use of performance charts, tables, and other data in determining performance in various phases of flight.
3. Effects of exceeding airplane limitations.
4. Effects of atmospheric conditions on performance.
5. Factors to be considered in determining that the required performance is within the airplane's capabilities.

E. TASK: AIRWORTHINESS REQUIREMENTS

REFERENCES: 14 CFR parts 23, 39, 43; FAA-S-8081-12, FAA-S-8081-14; POH/AFM.

Objective. To determine that the applicant exhibits instructional knowledge of the elements related to required airworthiness by explaining:

1. Required instruments and equipment for day/night VFR.
2. Procedures and limitations for determining airworthiness of the airplane with inoperative instruments and equipment with and without minimum equipment list (MEL).
3. Requirements and procedures for obtaining a special flight permit.
4. Airworthiness directives, compliance records, maintenance/inspection requirements, and appropriate records.
5. Procedures for deferring maintenance on aircraft without an approved MEL.

IV. AREA OF OPERATION: PREFLIGHT LESSON ON A MANEUVER TO BE PERFORMED IN FLIGHT

NOTE: Examiner shall select at least one maneuver TASK from AREAS OF OPERATION VII through XIII, and ask the applicant to present a preflight lesson on the selected maneuver as the lesson would be taught to a student.

A. TASK: MANEUVER LESSON

REFERENCES: FAA-H-8082-3, FAA-H-8083-9, FAA-H-8083-23, FAA-H-8083-25; FAA-S-8081-12, FAA-S-8081-14; POH/AFM.

Objective. To determine the applicant exhibits instructional knowledge of the selected maneuver by:

1. Stating the purpose.
2. Giving an accurate, comprehensive oral description, including the elements and common errors.
3. Using instructional aids, as appropriate.
4. Describing the recognition, analysis, and correction of common errors.

V. AREA OF OPERATION: PREFLIGHT PROCEDURES

NOTE: The examiner shall select at least one TASK.

A. TASK: PREFLIGHT INSPECTION

REFERENCES: AC 61-84; FAA-H-8083-3, FAA-H-8083-23; FAA-S-8081-12, FAA-S-8081-14; POH/AFM.

Objective. To determine that the applicant:

1. Exhibits instructional knowledge of the elements of a preflight inspection, as applicable to the airplane used for the practical test, by describing--

   a. reasons for the preflight inspection, items that should be inspected, and how defects are detected.
   b. importance of using the appropriate checklist.
   c. how to determine fuel and oil quantity and contamination.
   d. detection of fuel, oil, and hydraulic leaks.
   e. inspection of the oxygen system, including supply and proper operation (if applicable).
   f. inspection of the flight controls and water rudder (if applicable).
   g. detection of visible structural damage.
   h. removal of tie-downs, control locks, and wheel chocks.
   i. removal of ice and frost.
   j. importance of the proper loading and securing of baggage, cargo, and equipment.
   k. use of sound judgment in determining whether the airplane is airworthy and in condition for safe flight.

2. Exhibits instructional knowledge of common errors related to a preflight inspection by describing--

   a. failure to use or the improper use of a checklist.
   b. hazards which may result from allowing distractions to interrupt a visual inspection.
   c. inability to recognize discrepancies to determine airworthiness.
   d. failure to ensure servicing with the proper fuel and oil.
   e. failure to ensure proper loading and securing of baggage, cargo, and equipment.

3. Demonstrates and simultaneously explains a preflight inspection from an instructional standpoint.

B. TASK: SINGLE-PILOT RESOURCE MANAGEMENT

REFERENCES: FAA-H-8083-3; FAA-S-8081-12; POH/AFM.

Objective. To determine that the applicant:

1. Exhibits instructional knowledge of the elements of single-pilot resource management by describing--

   a. proper arranging and securing of essential materials and equipment in the cockpit.
   b. proper use and/or adjustment of cockpit items such as safety belts, shoulder harnesses, rudder pedals, and seats.
   c. occupant briefing on emergency procedures and use of safety belts.
   d. Proper utilization of all resources required to operate a flight safely, dispatchers, weather briefers, maintenance personnel, and air traffic control.

2. Exhibits instructional knowledge of common errors related to single-pilot crew resource management by describing--

   a. failure to place and secure essential materials and equipment for easy access during flight.
   b. failure to properly adjust cockpit items, such as safety belts, shoulder harnesses, rudder pedals, and seats.
   c. failure to provide proper adjustment of equipment and controls.
   d. failure to provide occupant briefing on emergency procedures and use of safety belts.
   e. failure to utilize all resources required to operate a flight safely.

3. Demonstrates and simultaneously explains single-pilot crew resource management from an instructional standpoint.

C. TASK: ENGINE STARTING

REFERENCES: FAA-H-8083-3, FAA-H-8083-23, FAA-H-8083-25; AC 91-13, AC 91-55; FAA-S-8081-12, FAA-S-8081-14; POH/AFM.

Objective. To determine that the applicant:

1. Exhibits instructional knowledge of the elements of engine starting, as appropriate to the airplane used for the practical test, by describing--

   a. safety precautions related to starting.
   b. use of external power.
   c. effect of atmospheric conditions on starting.
   d. importance of following the appropriate checklist.
   e. adjustment of engine controls during start.
   f. prevention of airplane movement during and after start.
   g. safety procedures for hand propping an airplane.
   h. carburetor fire hazard.

2. Exhibits instructional knowledge of common errors related to engine starting by describing--

   a. failure to properly use the appropriate checklist.
   b. failure to use safety precautions related to starting.
   c. improper adjustment of engine controls during start.
   d. failure to ensure proper clearance of the propeller.

3. Demonstrates and simultaneously explains engine starting from an instructional standpoint.

D. TASK: TAXIING—LANDPLANE

REFERENCES: FAA-H-8083-3; FAA-S-8081-12, FAA-S-8081-14; POH/AFM.

Objective. To determine that the applicant:

1. Exhibits instructional knowledge of the elements of landplane taxiing by describing--

   a. proper brake check and correct use of brakes.
   b. compliance with airport/taxiway surface marking, signals, and ATC clearances or instructions.
   c. how to control direction and speed.
   d. flight control positioning for various wind conditions.
   e. procedures used to avoid other aircraft and hazards.
   f. avoiding runway incursions.

2. Exhibits instructional knowledge of common errors related to landplane taxiing by describing--

   a. improper use of brakes.
   b. improper positioning of the flight controls for various wind conditions.
   c. hazards of taxiing too fast.
   d. failure to comply with airport/taxiway surface marking, signals, and ATC clearances or instructions.

3. Demonstrates and simultaneously explains landplane taxiing from an instructional standpoint.

4. Analyzes and corrects simulated common errors related to landplane taxiing.

G. TASK: BEFORE TAKEOFF CHECK

REFERENCES: FAA-H-8083-3, FAA-H-8083-23; FAA-S-8081-12, FAA-S-8081-14; POH/AFM.

Objective. To determine that the applicant:

1. Exhibits instructional knowledge of the elements of the before takeoff check by describing--

   a. positioning the airplane with consideration for other aircraft, surface conditions, and wind.
   b. division of attention inside and outside the cockpit.
   c. importance of following the checklist and responding to each checklist item.
   d. reasons for ensuring suitable engine temperatures and pressures for run-up and takeoff.
   e. method used to determine the airplane is in a safe operating condition.
   f. importance of reviewing takeoff performance airspeeds, expected takeoff distances, and emergency procedures.
   g. method used for ensuring that the takeoff area or path is free of hazards.
   h. method of avoiding runway incursions and ensuring no conflict with traffic prior to taxiing into takeoff position.

2. Exhibits instructional knowledge of common errors related to the before takeoff check by describing--

   a. failure to properly use the appropriate checklist.
   b. improper positioning of the airplane.
   c. improper acceptance of marginal engine performance.

   d. an improper check of flight controls.
   e. hazards of failure to review takeoff and emergency procedures.
   f. failure to avoid runway incursions and to ensure no conflict with traffic prior to taxiing into takeoff position.

3. Demonstrates and simultaneously explains a before takeoff check from an instructional standpoint.

4. Analyzes and corrects simulated common errors related to a before takeoff check.

VI. AREA OF OPERATION: AIRPORT OPERATIONS

NOTE: The examiner shall select at least one TASK.

A. TASK: RADIO COMMUNICATIONS AND ATC LIGHT SIGNALS

REFERENCES: FAA-H-8083-3; FAA-H-8083-25; FAA-S-8081-12, FAA-S-8081-14; AIM.

Objective. To determine that the applicant:

1. Exhibits instructional knowledge of the elements of radio communications and ATC light signals by describing--

   a. selection and use of appropriate radio frequencies.
   b. recommended procedure and phraseology for radio communications.
   c. receipt, acknowledgement of, and compliance with, ATC clearances and instructions.
   d. Interpretation of, and compliance with, ATC light signals.

2. Exhibits instructional knowledge of common errors related to radio communications and ATC light signals by describing--

   a. use of improper frequencies.
   b. improper procedure and phraseology when using radio communications.
   c. failure to acknowledge, or properly comply with, ATC clearances and instructions.
   d. failure to understand, or to properly comply with, ATC light signals.

3. Demonstrates and simultaneously explains radio communication procedures from an instructional standpoint.

4. Analyzes and corrects simulated common errors related to radio communications and ATC light signals.

B. TASK: TRAFFIC PATTERNS

REFERENCES: FAA-H-8083-3, FAA-H-8083-25; AC 90-42, AC 90-66; FAA-S-8081-12, FAA-S-8081-14; AIM.

Objective. To determine that the applicant:

1. Exhibits instructional knowledge of the elements of traffic patterns by describing--

   a. operations at airports and seaplane bases with and without operating control towers.
   b. adherence to traffic pattern procedures, instructions, and rules.
   c. how to maintain proper spacing from other traffic.
   d. how to maintain the desired ground track.
   e. wind shear and wake turbulence avoidance procedures.
   f. orientation with the runway or landing area in use.
   g. how to establish a final approach at an appropriate distance from the runway or landing area.
   h. use of checklist.

2. Exhibits instructional knowledge of common errors related to traffic patterns by describing--

 a. failure to comply with traffic pattern instructions, procedures, and rules.
 b. improper correction for wind drift.
 c. inadequate spacing from other traffic.
 d. poor altitude or airspeed control.

3. Demonstrates and simultaneously explains traffic patterns from an instructional standpoint.
4. Analyzes and corrects simulated common errors related to traffic patterns.

C. TASK: AIRPORT, RUNWAY AND TAXIWAY SIGNS, MARKINGS, AND LIGHTING

REFERENCES: FAA-H-8083-23, FAA-H-8083-25; FAA-S-8081-12, FAA-S-8081-14; AIM; AC 150/5340-1, AC 150/5340-18.

Objective. To determine that the applicant:

1. Exhibits instructional knowledge of the elements of airport, runway and taxiway signs, markings, and lighting by describing--

 a. identification and proper interpretation of airport, runway and taxiway signs and markings with emphasis on runway incursion avoidance.
 b. identification and proper interpretation of airport, runway and taxiway lighting with emphasis on runway incursion avoidance.

2. Exhibits instructional knowledge of common errors related to airport, runway and taxiway signs, markings, and lighting by describing--

 a. failure to comply with airport, runway and taxiway signs and markings.
 b. failure to comply with airport, runway and taxiway lighting.
 c. failure to use proper runway incursion avoidance procedures.

3. Demonstrates and simultaneously explains airport, runway and taxiway signs, markings, and lighting from an instructional standpoint.
4. Analyzes and corrects simulated common errors related to airport, runway and taxiway signs, markings, and lighting.

VII. AREA OF OPERATION: TAKEOFFS, LANDINGS, AND GO-AROUNDS

NOTE: The examiner shall select at least two takeoffs and two landing TASKs.

A. TASK: NORMAL AND CROSSWIND TAKEOFF AND CLIMB

REFERENCES: FAA-H-8083-3, FAA-H-8083-23; FAA-S-8081-12, FAA-S-8081-14; POH/AFM.

Objective. To determine that the applicant:

1. Exhibits instructional knowledge of the elements of a normal and crosswind takeoff and climb by describing--

 a. procedures before taxiing onto the runway or takeoff area to ensure runway incursion avoidance.
 b. normal and crosswind takeoff and lift-off procedures.
 d. proper climb attitude, power setting, and airspeed (V_Y).
 e. proper use of checklist.

2. Exhibits instructional knowledge of common errors related to a normal and crosswind takeoff and climb by describing--

 a. improper runway incursion avoidance procedures.

 b. improper use of controls during a normal or crosswind takeoff.
 c. inappropriate lift-off procedures.
 d. improper climb attitude, power setting, and airspeed (V_Y).
 e. improper use of checklist.

3. Demonstrates and simultaneously explains a normal or a crosswind takeoff and climb from an instructional standpoint.
4. Analyzes and corrects simulated common errors related to a normal or a crosswind takeoff and climb.

B. TASK: SHORT-FIELD TAKEOFF AND MAXIMUM PERFORMANCE CLIMB

REFERENCES: FAA-H-8083-3, FAA-H-8083-23; FAA-S-8081-12, FAA-S-8081-14; POH/AFM.

Objective. To determine that the applicant:

1. Exhibits instructional knowledge of the elements of a short-field takeoff and climb by describing--

 a. procedures before taxiing onto the runway or takeoff area to ensure runway incursion avoidance.
 b. short-field takeoff and lift-off procedures.
 c. initial climb attitude and airspeed (V_x) until obstacle is cleared (50 feet AGL).
 d. proper use of checklist.

2. Exhibits instructional knowledge of common errors related to a short-field takeoff and climb by describing--

 a. improper runway incursion avoidance procedures.
 b. improper use of controls during a short-field takeoff.
 c. improper lift-off procedures.
 d. improper initial climb attitude, power setting, and airspeed (V_x) to clear obstacle.
 e. improper use of checklist.

3. Demonstrates and simultaneously explains a short-field takeoff and climb from an instructional standpoint.
4. Analyzes and corrects simulated common errors related to a short-field takeoff and climb.

C. TASK: SOFT-FIELD TAKEOFF AND CLIMB

REFERENCES: FAA-H-8083-3; FAA-S-8081-12, FAA-S-8081-14; POH/AFM.

Objective. To determine that the applicant:

1. Exhibits instructional knowledge of the elements of a soft-field takeoff and climb by describing--

 a. procedures before taxiing onto the runway or takeoff area to ensure runway incursion avoidance.
 b. soft-field takeoff and lift-off procedures.
 c. initial climb attitude and airspeed, depending on if an obstacle is present.
 d. proper use of checklist.

2. Exhibits instructional knowledge of common errors related to a soft-field takeoff and climb by describing--

 a. improper runway incursion avoidance procedures.
 b. improper use of controls during a soft-field takeoff.
 c. improper lift-off procedures.
 d. improper climb attitude, power setting, and airspeed (V_Y) or (V_x).
 e. improper use of checklist.

3. Demonstrates and simultaneously explains a soft-field takeoff and climb from an instructional standpoint.
4. Analyzes and corrects simulated common errors related to a soft-field takeoff and climb.

F.   TASK: NORMAL AND CROSSWIND APPROACH AND LANDING

REFERENCES: FAA-H-8083-3, FAA-H-8083-23; FAA-S-8081-12, FAA-S-8081-14; POH/AFM.

Objective. To determine that the applicant:

1. Exhibits instructional knowledge of the elements of a normal and a crosswind approach and landing by describing--

   a. how to determine landing performance and limitations.
   b. configuration, power, and trim.
   c. obstructions and other hazards, which should be considered.
   d. a stabilized approach at the recommended airspeed to the selected touchdown area.
   e. course of action if selected touchdown area is going to be missed.
   f. coordination of flight controls.
   g. a precise ground track.
   h. wind shear and wake turbulence avoidance procedures.
   i. most suitable crosswind procedure.
   j. timing, judgment, and control procedure during roundout and touchdown.
   k. directional control after touchdown.
   l. use of brakes.
   m. use of checklist.

2. Exhibits instructional knowledge of common errors related to a normal and a crosswind approach and landing by describing--

   a. improper use of landing performance data and limitations.
   b. failure to establish approach and landing configuration at appropriate time or in proper sequence.
   c. failure to establish and maintain a stabilized approach.
   d. inappropriate removal of hand from throttle.
   e. improper procedure during roundout and touchdown.
   f. poor directional control after touchdown.
   g. improper use of brakes.

3. Demonstrates and simultaneously explains a normal or a crosswind approach and landing from an instructional standpoint.

4. Analyzes and corrects simulated common errors related to a normal or crosswind approach and landing.

G.   TASK: SLIP TO A LANDING

REFERENCES: FAA-H-8083-3, FAA-H-8083-23; FAA-S-8081-14; POH/AFM.

Objective. To determine that the applicant:

1. Exhibits instructional knowledge of the elements of a slip (forward and side) to a landing by describing--

   a. configuration, power, and trim.
   b. obstructions and other hazards, which should be considered.
   c. a stabilized slip at the appropriate airspeed to the selected touchdown area.
   d. possible airspeed indication errors.
   e. proper application of flight controls.
   f. a precise ground track.
   g. wind shear and wake turbulence avoidance procedures.
   h. timing, judgment, and control procedure during transition from slip to touchdown.
   i. directional control after touchdown.
   j. use of brakes.
   k. use of checklist.

2. Exhibits instructional knowledge of common errors related to a slip (forward and side) to a landing by describing--

   a. improper use of landing performance data and limitations.
   b. failure to establish approach and landing configuration at appropriate time or in proper sequence.
   c. failure to maintain a stabilized slip.
   d. inappropriate removal of hand from throttle.
   e. improper procedure during transition from the slip to the touchdown.
   f. poor directional control after touchdown.
   g. improper use of brakes.

3. Demonstrates and simultaneously explains a forward or sideslip to a landing from an instructional standpoint.

4. Analyzes and corrects simulated common errors related to a forward or sideslip to a landing.

H.   TASK: GO-AROUND/REJECTED LANDING

REFERENCES: FAA-H-8083-3, FAA-H-8083-23; FAA-S-8081-12, FAA-S-8081-14; POH/AFM.

Objective. To determine that the applicant:

1. Exhibits instructional knowledge of the elements of a go-around/rejected landing by describing--

   a. situations where a go-around/rejected landing is necessary.
   b. importance of making a prompt decision.
   c. importance of applying takeoff power immediately after the go-around/rejected landing decision is made.
   d. importance of establishing proper pitch attitude.
   e. wing flaps retraction.
   f. use of trim.
   g. landing gear retraction.
   h. proper climb speed.
   i. proper track and obstruction clearance.
   j. use of checklist.

2. Exhibits instructional knowledge of common errors related to a go-around/rejected landing by describing--

   a. failure to recognize a situation where a go-around/rejected landing is necessary.
   b. hazards of delaying a decision to perform a go-around/rejected landing.
   c. improper power application.
   d. failure to control pitch attitude.
   e. failure to compensate for torque effect.
   f. improper trim procedure.
   g. failure to maintain recommended airspeeds.
   h. improper wing flaps or landing gear retraction procedure.
   i. failure to maintain proper track during climb-out.
   j. failure to remain well clear of obstructions and other traffic.

3. Demonstrates and simultaneously explains a go-around/rejected landing from an instructional standpoint.

4. Analyzes and corrects simulated common errors related to a go-around/rejected landing.

I. TASK: SHORT-FIELD APPROACH AND LANDING

REFERENCES: FAA-H-8083-3, FAA-H-8083-23; FAA-S-8081-12, FAA-S-8081-14; POH/AFM.

Objective. To determine that the applicant:

1. Exhibits instructional knowledge of the elements of a short-field approach and landing by describing--

    a. how to determine landing performance and limitations.
    b. configuration and trim.
    c. proper use of pitch and power to maintain desired approach angle.
    d. obstructions and other hazards which should be considered.
    e. effect of wind.
    f. selection of touchdown and go-around points.
    g. a stabilized approach at the recommended airspeed to the selected touchdown point.
    h. coordination of flight controls.
    i. a precise ground track.
    j. timing, judgment, and control procedure during roundout and touchdown.
    k. directional control after touchdown.
    l. use of brakes.
    m. use of checklist.

2. Exhibits instructional knowledge of common errors related to a short-field approach and landing by describing--

    a. improper use of landing performance data and limitations.
    b. failure to establish approach and landing configuration at appropriate time or in proper sequence.
    c. failure to establish and maintain a stabilized approach.
    d. improper procedure in use of power, wing flaps, and trim.
    e. inappropriate removal of hand from throttle.
    f. improper procedure during roundout and touchdown.
    g. poor directional control after touchdown.
    h. improper use of brakes.

3. Demonstrates and simultaneously explains a short-field approach and landing from an instructional standpoint.
4. Analyzes and corrects simulated common errors related to a short-field approach and landing.

J. TASK: SOFT-FIELD APPROACH AND LANDING

REFERENCES: FAA-H-8083-3; FAA-S-8081-12, FAA-S-8081-14; POH/AFM.

Objective. To determine that the applicant:

1. Exhibits instructional knowledge of the elements of a soft-field approach and landing by describing--

    a. how to determine landing performance and limitations.
    b. configuration and trim.
    c. obstructions and other hazards which should be considered.
    d. effect of wind and landing surface.
    e. selection of touchdown area.
    f. a stabilized approach at the recommended airspeed to the selected touchdown area.
    g. coordination of flight controls.
    h. a precise ground track.
    i. timing, judgment, and control procedure during roundout and touchdown.
    j. touchdown in a nose-high pitch attitude at minimum safe airspeed.
    k. proper use of power.
    l. directional control after touchdown.
    m. use of checklist.

2. Exhibits instructional knowledge of common errors related to a soft-field approach and landing by describing--

    a. improper use of landing performance data and limitations.
    b. failure to establish approach and landing configuration at proper time or in proper sequence.
    c. failure to establish and maintain a stabilized approach.
    d. failure to consider the effect of wind and landing surface.
    e. improper procedure in use of power, wing flaps, or trim.
    f. inappropriate removal of hand from throttle.
    g. improper procedure during roundout and touchdown.
    h. failure to hold back elevator pressure after touchdown.
    i. closing the throttle too soon after touchdown.
    j. poor directional control after touchdown.
    k. improper use of brakes.

3. Demonstrates and simultaneously explains a soft-field approach and landing from an instructional standpoint.
4. Analyzes and corrects simulated common errors related to a soft-field approach and landing.

K. TASK: 180° POWER-OFF ACCURACY APPROACH AND LANDING

REFERENCES: FAA-H-8083-3; FAA-S-8081-12.

Objective. To determine that the applicant:

1. Exhibits instructional knowledge of the elements of a 180° power-off accuracy approach and landing by describing--

    a. configuration and trim.
    b. effects of wind and selection of a touchdown area.
    c. the key points in the pattern.
    d. a stabilized approach at the recommended airspeed to the selected touchdown area.
    e. coordination of flight controls.
    f. timing, judgment, and control procedure during roundout and touchdown.
    g. directional control after touchdown.
    h. use of checklist.

2. Exhibits instructional knowledge of common errors related to a 180° power-off accuracy approach and landing by describing--

    a. failure to establish approach and landing configuration at proper time or in proper sequence.
    b. failure to identify the key points in the pattern.
    c. failure to establish and maintain a stabilized approach.
    d. failure to consider the effect of wind and landing surface.
    e. improper use of power, wing flaps, or trim.
    f. improper procedure during roundout and touchdown.
    g. failure to hold back elevator pressure after touchdown.
    h. poor directional control after touchdown.
    i. improper use of brakes.

3. Demonstrates and simultaneously explains a 180° power-off accuracy approach and landing from an instructional standpoint.
4. Analyzes and corrects simulated common errors related to a 180° power-off accuracy approach and landing.

VIII. AREA OF OPERATION: FUNDAMENTALS OF FLIGHT

NOTE: The examiner shall select at least one TASK.

A. TASK: STRAIGHT-AND-LEVEL FLIGHT

REFERENCES: FAA-H-8083-3, FAA-H-8083-23; FAA-S-8081-14.

Objective. To determine that the applicant:

1. Exhibits instructional knowledge of the elements of straight-and-level flight by describing--

    a. effect and use of flight controls.
    b. the Integrated Flight Instruction method.
    c. outside and instrument references used for pitch, bank, and power control; the crosscheck and interpretation of those references; and the control procedure used.
    d. trim procedure.
    e. methods that can be used to overcome tenseness and overcontrolling.

2. Exhibits instructional knowledge of common errors related to straight-and-level flight by describing--

    a. failure to crosscheck and correctly interpret outside and instrument references.
    b. application of control movements rather than pressures.
    c. uncoordinated use of flight controls.
    d. faulty trim procedure.

3. Demonstrates and simultaneously explains straight-and-level flight from an instructional standpoint.
4. Analyzes and corrects simulated common errors related to straight-and-level flight.

B. TASK: LEVEL TURNS

REFERENCES: FAA-H-8083-3; FAA-S-8081-14.

Objective. To determine that the applicant:

1. Exhibits instructional knowledge of the elements of level turns by describing--

    a. effect and use of flight controls.
    b. the Integrated Flight Instruction method.
    c. outside and instrument references used for pitch, bank, and power control; the crosscheck and interpretation of those references; and the control procedure used.
    d. trim procedure.
    e. methods that can be used to overcome tenseness and overcontrolling.

2. Exhibits instructional knowledge of common errors related to level turns by describing--

    a. failure to crosscheck and correctly interpret outside and instrument references.
    b. application of control movements rather than pressures.
    c. uncoordinated use of flight controls.
    d. faulty altitude and bank control.

3. Demonstrates and simultaneously explains level turns from an instructional standpoint.
4. Analyzes and corrects simulated common errors related to level turns.

C. TASK: STRAIGHT CLIMBS AND CLIMBING TURNS

REFERENCES: FAA-H-8083-3; FAA-S-8081-14.

Objective. To determine that the applicant:

1. Exhibits instructional knowledge of the elements of straight climbs and climbing turns by describing--

    a. effect and use of flight controls.
    b. the Integrated Flight Instruction method.
    c. outside and instrument references used for pitch, bank, and power control; the crosscheck and interpretation of those references; and the control procedure used.
    d. trim procedure.
    e. methods that can be used to overcome tenseness and overcontrolling.

2. Exhibits instructional knowledge of common errors related to straight climbs and climbing turns by describing--

    a. failure to crosscheck and correctly interpret outside and instrument references.
    b. application of control movements rather than pressures.
    c. improper correction for torque effect.
    d. faulty trim procedure.

3. Demonstrates and simultaneously explains straight climbs and climbing turns from an instructional standpoint.
4. Analyzes and corrects simulated common errors related to straight climbs and climbing turns.

D. TASK: STRAIGHT DESCENTS AND DESCENDING TURNS

REFERENCES: FAA-H-8083-3; FAA-S-8081-14.

Objective. To determine that the applicant:

1. Exhibits instructional knowledge of the elements of straight descents and descending turns by describing--

    a. effect and use of flight controls.
    b. the Integrated Flight Instruction method.
    c. outside and instrument references used for pitch, bank, and power control; the crosscheck and interpretation of those references; and the control procedure used.
    d. trim procedure.
    e. methods that can be used to overcome tenseness and overcontrolling.

2. Exhibits instructional knowledge of common errors related to straight descents and descending turns by describing--

    a. failure to crosscheck and correctly interpret outside and instrument references.
    b. application of control movements rather than pressures.
    c. uncoordinated use of flight controls.
    d. faulty trim procedure.
    e. failure to clear engine and use carburetor heat, as appropriate.

3. Demonstrates and simultaneously explains straight descents and descending turns from an instructional standpoint.
4. Analyzes and corrects simulated common errors related to straight descents and descending turns.

IX. AREA OF OPERATION: PERFORMANCE MANEUVERS

NOTE: The examiner shall select at least TASKs A or B and C or D.

A. TASK: STEEP TURNS

REFERENCES: FAA-H-8083-3; FAA-S-8081-12, FAA-S-8081-14; POH/AFM.

Objective. To determine that the applicant:

1. Exhibits instructional knowledge of the elements of steep turns by describing--

   a. relationship of bank angle, load factor, and stalling speed.
   b. overbanking tendency.
   c. torque effect in right and left turns.
   d. selection of a suitable altitude.
   e. orientation, division of attention, and planning.
   f. entry and rollout procedure.
   g. coordination of flight and power controls.
   h. altitude, bank, and power control during the turn.
   i. proper recovery to straight-and-level flight.

2. Exhibits instructional knowledge of common errors related to steep turns by describing--

   a. improper pitch, bank, and power coordination during entry and rollout.
   b. uncoordinated use of flight controls.
   c. improper procedure in correcting altitude deviations.
   d. loss of orientation.

3. Demonstrates and simultaneously explains steep turns from an instructional standpoint.
4. Analyzes and corrects simulated common errors related to steep turns.

B. TASK: STEEP SPIRALS

REFERENCES: FAA-H-8083-3; FAA-S-8081-12.

Objective. To determine that the applicant:

1. Exhibits instructional knowledge of the elements of steep spirals by describing--

   a. selection of entry altitude.
   b. entry airspeed and power setting.
   c. selection of a proper ground reference point.
   d. division of attention and planning.
   e. coordination of flight and power controls.
   f. maintenance of constant radius around selected point.
   g. maintenance of constant airspeed throughout maneuver.

2. Exhibits instructional knowledge of common errors related to steep spiral by describing--

   a. improper pitch, bank, and power coordination during entry or completion.
   b. uncoordinated use of flight controls.
   c. improper planning and lack of maintenance of constant airspeed and radius.
   d. failure to stay orientated to the number of turns and the rollout heading.

3. Demonstrates and simultaneously explains a steep spiral from an instructional standpoint.
4. Analyzes and corrects simulated common errors related to steep spirals.

C. TASK: CHANDELLES

REFERENCES: FAA-H-8083-3; FAA-S-8081-12.

Objective. To determine that the applicant:

1. Exhibits instructional knowledge of the elements of chandelles by describing--

   a. selection of entry altitude.
   b. entry airspeed and power setting.
   c. division of attention and planning.
   d. coordination of flight controls.
   e. pitch and bank attitudes at various points during the maneuver.
   f. proper correction for torque effect in right and left turns.
   g. achievement of maximum performance.
   h. completion procedure.

2. Exhibits instructional knowledge of common errors related to chandelles by describing--

   a. improper pitch, bank, and power coordination during entry or completion.
   b. uncoordinated use of flight controls.
   c. improper planning and timing of pitch and bank attitude changes.
   d. factors related to failure in achieving maximum performance.
   e. a stall during the maneuver.

3. Demonstrates and simultaneously explains chandelles from an instructional standpoint.
4. Analyzes and corrects simulated common errors related to chandelles.

D. TASK: LAZY EIGHTS

REFERENCES: FAA-H-8083-3; FAA-S-8081-12.

Objective. To determine that the applicant:

1. Exhibits instructional knowledge of the elements of lazy eights by describing--

   a. selection of entry altitude.
   b. selection of suitable entry points.
   c. entry airspeed and power setting.
   d. entry procedure.
   e. orientation, division of attention, and planning.
   f. coordination of flight controls.
   g. pitch and bank attitudes at key points during the maneuver.
   h. importance of consistent airspeed and altitude control at key points during the maneuver.
   i. proper correction for torque effect in right and left turns.
   j. loop symmetry.

2. Exhibits instructional knowledge of common errors related to lazy eights by describing--

   a. poor selection of reference points.
   b. uncoordinated use of flight controls.
   c. unsymmetrical loops resulting from poorly planned pitch and bank attitude changes.
   d. inconsistent airspeed and altitude at key points.
   e. loss of orientation.
   f. excessive deviation from reference points.

3. Demonstrates and simultaneously explains lazy eights from an instructional standpoint.
4. Analyzes and corrects simulated common errors related to lazy eights.

X.   AREA OF OPERATION: GROUND REFERENCE MANEUVERS

NOTE: The examiner shall select TASK D and one other TASK.

A.   TASK: RECTANGULAR COURSE

REFERENCES: FAA-H-8083-3; FAA-S-8081-14.

Objective. To determine that the applicant:

1. Exhibits instructional knowledge of the elements of a rectangular course by describing--

    a. how to select a suitable altitude.
    b. how to select a suitable ground reference with consideration given to emergency landing areas.
    c. orientation, division of attention, and planning.
    d. configuration and airspeed prior to entry.
    e. relationship of a rectangular course to an airport traffic pattern.
    f. wind drift correction.
    g. how to maintain desired altitude, airspeed, and distance from ground reference boundaries.
    h. timing of turn entries and rollouts.
    i. coordination of flight controls.

2. Exhibits instructional knowledge of common errors related to a rectangular course by describing--

    a. poor planning, orientation, or division of attention.
    b. uncoordinated flight control application.
    c. improper correction for wind drift.
    d. failure to maintain selected altitude or airspeed.
    e. selection of a ground reference where there is no suitable emergency landing area within gliding distance.

3. Demonstrates and simultaneously explains a rectangular course from an instructional standpoint.
4. Analyzes and corrects simulated common errors related to a rectangular course.

B.   TASK: S-TURNS ACROSS A ROAD

REFERENCES: FAA-H-8083-3; FAA-S-8081-14.

Objective. To determine that the applicant:

1. Exhibits instructional knowledge of the elements of S-turns across a road by describing--

    a. how to select a suitable altitude.
    b. how to select a suitable ground reference line with consideration given to emergency landing areas.
    c. orientation, division of attention, and planning.
    d. configuration and airspeed prior to entry.
    e. entry procedure.
    f. wind drift correction.
    g. tracking of semicircles of equal radii on either side of the selected ground reference line.
    h. how to maintain desired altitude and airspeed.
    i. turn reversal over the ground reference line.
    j. coordination of flight controls.

2. Exhibits instructional knowledge of common errors related to S-turns across a road by describing--

    a. poor entry procedure.
    b. poor planning, orientation, or division of attention.
    c. uncoordinated flight control application.
    d. improper correction for wind drift.
    e. an unsymmetrical ground track.
    f. failure to maintain selected altitude or airspeed.
    g. selection of a ground reference where there is no suitable emergency landing area within gliding distance.

3. Demonstrates and simultaneously explains S-turns across a road from an instructional standpoint.
4. Analyzes and corrects simulated common errors related to S-turns across a road.

C.   TASK: TURNS AROUND A POINT

REFERENCES: FAA-H-8083-3; FAA-S-8081-14.

Objective. To determine that the applicant:

1. Exhibits instructional knowledge of the elements of turns around a point by describing--

    a. how to select a suitable entry altitude.
    b. how to select a suitable ground reference point with consideration given to emergency landing areas.
    c. orientation, division of attention, and planning.
    d. configuration and airspeed prior to entry.
    e. entry procedure.
    f. wind drift correction.
    g. how to maintain desired altitude, airspeed, and distance from reference point.
    h. coordination of flight controls.

2. Exhibits instructional knowledge of common errors related to turns around a point by describing--

    a. faulty entry procedure.
    b. poor planning, orientation, or division of attention.
    c. uncoordinated flight control application.
    d. improper correction for wind drift.
    e. failure to maintain selected altitude or airspeed.
    f. selection of a ground reference point where there is no suitable emergency landing area within gliding distance.

3. Demonstrates and simultaneously explains turns around a point from an instructional standpoint.
4. Analyzes and corrects simulated common errors related to turns around a point.

D.   TASK: EIGHTS ON PYLONS

REFERENCES: FAA-H-8083-3; FAA-S-8081-12.

Objective. To determine that the applicant:

1. Exhibits instructional knowledge of the elements of eights on pylons by describing--

    a. how to determine the approximate pivotal altitude.
    b. how to select suitable pylons with consideration given to emergency landing areas.
    c. orientation, division of attention, and planning.
    d. configuration and airspeed prior to entry.
    e. relationship of groundspeed change to the performance of the maneuver.
    f. pilot's "line-of-sight" reference to the pylon.
    g. entry procedure.
    h. procedure for maintaining "line-of-sight" on the pylon.
    i. proper planning for turn entries and rollouts.
    j. how to correct for wind drift between pylons.
    k. coordination of flight controls.

2. Exhibits instructional knowledge of common errors related to eights on pylons by describing--

    a. faulty entry procedure.
    b. poor planning, orientation, and division of attention.
    c. uncoordinated flight control application.
    d. use of an improper "line-of-sight" reference.
    e. application of rudder alone to maintain "line-of-sight" on the pylon.
    f. improper planning for turn entries and rollouts.
    g. improper correction for wind drift between pylons.
    h. selection of pylons where there is no suitable emergency landing area within gliding distance.

3. Demonstrates and simultaneously explains eights on pylons from an instructional standpoint.
4. Analyzes and corrects simulated common errors related to eights on pylons.

XI. AREA OF OPERATION: SLOW FLIGHT, STALLS, AND SPINS

NOTE: The examiner must select at least one proficiency stall (TASK B or C). At least one demonstration stall (TASK D, E, F, or H) and TASK G.

A. TASK: MANEUVERING DURING SLOW FLIGHT

REFERENCES: FAA-H-8083-3; FAA-S-8081-12, FAA-S-8081-14; POH/AFM.

Objective. To determine that the applicant:

1. Exhibits instructional knowledge of the elements of maneuvering during slow flight by describing--

   a. relationship of configuration, weight, center of gravity, maneuvering loads, angle of bank, and power to flight characteristics and controllability.
   b. relationship of the maneuver to critical flight situations, such as go-around.
   c. performance of the maneuver with selected landing gear and flap configurations in straight-and-level flight and level turns.
   d. specified airspeed for the maneuver.
   e. coordination of flight controls.
   f. trim technique.
   g. re-establishment of cruise flight.

2. Exhibits instructional knowledge of common errors related to maneuvering during slow flight by describing--

   a. failure to establish specified gear and flap configuration.
   b. improper entry technique.
   c. failure to establish and maintain the specified airspeed.
   d. excessive variations of altitude and heading when a constant altitude and heading are specified.
   e. rough or uncoordinated control technique.
   f. improper correction for torque effect.
   g. improper trim technique.
   h. unintentional stalls.
   i. inappropriate removal of hand from throttles.

3. Demonstrates and simultaneously explains maneuvering during slow flight from an instructional standpoint.
4. Analyzes and corrects simulated common errors related to maneuvering during slow flight.

B. TASK: POWER-ON STALLS (PROFICIENCY)

REFERENCES: AC 61-67; FAA-H-8083-3; FAA-S-8081-12, FAA-S-8081-14; POH/AFM.

Objective. To determine that the applicant:

1. Exhibits instructional knowledge of the elements of power-on stalls, in climbing flight (straight or turning), with selected landing gear and flap configurations by describing--

   a. aerodynamics of power-on stalls.
   b. relationship of various factors such as landing gear and flap configuration, weight, center of gravity, load factor, and bank angle to stall speed.
   c. flight situations where unintentional power-on stalls may occur.
   d. entry technique and minimum entry altitude.
   e. performance of power-on stalls in climbing flight (straight or turning).
   f. coordination of flight controls.
   g. recognition of the first indications of power-on stalls.
   h. recovery technique and minimum recovery altitude.

2. Exhibits instructional knowledge of common errors related to power-on stalls, in climbing flight (straight or turning), with selected landing gear and flap configurations by describing--

   a. failure to establish the specified landing gear and flap configuration prior to entry.
   b. improper pitch, heading, and bank control during straight ahead and turning stalls.
   c. improper pitch and bank control during turning stalls.
   d. rough or uncoordinated control procedure.
   e. failure to recognize the first indications of a stall.
   f. failure to achieve a stall.
   g. improper torque correction.
   h. poor stall recognition and delayed recovery.
   i. excessive altitude loss or excessive airspeed during recovery.
   j. secondary stall during recovery.

3. Demonstrates and simultaneously explains power-on stalls, in climbing flight (straight or turning), with selected landing gear and flap configurations, from an instructional standpoint.
4. Analyzes and corrects simulated common errors related to power-on stalls, in climbing flight (straight or turning), with selected landing gear and flap configurations.

C. TASK: POWER-OFF STALLS (PROFICIENCY)

REFERENCES: FAA-H-8083-3; FAA-S-8081-12, FAA-S-8081-14; POH/AFM.

Objective. To determine that the applicant:

1. Exhibits instructional knowledge of the elements of power-off stalls, in descending flight (straight or turning), with selected landing gear and flap configurations by describing--

   a. aerodynamics of power-off stalls.
   b. relationship of various factors, such as landing gear and flap configuration, weight, center of gravity, load factor, and bank angle to stall speed.
   c. flight situations where unintentional power-off stalls may occur.
   d. entry technique and minimum entry altitude.
   e. performance of power-off stalls in descending flight (straight or turning).
   f. coordination of flight controls.
   g. recognition of the first indications of power-off stalls.
   h. recovery technique and minimum recovery altitude.

2. Exhibits instructional knowledge of common errors related to power-off stalls, in descending flight (straight or turning), with selected landing gear and flap configurations by describing--

   a. failure to establish the specified landing gear and flap configuration prior to entry.
   b. improper pitch, heading, and bank control during straight-ahead stalls.
   c. improper pitch and bank control during turning stalls.
   d. rough or uncoordinated control technique.
   e. failure to recognize the first indications of a stall.
   f. failure to achieve a stall.
   g. improper torque correction.
   h. poor stall recognition and delayed recovery.
   i. excessive altitude loss or excessive airspeed during recovery.
   j. secondary stall during recovery.

3. Demonstrates and simultaneously explains power-off stalls, in descending flight (straight or turning), with selected landing gear and flap configurations, from an instructional standpoint.

4. Analyzes and corrects simulated common errors related to power-off stalls, in descending flight (straight or turning), with selected landing gear and flap configurations.

D. TASK: CROSSED-CONTROL STALLS (DEMONSTRATION)

REFERENCES: FAA-H-8083-3; FAA-S-8081-12, FAA-S-8081-14; POH/AFM.

Objective. To determine that the applicant:

1. Exhibits instructional knowledge of the elements of crossed-control stalls, with the landing gear extended by describing--

    a. aerodynamics of crossed-control stalls.
    b. effects of crossed controls in gliding or reduced airspeed descending turns.
    c. flight situations where unintentional crossed-control stalls may occur.
    d. entry procedure and minimum entry altitude.
    e. recognition of crossed-control stalls.
    f. recovery procedure and minimum recovery altitude.

2. Exhibits instructional knowledge of common errors related to crossed-control stalls, with the landing gear extended by describing--

    a. failure to establish selected configuration prior to entry.
    b. failure to establish a crossed-control turn and stall condition that will adequately demonstrate the hazards of a crossed-control stall.
    c. improper or inadequate demonstration of the recognition and recovery from a cross-control stall.
    d. failure to present simulated student instruction that emphasizes the hazards of a cross-control condition in a gliding or reduced airspeed condition.

3. Demonstrates and simultaneously explains a crossed-control stall, with the landing gear extended, from an instructional standpoint.

4. Analyzes and corrects simulated common errors related to a crossed-control stall with the landing gear extended.

E. TASK: ELEVATOR TRIM STALLS (DEMONSTRATION)

REFERENCES: FAA-H-8083-3; FAA-S-8081-12, FAA-S-8081-14; POH/AFM.

Objective. To determine that the applicant:

1. Exhibits instructional knowledge of the elements of elevator trim stalls, in selected landing gear and flap configurations by describing--

    a. aerodynamics of elevator trim stalls.
    b. hazards of inadequate control pressures to compensate for thrust, torque, and up-elevator trim during go-around and other related maneuvers.
    c. entry procedure and minimum entry altitude.
    d. recognition of elevator trim stalls.
    e. importance of recovering from an elevator trim stall immediately upon recognition.

2. Exhibits instructional knowledge of common errors related to elevator trim stalls, in selected landing gear and flap configurations by describing--

    a. failure to present simulated student instruction that adequately emphasizes the hazards of poor correction for torque and up-elevator trim during go-around and other maneuvers.
    b. failure to establish selected configuration prior to entry.
    c. improper or inadequate demonstration of the recognition and of the recovery from an elevator trim stall.

3. Demonstrates and simultaneously explains elevator trim stalls, in selected landing gear and flap configurations, from an instructional standpoint.

4. Analyzes and corrects simulated common errors related to elevator trim stalls in selected landing gear and flap configurations.

F. TASK: SECONDARY STALLS (DEMONSTRATION)

REFERENCES: FAA-H-8083-3; FAA-S-8081-12, FAA-S-8081-14; POH/AFM.

Objective. To determine that the applicant:

1. Exhibits instructional knowledge of the elements of secondary stalls, in selected landing gear and flap configurations by describing--

    a. aerodynamics of secondary stalls.
    b. flight situations where secondary stalls may occur.
    c. hazards of secondary stalls during normal stall or spin recovery.
    d. entry procedure and minimum entry altitude.
    e. recognition of a secondary stall.
    f. recovery procedure and minimum recovery altitude.

2. Exhibits instructional knowledge of common errors related to secondary stalls, in selected landing gear and flap configurations by describing--

    a. failure to establish selected configuration prior to entry.
    b. improper or inadequate demonstration of the recognition of and recovery from a secondary stall.
    c. failure to present simulated student instruction that adequately emphasizes the hazards of poor procedure in recovering from a primary stall.

3. Demonstrates and simultaneously explains secondary stalls, in selected landing gear and flap configurations, from an instructional standpoint.

4. Analyzes and corrects simulated common errors related to secondary stalls in selected landing gear and flap configurations.

G. TASK: SPINS

NOTE: At the discretion of the examiner, a logbook record attesting applicant instructional competency in spin entries, spins, and spin recoveries may be accepted in lieu of this TASK. The flight instructor who conducted the spin instruction shall certify the logbook record.

REFERENCES: 14 CFR part 23; Type Certificate Data Sheet; AC 61-67, FAA-H-8083-3; POH/AFM.

Objective. To determine that the applicant:

1. Exhibits instructional knowledge of the elements of spins by describing--

    a. anxiety factors associated with spin instruction.
    b. aerodynamics of spins.
    c. airplanes approved for the spin maneuver based on airworthiness category and type certificate.
    d. relationship of various factors such as configuration, weight, center of gravity, and control coordination to spins.
    e. flight situations where unintentional spins may occur.
    f. how to recognize and recover from imminent, unintentional spins.
    g. entry procedure and minimum entry altitude for intentional spins.
    h. control procedure to maintain a stabilized spin.
    i. orientation during a spin.
    j. recovery procedure and minimum recovery altitude for intentional spins.

2. Exhibits instructional knowledge of common errors related to spins by describing--

    a. failure to establish proper configuration prior to entry.
    b. failure to achieve and maintain a full stall during spin entry.
    c. failure to close throttle when a spin entry is achieved.
    d. failure to recognize the indications of an imminent, unintentional spin.
    e. improper use of flight controls during spin entry, rotation, or recovery.
    f. disorientation during a spin.
    g. failure to distinguish between a high-speed spiral and a spin.
    h. excessive speed or accelerated stall during recovery.
    i. failure to recover with minimum loss of altitude.
    j. hazards of attempting to spin an airplane not approved for spins.

3. Demonstrates and simultaneously explains a spin (one turn) from an instructional standpoint.
4. Analyzes and corrects simulated common errors related to spins.

H. TASK: ACCELERATED MANEUVER STALLS (DEMONSTRATION)

NOTE: This TASK shall be completed by oral examination or demonstration at discretion of examiner.

REFERENCES: FAA-H-8083-3; POH/AFM.

Objective. To determine that the applicant:

1. Exhibits instructional knowledge of the elements of accelerated maneuver stalls by describing--

    a. aerodynamics of accelerated maneuver stalls.
    b. flight situations where accelerated maneuver stalls may occur.
    c. hazards of accelerated stalls during stall or spin recovery.
    d. entry procedure and minimum entry altitude.
    e. recognition of the accelerated stall.
    f. recovery procedure and minimum recovery altitude.

2. Demonstrates and simultaneously explains accelerated maneuver stall, from an instructional standpoint--
3. Exhibits instructional knowledge of common errors related to accelerated maneuver stalls by describing--

    a. failure to establish proper configuration prior to entry.
    b. improper or inadequate demonstration of the recognition of and recovery from an accelerated maneuver stall.
    c. Failure to present simulated student instruction that adequately emphasizes the hazards of poor procedures in recovering from an accelerated stall.

4. Analyzes and corrects simulated common errors related to accelerated stalls.

XII. AREA OF OPERATION: BASIC INSTRUMENT MANEUVERS

NOTE: The examiner shall select at least one TASK.

A. TASK: STRAIGHT-AND-LEVEL FLIGHT

REFERENCES: FAA-H-8083-3, FAA-H-8083-15; FAA-S-8081-14.

Objective. To determine that the applicant:

1. Exhibits instructional knowledge of the elements of straight-and-level flight solely by reference to instruments by describing--

    a. instrument crosscheck, instrument interpretation, and aircraft control.
    b. instruments used for pitch, bank, and power control, and how those instruments are used to maintain altitude, heading, and airspeed.
    c. trim procedure.

2. Exhibits instructional knowledge of common errors related to straight-and-level flight solely by reference to instruments by describing--

    a. "fixation," "omission," and "emphasis" errors during instrument crosscheck.
    b. improper instrument interpretation.
    c. improper control applications.
    d. failure to establish proper pitch, bank, or power adjustments during altitude, heading, or airspeed corrections.
    e. faulty trim procedure.

3. Demonstrates and simultaneously explains straight-and-level flight, solely by reference to instruments, from an instructional standpoint.
4. Analyzes and corrects simulated common errors related to straight-and-level flight, solely by reference to instruments.

B. TASK: CONSTANT AIRSPEED CLIMBS

REFERENCES: FAA-H-8083-3, FAA-H-8083-15; FAA-S-8081-14.

Objective. To determine that the applicant:

1. Exhibits instructional knowledge of the elements of straight and turning, constant airspeed climbs, solely by reference to instruments by describing--

    a. instrument crosscheck, instrument interpretation, and aircraft control.
    b. instruments used for pitch, bank, and power control during entry, during the climb, and during level off, and how those instruments are used to maintain climb heading and airspeed.
    c. trim procedure.

2. Exhibits instructional knowledge of common errors related to straight and turning, constant airspeed climbs, solely by reference to instruments by describing--

    a. "fixation," "omission," and "emphasis" errors during instrument crosscheck.
    b. improper instrument interpretation.
    c. improper control applications.
    d. failure to establish proper pitch, bank, or power adjustments during heading and airspeed corrections.
    e. improper entry or level-off procedure.
    f. faulty trim procedure.

3. Demonstrates and simultaneously explains straight and turning, constant airspeed climb, solely by reference to instruments, from an instructional standpoint.
4. Analyzes and corrects simulated common errors related to straight and turning, constant airspeed climbs, solely by reference to instruments.

C. TASK: CONSTANT AIRSPEED DESCENTS

REFERENCES: FAA-H-8083-3, FAA-H-8083-15; FAA-S-8081-14.

Objective. To determine that the applicant:

1. Exhibits instructional knowledge of the elements of straight and turning, constant airspeed descents, solely by reference to instruments by describing--

    a. instrument crosscheck, instrument interpretation, and aircraft control.

    b. instruments used for pitch, bank, and power control during entry, during the descent, and during level off, and how those instruments are used to maintain descent heading and airspeed.

    c. trim procedure.

2. Exhibits instructional knowledge of common errors related to straight and turning, constant airspeed descents, solely by reference to instruments by describing--

    a. "fixation," "omission," and "emphasis" errors during instrument crosscheck.

    b. improper instrument interpretation.

    c. improper control applications.

    d. failure to establish proper pitch, bank, or power adjustments during heading and airspeed corrections.

    e. improper entry or level-off procedure.

    f. faulty trim procedure.

3. Demonstrates and simultaneously explains a straight and turning, constant airspeed descent, solely by reference to instruments, from an instructional standpoint.

4. Analyzes and corrects simulated common errors related to straight and turning, constant airspeed descents, solely by reference to instruments.

D. TASK: TURNS TO HEADINGS

REFERENCES: FAA-H-8083-3, FAA-H-8083-15; FAA-S-8081-14.

Objective. To determine that the applicant:

1. Exhibits instructional knowledge of the elements of turns to headings, solely by reference to instruments by describing--

    a. instrument crosscheck, instrument interpretation, and aircraft control.

    b. instruments used for pitch, bank, and power control during turn entry, during the turn, and during the turn rollout, and how those instruments are used.

    c. trim procedure.

2. Exhibits instructional knowledge of common errors related to turns to headings, solely by reference to instruments by describing--

    a. "fixation," "omission," and "emphasis" errors during instrument crosscheck.

    b. improper instrument interpretation.

    c. improper control applications.

    d. failure to establish proper pitch, bank, and power adjustments during altitude, bank, and airspeed corrections.

    e. improper entry or rollout procedure.

    f. faulty trim procedure.

3. Demonstrates and simultaneously explains a turn to a heading, solely by reference to instruments, from an instructional standpoint.

4. Analyzes and corrects simulated common errors related to turns to headings, solely by reference to instruments.

E. TASK: RECOVERY FROM UNUSUAL FLIGHT ATTITUDES

REFERENCES: FAA-H-8083-3, FAA-H-8083-15; FAA-S-8081-14.

Objective. To determine that the applicant:

1. Exhibits instructional knowledge of the elements of recovery from unusual flight attitudes by describing--

    a. conditions and situations that may result in unusual flight attitudes.

    b. the two basic unusual flight attitudes—nose-high (climbing turn) and nose-low (diving spiral).

    c. how unusual flight attitudes are recognized.

    d. control sequence for recovery from a nose-high attitude and the reasons for that sequence.

    e. control sequence for recovery from a nose-low attitude and the reasons for that sequence.

    f. reasons why the controls should be coordinated during unusual flight attitude recoveries.

2. Exhibits instructional knowledge of common errors related to recovery from unusual flight attitudes by describing--

    a. failure to recognize an unusual flight attitude.

    b. consequences of attempting to recover from an unusual flight attitude by "feel" rather than by instrument indications.

    c. inappropriate control applications during recovery.

    d. failure to recognize from instrument indications when the airplane is passing through a level flight attitude.

3. Demonstrates and simultaneously explains a recovery from nose-high and a nose-low flight attitude from an instructional standpoint.

4. Analyzes and corrects simulated common errors related to recovery from unusual flight attitudes.

XIII. AREA OF OPERATION: EMERGENCY OPERATIONS

NOTE: The examiner shall select at least TASKs A and B.

A. TASK: EMERGENCY APPROACH AND LANDING (SIMULATED)

NOTE: The examiner shall NOT simulate a power failure by placing the fuel selector to the "off" position or by placing the mixture control in the "idle-cutoff" position. No simulated emergency approach shall be continued below 500 feet AGL, unless over an area where a safe landing can be accomplished in compliance with 14 CFR section 91.119.

REFERENCES: FAA-H-8083-3; FAA-S-8081-12, FAA-S-8081-14; POH/AFM.

Objective. To determine that the applicant:

1. Exhibits instructional knowledge of the elements related to an emergency approach and landing by describing--

    a. prompt establishment of the best glide airspeed and the recommended configuration.

    b. how to select a suitable emergency landing area.

    c. planning and execution of approach to the selected landing area.

    d. use of emergency checklist.

    e. importance of attempting to determine reason for the malfunction.

    f. importance of dividing attention between flying the approach and accomplishing emergency checklist.

    g. procedures that can be used to compensate for undershooting or overshooting selected emergency landing area.

2. Exhibits instructional knowledge of common errors related to an emergency approach and landing by describing--

   a. improper airspeed control.
   b. poor judgment in the selection of an emergency landing area.
   c. failure to estimate the approximate wind speed and direction.
   d. failure to fly the most suitable pattern for existing situation.
   e. failure to accomplish the emergency checklist.
   f. undershooting or overshooting selected emergency landing area.

3. Demonstrates and simultaneously explains an emergency approach with a simulated engine failure from an instructional standpoint.

4. Analyzes and corrects simulated common errors related to an emergency approach with a simulated engine failure.

B. TASK: SYSTEMS AND EQUIPMENT MALFUNCTIONS

REFERENCES: FAA-H-8083-3; FAA-S-8081-12, FAA-S-8081-14; POH/AFM.

NOTE: The examiner shall not simulate a system or equipment malfunction in a manner that may jeopardize safe flight or result in possible damage to the airplane.

Objective. Exhibits instructional knowledge of at least five (5) of the systems and equipment malfunctions, appropriate to the airplane used for the practical test by describing recommended pilot action for:

1. Smoke, fire, or both, during ground or flight operations.
2. Rough running engine or partial power loss.
3. Loss of engine oil pressure.
4. Fuel starvation.
5. Engine overheat.
6. Hydraulic malfunction.
7. Electrical malfunction.
8. Carburetor or induction icing.
9. Door or window opening in flight.
10. Inoperative or "runaway" trim.
11. Landing gear or flap malfunction.
12. Pressurization malfunction.

C. TASK: EMERGENCY EQUIPMENT AND SURVIVAL GEAR

REFERENCES: FAA-H-8083-3, FAA-H-8083-23; FAA-S-8081-12, FAA-S-8081-14; POH/AFM.

Objective. To determine that the applicant exhibits instructional knowledge of the elements related to emergency equipment and survival gear appropriate to the airplane used for the practical test by describing:

1. Equipment and gear appropriate for operation in various climates, over various types of terrain, and over water.
2. Purpose, method of operation or use, servicing and storage of appropriate equipment.

D. TASK: EMERGENCY DESCENT

REFERENCES: FAA-H-8083-3; FAA-S-8081-12, FAA-S-8081-14; POH/AFM.

Objective. To determine that the applicant exhibits instructional knowledge of the elements related to emergency descents appropriate to the airplane flown by describing:

1. Exhibits instructional knowledge of the elements related to an emergency descent by describing--

   a. situations that require an emergency descent.
   b. proper use of the prescribed emergency checklist to verify accomplishment of procedures for initiating the emergency descent.
   c. proper use of clearing procedures before initiating and during the emergency descent.
   d. procedures for recovering from an emergency descent.
   e. manufacturer's procedures.

2. Exhibits instructional knowledge of common errors related to an emergency descent by describing--

   a. the consequences of failing to identify reason for executing an emergency descent.
   b. improper use of the prescribed emergency checklist to verify accomplishment of procedures for initiating the emergency descent.
   c. improper use of clearing procedures before initiating and during the emergency descent.
   d. improper procedures for recovering from an emergency descent.

3. Demonstrates and simultaneously explains an approach and landing with a simulated inoperative engine from an instructional standpoint.

4. Analyzes and corrects simulated common errors related to an approach and landing with an inoperative engine.

XIV. AREA OF OPERATION: POSTFLIGHT PROCEDURES

A. TASK: POSTFLIGHT PROCEDURES

REFERENCES: FAA-H-8083-3, FAA-H-8083-23; FAA-S-8081-12, FAA-S-8081-14; POH/AFM.

Objective. To determine that the applicant:

1. Exhibits instructional knowledge of the elements of postflight procedures by describing--

   a. parking procedure.
   b. engine shutdown and securing cockpit.
   c. deplaning passengers.
   d. securing airplane.
   e. postflight inspection.
   f. refueling.

2. Exhibits instructional knowledge of common errors related to postflight procedures by describing--

   a. hazards resulting from failure to follow recommended procedures.
   b. poor planning, improper procedure, or faulty judgment in performance of postflight procedures.

APPENDIX B
ORAL EXAM GUIDE

Most flight schools and many CFIs recommend that pilots preparing for their practical test study an "Oral Exam Guide." We agree: This book is both an "oral exam guide" and a "flight test guide." With the Gleim system, you are well prepared. This book has **everything** you need to know to pass your FAA practical test with confidence. Three other Gleim books that may be applicable to the flight instructor oral exam and the entire FAA practical test are

Aviation Weather and Weather Services (AWWS),
Flight/Ground Instructor FAA Knowledge Test (FIGI), and
Pilot Handbook (PH)

These books contain all the information you need to do well on your flight instructor practical test.

Consider this appendix your **ORAL EXAM GUIDE**.

1. Review the requirements to obtain a flight instructor certificate on page 2 of FIFM (this book).

2. Essential reading: "Oral Portion of the Practical Test" on page 49 (FIFM).

3. Relatedly, read the following on pages 42 through 49 (FIFM):

    "Airplane and Equipment Requirements"
    "What to Take to Your Practical Test"
    "Practical Test Application Form"
    "Authorization to Take the Practical Test"

4. FARs: In *Flight/Ground Instructor FAA Knowledge Test* (FIGI), read the 14-page outline in Study Unit 7 for a condensed review of the most important FARs related to flight instructing.

FAR PART 61
61.3   Requirements for Certificates, Ratings, and Authorizations
61.57   Recent Flight Experience: Pilot in Command
61.133   Commercial Pilot Privileges and Limitations

FAR PART 91
91.3   Responsibility and Authority of the Pilot in Command
91.103   Preflight Action
91.123   Compliance with ATC Clearances and Instructions
91.129   Operations in Class D Airspace
91.131   Operations in Class B Airspace
91.135   Operations in Class A Airspace
91.155   Basic VFR Weather Minimums
91.157   Special VFR Weather Minimums
91.211   Supplemental Oxygen
91.215   ATC Transponder and Altitude Reporting Equipment and Use

NTSB PART 830, Notification and Reporting of Aircraft Accidents or Incidents and Overdue Aircraft, and Preservation of Aircraft Wreckage, Mail, Cargo, and Records

In *Pilot Handbook* (PH), Chapter 4, read the following FARs:

91.7	Civil Aircraft Airworthiness
91.9	Civil Aircraft Flight Manual, Marking, and Placard Requirements
91.126-91.135	Airspace FARs
91.167-91.187	Private Pilot FARs
91.403-91.413	Maintenance and inspection FARs

NOTE: FARs ARE NOW REFERRED TO AS CFRs. The FAA has recently begun to abbreviate Federal Aviation Regulations as "14 CFR" rather than "FARs." CFR stands for Code of Federal Regulations, and the Federal Aviation Regulations are in Title 14. For example, FAR Part 1 and FAR 61.109 are now referred to as 14 CFR Part 1 and 14 CFR Sec. 61.109, respectively. CFIs and pilots continue to use the acronym FAR.

Examiner Questions:

You will be ready for your practical test. A summary of preparatory tasks is given on pages 39 through 41. The following pages contain questions previously asked by designated examiners and FAA inspectors on flight instructor practical tests. In addition to our brief answers, there are cross references to Gleim books if you wish to research the answers to these questions.

Part I Fundamentals of instructing

1. The learning process

1. Briefly define learning.

Learning can be defined as a change in behavior as a result of experience. (FIFM p. 65)

2. What are the characteristics of learning?

Learning is purposeful, comes through experience, is multifaceted, and is an active process. (FIFM p. 66)

3. What is the principle of readiness?

Individuals learn best when they are ready and eager to learn. Students with a strong purpose, a clear objective, and a well-fixed reason for learning tend to progress better than those who lack motivation. (FIFM p. 66)

4. What is the principle of exercise?

Things most often repeated are best remembered. This is the basis for practice and drill. (FIFM p. 67)

5. What is the principle of effect?

Learning is strengthened when accompanied by a pleasant or satisfying feeling. Whatever the learning situation, it should contain elements that affect students positively. (FIFM p. 67)

6. What is the principle of primacy?

Things learned first often create a strong, almost unshakable impression. Therefore, it is better to teach the right way, the first time. (FIFM p. 67)

7. What is the principle of intensity?

A student will learn more from the real thing than from a substitute. (FIFM p. 67)

8. What is the principle of recency?

Things most recently learned are best remembered. Instructors should repeat important matters at the end of a lesson to make sure the student remembers them. (FIFM p. 67)

9. What are the various levels of learning?

Rote, understanding, application, and correlation. (FIFM p. 68)

10. What are the three domains of learning?

Cognitive (knowledge), affective (attitudes, beliefs, and values), and psychomotor (physical skills). (FIFM p. 69)

11. What is the most effective way for an instructor to teach a new maneuver?

With a clear, step-by-step example. The instructor provides a demonstration, emphasizing the steps and techniques, to provide the student with a clear impression of what to do. (FIFM p. 69)

12. What does the learning curve demonstrate? How is it useful?

The learning curve provides the instructor with a pattern of the student's progression and informs the instructor when to expect a plateau, followed by further learning. (FIFM p. 70)

13. Why is the length of a lesson important?

A lesson that is too long may result in increased errors and decreased motivation. (FIFM p. 70)

14. What are the three processes of memory?

The sensory register, working or short-term memory, and long-term memory. (FIFM pp. 70-71)

15. What are the three theories of forgetting?

The theories of disuse, interference, and repression. (FIFM p. 71)

16. How can an instructor decrease the amount of knowledge forgotten?

An instructor can decrease the amount of knowledge forgotten by providing positive feedback and meaningful repetition, promoting favorable attitudes toward learning, emphasizing associations between bits of information or actions, and facilitating the use of all the student's senses in the learning process. (FIFM p. 71)

17. What is a positive transfer of learning? What is a negative transfer of learning?

Positive transfer occurs when the learning of one skill aids in the learning of another skill. Negative transfer occurs when a previously learned skill interferes with the learning of a new skill. (FIFM p. 72)

18. How can an instructor facilitate the transfer of learning?

An instructor can facilitate the transfer of learning by planning for learning transfer, making sure the student understands the material, maintaining high-order learning standards, and providing meaningful learning experiences. (FIFM p. 72)

2. The teaching process

19. How should a flight instructor prepare for a productive flight lesson?

An instructor should determine the material to be covered, the objectives of the lesson, and the goals to be attained. (FIFM p. 77)

20. What are performance-based objectives?

Performance based objectives are lesson plans that set measurable, reasonable standards that describe the desired performance of the student. (FIFM p. 77)

21. When should the lecture method of presentation be used? When should the demonstration method be used?

The lecture method should be used primarily to introduce students to new material. The demonstration method should be used extensively during flight training. (FIFM p. 78)

22. Why should an instructor always conduct reviews and evaluations after student demonstrations?

Reviews and evaluations make students aware of their progress and prepare them for future lessons. (FIFM p. 79)

23. Is it important for students to be aware of their progress? Why?

Yes. A lack of awareness may form a barrier between the student and instructor. (FIFM p. 79)

3. Teaching methods

24. How should an effective lesson be organized?

Every lesson should be organized into three phases: introduction, development, and conclusion. (FIFM p. 80)

25. What are the four types of lectures?

The four types of lectures are the illustrated talk, the briefing, the formal speech, and the teaching lecture. (FIFM p. 81)

26. Describe the cooperative learning method.

Students are organized into groups to maximize each other's learning. (FIFM p. 81)

27. What is the effect of the skillful use of questions in the guided discussion method?

Skillful use of questions brings about discussion and develops an understanding. (FIFM p. 82)

28. What questions should be avoided in the guided discussion?

Questions that only require a short, categorical answer, e.g., yes/no questions, should be avoided in a guided discussion. (FIFM p. 82)

29. How should instructors address their own mistakes made during demonstrations (i.e., the demonstration method)?

Instructors should point out their own mistakes and correct them as necessary, but should not downplay or ignore mistakes as this would create confusion. (FIFM p. 82)

30. What is the main advantage to the computer-based training method?

Students can work and progress at their own pace. (FIFM p. 83)

4. Evaluation

31. What is the purpose of a critique?

A critique is provided to a student as a means to better their performance. Critiques must be given for every lesson so that the student knows where they stand, and how they can improve upon their performance. (FIFM p. 84)

32. What are the characteristics of an effective oral question?

An effective oral question has only one correct answer; applies to the subject being taught; is brief, concise, and clear; centers on one idea; and presents a challenge. (FIFM p. 85)

33. What types of oral questions should be avoided?

Trick, irrelevant, or too general questions should be avoided. (FIFM p. 85)

34. How should the instructor respond to a question to which (s)he does not know the answer?

(S)he should admit to not knowing the answer and volunteer to help the student find the answer. (FIFM p. 86)

35. What are the attributes of a good test question?

A good test question is reliable, valid, usable, comprehensive, and discriminating. (FIFM pp. 86-87)

36. What is the instructor's responsibility regarding the PTS?

The instructor should teach to the Practical Test Standards so students will be prepared when they are tested to them. (FIFM p. 87)

5. Flight instructor characteristics and responsibilities

37. What are some qualities of a flight instructor who seeks to be a professional?

Sincerity, a good personal appearance, a professional image, the exercise of safe and good judgment, and the desire to better oneself through learning. (FIFM p. 90)

38. Why should a flight instructor not give a student a solo endorsement if the student is not ready to solo?

It is a breach of faith between the instructor and student, and more importantly, it is unsafe for everyone involved. (FIFM p. 89)

39. What is the flight instructor's responsibility regarding the flight review?

CFIs should not sign off anyone who has not satisfactorily completed the flight review. Instead, the training should only appear as dual in the logbook. (FIFM p. 89)

6. Human factors

40. How should the instructor control the behavior of his/her student?

In a way that encourages a goal and puts the instructor in a position of authority. (FIFM p. 73)

41. What is the first of the basic human needs?

The first basic human need is physical (water, shelter, rest, exercise). (FIFM p. 73)

42. What is the last of the basic human needs?

The last basic human need is self-fulfillment (realizing a potential). (FIFM p. 74)

43. In what order are human needs fulfilled?

Human needs are fulfilled in ascending order. A higher need cannot be fulfilled unless the lower needs already have been fulfilled. (FIFM pp. 73-74)

44. Why do humans use defense mechanisms?

Humans use defense mechanisms to soften feelings of failure, alleviate feelings of guilt, and protect feelings of worth. (FIFM p. 74)

45. What are some of the common defense mechanisms?

Some common defense mechanisms are rationalization, flight, aggression, resignation, compensation, and denial. (FIFM pp. 74-75)

46. What happens to effective pilot decision making when a defense mechanism is involved?

Effective pilot decision making is decreased because defense mechanisms create a distortion of reality. (FIFM p. 75)

47. Why should instructors approach students as individuals?

Instructors who limit their thinking to an entire group are generalizing and cannot give particular students the individual attention they need. (FIFM p. 75)

48. Why is it important for instructors to praise and give constructive criticism?

Praise and constructive criticism deflect study frustration and make students feel that they are accomplishing something. (FIFM p. 75)

7. Planned instructional activity

49. What is the first step in planning a lesson?

The first step is to determine the objectives and completion standards for the lesson. (FIFM p. 91)

50. What are the contents of a training syllabus?

A training syllabus is an abstract or digest with blocks of learning organized in an efficient order. A syllabus should be flexible to allow an instructor to change it to the most effective training possible. (FIFM p. 92)

51. What is the purpose of a lesson plan?

A lesson plan should provide the best instruction possible within the given period, and make sure the instructor keeps a constant check on his/her own activity. (FIFM p. 92)

52. What are the characteristics of a well-planned lesson?

A well-planned lesson has unity, good content, scope, practicality, instructional steps, and flexibility. (FIFM p. 93)

53. How should the lesson plan be used?

The lesson plan should be used as a guide, but not as a substitute to thinking. (FIFM p. 92)

54. Why should a new lesson be made for every student and every instructional period?

Not every student and training environment is the same. Failure to create a new lesson plan can result in an ineffective lesson. (FIFM p. 92)

55. What are the elements of a lesson plan?

The elements of a lesson plan are: objective, elements involved, schedule, equipment, instructor's actions, student's actions, and completion standards. (FIFM p. 93)

56. What is the purpose of the "elements involved" section of the lesson plan?

The "elements involved" section is a statement of the elements of knowledge and skill necessary for the fulfillment of the lesson objective. It can include new or previously learned material. (FIFM p. 93)

Part II Technical subject areas

1. Aeromedical factors

1. What is hypoxia? How can it be prevented?

Hypoxia is insufficient oxygen in the blood. It can be prevented by flying at a lower altitude or by using supplemental oxygen. (PH Ch. 6)

2. What is hyperventilation? How can it be treated?

Hyperventilation is insufficient carbon dioxide in the blood. It can be treated by taking slow, deep breaths or breathing into a bag. (PH Ch. 6)

3. How can stress affect your flying?

Stress slows your reactions and degrades decision making ability. (PH Ch. 6)

4. What is a good rule for flying if taking medication?

Do NOT fly if you are taking medication unless the medication is approved by the FAA. (PH Ch. 6)

5. What is carbon monoxide poisoning? What are some sources of it in airplanes?

Carbon monoxide poisoning occurs when carbon monoxide enters the blood, thereby causing hypoxia. The most common source of carbon monoxide in aircraft cockpits is exhaust fumes from a defective heater or other source. (PH Ch. 6)

6. How long should pilots and passengers wait to fly after SCUBA diving?

If a controlled ascent was required during the dive, wait 24 hours before flying. If a controlled ascent was not required, wait 12 hours before flying up to 8,000 ft. and 24 hours for any altitude above 8,000 ft. (PH Ch. 6)

7. What are a few illusions that lead to spatial disorientation?

Some illusions that lead to spatial disorientation are the leans, Coriolis illusion, graveyard spiral and spin, somatogravic illusion, inversion illusion, elevator illusion, false horizon, autokinesis, and size-distance illusion. (PH Ch. 6)

8. How can you recover from spatial disorientation?

The best way to recover from spatial disorientation is to focus on the flight instruments and rely on their indications. (PH Ch. 6)

9. What are some illusions that lead to landing error?

Illusions that can lead to landing error include runway width, runway terrain, featureless terrain, and atmospheric illusions. (PH Ch. 6)

2. Visual scanning and collision avoidance

10. What are a few environmental situations that will degrade your vision?

Some environmental situations that will degrade your vision include low levels of light in the cockpit, excessive light or glare, dirty windshields, smoke or haze, and carbon monoxide for night vision. (FIFM p. 102)

11. When do most mid-air accidents occur?

Most mid-air accidents occur during day VFR. (FIFM p. 104)

12. What is the see and avoid concept?

It is the pilot's responsibility to make and maintain visual contact with other aircraft and avoid them as necessary. (FIFM p. 104)

13. What is the proper way to scan for traffic during the day? What is the proper way to scan at night?

Scan in small (10°) increments during the day. Scan with peripheral vision at night. (FIFM pp. 104-105)

14. What responsibility does the instructor have regarding collision avoidance?

The instructor must make sure that each student understands the importance of proper clearing technique and procedures and that each student uses them regularly. (FIFM pp. 102-106)

3. Principles of flight

15. What are the four basic forces acting on an airplane in flight?

Lift, weight, thrust, and drag are the four forces acting on an airplane in flight. (PH Ch. 1)

16. What is static stability?

Static stability is the airplane's initial tendency to return to its original attitude after its equilibrium is disturbed. If an airplane flies through turbulence, and its immediate tendency is to return to its original attitude, it has positive static stability. (FIFM p. 108)

17. What is dynamic stability?

Dynamic stability is the airplane's tendency to enter increasing oscillations OR decreasing oscillations after its equilibrium is disrupted. For example, if an airplane flies through turbulence and the oscillations it experiences diverge from its original trimmed attitude, the airplane exhibits negative dynamic stability; or if the oscillations decrease in size, the airplane has positive dynamic stability. (FIFM p. 108)

18. What is the most desirable stability to have?

Positive static stability and positive dynamic stability are most desirable. (FIFM p. 108)

19. What are the four elements of torque?

The four elements are: torque from the engine and propeller, spiraling slipstream, gyroscopic action of the propeller, and asymmetrical loading of the propeller. (FIFM p. 108)

20. What is load factor?

Load factor is the ratio of the total load supported by the wings to the actual weight of the airplane. (FIFM p. 109)

21. What is the limit load factor?

The limit load factor is the highest load factor that can be expected in normal operations. It can be sustained without any structural damage. (FIFM p. 109)

22. How does lift develop?

Lift develops as a result of differential pressure between the upper and lower surfaces of the wing. The air that travels over the top of the wing must go farther and faster, and thus has a lower pressure than the air that travels under the bottom side of the wing. The pressure difference pulls the wing upward. (FIFM p. 110)

23. In what situation are wingtip vortices the strongest?

Wingtip vortices are the strongest when the airplane is heavy, clean, and slow. (FIFM p. 110)

24. What could happen if wake turbulence is encountered?

Encountering wake turbulence can result in induced rolling, structural failure, and a general inability to control the aircraft. (FIFM p. 110)

4. Elevators, ailerons, and rudder

25. How does the elevator initiate an increase in pitch?

Increased back pressure on the yoke raises the elevator. This increases the negative angle of attack and increases the down force of the tail, thus raising the pitch attitude. (FIFM p. 112)

26. What are antiservo tabs?

Antiservo tabs are found on the trailing edge of a stabilator. They move in the same direction as the stabilator and increase control pressure. (FIFM p. 112)

27. What causes an airplane to turn to the right?

The yoke is deflected to the right, causing the right aileron to move up and the left aileron to move down. The lowered aileron causes an increase in the angle of attack and therefore creates more lift, thus raising the left wing and transferring a portion of the vertical component of lift to the rightward horizontal component. (FIFM pp. 112-113)

28. How does the rudder cause an airplane to yaw?

The rudder pedals are deflected in the desired direction of yaw. This pushes the rudder into the airflow, causing a horizontal force to be exerted in the opposite direction. (FIFM p. 113)

5. Trim devices and wing flaps

29. In which direction do trim tabs move?

Trim tabs move in the opposite direction of the elevator. (FIFM p. 114)

30. What is the correct use of trim?

Trim should be used to relieve control pressure, not as the primary means to control pitch. (FIFM p. 114)

31. What is the purpose of wing flaps?

Wing flaps permit a slower landing speed, a steeper angle of descent without an increase in airspeed, and a shorter takeoff roll. (FIFM p. 115)

6. Airplane weight and balance

32. What are some of the few effects that being over gross weight can have on an airplane?

Being over gross weight is illegal and very dangerous. It can increase the takeoff and landing distance, decrease climb performance, cause structural damage, and accelerate metal fatigue. (PH Ch. 5)

33. Where are the measurements taken for the computation of the CG?

The CG is computed from the reference datum which is defined by the manufacturer. It is frequently the firewall or the leading edge of the wing. (PH Ch. 5)

34. What are the handling characteristics of an airplane with a forward CG?

An airplane with a forward CG is generally more stable than one with an aft CG. It will fly at a slower airspeed (more drag), and it will stall at a higher indicated stall speed. (PH Ch. 5)

35. What are the handling characteristics of an airplane with an aft CG?

An airplane with an aft CG is generally unstable. It will fly at a higher airspeed, but it may be impossible to recover from a stall or spin. (PH Ch. 5)

36. How is the CG calculated?

The center of gravity (CG) is calculated by dividing the airplane's total weight by the total moment. Total moment is the sum of the products of the weight of each item in the airplane, e.g., occupant, baggage, fuel, etc., and the arm. Arm is the distance of the item from the datum. (PH Ch. 5)

37. Why should a weight and balance be calculated with zero fuel under the given conditions before departing?

As fuel burns, the CG moves. Calculating a zero fuel CG will ensure the entire flight stays within CG envelope. (PH Ch. 5)

38. What is the difference between a moment and an arm?

The arm is the distance from the reference datum to the center of gravity of an item. The moment is any force that causes an object to rotate; it is the product of weight and distance (arm). (PH Ch. 5)

7. Navigation and flight planning

39. What are some considerations when planning a cross-country flight?

A pilot should consider fuel requirements, available alternate airfields, the terrain, and ensuring the flight occurs outside of restricted or prohibited airspace. (FIFM p. 123)

40. What is pilotage? What is dead reckoning?

Pilotage is the use of a sectional chart to fly from one landmark to another. Dead reckoning is the navigation of the aircraft using computations based on airspeed, heading, etc. (FIFM p. 123)

41. What is the procedure for diverting to an alternate airport while on a cross-country flight?

Confirm your present position, select the alternate airport and turn toward it, draw a new course line, then calculate actual heading, time, and fuel consumption to the alternate. (FIFM p. 124)

42. What are some precautions you can take to avoid getting lost on a cross-country?

Always know where you are, plan ahead to find the next checkpoint, take action if checkpoint is not seen or reached. (FIFM p. 124)

43. What is the procedure you should follow if you become lost while on a cross-country flight?

Select the best course of action, maintain original heading, and climb if necessary. Attempt to identify a known landmark, use all available radio navigation, or contact ATC. Finally, plan a precautionary landing if needed. (FIFM p. 124)

44. What is the reason for using a properly prepared flight log?

A properly prepared flight log assists in planning, as well as reduces workload and tracks one's progress while on the flight. (FIFM p. 124)

45. What resources should be used to make a good "go/no-go" decision?

All available weather information should be considered, especially FSS. (FIFM p. 125)

46. What is the purpose of filing a VFR flight plan?

Filing a VFR flight plan is a good safety precaution. (FIFM p. 125)

8. Night operations

47. How long does it take your eyes to fully adapt to darkness?

It takes your eyes 30 minutes to fully adapt to darkness. If you accidentally view a bright light, your night vision will be lost instantly. (FIFM p. 126)

48. What is the recommended altitude to begin using oxygen at night to reduce the risk of hypoxia?

It is recommended that supplemental oxygen be used above 5,000 ft. at night to avoid the effects of hypoxia and impaired vision. (PH Ch. 6)

49. What portion of the eye is used for night flight? Is it more or less sensitive than the portion used for day?

Rods are used for night vision. They are less sensitive than the cones (used for day/color vision), but there are many more of them. (FIFM p. 126)

50. What additional equipment is your airplane required to have for night flight?

In addition to the equipment required for day VFR flight, your airplane is also required to have approved position and anti-collision lights, an electrical source, a spare set of fuses, and one electric landing light if being operated on a "for hire" basis. (PH Ch. 4)

51. What makes the preflight at night different from the preflight during the day?

The night preflight is difficult to perform because you cannot see as well at night. In addition to the normal checks and inspections, your night preflight should include a check of all lights and lighting systems. (FIFM p. 128)

52. What additional equipment should you bring with you on a night flight?

You should bring two flashlights and spare batteries. (FIFM p. 127)

53. How much fuel reserve is required for night flight?

The FARs require a VFR fuel reserve of 45 minutes at normal cruise. More is better. (PH Ch. 4)

54. What color are runway lights?

Runway lights are white. (They turn to yellow on an instrument runway with 2,000 feet left). (PH Ch. 3)

55. What color are taxiway edge lights? What color are taxiway centerline lights?

Taxiway edge lights are blue, and taxiway centerline lights are green. (PH Ch. 3)

56. Why is an engine failure especially dangerous at night?

It is harder to see safe landing areas in darkness, and there is little or no outside reference to help maintain directional control of the aircraft. (FIFM p. 133)

9. High-altitude operations

57. Explain how cabin pressurization is accomplished in most light airplanes?

Pressurization in most light airplanes is accomplished when the turbocharger's compressor pumps air into the cabin. (FIFM p. 138)

58. What happens when an airplane reaches its maximum differential pressure?

If a pressurized airplane reaches its maximum differential cabin pressure, any increase in altitude will result in an increase in cabin altitude. (FIFM p. 138)

59. What is the purpose of the cabin differential pressure gauge?

The cabin differential pressure gauge indicates the difference between cabin pressure and ambient (atmospheric) pressure. (FIFM p. 139)

60. What is the purpose of the cabin altimeter?

The cabin altimeter provides a check on the performance of the pressurization of the system by indicating cabin altitude. (FIFM p. 139)

61. How can a sudden supply of pure oxygen affect hypoxia symptoms?

A sudden supply of pure oxygen can aggravate the symptoms of hypoxia. (FIFM p. 137)

62. What action should take place following any sort of decompression?

Oxygen masks should be donned, followed by a rapid (emergency) descent to avoid the onset of hypoxia. (FIFM p. 141)

63. What is the most dangerous and harmful type of decompression?

An explosive decompression is the most dangerous and harmful type. (FIFM p. 141)

64. Why should all passengers wear their safety belts at all times during a flight?

In the event of a rapid or explosive decompression, there is less likelihood that passengers will be thrown out of the airplane if they are wearing safety belts. (FIFM p. 141)

10. FARs

65. Is any action required with regard to change of address when a pilot moves residence?

Yes. A pilot must notify the FAA in writing within 30 days of a move. (FIGI p. 203)

66. What is the pilot in command's responsibility?

The pilot in command (PIC) has final authority of the aircraft, and is directly responsible for the safety of crew and passengers. (PH Ch. 4)

67. How long must you wait after consuming alcohol before acting as a crewmember of a civil airplane?

You must wait 8 hours after consuming alcohol before acting as a required crewmember on a civil airplane. (FIGI p. 206)

68. What is the maximum allowable blood alcohol level for acting as a crewmember?

You may not act as a required crewmember on a civil aircraft while having .04% by weight or more blood alcohol content. (FIGI p. 206)

69. When must every passenger be provided supplemental oxygen?

All passengers must be provided with supplemental oxygen when the cabin altitude is above 15,000 ft. MSL. (FIGI p. 210)

70. When is an accident report required to be submitted to the NTSB?

Upon serious injury to occupants or substantial damage to the aircraft. (FIGI p. 212)

71. Is damage to the landing gear or tires considered substantial damage?

Damage to the landing gear or tires is not considered substantial damage and no report to the NTSB is required. (FIGI p. 212)

72. What other types of situations require a report to the NTSB?

The NTSB requires reports for the following situations: inability of required crewmember to perform duties, inflight fire, flight control system malfunction, mid-air collision, or an overdue aircraft believed to be involved in an accident. (FIGI p. 212)

73. Is a written incident report always required to the NTSB?

A written report is only required when requested. (FIGI p. 212)

74. How is a serious injury defined by the NTSB?

A serious injury is defined by NTSB Part 830 as an injury requiring hospitalization for more than 48 hours, most bone fractures, muscle or nerve damage, internal organ damage, or second or third degree burns affecting more than 5% of the body. (FIGI p. 212)

75. As PIC, what is your responsibility to your passengers with regard to safety belts?

Each passenger must be briefed on safety belt operation and notified when safety belts must be worn. (FIGI p. 207)

76. What is a NASA ASRP report? When should they be filed?

It is a voluntary program designed to gather information about deficiencies in the aviation system. A NASA report form should be filed when a FAR is violated inadvertently without any criminal action. (PH Ch. 4)

77. What are the prerequisites for a practical test?

The prerequisites for a practical test are passing of the appropriate knowledge test (if required), an instructor endorsement, and a third class medical certificate. (FIGI p. 202)

78. What must the flight review consist of?

At least one hour of ground and one hour of flight instruction. (FIGI p. 203)

79. How much flight training can one give in a 24 hr. period?

Eight hours of flight training can be given in a 24 hour period. (FIGI p. 205)

80. How can a flight instructor certificate be renewed?

A flight instructor certificate can be renewed by taking a practical test, providing a record of instruction, or completing a flight instructor refresher course within 90 days of renewal. (FIGI p. 206)

11. Use of minimum equipment list

81. What is a minimum equipment list?

A minimum equipment list is a specific inoperative equipment document for a particular make and model aircraft. (FIFM p. 193)

82. Can you take off with inoperative equipment without an MEL?

Yes, with conditions. (FIFM p. 193)

83. What is an STC? When are they used?

An STC is a supplemental type certificate. An STC is required when a major change in type design is made to the airplane (not always required with a minimum equipment list). (14 CFR Part 21.113)

12. Publications

84. What information does an *Airport/Facilities Directory* provide?

Airport/Facilities Directories (A/FD) provide all the information needed for an airport or radio navigation aid (NAVAID). *A/FDs* also provide published NOTAMs and areas of parachute and aerobatic activity. (FIFM p. 144)

85. What information does the Airmen's Information Manual provide?

The AIM provides information regarding airport operations, navigation aids, airspace, flight operations, and ATC procedures. (FIFM p. 145)

86. What is an Advisory Circular (AC)?

ACs are used by the FAA as a means of issuing nonregulatory information to pilots, mechanics, and manufacturers. (FIFM p. 147)

87. What is a NOTAM?

A NOTAM is a notice to airmen. This is a system that disseminates time-critical aeronautical information. (FIFM p. 147)

88. What are the three types of NOTAMs?

The three types of NOTAMs are D or distant, L or local, and FDC or flight data center NOTAMs (AC). (FIFM pp. 147-148)

89. What is the PTS? What is your responsibility as an instructor with regard to the PTS?

PTS stands for Practical Test Standards. They are the FAA-issued standards of knowledge and skill that must be demonstrated before a flight certificate is issued. The flight instructor is responsible for training the student to the acceptable standards of each task within the appropriate PTS. (FIFM p. 150)

13. National airspace system

90. What are the minimum VFR cloud clearance and visibility requirements for Class E airspace below 10,000 ft. MSL? What are they above 10,000 ft. MSL?

The minimum VFR cloud clearance and visibility requirements in Class E airspace below 10,000 ft. MSL are 500 ft. below clouds, 1,000 ft. above clouds, 2,000 ft. horizontally from clouds, and 3 miles visibility. The minimum VFR cloud clearance and visibility requirements in Class E airspace above 10,000 ft. MSL are 1,000 ft. below clouds, 1,000 ft. above clouds, 1 SM horizontally from clouds, and 5 miles visibility. (PH Ch. 3)

91. What are the minimum VFR cloud clearance and visibility requirements for Class G airspace below 1,200 ft. AGL during the day? What are they at night?

The minimum VFR cloud clearance and visibility requirements in Class G airspace below 1,200 ft. AGL during the day are clear of clouds and 1 mile visibility. The minimum VFR cloud clearance and visibility requirements in Class G airspace below 1,200 ft. AGL at night are 500 ft. below clouds, 1,000 ft. above clouds, 2,000 ft. horizontally from clouds, and 3 miles visibility. (PH Ch. 3)

92. What are the minimum VFR cloud clearance and visibility requirements for Class G airspace above 1,200 ft. AGL but below 10,000 ft. MSL during the day? What are they at night?

The minimum VFR cloud clearance and visibility requirements in Class G airspace above 1,200 ft. AGL but below 10,000 ft. MSL during the day are 500 ft. below clouds, 1,000 ft. above clouds, 2,000 ft. horizontally from clouds, and 1 mile visibility. The minimum VFR cloud clearance and visibility requirements in Class G airspace above 1,200 ft. AGL but below 10,000 ft. MSL at night are 500 ft. below clouds, 1,000 ft. above clouds, 2,000 ft. horizontally from clouds, and 3 miles visibility. (PH Ch. 3)

93. What are the minimum VFR cloud clearance and visibility requirements for Class B airspace? What are they for Class C airspace? What are they for Class D airspace?

The minimum VFR cloud clearance and visibility requirements in Class B airspace are clear of clouds and 3 miles visibility. The minimum VFR cloud clearance and visibility requirements in Class C and Class D airspace are 500 ft. below clouds, 1,000 ft. above clouds, 2,000 ft. horizontally from clouds, and 3 miles visibility. (PH Ch. 3)

94. What is required for a student pilot to fly solo in Class B airspace?

To fly solo in Class B airspace, a student pilot must have received all of the standard pre-solo endorsements, as well as an endorsement stating that (s)he has received training for the specific Class B area in which the solo flight is to be conducted. (PH Ch. 3)

95. What class of airspace requires a clearance prior to entry? What classes of airspace require that 2-way radio communication be established prior to entry?

Class B airspace requires a clearance prior to entry, and Class C and Class D airspace require that 2-way radio communication be established prior to entry. (PH Ch. 3)

96. How do you determine when 2-way radio communication has been established?

Two-way radio communication has been established when ATC responds with your correct call sign. (PH Ch. 3)

97. When is a transponder with Mode C required for VFR flight?

A working transponder with Mode C is required any time you are above 10,000 ft. MSL, inside Class B or Class C airspace, and above Class B or Class C airspace up to 10,000 ft. MSL, or within 30 NM of a Class B primary airport. (PH Ch. 3)

98. What are the typical dimensions of Class D airspace?

Class D airspace typically extends upward from the surface to 2,500 ft. AGL and outward to a 5 SM radius from the primary airport. Airspace dimensions may vary according to local requirements, however. (PH Ch. 3)

99. What are the typical dimensions of Class C airspace?

Class C airspace is typically composed of two sections that are referred to as the surface area and the shelf area. The surface area typically extends upward from the surface to 4,000 ft. MSL and outward to a 5 NM radius upward from the primary airport. The shelf area typically extends upward from 1,200 ft. MSL to 4,000 ft. MSL and outward to a 10 NM radius from the primary airport. Airspace dimensions may vary according to local requirements, however. (PH Ch. 3)

100. What is a TRSA?

TRSA stands for Terminal Radar Service Area. TRSAs are established around Class D airports that have radar service capability, but do not meet all of the criteria to be designated as Class C airspace. Participation in TRSA service is voluntary (though it is recommended), but 2-way radio communication must still be established prior to entering Class D airspace. (PH Ch. 3)

101. What are the minimum cloud clearance and visibility requirements to obtain a special VFR clearance?

To obtain a special VFR clearance, you must be able to remain clear of clouds and have at least 1 mile visibility. (PH Ch. 3)

102. What is a Prohibited Area? What is a Restricted Area? What is a Military Operations Area? What is an Alert Area? What is a Warning Area?

Prohibited Areas are established for reasons of national security; flight is prohibited at all times within them. Restricted Areas are established to contain unusual, often invisible hazards to aircraft such as aerial gunnery or missile tests. Flight is restricted within a Restricted Area when that area is active. Military Operations Areas (MOAs) are established to separate IFR and military traffic. VFR flight is always permitted within MOAs. Alert Areas are established to notify pilots of unusual aerial activity such as a high volume of flight training but flight is always permitted within them. Warning areas are located offshore and are established to alert pilots. (PH Ch. 3)

103. When should you contact ATC after leaving from an uncontrolled satellite airport located in Class C or Class D airspace?

After departing an uncontrolled satellite airport in Class C or Class D airspace, contact ATC as soon as practicable. (FIGI p. 107)

104. What is a Military Training Route?

Military Training Routes are depicted on sectional charts to alert pilots to establish flight paths used for military training, usually occurring at high speeds and low altitudes. (PH Ch. 3)

14. Logbook entries and certificate endorsements

105. What record of student endorsements must an instructor keep? How long must they be kept?

An instructor must keep the name of the student and the date, type, and result of a practical or knowledge test for the record of student endorsements. Additionally, an instructor must keep a record of the name of each student endorsed for solo. All endorsement records must be kept for three years. (FIFM pp. 162-174)

106. What is required for a student to fly solo?

A logbook endorsement within the preceding 90 days, a student pilot certificate endorsement, and a passing score on a knowledge test given by the instructor are required for a student to fly solo. (FIGI p. 204)

107. What is required for a student to fly solo cross-country?

A logbook endorsement and a student pilot certificate endorsement are required for a student to fly solo cross-country. (FIGI p. 204)

Part III Flight operations

1. Certificates and documents

1. How long is a second-class medical certificate valid? How long is a third-class medical certificate valid?

A second-class medical certificate is valid for 12 months following the date of the examination, regardless of age. A third-class medical certificate is valid until the end of the 36th calendar month following the date of the examination if you were under 40 years of age on the day of the medical exam. If you were 40 years old or older on the day of the exam, then the certificate is valid until the end of the 24th calendar month following the date of the examination. (FIFM p. 177)

2. Is a flight instructor required to hold a medical certificate?

If the student is appropriately certified and rated for the aircraft, and a safety pilot is not required for the flight, a flight instructor does not have to hold a medical certificate. (FIFM p. 177)

3. What certificates and documents are required to be aboard an aircraft during a Part 91 flight for it to be considered legal?

Remember A.R.R.O.W: Airworthiness certificate, Registration, Radio station license (if you are flying outside the U.S.), Operating limitations, and Weight and balance. (FIFM p. 196)

4. Is an AD required to be complied with?

Yes, unless there is a specific exemption. (FIFM p. 194)

5. How often must the transponder be tested and inspected to be considered legal? How often must the ELT be tested and inspected to be considered legal?

A transponder must be tested every 24 calendar months. An ELT must be tested every 12 calendar months. (FIFM p. 195)

6. How often must the ELT battery be replaced or recharged?

The ELT battery must be replaced or recharged after half of its useful life or 1 hour of cumulative use. (FIFM p. 195)

7. Who must sign off for an annual or 100-hr. inspection?

An A&P with an IA (inspection authorization) must sign off for an annual inspection. A mechanic holding an A&P certificate may sign off for a 100-hour inspection. (FIFM p. 195)

8. What is a way to ensure compliance with all ADs?

To ensure compliance with all ADs, compile a list of all current ADs for your aircraft. Compare this ADs list with the AD compliance record maintained by the mechanic and ensure all applicable ADs have been complied with. (14 CFR Part 39)

9. How long is an airworthiness certificate valid?

An airworthiness certificate is valid as long as all maintenance, ADs, and equipment FARs have been complied with. (PH Ch. 4)

2. Weather information

10. Where is weather information available on the ground?

Weather information is available on the ground from a Flight Service Station (FSS), Direct User Access Terminal System (DUATS), and Telephone Information Briefing Service (TIBS). You can speak to a pre-flight briefer at FSS and/or receive TIBS by calling 1-800-WX-BRIEF anywhere in the country. DUATS is a free service available to pilots on the Internet. Here you can receive weather information, and file a flight plan. TIBS is recorded weather information that can be obtained by calling 1-800-WX-BRIEF. (AWWS PIII-Ch. 1)

11. Where is weather information available in-flight?

Weather information is available in-flight with

EFAS   En-route Flight Advisory Service
HIWAS   Hazardous In-Flight Weather Advisory Service
FSS   Flight Service Station
TWEB   Transcribed Weather Broadcast
ATIS   Automatic Terminal Information Service
ASOS   Automated Surface Observation Service
AWOS   Automatic Weather Observation Service

EFAS is also known as "Flight Watch" and is available almost anywhere in the country on 122.0. You can file a PIREP and obtain numerous types of weather information with EFAS. HIWAS is a recorded briefing of hazardous weather over select VOR frequencies. FSS frequencies are shown on navigational charts and are usually available for ATC. TWEB is a recorded broadcast of current and adverse weather conditions over select VOR and NDB frequencies. ATIS is recorded weather information for a terminal area. AWOS and ASOS are automated weather reporting stations found at many airports. (AWWS PIII-Ch. 1)

12. What is a METAR?

A METAR is a current weather observation that is updated at a regular interval, and applies for a 5-mile radius around the observation point (usually at any airport). (AWWS PIII-Ch. 2)

13. What is a TAF? How often are TAFs updated?

A TAF is a forecast of conditions for the next 24 hours that applies to a 5-mile radius around the place of the report. TAFs are updated four times a day. (AWWS PIII-Ch. 7)

14. How do TAFs indicate wind shear?

TAFs indicate wind shear after sky conditions, with the abbreviation WS. (FIFM pp. 179-180)

15. What is a PIREP? How is one submitted? How can a pilot receive one?

A PIREP is a Pilot Weather Report. They are important sources of observed weather aloft. PIREPs are submitted by pilots on EFAS, and can be received by EFAS or a FSS. (AWWS PIII-Ch. 3)

16. What is a SIGMET? What is a convective SIGMET? What is an AIRMET?

SIGMETs are issued for all aircraft and may include severe icing not associated with thunderstorms, clear air turbulence, dust storms, and volcanic eruptions. Convective SIGMETs are issued for severe thunderstorms, embedded thunderstorms, lines of thunderstorms, and tornados, all of which imply severe or greater turbulence, severe icing, and low-level wind shear. AIRMETs are issued for moderate icing, moderate turbulence, IFR conditions over 50% of an area, sustained surface winds of 30 kt. or greater, nonconvective low-level windshear, and mountain obscuration. (AWWS PIII-Ch. 10)

17. How are SIGMETS and AIRMETS received?

FSS, DUATS, ARTCC. (FIFM pp. 182-183)

18. What are some forecast conditions that may lead to a no-go decision?

Thunderstorms, lines of thunderstorms, embedded thunderstorms, fast-moving squall lines, moderate or greater turbulence, icing, and fog may lead to a no-go decision. (FIFM p. 184)

19. What are some personal factors that relate to the "go/no-go" decision?

Personal factors relating to the "go/no-go" decision include experience, personal limitations, physical and mental health, and the condition of the airplane and its equipment. (FIFM p. 184)

20. Why is it important to obtain a thorough check of the weather before departing for any flight?

A weather briefing will alert you to any potential hazards along the route of flight and may lead to a better "go/no-go" decision. It is also required if flying away from the vicinity of the departure airport. (FIFM p. 180)

3. Operation of systems

21. Which instruments utilize the pitot-static system?

The airspeed indicator, the altimeter, and the vertical speed indicator utilize the pitot-static indicator. (FIFM p. 187)

22. Which instruments operate off the vacuum system?

The attitude indicator and the heading indicator operate off the vacuum system. (FIFM p. 187)

23. What is the purpose of slots?

The slots direct airflow to the upper surface of a wing, which delays airflow separation, thus making the wing stall at a higher angle of attack. (FIFM p. 186)

24. What is the purpose of slats?

Slats are leading edge devices that move forward on tracks at higher angles of attack to delay airflow separation. (FIFM p. 186)

25. What is the purpose of spoilers?

Spoilers are on the upper surface of a wing and "spoil" the smooth airflow to reduce the wing's lifting force, thus increasing the rate of descent without increasing airspeed. (FIFM p. 186)

26. How does the propeller work, in basic terms?

The propeller is essentially a wing that pulls the airplane through the air. (FIFM p. 186)

Make sure you know how to explain and describe everything about the powerplant; and the fuel, oil, hydraulic, electrical, environmental, deicing, and avionics systems in your airplane.

4. Performance and limitations

27. What is the relationship between air density and airplane performance?

There is a direct relationship between air density and airplane performance, i.e., airplane performance decreases as air density decreases. (FIFM pp. 190-191)

28. What is your responsibility as a flight instructor regarding airplane performance?

You must instruct students about the hazards of flying the airplane outside of its limitations. Also, you must develop their judgment with regard to the general safety of the flight. (FIFM p. 191)

Make sure you are familiar with all the airplane's performance and limitations tables. Also, make sure you know how to work through a weight and balance and are able to teach it.

5. Airport markings and ATC light gun signals

29. What is the signal indicating you are cleared to land? What signal indicates that you are cleared for takeoff?

Steady green indicates cleared to land. Steady green also indicates cleared for takeoff. (PH Ch. 3)

30. What does steady red indicate on the ground? In the air?

On the ground, steady red means stop. In the air, it means give way to other aircraft and continue circling. (PH Ch. 3)

31. What is the procedure if you have lost communication at a tower-controlled field?

Observe the flow of traffic, enter the pattern at 45° to the downwind, look for light gun signals, rock wings to acknowledge receipt of the signal. (PH Ch. 3)

32. After landing without a radio at a controlled field, must you stop and wait after clearing the runway for another light gun signal?

After landing without a radio, you must wait for flashing green to taxi after landing. (FIFM p. 242)

33. What is a displaced threshold used for? How is it indicated?

A displaced threshold is used for taxi, takeoff, and landing rollout. It is indicated by white arrows pointing to the runway threshold bar. (PH Ch. 3)

34. What are hold short lines? How are they indicated?

Hold short lines indicate where the airplane is supposed to stop unless cleared to cross. They consist of 4 yellow lines, 2 solid and 2 dashed. (PH Ch. 3)

35. What color are mandatory instruction signs? What color are taxiway location signs? What color are direction signs?

Mandatory instruction signs consist of white letters on a red background. Taxiway location signs consist of yellow letters on a black background. Direction signs consist of black characters on a yellow background. (PH Ch. 3)

36. What color are taxiway markings? What color are runway markings?

There are yellow centerline and shoulder markings on taxiways. There are white centerline and shoulder markings on runways. (PH Ch. 3)

37. What is a non-movement area?

A non-movement area is an area where aircraft can move without being in contact with ATC, i.e., the ramp of an FSO. (PH Ch. 3)

6. Spins

38. What is a spin?

An aggravated stall that results in autorotation. (FIFM p. 445)

39. In what situations is a spin likely to occur?

A spin is likely to occur in a stall while executing a turn with improper rudder technique. Takeoff and departure, approach to landing and go-arounds, and engine failure are the common spin situations. (FIFM p. 446)

40. What is the difference between a steep spiral and a spin?

The airplane must be fully stalled for a spin, but not so for a steep spiral. (FIFM p. 449)

41. What is the recommended spin recovery procedure for your airplane?

The recommended spin recovery procedure is usually throttle to idle, neutralize aileron, rudder in opposite direction of turn, break the stall, recover from the steep descent. (Check your POH for your airplane's specific spin recovery procedure.) (PH Ch. 1)

7. System and equipment malfunctions

The following questions are airplane specific. The answers given here are a general guide only. To fully understand what the correct procedure is for every situation in your airplane, refer to the POH.

42. What would happen if the vacuum pump failed in your airplane? Is it considered an emergency situation in VFR conditions?

A failed vacuum pump will result in the loss of the attitude indicator and the heading indicator. A vacuum pump failure often goes unnoticed by pilots because the gyros take a long time to wind down. This is not considered an emergency situation in VFR conditions because these instruments are not necessary for safe flight. (PH Ch. 2)

43. What would happen to the altimeter if the pitot-static system were obstructed? What would happen to the VSI? What would happen to the airspeed indicator?

If the pitot-static system was blocked, the altimeter would indicate the altitude where the system became blocked. The VSI would give no indication in a climb or descent. The airspeed indicator would read zero. (PH Ch. 2)

44. What error does the altimeter indicate if you are using an alternate static source?

Because alternate static sources are usually located inside the cockpit where the pressure is lower than it is outside the airplane, the altimeter indicates higher than normal. (PH Ch. 2)

45. What does a low voltage light indicate when it is illuminated? How would you address this situation?

A low voltage light indicates that the alternator or generator is not working and the electrical equipment is running from the battery. To effectively deal with the situation, turn off all non-essential electrical equipment and land as soon as practical. (PH Ch. 2)

46. What is the procedure if a circuit breaker pops?

If a circuit breaker pops and it is resettable, push it back in once. If it pops out again, leave it out and determine the next course of action. Pushing the breaker in more than once may lead to an electrical fire. (PH Ch. 2)

47. What are some common causes of a rough running engine in flight?

Though there are numerous factors that can cause an engine to run rough, the most common causes are carburetor ice, a problematic magneto, or fouled spark plugs. (PH Ch. 2)

48. What is the necessary action if you lose a magneto in flight?

If a magneto is lost in flight, determine the bad magneto by switching from both to either left or right. The bad magneto will be evident when the engine will not run on that particular magneto. After you identify the bad magneto, run on the good one and land as soon as practical. (PH Ch. 2)

49. Is using all the fuel from one tank before switching to another a good practice? Why?

Using all the fuel from one tank is not a good practice because it may cause vapor lock in the fuel line. The result may be an engine failure without the ability to restart it inflight. (PH Ch. 2)

50. What should you do if you notice that you are losing oil pressure?

A loss of oil pressure should be viewed as an emergency situation. The engine will not continue to run with low oil pressure. Be prepared for a forced landing. Attempt to fly to the nearest airport. (PH Ch. 2)

51. What is the procedure if you have an engine failure in flight?

If you have an engine failure in flight, establish best glide speed, determine landing site and fly towards it, and go through forced landing checklist if there is time. Put 7700 in the transponder and declare an emergency. (FIFM pp. 490-497)

Be confident; you will do fine. You will never feel totally prepared. If you have studied this book, you will pass with confidence. This book contains the answer to virtually every question, issue, and requirement that is possible on the flight instructor practical test. GOOD LUCK!

The questions contained in this oral exam guide only cover the oral tasks listed in the PTS. Some tasks that can be tested by either the oral exam or flight portion of the exam are also included. Tasks that are considered only testable in flight have not been included. Your instructor should prepare you well for this portion of the exam.

LESSON:

STUDENT: _____ **DATE:** _____

OBJECTIVE

CONTENT

SCHEDULE

EQUIPMENT

INSTRUCTOR'S ACTIONS

STUDENT'S ACTIONS

COMPLETION STANDARDS

ABBREVIATIONS AND ACRONYMS IN FLIGHT INSTRUCTOR FLIGHT MANEUVERS AND PRACTICAL TEST PREP

AD	airworthiness directive
A/FD	*Airport/Facility Directory*
AGL	above ground level
AI	attitude indicator
ALT	altimeter
ASI	airspeed indicator
ATC	air traffic control
ATP	airline transport pilot
CFI	certificated flight instructor
CFII	certificated flight instructor--instrument
ELT	emergency locator transmitter
FAA	Federal Aviation Administration
FAF	final approach fix
FAR	Federal Aviation Regulations
FCC	Federal Communications Commission
FOI	fundamentals of instructing
FSDO	Flight Standards District Office
FSS	Flight Service Station
HI	heading indicator
ICAO	International Civil Aviation Organization
IMC	instrument meteorological conditions
LLWAS	low-level wind shear alert system
MP	manifold pressure
NM	nautical mile
NOTAM	notice to airmen
NWS	National Weather Service
PIREP	pilot weather report
POH	Pilot's Operating Handbook
PTS	Practical Test Standards
SM	statute mile
TC	turn coordinator
V_A	design maneuvering speed
V_{FE}	maximum flap extended speed
VFR	visual flight rules
V_{LE}	maximum landing gear extended speed
V_{LO}	maximum landing gear operating speed
VMC	visual meteorological conditions
V_{NE}	never-exceed speed
V_R	rotation speed
V_{SO}	stalling speed in the landing configuration
V_{S1}	stalling speed in a specified configuration
VSI	vertical speed indicator
V_X	best angle of climb speed
V_Y	best rate of climb speed

AUTHORS' RECOMMENDATIONS

The Experimental Aircraft Association, Inc. is a very successful and effective nonprofit organization that represents and serves those of us interested in flying, in general, and in sport aviation, in particular. We personally invite you to enjoy becoming a member. Visit their website at www.eaa.org.

For 1-year memberships:

$60 (U.S. and Canada) with both magazines
$40 (U.S. and Canada) with one magazine
Add $16 per magazine for international
$23 for students under 19 years old or schools/libraries
Family membership available for $70 with both magazines or $50 with one magazine

> Membership includes choices between the monthly magazines *EAA Sport Aviation* and *EAA Sport Pilot*.

Write: EAA Aviation Center
P.O. Box 3086
Oshkosh, Wisconsin 54903-3086

Call: (920) 426-4800
(800) 843-3612
E-mail: membership@eaa.org

The annual EAA Oshkosh AirVenture is an unbelievable aviation spectacular with over 12,000 airplanes at one airport! Virtually everything aviation-oriented you can imagine! Plan to spend at least 1 day (not everything can be seen in a day) in Oshkosh (100 miles northwest of Milwaukee).

Convention dates: 2007 -- July 23 through July 29
2008 -- July 28 through August 3
2009 -- July 27 through August 2

The annual Sun 'n Fun EAA Fly-In is also highly recommended. It is held at the Lakeland, FL (KLAL) airport (between Orlando and Tampa). Visit the Sun 'n Fun website at www.sun-n-fun.org.

Convention dates: 2007 -- April 17 through April 23

The National Association of Flight Instructors (NAFI) is a nonprofit organization dedicated to raising and maintaining the professional standing of the flight instructor in the aviation community. Members accept the responsibility to practice its profession according to the highest ethical standards. We personally invite you to become a member:

Individual Members - Certified flight instructors (civilian and military)
Annual NAFI Membership Dues = $39

Corporate Members - Organizations in support of professional flight education
Annual Corporate Membership Dues = $250

Membership includes the monthly NAFI magazine *Mentor*, flight instructor liability insurance, and representation with the FAA in Washington, DC. Visit the NAFI website at www.nafinet.org.

Write: NAFI
EAA Aviation Center
P.O. Box 3086
Oshkosh, Wisconsin 54903-3086

Call: (920) 426-6801
Fax: (920) 426-6865
E-mail: nafi@eaa.org

LOGBOOK ENDORSEMENTS FOR PRACTICAL TEST

The following endorsements must be in your logbook and be presented to your FAA inspector/examiner at your practical test.

1. Endorsement for ground flight proficiency/practical test: FAR 61.183(g) and 61.187(a) and (b)

I certify that (First name, MI, Last name) has received the required training of Sec. 61.187(a) and (b). I have determined he/she is prepared for the CFI-(aircraft category and class) practical test.

Signed	Date	Name	CFI Number	Exp. Date

2. Endorsement for spin training: FAR 61.183(i)(1)

I certify that (First name, MI, Last name) has received the required training of Sec. 61.183(i). I have determined he/she is competent and proficient on instructional skills for training stall awareness, spin entry, spins, and spin recovery procedures.

Signed	Date	Name	CFI Number	Exp. Date

3. Endorsement for fundamentals of instructing: FAR 61.183(d) and 61.185(a)(1)

I certify that (First name, MI, Last name) has received the required FOI training of Sec. 61.185(a)(1).

Signed	Date	Name	CFI Number	Exp. Date

INDEX

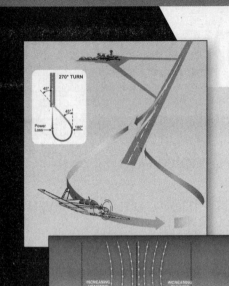

Gleim's Complete Pilot Kits

All kits include: Knowledge Test Book and Test Prep CD-Rom, Flight Maneuvers Book, Online Ground School, Syllabus, and Flight Bag (*Includes audios on CD*)

Sport Pilot (Sport and Private Pilot Kits also include FAR/AIM, Pilot Handbook, Pilot Logbook, Flight Computer, Navigational Plotter) .... $149.95 _____

Private Pilot* .... $219.95 _____

Instrument Pilot* (also includes Aviation Weather & Weather Services) .... $216.95 _____

Commercial Pilot .... $154.95 _____

Instrument*/Commercial Pilot (also includes Aviation Weather & Weather Services) $341.95 _____

Sport Pilot Flight Instructor (Sport Pilot Knowledge Test Book, Syllabus not included) $154.95 _____

Flight Instructor (also includes FAR/AIM and Pilot Handbook; Syllabus not included) $154.95 _____

ATP (also includes Pilot Handbook and Aviation Weather & Weather Services, FAR/AIM; Syllabus, Flight Maneuvers book, Flight Bag not included) $169.95 _____

Also Available:

Flight Engineer Test Prep CD-Rom .... $59.95 _____

Shipping (nonrefundable): **$20 per kit** $ _____
(Alaska and Hawaii please call for shipping details)
Add applicable sales tax for shipments within the state of Florida. $ _____
For orders outside the United States, please visit our website at
www.gleim.com/aviation/products.php to place your order. TOTAL $ _____

Reference Materials and Other Accessories Available by Contacting Gleim.

TOLL FREE: (800) 874-5346
INTERNET: gleim.com

LOCAL: (352) 375-0772
FAX: (352) 375-6940
E-MAIL: sales@gleim.com

Gleim Publications, Inc.
P.O. Box 12848
Gainesville, FL 32604

NAME (please print) _____

ADDRESS _____ Apt. _____
(street address required for UPS)

_____

CITY _____ STATE ____ ZIP _____

_____ MC/VISA/DISC _____ Check/M.O. Daytime Telephone (____) ___ - ____

Credit Card # _____ - _____ - _____

Exp. ____ / ____ Signature _____
Mo./Yr.

E-mail Address _____

1. We process and ship orders daily, within one business day, over 99.8% of the time. Call by 4PM Eastern for same-day service!
2. Please PHOTOCOPY this order form for others.
3. No CODs. Orders from individuals must be prepaid. Volume orders may be purchased on account.
4. Gleim Publications, Inc. guarantees the immediate refund of all resalable texts and unopened software and audios if returned within 30 days. Applies only to items purchased direct from Gleim Publications, Inc. Our shipping charge is nonrefundable.
5. Components of specially priced package deals are nonreturnable.
   Prices subject to change without notice.
   January 2007

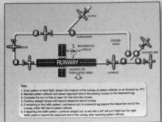

Please forward your suggestions, corrections, and comments concerning typographical errors, etc., to **Irvin N. Gleim • c/o Gleim Publications, Inc. • P.O. Box 12848 • University Station • Gainesville, Florida • 32604.** Please include your name and address so we can properly thank you for your interest.

1. _____

2. _____

3. _____

4. _____

5. _____

6. _____

7. _____

8. _____

9. _____

10. _____

11. _____

12. _____

13. _____

14. _____

15. _____

16. _____

17. _____

18. _____

Remember, for superior service: <u>Mail</u>, <u>e-mail</u>, or <u>fax</u> questions about our materials.
<u>Telephone</u> questions about orders, prices, shipments, or payments.

Name: _____

Address: _____

City/State/Zip: _____

Telephone: Home: _____ Work: _____ Fax: _____

E-mail: _____